MACMILLAN ENCYCLOPEDIA OF

ENERGY

EDITORIAL BOARD

MACMILLAN ENCYCLOPEDIA OF

ENERGY

JOHN ZUMERCHIK

Editor in Chief

VOLUME 1

Macmillan Reference USA

an imprint of the Gale Group

New York • Detroit • San Francisco • London • Boston • Woodbridge, CT

Macmillan Encyclopedia of Energy

Copyright © 2001 Macmillian Reference USA, an imprint of Gale Group

Macmillan Reference USA
1633 Broadway
New York, NY 10019

Macmillan Reference USA
27500 Drake Rd.
Farmington Hills, MI 48331-3535

Library of Congress Catalog Card Number: 00-062498

Printed in the United States of America
Printing number
1 2 3 4 5 6 7 8 9 10

Library of Congress Cataloging-in-Publication Data
Macmillan encyclopedia of energy / John Zumerchik, editor in chief.
    p. cm.
    Includes bibliographical references and index.
    ISBN 0-02-865021-2 (set : hc). — ISBN 0-02-865018-2 (vol. 1). — ISBN 0-02-865019-0 (vol. 2). — ISBN 0-02-865020-4 (vol. 3).
    1. Power resources—Encyclopedias. I. Title: Encyclopedia of Energy. II. Zumerchik, John
TJ163.25 .M33 2000

              00-062498

621.042'03 dc21

CONTENTS

PREFACE

In the mid-1970s, many environmental organizations and energy experts predicted a bleak future of widespread economic and social disruption caused by energy scarcities and continually rising costs of fossil fuels. These predictions never came true. In fact, the opposite occurred—fossil fuel prices in the year 2000 are actually lower than the brief peaks that were reached after the OPEC oil embargo of 1973. Despite this reality, there continue to be long-term energy problems for which there are no easy answers. Foremost is availability, related to the unsustainability of ever more people demanding an ever-declining supply of fossil fuels. Second is growth in electricity demand, which requires more power plants in an era of growing political opposition to the greater use of fossil fuels and nuclear power. Third is pollution, because the more power plants and vehicles there are, the harder it is to reduce harmful emissions. And finally, there is global warming, the contentiously debated link between energy-related carbon dioxide emissions and climate change.

The three-volume *Macmillan Encyclopedia of Energy* is an interdisciplinary work with a very ambitious scope that addresses these continuing problems. Entries represent the fields of physics, chemistry, biology, geophysics, and engineering, as well as history, economics, psychology, sociology, and ethics. Traditional energy concepts such as electricity, thermodynamics, and conservation of energy are included, but so are less obvious concepts such as combustion, catalysts, propulsion, and matter and energy. The traditional fuels such as gasoline, diesel, and jet fuel are included, but so are the alternative fuels such as methanol, synthetic fuel, and hydrogen. The traditional energy sources such as coal, natural gas, petroleum, and nuclear fission are included, but so are the more forward-looking sources such as wind, solar, fuel cells, and nuclear fusion.

The *Encyclopedia* covers the major energy-using technologies and examines the advances that have been made in terms of performance and efficiency. This is an important approach because the unforeseen technology of tomorrow might be far different from what exists today, leading to an alteration in patterns of energy consumption. Only by improving how experts predict how technological innovation will alter energy consumption will we be able to avoid making erroneous energy predictions similar to those made in the 1970s. A large number of entries related to transportation are included because this sector of the economy faces the greatest challenges in the future with regard to energy use. The reliance on imported petroleum and the difficulty in switching to another energy source make the transportation sector particularly vulnerable to supply disruptions and wild price fluctuations. Subjects related to electricity are also well represented because of the rapid pace of growth in the field of electric technology.

The interdisciplinary approach of the *Encyclopedia* led to different approaches for different types of articles. Besides the historical articles covering the sources, uses, and social views of energy, many technology articles take a historical approach to discussing the invention and innovation of the technology, especially the pace of innovation and energy efficiency improvements between 1970 and 2000. Sixty biographies provide information on specific scientists and engineers who made major contributions to the understanding of energy and the related technologies. Since environmental and safety issues are important to the story of energy, these issues are addressed throughout the *Encyclopedia*.

The *Encyclopedia* includes 253 alphabetically arranged entries written by 170 authors. The text is supplemented with more than 600 photographs, illustrations, sidebars, and maps. Entries contain a set of cross-references to related entries within the set, as well as a bibliography of related books and journal articles to guide readers who want to learn more about a given topic. The front matter in Volume 1 includes a list of entry topics, and the back matter in Volume 3 contains both an extensive timeline of important dates in energy history and a comprehensive subject index.

The purpose of this publication is to provide an up-to-date reference guide for people who want full, current, trustworthy information about all aspects of energy. This includes individuals who want more information than they read in newspapers or magazines, high school and college students who want information for class discussions and papers, people looking for expert information on which to base their everyday energy-related decisions, and scholars who want a review of interdisciplinary research.

The *Encyclopedia* will help to foster a greater awareness of the historical significance of energy and the relationship between energy and technology. Decisions that are related to technology and that affect energy consumption are made on a regular basis. However, since energy is mostly invisible (one cannot see a kilowatt of electricity or a cubic foot of natural gas), these decisions are not commonly thought of as energy decisions even when the energy aspect is an essential aspect. Choices related to transportation, shelter, comfort, and recreation all entail energy decisions, but they are not themselves considered to be energy decisions. For example, driving to work instead of walking, purchasing bigger homes and vehicles instead of smaller ones, and acquiring the latest plug-in appliance all have short- and long-term energy-related consequences. Although each individual decision has an insignificant effect on overall consumption, once the behavior becomes a norm, the effect can be very significant for society since reversal is often impossible.

Production of this three-volume set has involved the efforts of many people. In particular, I want to thank my associate editors, Herm Bieber, Alan Chachich, Fred Denny, Barney Finn, John Firor, Howard Geller, David Greene, Joseph Priest, Rosalie Ruegg, Harold Wallace, and Ellen Zeman for their suggestions with regard to the organization and content of the work and for lending their expertise in evaluating and reviewing manuscripts. I would also like to thank the contributors for their cooperation in ensuring that the content and level of presentation of their entries was appropriate for our audience.

I am equally indebted to the staff at Macmillan Reference USA for their prodding and pushing that helped speed the completion of this work. In particular, I want to thank Charlie Montney and Brian Kinsey for their day-to-day management of the project and Elly Dickason, the publisher of Macmillan, for her support in producing the work you see before you.

John Zumerchik
New York City
2000

LIST OF ARTICLES

LIST OF CONTRIBUTORS

Tyno Abdul-Redah
Technical University, Berlin, Germany
Nernst, Walther Hermann (1864-1941)

Albert Abramson
Las Vegas, NV
Alexanderson, Ernst Frederik Werner

Samuel N. Addy
University of Alabama, Tuscaloosa
Economic Externalities

Thomas S. Ahlbrandt
U.S. Geological Survey, Denver
Fossil Fuels
Reserves and Resources

Hashem Akbari
Lawrence Berkeley National Laboratory, Berkeley, CA
Control Systems, History of (with Anibal T. De Almeida)
Cool Communities (with Arthur H. Rosenfeld)
Energy Management Control Systems (with Anibal T. De Almeida)

James D. Allan
Cypress Semiconductor Corporation, Colorado Springs
Tesla, Nikola (1856-1943)

Charles A. Amann
KAB Engineering, Bloomfield Hills, MI
Automobile Performance
Drivetrains
Engines
Gasoline Engines

John D. Anderson
National Air and Space Museum, Smithsonian Institution, Washington, DC
Aerodynamics
Aircraft

Brian S. Baigrie
University of Toronto, Ontario, Canada
Newton, Isaac (1642-1727)

Stuart E. Baird
International Council for Local Environmental Initiatives, Toronto, Ontario, Canada
Ocean Energy

Dennis G. Baker
University of Michigan, Ann Arbor
Atmosphere (with Anita Baker-Blocker)

Anita Baker-Blocker
Ann Arbor, MI
Atmosphere (with Dennis G. Baker)

Ronald L. Bannister
American Society of Engineers, Winter Springs, FL
Turbines, Gas
Turbines, Steam

Gary C. Barber
Oakland University
Tribology (with Barbara A. Oakley)

Dennis Barnaal
Luther College
Electricity

Thomas A. Barthold
Joint Committee on Taxation, U.S. Congress
Taxation of Energy (wth Mark Mazur)

Stephanie J. Battles
U.S. Energy Information Administration, Washington, DC
Energy Intensity Trends

Donald A. Bender
Trinity Flywheel Power, Livermore, CA
Flywheels

Herman Bieber
Engineering Consultant, Kenilworth, NJ
  Charcoal
  Explosives and Propellants
  Kerosene
  Natural Gas, Processing and Conversion of
    (with John Zumerchik)
  Synthetic Fuel (with John Zumerchik)

Roy Billinton
University of Saskatchewan, Saskatoon, Canada
  Electric Power, System Reliability and

Christopher Borroni-Bird
General Motors, Oakland Township, MI
  Fuel Cells
  Fuel Cell Vehicles

Brian Bowers
Science Museum, London, United Kingdom
  Wheatstone, Charles (1802-1875)

Robert L. Bradley, Jr.
Institute for Energy Research, Houston, TX
  Green Energy

Robert N. Brady
Vancouver Community College, Vancouver, Canada
  Diesel Cycle Engines
  Diesel Fuel

Kenneth Brezinsky
University of Illinois, Chicago
  Combustion

Allison Brody
Oklahoma City, OK
  Biological Energy Use, Ecosystem
    Functioning of

Raymon L. Brown
Oklahoma Geological Survey, Norman, OK
  Oil and Gas, Exploration for

Shannon A. Brown
Washington, DC
  Military Energy Use, Historical Aspects of

Andrew F. Burke
Institute of Transportation Studies, University of California, Davis
  Hybrid Vehicles

Tom Butler
Institute of Ecosystem Studies, Cornell University
  Acid Rain

A. Douglas Carmichael
Massachusetts Institute of Technology
  Ships (with Clifford A. Whitcomb)

David Claridge
Texas A&M University
  Heat Transfer (with W. Dan Turner)

David Conover
National Evaluation Services, Inc., Falls Church, VA
  Building Design, Energy Codes and (with
    Jeffrey A. Johnson)

Roger Conway
Office of Energy Policy and New Uses, U.S. Department of Agriculture, Washington, DC
  Agriculture (with Hosein Shapouri)

Christopher Cousins
University of Exeter, Exeter, United Kingdom
  Thomson, Joseph John (1856-1940) (with Leif
    Gerward)

Carol Dahl
Colorado School of Mines
  Supply and Demand and Energy Prices

Burtron H. Davis
University of Kentucky, Lexington
  Catalysts (with Joe W. Hightower)

Anibal T. De Almeida
University of Coimbra, Coimbra, Portugal
  Control Systems, History of (with Hashem
    Akbari)
  Energy Management Control Systems (with
    Hashem Akbari)

Robert J. Deltete
Seattle University
  Gibbs, Josiah Willard (1839-1903)

Fred I. Denny
Louisana State University
  Electric Vehicles (with John Zumerchik)
  Environmental Problems and Energy Use

Joseph R. DesJardins
St. John's University
  Ethical and Moral Aspects of Energy Use

Dennis R. Diehl
Newcomen Society of the United States, Exton, PA
  Fulton, Robert (1765-1815)
  Newcomen, Thomas (1663-1729)

Rankine, William John Macquorn (1820-1872)
Savery, Thomas (1650-1715)
Stephenson, George (1781-1848)

David E. Dismukes
Lousiana State University
Electric Power, Generation of

Linda Doman
U.S. Energy Information Administration, Washington, DC
Hydroelectric Energy

Robert H. Doremus
Rensselaer Polytechnic Institute
Materials

Joseph Eto
Lawrence Berkeley National Laboratory, Berkeley, CA
Demand-Side Management

C. W. F. Everitt
Stanford University
Maxwell, James Clerk (1831-1879)
Scientific and Technical Understanding

Anthony Fainberg
Bethesda, MD
Coal, Consumption of

Bernard Finn
National Museum of American History, Smithsonian Institution, Washington, DC
Electricity, History of

Barbara Flume-Gorczyca
Bonn, Germany
Clausius, Rudolf Julius Emmanuel (1822-1888)
Seebeck, Thomas Johann (1770-1831)
Siemens, Ernst Werner von (1816-1892)

Robert E. Gallamore
Transportation Technology Center, Inc., Pueblo, CO
Locomotive Technology

James H. Gary
Colorado School of Mines
Refineries

Leif Gerward
Technical University of Denmark, Kongens Lyngby, Denmark
Thomson, Joseph John (1856-1940) (with Christopher Cousins)

David B. Goldstein
Natural Resources Defense Council, San Francisco, CA
Appliances

Robin B. Goodale
Houston, TX
Sperry, Elmer Ambrose (1860-1930)

Richard L. Gordon
Pennsylvania State University
Energy Economics
Subsidies and Energy Costs

Peter A. Gorin
National Air and Space Museum, Smithsonian Institution, Washington, DC
Rocket Propellants
Spacecraft Energy Systems

Kenneth Green
Reason Public Policy Institute, Austin, TX
Air Pollution
Climatic Effects

Randall Guensler
Georgia Institute of Technology
Emission Control, Vehicle
Traffic Flow Management

J. Storrs Hall
Institute for Molecular Manufacturing, Palo Alto, CA
Nanotechnologies

M. Maureen Hand
National Renewable Energy Laboratory, Golden, CO
Turbines, Wind

Mark E. Hanson
Energy Center of Wisconsin, Madison
Market Transformation (with Dan W. York)

David E. Harris
University of Southern Maine
Biological Energy Use, Cellular Processes of

Alan S. Heather
Radio Society of Great Britain, Potters Bar, United Kingdom
Heaviside, Oliver (1850-1925)

Karl J. Hejlik
Orlando, FL
Lasers
Townes, Charles Hard (1915-)

Gilbert Held
Macon, GA
Communications and Energy

Jamie N. Heller
PHB Hagler Bailly, Inc., Washington, DC
Coal, Transportation and Storage of (with Stan M. Kaplan)

Ken Helmick
General Motors, Warren, MI
Steam Engines

P. C. Hemmer
Norwegian University of Science and Technology, Trondheim, Norway
Onsager, Lars (1903-1976)

Charles E. Hickman
Tennessee Technological University (ret.)
Emission Control, Power Plant

Joe W. Hightower
Rice University
Catalysts (with Burtron H. Davis)

Lawrence J. Hill
Honolulu, HI
Regulation and Rates for Electricity

George W. Hinman
Washington State University, Pullman
Oil and Gas, Production of (with Nancy W. Hinman)

Nancy W. Hinman
University of Montana, Missoula
Oil and Gas, Production of (with George W. Hinman)

Allan R. Hoffman
U.S. Department of Energy, Washington, DC
Solar Energy

Don C. Hopkins
South Dakota School of Mines and Technology
Mechanical Transmission of Energy
Perpetual Motion
Propulsion

Stanley H. Horowitz
Consultant, Columbus, OH
Electric Power, System Protection, Control, and Monitoring of

J. Benjamin Horvay
Consulting Engineer, Isle of Palms, SC
Refrigerators and Freezers

D. W. Ignat
International Atomic Energy Agency, Vienna, Austria
Nuclear Fusion

Jeffrey A. Johnson
New Buildings Institute, White Salmon, WA
Building Design, Commercial (with William Steven Taber, Jr.)
Building Design, Energy Codes and (with David Conover)

Tina M. Kaarsberg
Office of Power Technologies, U.S. Department of Energy
Efficiency of Energy Use (with Arthur H. Rosenfeld and Joseph J. Romm)

Stan M. Kaplan
PHB Hagler Bailly, Inc., Washington, DC
Coal, Transportation and Storage of (with Jamie N. Heller)

David Keston
Siemens UK, Bracknell, United Kingdom
Fourier, Jean Baptiste Joseph (1768-1830)

Jeremy Kinney
Auburn University
Propellers

Ahmet E. Kocagil
Pensylvania State University, University Park, PA
Risk Assessment and Management

Jonathan G. Koomey
Lawrence Berkeley National Laboratory, Berkeley, CA
True Energy Costs

Japhet Koteen
U.S. Environmental Protection Agency, Washington, DC
Building Design, Residential (with Sam Rashkin)

Satish Kumar
Lawrence Berkeley National Laboratory, Berkeley, CA
Air Quality, Indoor

Esher Kweller
U.S. Department of Energy, Washington, DC
Furnaces and Boilers (with Roger McDonald)

Jack Lanigan, Sr.
Mi-Jack Products, Hazel Crest, IL
Freight Movement (with John Zumerchik)

Gary Libecap
University of Arizona, Tucson
Property Rights

Don Lichtenberg
Indiana University, Bloomington
Big Bang Theory
Matter and Energy
Particle Accelerators

Jiang Long
Global Bearings, Inc., Tulsa, OK
Bearings

James D. Lutz
Lawrence Berkeley National Laboratory, Berkeley, CA
Water Heating

James N. Maughn
Engineering Consultant, Birmingham, AL
Utility Planning

Mark Mazur
*U.S. Energy Information Administration, U.S.
Department of Energy, Washington, DC*
Taxation of Energy (with Thomas A. Barthold)

Robert V. McCormick
Troy, MI
Natural Gas, Transportation, Distribution, and
Storage of

Roger McDonald
Brookhaven National Laboratory, Upton, NY
Furnaces and Boilers (with Esher Kweller)

Alan B. McEwen
Covalent Associates, Inc., Woburn, MA
Capacitors and Ultracapacitors

Ernest L. McFarland
University of Guelph, Ontario, Canada
Geothermal Energy
Nuclear Energy

Sue McNeil
University of Illinois, Chicago
Capital Investment Decisions

Alan K. Meier
Lawrence Berkeley National Laboratory, Berkeley, CA
Conservation Supply Curves (with Arthur H.
Rosenfeld)

Paul H. E. Meijer
Catholic University of America
Kamerlingh Onnes, Heike (1853-1926)

Elena Subia Melchert
U.S. Department of Energy, Washington, DC
Oil and Gas, Drilling for (with John
Zumerchik)

Hugo Karl Messerle
Turramurra, Australia
Magnetohydrodynamics

James G. Mills
U.S. Federal Trade Commission, Washington, DC
Efficiency of Energy Use, Labeling of

J. Bernard Moore
Hyattsville, MD
Solar Energy, Historical Evolution of
the Use of

Sanford L. Moskowitz
Chemical Heritage Foundation, Philadelphia, PA
Gasoline and Additives
Ipatieff, Vladimir Nikolaevitch (1867-1952)
Lewis, Warren K. (1882-1975)
Petroleum Consumption
Refining, History of

Deborah L. Mowery
Sandia National Laboratories, Albuquerque, NM
Biofuels
Nitrogen Cycle

David Mulcahy
Polytechnic University, Brooklyn, NY
Animal and Human Energy

Michael R. Muller
Rutgers University
Auditing of Energy Use

Joseph F. Mulligan
University of Maryland, Baltimore County
Carnot, Nicolas Leonard Sadi (1796-1832)
Fermi, Enrico (1901-1954)
Helmholtz, Hermann von (1821-1894)
Hertz, Henrich Rudolph (1857-1894)

Joule, James Prescott (1818-1889)
Mayer, Julius Robert (1814-1878)
Meitner, Lise (1878-1968)
Thompson, Benjamin (Count Rumford) (1753-1814)

Michael L. Murphy
Energy Products of Idaho, Coeur d'Alene, ID
Waste-to-Energy Technology

Laura Nader
University of California, Berkeley
Culture and Energy Usage

Bernard A. Nagengast
Consulting Engineer, Sidney, OH
Air Conditioning
Heat Pumps

Bruce Nordman
Lawrence Berkeley National Laboratory, Berkeley, CA
Office Equipment (with Mary Ann Piette)

Barbara A. Oakley
Oakland University
Tribology

Joan M. Ogden
Center for Energy and Environmental Studies, Princeton University
Hydrogen

Abraham Pais
Rockefeller University
Einstein, Albert (1879-1955)

John A. Palmer
Colorado School of Mines
Electric Power Substations
Electric Power Transmission and Distribution Systems

Giuliano Pancaldi
University of Bologna, Italy
Volta, Alessandro (1745-1827)

William S. Peirce
Case Western Reserve University
Economically Efficient Energy Choices

Karl A. Petersen
Steam Automobile Club of America, Washington, IL
Trevithick, Richard (1771-1833)
Watt, James (1736-1819)

Morris A. Pierce
University of Rochester
Cogeneration Technologies
District Heating and Cooling
Domestic Energy Use

Mary Ann Piette
Lawrence Berkeley National Laboratory, Berkeley, CA
Office Equipment (with Bruce Nordman)

Frank R. Power
National Institute of Standards and Technology, Washington, DC
Futures

Joseph Priest
Miami University, OH
Cogeneration
Conservation of Energy
Elastic Energy
Gravitational Energy
Heat and Heating
Kinetic Energy
Magnetism and Magnets
Nuclear Fission
Piezoelectric Energy
Potential Energy
Power
Pressure
Renewable Energy
Storage
Storage Technology
Thermal Energy
Units of Energy
Waves
Work and Energy

Douglas Quinney
Keele University, Keele, United Kingdom
Archimedes (287-212 BC)
Bernoulli, Daniel (1700-1782)

Sam Rashkin
U.S. Environmental Protection Agency, Washington, DC
Building Design, Residential (with Japhet Koteen)

John R. Ray
Clemson University
Thermodynamics

Mark S. Rea
Lighting Research Center, Rensselaer Polytechnic Institute
Lighting

Douglas B. Reynolds
University of Alaska, Fairbanks
Import/Export Market for Energy
Industry and Business, Energy as a Factor of Production in

Terry S. Reynolds
Michigan Technological University
Kinetic Energy, Historical Evolution of the Use of

Joseph J. Romm
Center for Energy and Climate Solutions, Annandale, VA
Efficiency of Energy Use (with Arthur H. Rosenfeld and Tina M. Kaarsberg)
Industry and Business, Productivity and Energy Efficiency in

Arthur H. Rosenfeld
Commissioner, California Energy Commission
Conservation Supply Curves (with Alan K. Meier)
Cool Communities (with Hashem Akbari)
Efficiency of Energy Use (with Tina M. Kaarsberg and Joseph J. Romm)

Colin A. Russell
The Open University, Milton Keynes, United Kingdom
Lyell, Charles (1797-1875)

Alan H. Sanstad
Lawrence Berkeley National Laboratory, Berkeley, CA
Market Imperfections

Linda Sargent Wood
University of Maryland, College Park
Carson, Rachel
Fuller, R. Buckminster, Jr. (1895-1983)

Charles K. Scharnberger
Millersville University
Seismic Energy

Paul M. Schimek
U.S. Department of Transportation, Cambridge, MA
Mass Transit

John Schubert
Adventure Cycling Association
Bicycling (with John Zumerchik)

Silvan S. Schweber
Brandeis University
Bethe, Hans Albrecht (1906-   )

Andrew M. Sessler
Lawrence Berkeley National Laboratory, Berkeley, CA
Sakharov, Andrei Dmitrievich (1921-1989)

Hosein Shapouri
Office of Energy Policy and New Uses, U.S. Department of Agriculture, Washington, DC
Agriculture (with Roger Conway)

David B. Sicilia
University of Maryland, College Park
Diesel, Rudolph (1858-1913)
Nuclear Energy, Historical Evolution of the Use of

Charles W. Siegmund
Morris Plains, NJ
Residual Fuels

Robert Sier
Chelmsford, United Kingdom
Ampère, André-Marie (1775-1836)
Black, Joseph (1728-1799)
Ericsson, John (1803-1899)
Otto, Nikolaus August (1832-1891)
Smeaton, John (1724-1792)
Stirling, Robert (1790-1878)

Scott B. Sitzer
U.S. Energy Information Administration, Washington, DC
Economic Growth and Energy Consumption

Vaclav Smil
University of Manitoba, Winnipeg, Canada
Geography and Energy Use
Historical Perspectives and Social Consequences

Crosbie Smith
University of Kent, Canterbury, United Kingdom
Thomson, William (Lord Kelvin) (1824-1907)

Stig Steenstrup
Niels Bohr Institute, Copenhagen, Denmark
Becquerel, Alexandre-Edmund (1820-1891)

Richard L. Stroup
Montana State University, Bozeman
Environmental Economics

Robert Strunz
University of Limerick, Ireland
Parsons, Charles Algernon (1854-1931)

Ronald J. Sutherland
Center for the Advancement of Energy Markets, Burke, VA
Efficiency of Energy Use, Economic Concerns and

William Steven Taber, Jr.
Princeton Development Corporation, Sausalito, CA
Building Design, Commercial (with Jeffrey A. Johnson)

Barry L. Tarmy
TBD Technology, Berkeley Heights, NJ
Chemical Energy, Historical Evolution of the Use of
Houdry, Eugene Jules (1892-1962)

Leonard S. Taylor
University of Maryland, College Park
Edison, Thomas Alva (1847-1931)
Faraday, Michael (1791-1867)
Laplace, Pierre Simon (1749-1827)
Oersted, Hans Christian (1777-1851)

William F. Taylor
Consultant, Mountainsaide, NJ
Aviation Fuel

Louis S. Thompson
The World Bank, Washington, DC
Railway Passenger Service

Richard D. Thornton
Concord, MA
Magnetic Levitation

Brian F. Thumm
St. Francisville, LA
Nuclear Fission Fuel
Transformers

Jerry C. Tien
University of Missouri, Rolla
Coal, Production of

W. Dan Turner
Texas A&M University
Heat Transfer (with David Claridge)

Arlon R. Tussing
University of Alaska, Anchorage
Natural Gas, Consumption of

Brent H. Van Arsdell
American Stirling Company
Stirling Engines

F. Robert van der Linden
National Air and Space Museum, Smithsonian Institution, Washington, DC
Air Travel

Margaret H. Venable
Georgia Perimeter College
Curie, Marie Sklodowska (1867-1934) (with T. Leon Venable)

T. Leon Venable
Agnes Scott College
Curie, Marie Sklodowska (1867-1934) (with Margaret H. Venable)

Joseph D. Walter
University of Akron
Tires

Alecia Ward
Midwest Energy Efficiency Alliance, Chicago, IL
Windows

Clifford A. Whitcomb
Massachusetts Institute of Technology
Ships (with A. Douglas Carmichael)

Donald Williams
Hope College
Nuclear Waste

Roy W. Willis
Propane Education and Research Council, Washington, DC
Liquefied Petroleum Gas

William L. Withuhn
Smithsonian Institution, Washington, DC
Transportation, Evolution of Energy Use and

Ernst Worrell
Lawrence Berkeley National Laboratory, Berkeley, CA
Manufacturing

David W. Yarbrough
Tennessee Technological University
Insulation

Dan W. York
Energy Center of Wisconsin, Madison
Market Transformation (with Mark E. Hanson)

Ellen J. Zeman
Burlington, VT
Molecular Energy

John Zumerchik
New York, NY
Alternative Fuels and Vehicles
Batteries
Behavior
Bicycling (with John Schubert)
Consumption
Electric Motor Systems
Electric Vehicles (with Fred I. Denny)
Freight Movement (with Jack Lanigan, Sr.)
Government Agencies
Government Intervention in Energy Markets
Methane
Methanol
National Energy Laboratories
Natural Gas, Processing and Conversion of (with Herman Bieber)
Oil and Gas, Drilling for (with Elena Subia Melchert)
Sustainable Resources
Synthetic Fuel (with Herman Bieber)

COMMON ABBREVIATIONS AND MATHEMATICAL SYMBOLS

=	equals; double bond	"	second
≠	not equal to	'	minute
≡	identically equal to; equivalent to; triple bond	ε_0	electric constant
~	approximately	μ	micro-
≈	approximately equal to	μ_0	magnetic constant
≅	congruent to; approximately equal to	μA	microampere
∝	proportional to	$\mu A\,h$	microampere hour
<	less than	μC	microcoulomb
>	greater than	μF	microfarad
<<	much less than	μg	microgram
>>	much greater than	μK	microkelvin
≤	less than or equal to	μm	micrometer
≥	greater than or equal to	μm	micron
→	approaches, tends to; yields; is replaced by	$\mu m\,Hg$	microns of mercury
⇒	implies; is replaced by	μmol	micromole
⇐	is implied by	$\mu s, \mu sec$	microsecond
⇓	mutually implies	μu	microunit
⇔	if and only if	$\mu\Omega$	microhm
⇌	reversible reaction	σ	Stefan-Boltzmann constant
⊥	perpendicular to	Ω	ohm
‖	parallel to	Ωcm	ohm centimeter
| |	absolute value of	$\Omega cm/(cm/cm^3)$	ohm centimeter per centimeter per cubic centimeter
+	plus	A	ampere
−	minus	Å	angstrom
/	divided by	a	atto-
×	multiplied by	A_S	atmosphere, standard
±	plus or minus	abbr.	abbreviate; abbreviation
√	radical	abr.	abridged; abridgment
∫	integral	Ac	Actinium
Σ	summation	ac	alternating-current
Π	product	aF	attofarad
∂	partial derivative	af	audio-frequency
°	degree	Ag	silver
°B	degrees Baumé	A h	ampere hour
°C	degrees Celsius (centigrade)	AIP	American Institute of Physics
°F	degrees Fahrenheit	Al	aluminum
!	factorial	alt	altitude

Am	americium
AM	amplitude-modulation
A.M.	ante meridiem
amend.	amended; amendment
annot.	annotated; annotation
antilog	antilogarithm
app.	appendix
approx	approximate (in subscript)
Ar	argon
arccos	arccosine
arccot	arccotangent
arccsc	arccosecant
arc min	arc minute
arcsec	arcsecant
arcsin	arcsine
arg	argument
As	arsenic
At	astatine
At/m	ampere turns per meter
atm	atmosphere
at. ppm	atomic parts per million
at. %	atomic percent
atu	atomic time unit
AU	astronomical unit
a.u.	atomic unit
Au	gold
av	average (in subscript)
b	barn
B	boron
Ba	barium
bcc	body-centered-cubic
B.C.E.	before the common era
Be	beryllium
Bi	biot
Bi	bismuth
Bk	berkelium
bp	boiling point
Bq	becquerel
Br	bromine
Btu, BTU	British thermal unit
C	carbon
c	centi-
c.	circa, about, approximately
C	coulomb
c	speed of light
Ca	calcium
cal	calorie
calc	calculated (in subscript)
c.c.	complex conjugate
CCD	charge-coupled devices

Cd	cadmium
cd	candela
CD	compact disc
Ce	cerium
C.E.	common era
CERN	European Center for Nuclear Research
Cf	californium
cf.	confer, compare
cgs	centimeter-gram-second (system)
Ci	curie
Cl	chlorine
C.L.	confidence limits
c.m.	center of mass
cm	centimeter
Cm	curium
cm^3	cubic centimeter
Co	cobalt
Co.	Company
coeff	coefficient (in subscript)
colog	cologarithm
const	constant
Corp.	Corporation
cos	cosine
cosh	hyperbolic cosine
cot	cotangent
coth	hyperbolic cotangent
cp	candlepower
cP	centipoise
cp	chemically pure
cpd	contact potential difference
cpm	counts per minute
cps	cycles per second
Cr	chromium
cS	centistoke
Cs	cesium
csc	cosecant
csch	hyperbolic cosecant
Cu	copper
cu	cubic
cw	continuous-wave
D	Debye
d	deci-
da	deka-
dB, dBm	decibel
dc	direct-current
deg	degree
det	determinant
dev	deviation
diam	diameter
dis/min	disintegrations per minute

dis/s	disintegrations per second
div	divergence
DNA	deoxyribose nucleic acid
DOE, DoE	U.S. Department of Energy
Dy	dysprosium
dyn	dyne
E	east
e	electronic charge
E	exa-
e, exp	exponential
e/at.	electrons per atom
e b	electron barn
e/cm³	electrons per cubic centimeter
ed. (pl., eds.)	editor
e.g.	exempli gratia, for example
el	elastic (in subscript)
emf, EMF	electromotive force
emu	electromagnetic unit
Eng.	England
EPA	U.S. Environmental Protection Agency
Eq. (pl., Eqs.)	equation
Er	erbium
EREN	Energy Efficiency and Renewable Energy Network
erf	error function
erfc	error function (complement of)
Es	einsteinium
e.s.d.	estimated standard deviation
esu	electrostatic unit
et al.	et alii, and others
etc.	et cetera, and so forth
e.u.	electron unit
eu	entropy unit
Eu	europium
eV	electron volt
expt	experimental (in subscript)
F	farad
F	Faraday constant
f	femto-
F	fermi
F	fluorine
fc	foot-candle
fcc	face-centered-cubic
Fe	iron
FERC	Federal Energy Regulatory Commission (U.S.)
fF	femtofarad
Fig. (pl., Figs.)	figure
fL	foot-lambert
fm	femtometer

Fm	fermium
FM	frequency-modulation
f. (pl., ff.)	following
fpm	fissions per minute
Fr	francium
Fr	franklin
fs	femtosecond
ft	foot
ft lb	foot-pound
ft lbf	foot-pound-force
f.u.	formula units
g	acceleration of free fall
G	gauss
G	giga-
g	gram
G	gravitational constant
Ga	gallium
Gal	gal (unit of gravitational force)
gal	gallon
g-at.	gram-atom
g.at. wt	gram-atomic-weight
Gc/s	gigacycles per second
Gd	gadolinium
Ge	germanium
GeV	giga-electron-volt
GHz	gigahertz
Gi	gilbert
grad	gradient
GV	gigavolt
Gy	gray
h	hecto-
H	henry
h	hour
H	hydrogen
h	Planck constant
H.c.	Hermitian conjugate
hcp	hexagonal-close-packed
He	helium
Hf	hafnium
hf	high-frequency
hfs	hyperfine structure
hg	hectogram
Hg	mercury
Ho	holmium
hp	horsepower
Hz	hertz
I	iodine
IAEA	International Atomic Energy Agency
ICT	International Critical Tables
i.d.	inside diameter

i.e.	id est, that is
IEA	International Energy Agency
IEEE	Institute of Electrical and Electronics Engineers
if	intermediate frequency
Im	imaginary part
in.	inch
In	indium
Inc.	Incorporated
inel	inelastic (in subscript)
ir, IR	infrared
Ir	iridium
J	joule
Jy	jansky
k, k_b	Boltzmann's constant
K	degrees Kelvin
K	kayser
k	kilo-
K	potassium
kA	kiloamperes
kbar	kilobar
kbyte	kilobyte
kcal	kilocalorie
kc/s	kilocycles per second
kdyn	kilodyne
keV	kilo-electron-volt
kG	kilogauss
kg	kilogram
kgf	kilogram force
kg m	kilogram meter
kHz	kilohertz
kJ	kilojoule
kK	kilodegrees Kelvin
km	kilometer
kMc/s	kilomegacycles per second
kn	knot
kOe	kilo-oersted
kpc	kiloparsec
Kr	krypton
ks, ksec	kilosecond
kt	kiloton
kV	kilovolt
kV A	kilovolt ampere
kW	kilowatt
kWh	kilowatt hour
kΩ	kilohm
L	lambert
L	langmuir
l, L	liter
La	lanthanum
LA	longitudinal-acoustic
lab	laboratory (in subscript)
lat	latitude
lb	pound
lbf	pound-force
lbm	pound-mass
LED	light emitting diode
Li	lithium
lim	limit
lm	lumen
lm/W	lumens per watt
ln	natural logarithm (base e)
LO	longitudinal-optic
log	logarithm
Lr	lawrencium
LU	Lorentz unit
Lu	lutetium
lx	lux
ly, lyr	light-year
M	Mach
M	mega-
m	meter
m	milli-
m	molal (concentration)
M	molar (concentration)
m_e	electronic rest mass
m_n	neutron rest mass
m_p	proton rest mass
MA	megaamperes
mA	milliampere
ma	maximum
mb	millibarn
mCi	millicurie
Mc/s	megacycles per second
Md	mendelevium
MeV	mega-electron-volt; million electron volt
Mg	magnesium
mg	milligram
mH	millihenry
mho	reciprocal ohm
MHz	megahertz
min	minimum
min	minute
mK	millidegrees Kelvin; millikelvin
mks	meter-kilogram-second (system)
mksa	meter-kilogram-second ampere
mksc	meter-kilogram-second coulomb
ml	milliliter
mm	millimeter
mmf	magnetomotive force

mm Hg	millimeters of mercury
Mn	manganese
MO	molecular orbital
Mo	molybdenum
MOE	magneto-optic effect
mol	mole
mol %, mole %	mole percent
mp	melting point
Mpc	megaparsec
mpg	miles per gallon
mph	miles per hour
MPM	mole percent metal
Mrad	megarad
ms, msec	millisecond
mt	metric tonne
Mt	million metric tonnes
mu	milliunit
MV	megavolt; million volt
mV	millivolt
MW	megawatt
MWe	megawatts of electric power
Mx	maxwell
mμm	millimicron
MΩ	megaohm
n	nano-
N	newton
N	nitrogen
N	normal (concentration)
N	north
N, N_A	Avogadro constant
Na	sodium
NASA	National Aeronautics and Space Administration
nb	nanobarn
Nb	niobium
Nd	neodymium
N.D.	not determined
NDT	nondestructive testing
Ne	neon
n/f	neutrons per fission
Ni	nickel
N_L	Loschmidt's constant
nm	nanometer
No	nobelium
No.	number
NOx, NO$_x$	oxides of nitrogen
Np	neper
Np	neptunium
NRC	Nuclear Regulatory Commission (U.S.)
ns, nsec	nanosecond

n/s	neutrons per second
n/s cm^2	neutrons per second per square centimeter
NSF	National Science Foundation (U.S.)
ns/m	nanoseconds per meter
O	oxygen
$o()$	of order less than
$O()$	of the order of
obs	observed (in subscript)
o.d.	outside diameter
Oe	oersted
ohm^{-1}	mho
Os	osmium
oz	ounce
P	peta-
P	phosphorus
p	pico-
P	poise
Pa	pascal
Pa	protactinium
Pb	lead
pc	parsec
Pd	palladium
PD	potential difference
pe	probable error
pF	picofarad
pl.	plural
P.M.	post meridiem
Pm	promethium
Po	polonium
ppb	parts per billion
p. (pl., pp.)	page
ppm	parts per million
Pr	praseodymium
psi	pounds per square inch
psi (absolute)	pounds per square inch absolute
psi (gauge)	pounds per square inch gauge
Pt	platinum
Pu	plutonium
Q, quad	quadrillion Btus
R	roentgen
Ra	radium
rad	radian
Rb	rubidium
Re	real part
Re	rhenium
rev.	revised
rf	radio frequency
Rh	rhodium
r.l.	radiation length
rms	root-mean-square

Rn	radon
RNA	ribonucleic acid
RPA	random-phase approximation
rpm	revolutions per minute
rps, rev/s	revolutions per second
Ru	ruthenium
Ry	rydberg
s, sec	second
S	siemens
S	south
S	stoke
S	sulfur
Sb	antimony
Sc	scandium
sccm	standard cubic centimeter per minute
Se	selenium
sec	secant
sech	hyperbolic secant
sgn	signum function
Si	silicon
SI	Systäme International
sin	sine
sinh	hyperbolic sine
SLAC	Stanford Linear Accelerator Center
Sm	samarium
Sn	tin
sq	square
sr	steradian
Sr	strontium
STP	standard temperature and pressure
Suppl.	Supplement
Sv	sievert
T	tera-
T	tesla
t	tonne
Ta	tantalum
TA	transverse-acoustic
tan	tangent
tanh	hyperbolic tangent
Tb	terbium
Tc	technetium

Td	townsend
Te	tellurium
TE	transverse-electric
TEM	transverse-electromagnetic
TeV	tera-electron-volt
Th	thorium
theor	theory, theoretical (in subscript)
THz	tetrahertz
Ti	titanium
Tl	thallium
Tm	thulium
TM	transverse-magnetic
TO	transverse-optic
tot	total (in subscript)
TP	temperature-pressure
tr, Tr	trace
trans.	translator, translators; translated by; translation
u	atomic mass unit
U	uranium
uhf	ultrahigh-frequency
uv, UV	ultraviolet
V	vanadium
V	volt
Vdc	Volts direct current
vol. (pl., vols.)	volume
vol %	volume percent
vs.	versus
W	tungsten
W	watt
W	West
Wb	weber
Wb/m^2	webers per square meter
wt %	weight percent
W.u.	Weisskopf unit
Xe	xenon
Y	yttrium
Yb	ytterbium
yr	year
Zn	zinc
Zr	zirconium

ACID RAIN

"Acid rain" is a popular term that can include all forms of precipitation (as well as fog and cloudwater) that is more acidic than expected from natural causes. Measurement of precipitation acidity at several remote sites around the world show natural background levels of acidity to be around pH of 5.1 to 5.2 (8 to 6 ueq/l H^+ or hydrogen in concentration, respectively). This compares with present annual average values of pH 4.3 to 4.4 (50 to 40 ueq/l H^+) for most of the northeastern and midwestern United States. Note that as pH decreases, H^+ concentration or acidity increases exponentially. Individual storms, especially in the summer, can often produce pH values below 3.5 (>300 ueq/l H^+). Cloudwater and fog often show even higher concentrations of acidity and this has major implications for high-elevation ecosystems such as mountain forests and water bodies. A more appropriate term than acid rain is "acid deposition," which includes both wet and dry deposition of acidic sulfur and nitrogen compounds to the earth's surface from the atmosphere.

Acid deposition is of greatest concern wherever there are large amounts of fossil fuel combustion upwind of an area. Eastern North America, large areas of Europe, and eastern Asia all receive acidic deposition. Acidic deposition is especially a concern when poorly buffered soils, with little acid-neutralizing capacity, are impacted. In North America, large areas of eastern Canada, the Adirondack Mountains of upstate New York, and sections of New England all are considered "acid sensitive " areas, where resistant bedrocks and thin soils prevent significant neutralization of acidity.

HISTORICAL PERSPECTIVE

Acidic deposition is not a new phenomena, as E. B. Cowling (1982) has noted. In 1872, the term "acid rain" was first known to be used by Angus Smith to describe the precipitation around Manchester, England. Smith analyzed the chemistry of the rain and attributed this acid rain to combustion of coal. He also noted damage from acid rain to plants and materials. C. Crowther and H. G. Ruston (1911) demonstrated gradients in rainfall acidity decreasing from the center of Leeds, England and associated the acidity with coal combustion. E. Gorham (1957, 1958) established that acid precipitation affects the acid-neutralizing capacity of lakes and bogs. A. Dannevig (1959) of Norway recognized the relationship between acid precipitation, lake and stream acidity, and the disappearance of fish. S. Oden (1968) used trajectory analysis to demonstrate that acid precipitation in Sweden was the result of long-range transport and transformation of sulfur emissions from England and central Europe. In 1972, Likens et al. identified acid precipitation in eastern North America. G. E. Likens and F. H. Bormann (1974) demonstrated its regional distribution in the eastern United States and indicated that the transformation of nitrogen oxides (NO_x), as well as sulfur dioxide (SO_2), adds to precipitation acidity. D. W. Schindler and his colleagues (1985) performed a whole lake acidification in Canada and documented the adverse decline of the lake food web at pH levels as high as 5.8. The issue of acid rain or deposition has generated a vast amount of knowledge and understanding of atmospheric and watershed processes, and research in the field continues today.

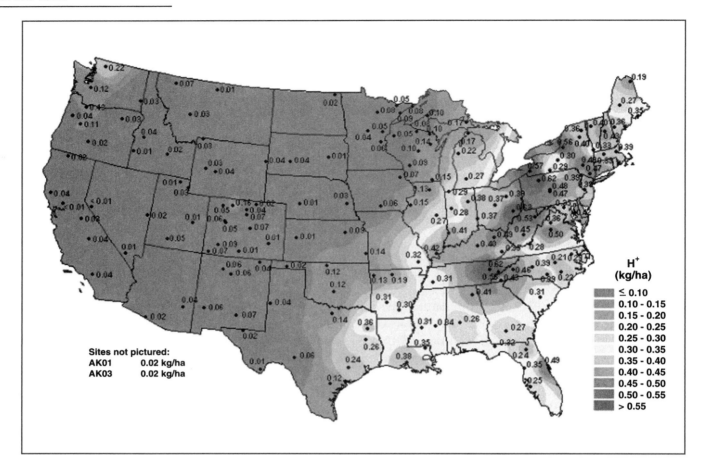

Figure 1.
Acid deposition is clearly a widespread phenomenon in the eastern third of the country.

FORMATION AND TRENDS

The formation of acidic deposition is largely from the combustion of fossil fuels and the smelting of sulfide ores. Minor natural sources exist such as the formation of hydrochloric and sulfuric acid from gaseous volcanic eruptions.

There are well over 100 gaseous and aqueous phase reactions that can lead to acid formation and more than fifty oxidizing agents and catalysts may be involved. However, in the simplest terms sulfur in fuels is oxidized to SO_2, and SO_2 in the atmosphere is further oxidized and hydrolyzed to sulfuric acid. Most nitric acid is formed by the fixation of atmospheric nitrogen gas (N_2) to NO_x (NO and NO_2) during high temperature combustion, followed by further oxidation and hydrolysis that produces nitric acid in the atmosphere. These materials can be dry-deposited onto surfaces, or be removed from the atmosphere by precipitation. The acid-generating reactions can take from hours to days depending on a wide range of atmospheric parameters such as temperature, humidity, and the presence of oxidizing agents such as hydroxl (OH) radicals, ozone (O_3) and hydrogen peroxide (H_2O_2). Depending on these conditions, and other factors such as height of release and wind speed, sulfur and nitrogen oxides can be transformed and deposited as acid deposition anywhere from a few kilometers to thousands of kilometers from their original source. Figure 1 shows the geographic distribution of acid deposition from precipitation for the United States.

The U.S. trends in emissions SO_2 and NO_x from 1900 to 1997 are shown in Figure 2. The pattern for SO_2 emissions since 1900 has shown three peaks. From 1900 to the 1920s there was a general increase

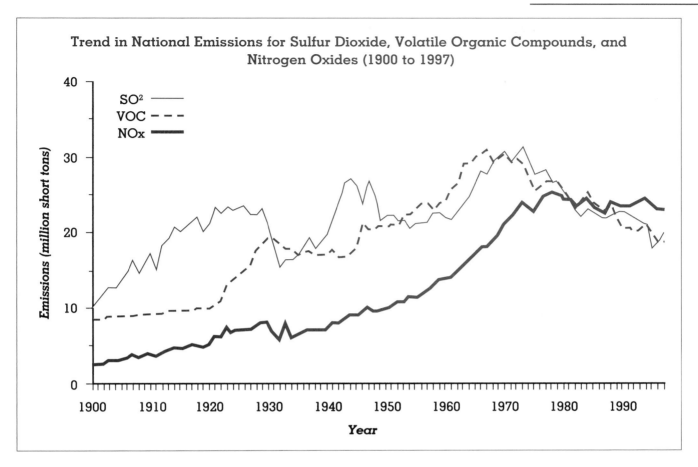

Trend in National Emissions for Sulfur Dioxide, Volatile Organic Compounds, and Nitrogen Oxides (1900 to 1997)

Figure 2.
Emissions of SO_2, NO_x, and volatile organic compounds in the United States, 1900–1997.

followed by a sharp decline during the Great Depression of the 1930s. World War II produced another peak, followed by a significant decline at the end of World War II. SO_2 emissions steadily rose from the early 1950s to the early 1970s. The Clean Air Act of 1970, which was directed at other air pollution concerns and not directly at acid depositon, was largely responsible for the decline in SO_2 emissions. At the time there was essentially no awareness that an "acid rain" problem existed in North America. By 1995, the implementation of the Clean Air Act Amendments (CAAA) of which specifically targeted SO_2 reductins to reduce acid deposition began to further decrease SO_2 emissions in a large part of the eastern United States, where acid deposition is most acute. NO_x emissions rose steadily until the 1970s when emissions leveled off and then showed a very slight decline.

SOURCES OF ACID DEPOSITION

Major sources for emissions of SO_2 and NO_x in the United States are presented in Figures 3 and 4 respectively. Approximately two-thirds of the SO_2 emissions are from electric utilities. Efforts to reduce SO_2 emissions both nationally and regionally have focused on electric utilities. The CAAA of 1990 have stipulated a reduction of 9.1 million metric tons (10 million short tons) of SO_2 below 1980 levels, with most of this reduction coming from coal-fired power plants. Implementation of Phase I reductions (1995–2000) has been successful and has resulted in an 18 percent decline in SO_2 emissions from electric utilities, compared with 1990 SO_2 emissions. There has been a 16 percent decline in SO_2 for this time period when all sources are considered. Phase 2 of the CAAA, which is designed to reduce SO_2 emissions from electric utilities by another 20 percent

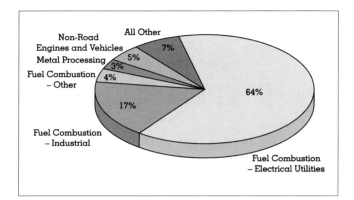

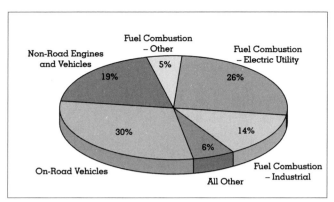

Figure 3.
National sulfur dioxide emissions by source category, 1997. Electric utilities account for almost two-thirds of SO_2 emissions, even after initial implementation of the CAAA of 1990. Total SO_2 emissions for 1997 were 18.5 million metric tons.

Figure 4.
National nitrogen oxide emissions by source category, 1997. Electric utilities and "on-road vehicles" account for more than half of the NO_x emissions. Total NO_x emissions for 1997 were 21.4 million metric tons.

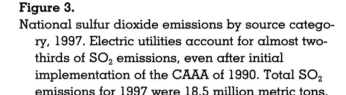

(compared to 1990), will go into effect from 2000 to 2005.

In the United States, recent reductions in emissions of SO_2 have been achieved by a shift to burning low sulfur coal and by the introduction of SO_2 scrubbers that remove SO_2 gases from power plant stacks. Most of the reductions in SO_2, mandated by the CAAA have come from the shift to burning low sulfur coal from the western United States. Electric utilities account for about one-fourth of the NO_x emissions. However, the largest single sources of NO_x emissions are "on-road vehicles," mainly cars and trucks, which account for 30 percent of the NO_x emissions (Figure 4). Control of NO_x emissions from vehicles is technically more difficult to achieve. Utilities can meet these targets for coal-fired boilers by using low NO_x burner technology (LNBT) or by "emissions averaging." Emissions averaging for a utility requires over-control by one boiler to make up for higher emissions at another boiler.

EFFECTS OF ACID DEPOSITION

Acid deposition and the associated particulate nitrates and sulfates are implicated in the deterioration of certain sensitive ecosystems, decreased visibility, negative human health effects, and increased degradation of certain stone building materials and cultural resources, especially those made of limestone and marble. Fine particulate nitrate and sulfate particles associated with acid deposition are implicated in aggravating cardiorespiratory diseases such as asthma and chronic bronchitis, especially in urban areas.

In many cases estimating the impact of acid deposition on various ecosystems can be a difficult process because acid deposition is only one of many impacts that can effect a response. However, wet and dry acid deposition has been documented as a major factor in the following ecosystem responses.

Aquatic Effects

In both Europe and eastern North America the negative impacts of acid deposition were first documented in lakes and streams found in acid-sensitive areas. In the early 1970s the loss of fish populations and increasing acidity in rural lakes and streams were documented both in Scandinavia and North America. In the United States, studies showed increasing acidification of lakes and loss of fish populations in the Adirondack Mountains of New York. The increased dissolved inorganic aluminum leaching from watersheds due to increased acidity proved toxic to fish in this region. In addition to dissolved aluminum toxicity, increased acidification leads to a large-scale disruption of lake food webs. For example, the experimental acidification of an entire Canadian lake (pH 6.8 to pH 5.09 from 1976 to 1983) led progressively to a loss of freshwater shrimp, all minnow species, and crayfish. These were important food sources for the lake trout population. By the end of the experiment all fish reproduction had ceased. There were also large changes in the species compositon of smaller organisms (insects and crustaceans) lower in the food chain.

Another aquatic impact of acid deposition is episodic acidification. For example, one form of episodic acidification occurs during spring snow melt. When a winter snowpack first melts, acids and other soluble material are concentrated and released, causing an initial "acid pulse" of meltwater, with acidity levels that may be higher than any of the original snowfall. These highly acid episodes, which are also often associated with high dissolved aluminum concentration in runoff, can be especially damaging in streams where fish and other organisms cannot seek refuge in less acid waters. Large storms, which produce high amounts of runoff during other seasons, can also produce episodic acidification.

Historical evidence has linked acidic deposition to the acidification of surface waters in many regions of eastern North America and Europe. Thousands of lakes and streams in these areas are significantly more acid than they were a few decades ago. Large regions of eastern Canada lying on the resistant bedrock of the Precambrian Shield are sensitive to ongoing acidification. In the eastern United States, surface water acidification has occurred in the Adirondack Mountains, the Pocono/Catskill region, the mid-Appalachians, the eastern portion of the Upper Midwest, the New Jersey Pine Barrens, and to a lesser extent, the Florida panhandle. Even with reduced emissions, acidification continues today in many regions. One reason improvements have been smaller than expected is that declines in sulfur emissions have also been accompanied by declines in emissions and deposition of acid-neutralizing agents found in atmospheric dust, both in North America and Europe. The most likely causes for the declines in atmospheric dust are cleaner combustion processes, controls on particulate emissions from smokestacks, changing agricultural practices (no-till), and fewer unpaved roads.

Controlling the effects of acid deposition by the use of lime or other acid-neutralizing compounds has been tried, but mainly on an experimental basis. Adding lime to lakes usually has only a short-term effect in terms of neutralizing lake acidity. The longevity of the effect is directly related to lake's

The pollution rising from the smokestacks of this power plant can cause acid rain. (Photo Researchers Inc.)

water residence time, or how long it takes for the lake volume to be replaced with new water. Another experimental control is the liming of a lake or stream watershed. While such an approach can improve forest health, as well as reduce lake acidity for a more extended time, it is prohibitively expensive as a widespread solution to acid deposition.

Terrestial Effects

The documentation of regional level terrestrial consequences of acid deposition is complicated. For example, forested ecosystems in eastern North America can be influenced by other factors such as high atmospheric ozone concentrations, drought, insect outbreaks and disease, sometimes from non-native sources. However there is a general consensus on some impacts of acidic depositon on both soils and forests in sensitive regions.

In the eastern United States, high elevation red spruce and fir forests found in the Northeast have suffered significant injury and mortality. Significant but lesser amounts of damage have also been found in high elevation spruce-fir forests in the southern Appalachians. Damage can occur directly to trees. For example, foliar leaching of plant nutrients such as calcium, and susceptability to winter cold damage thought to be the result of exposure to highly acid cloudwater can be direct impacts. Besides direct effects, acid deposition on poorly buffered, nutrient deficient soils has caused the leaching of valuable plant nutrients such as calcium, magnesium and other base cations and the release of aluminum, which can be injurious to plants and, as mentioned earlier, toxic to aquatic life. The loss of soil base cations can have long-term deleterious effects and may delay recovery of stressed ecosystems for decades or even longer. Such long-term soil nutrient losses also occur in sensitive low elevation forested ecosystems.

The role of nitrogen in the form of nitrate (NO_3^-) from both wet and dry nitric acid deposition can have both positive and negative effects on ecosystems. NO_3^- is an important plant nutrient and in nitrogen-poor soils can lead to increased plant growth and vigor. However in many acid-sensitive soils receiving high acidic deposition, nitrogen in soils is at or near saturation and can lead to the leaching of other important plant nutrients such as base cations or the release of aluminum into solution. The relative importance of nitric acid deposition will continue to grow as substantial reductions in SO_2 emissions occur and emissions of NO_x do not appreciably decline.

CONCLUSIONS

Acid deposition is a regional problem wherever large amounts of fossil fuels are consumed. There have been significant efforts in both Europe and North America to reduce acid deposition because of its many deleterious effects. This effort has focused mainly on the reduction of SO_2 emissions. In the future acid deposition will have to be addressed in eastern Asia, where rapid industrialization and increased use of fossil fuels is likely.

In the United States the passage of the CAAAs of 1990, and their implementation starting in 1995, was an attempt to "solve" the acid rain problem mainly by reducing SO_2 emissions from electric power plants. While significant reductions in SO_2 emissions have occurred, and there already has been

a reduction in deposition of sulfur and acidity, there have not been significant improvements in some sensitive regions such as the Adirondack Mountains of New York, the Green and White Mountains of New England, and the southern Appalachians. However, further reductions in SO_2 (and less so for NO_x) are expected until the year 2005. However, this too may not be enough. It is likely that further reductions in emissions of both SO_2 and NO_x, beyond those required by the CAAA, will be necessary if the goal is to protect sensitive ecosystems and public health. One reason recovery has been very limited is that acid deposition over decades has removed base cations from watersheds that are crucial to maintaining proper soil chemistry for plant growth and acid-neutralizing capacity for aquatic ecosystems. Recovery may be a slow process. Another concern is the very limited reduction in NO_x emissions over the last decade. Nitric acid is becoming a major component of acid deposition and significant reductions in NO_x emissions will probably be necessary to "solve" the acid rain problem.

Tom Butler

See also: Air Pollution; Atmosphere.

BIBLIOGRAPHY

Cowling, E. B. (1982). "Acid Precipitation in Historical Perspective." *Environmental Science and Technology* 16: 110A–123A.

Crowther, C., and Ruston, H. G. (1911). *Journal of Agricultural Science* 4:25–55.

Dannevig, A. (1959). *Jager og Fisker* 3:116–118.

Gorham, E. (1957). *Limnology and Oceanography* 2:22

Gorham, E (1958). *Phil. Trans. Royal Soc. London, Ser. B* 247:147–178.

Likens, G. E., and Bormann, F. H. (1974). "Acid Rain: A Serious Environmental Problem." *Science* 184:1176–1179

Likens, G. E.; Bormann, F. H.; and Johnson, N. M. (1972). "Acid rain." *Environment* 14:33–40.

Oden, S. (1968). "The Acidification of Air and Precipitation and Its Consequences in the Natural Environment." *Ecology Committee Bulletin No. 1*, Swedish National Science Research Council, Stockholm. Arlington, VA, Translation Consultants Ltd.

Schindler, D. W.; Mills, K. H.; Malley, D. F.; Findlay, D. L.; Shearer, J.A.; Davies, I. J.; Turner, M. A.; Linsey, G. A.; Cruikshank, D. R. (1985). "Long Term Ecosystem Stress: The Effects of Years of Experimental Acidification on a Small Lake." *Science* 228:1395–1401.

Smith, R. A. (1872). *Air and Rain: The Beginnings of a Chemical Climatology*. London: Longmans Greene.

ACOUSTICAL ENERGY

See: Waves

ADVANCED TRAFFIC MANAGEMENT SYSTEMS

See: Traffic Flow Management

AERODYNAMICS

The next time you hear an airplane flying overhead, look up, and pause for a moment. What you see is a machine that is heavier than air, but which is somehow being sustained in the air. This is due to the airflow over the airplane. This airflow exerts a lift force which counteracts the weight of the airplane and sustains it in the air—a good thing. The airflow also exerts a drag force on the airplane which retards its motion—a bad thing. The drag must be counteracted by the thrust of the engine in order to keep the airplane going. The production of thrust by the engine consumes energy. Hence, the energy efficiency of the airplane is intimately related to aerodynamic drag. This is just one of many examples where the disciplines of aerodynamics and energy interact.

DEFINITION

Aerodynamics deals with the flow of gases, particularly air, and the interaction with objects immersed in the flow. The interaction takes the form of an aerodynamic force and moment exerted on the object by the flow, as well as heat transfer to the object (aerodynamic heating) when the flow velocities exceed several times the speed of sound.

SOURCES OF AERODYNAMIC FORCE

Stop for a moment, and lift this book with your hands. You are exerting a force on the book; the book is feeling this force by virtue of your hands being in contact with it. Similarly, a body immersed in a liquid or a gas (a fluid) feels a force by virtue of the body surface being in contact with the fluid. The forces exerted by the fluid on the body surface derive from two sources. One is the pressure exerted by the fluid on every exposed point of the body surface. The net force on the object due to pressure is the integrated effect of the pressure summed over the complete body surface. In the aerodynamic flow over a body, the pressure exerted by the fluid is different at different points on the body surface (i.e., there is a distribution of variable values of pressure over the surface). At each point, the pressure acts locally perpendicular to the surface. The integrated effect of this pressure distribution over the surface is a net force—the net aerodynamic force on the body due to pressure. The second source is that due to friction between the body surface and the layer of fluid just adjacent to the surface. In an aerodynamic flow over a body, the air literally rubs over the surface, creating a frictional shear force acting on every point of the exposed surface. The shear stress is tangential to the surface at each point, in contrast to the pressure, which acts locally perpendicular to the surface. The value of the shear stress is different at different points on the body surface. The integrated effect of this shear stress distribution is a net integrated force on the body due to friction.

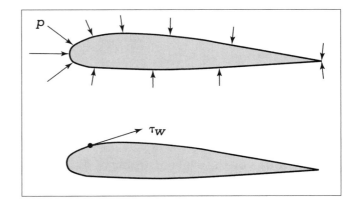

Figure 1.
Schematic of pressure and shear stress distribution over a body surface.

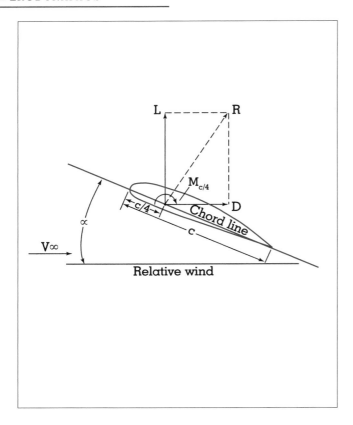

Figure 2.
Resolution of aerodynamic force into lift and drag.

The pressure (p) and shear stress (τ_ω) distributions over an airfoil-shaped body are shown schematically in Figure 1. The pressure and shear stress distributions exerted on the body surface by the moving fluids are the two hands of nature that reach out and grab the body, exerting a net force on the body—the aerodynamic force.

RESOLUTION OF THE AERODYNAMIC FORCE

The net aerodynamic force exerted on a body is illustrated in Figure 2 by the arrow labeled R. The direction and speed of the airflow ahead of the body is denoted by V_∞, called the relative wind. The body is inclined to V_∞. by the angle of attack, α. The resultant aerodynamic force R can be resolved into two components; lift, L, perpendicular to V_∞; and drag, D, parallel to V_∞. In Figure 2, R is shown acting through a point one-quarter of the body length from the nose, the quarter-chord point. Beacuse the aerodynamic

force derives from a distributed load due to the pressure and shear stress distributions acting on the surface, its mechanical effect can be represented by a combination of the net force vector drawn through any point and the resulting moment about that point. Shown in Figure 2 is R located (arbitrarily) at the quarter-chord point and the moment about the quarter-chord point, $M_{c/4}$.

The aerodynamic force varies approximately as the square of the flow velocity. This fact was established in the seventeenth century—experimentally by Edme Marione in France and Christiaan Huygens in Holland, and theoretically by Issac Newton. Taking advantage of this fact, dimensionless lift and drag coefficients, C_L and C_D respectively, are defined as

$$L = \tfrac{1}{2}\rho_\infty V_\infty^2 \, S \, C_L$$
$$D = \tfrac{1}{2}\rho_\infty V_\infty^2 \, S \, C_D$$

where ρ_∞, is the ambient density in the freestream, and S is a reference area, which for airplanes is usually chosen to be the planform area of the wings (the projected wing area you see by looking at the wing directly from the top or bottom), and for projectile-like bodies is usually chosen as the maximum cross-sectional area perpendicular to the axis ofthe body (frontal area).

At flow speeds well below the speed of sound, the lift coefficient depends only on the shape and orientation (angle of attack) of the body:

$$C_L = f \,(\text{shape}, \alpha)$$

The drag coefficient also depends on shape and α, but in addition, because drag is partially due to friction, and frictional effects in a flow are governed by a powerful dimensionless quantity called Reynolds number, then C_D is also a function of the Reynolds number, Re:

$$C_D = f \,(\text{shape}, \alpha, \text{Re})$$

where $\mathrm{Re} \cong \rho_\infty V_\infty$, c/μ_∞. Here, c is the reference length of the body and μ_∞ is the viscosity coefficient of the fluid. At speeds near and above the speed of sound, these coefficients also become functions of Mach number, $M_\infty \cong V_\infty/a_\infty$, where a_∞, is the speed of sound in the freestream:

$$C_L = f \,(\text{shape}, \alpha, M_\infty)$$
$$C_D = f \,(\text{shape}, \alpha, \text{Re}, M_\infty)$$

The lift and drag characteristics of a body in a flow are almost always given in terms of C_L and C_D rather than the forces themselves, because the force coefficients are a more fundamental index of the aerodynamic properties.

DRAG

One of the most important aerodynamic effects on the consumption of energy required to keep a body moving through a fluid is the aerodynamic drag. The drag must be overcome by the thrust of a propulsion mechanism, which in turn is consuming energy. Everything else being equal, the higher the drag, the more energy is consumed. Therefore, for energy efficiency, bodies moving through a fluid should be low-drag bodies. To understand how to obtain low drag, we have to first understand the nature of drag, and what really causes it.

The influence of friction on the generation of drag is paramount. In most flows over bodies, only a thin region of the flow adjacent to the surface is affected by friction. This region is called the boundary layer (Figure 3). Here, the thickness of the boundary layer is shown greatly exaggerated; in reality, for ordinary flow conditions, the boundary layer thickness, δ, on the scale of Figure 3 would be about the thickness of a sheet of paper. However, the secrets of drag production are contained in this very thin region. For example, the local shear stress at the wall, labeled in Figure 3 as τ_ω, when integrated over the entire surface creates the skin friction drag D_f on the body. The magnitude of τ_ω, hence, D_f, is determined by the nature of the velocity profile through the boundary layer, (i.e., the variation of the flow velocity as a function of distance y normal to the surface at a given station, x, along the surface). This velocity variation is quite severe, ranging from zero velocity at the surface (due to friction, the molecular layer right at the wall is at zero velocity relative to the wall) to a high velocity at the outer edge of the boundary layer. For most fluids encountered in aerodynamics, the shear stress at the surface is given by the Newtonian shear stress law:

$$\tau_\omega = \mu \, (dV/dy)_{y=0}$$

where μ is the viscosity coefficient, a property of the fluid itself, and $(dV/dy)_{y=0}$ is the velocity gradient at the wall. The more severe is the velocity variation in the boundary layer, the larger is the velocity gradient

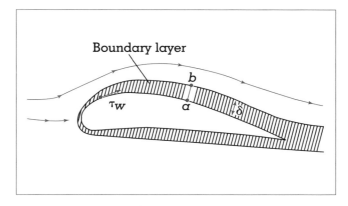

Figure 3.
The boundary layer.

at the wall, and the greater is the shear stress at the wall.

The above discussion has particular relevance to drag when we note that the flow in the boundary layer can be of two general types: laminar flow, in which the streamlines are smooth and regular, and an element of the fluid moves smoothly along a streamline; and turbulent flow, in which the streamlines break up and a fluid element moves in a random, irregular, and tortuous fashion. The differences between laminar and turbulent flow are dramatic, and they have a major impact on aerodynamics. For example, consider the velocity profiles through a boundary layer, as sketched in Figure 4. The profiles are different, depending on whether the flow is laminar or turbulent. The turbulent profile is "fatter," or fuller, than the laminar profile. For the turbulent profile, from the outer edge to a point near the surface, the velocity remains reasonably close to the freestrearn velocity; it then rapidly decreases to zero at the surface. In contrast, the laminar velocity profile gradually decreases to zero from the outer edge to the surface. Now consider the velocity gradient at the wall, $(dV/dy)_{y=0}$, which is the reciprocal of the slope of the curves shown in Figure 4 evaluated at y = 0. It is clear that $(dV/dy)_{y=0}$ for laminar flow is less than $(dV/dy)_{y=0}$ for turbulent flow. Recalling the Newtonian shear stress law for τ_ω leads us to the fundamental and highly important fact that laminar stress is less than turbulent shear stress:

$$\tau_{\omega \text{ laminar}} < \tau_{\omega \text{ turbulent}}$$

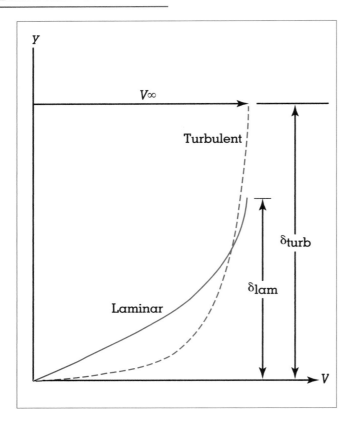

Figure 4.
Comparison of the velocity profiles for laminar and turbulent boundary layers.

This obviously implies that the skin friction exerted on an airplane wing or body will depend on whether the boundary layer on the surface is laminar or turbulent, with laminar flow yielding the smaller skin friction drag.

It appears to be almost universal in nature that systems with the maximum amount of disorder are favored. For aerodynamics, this means that the vast majority of practical viscous flows are turbulent. The boundary layers on most practical airplanes, missiles, and ship hulls, are turbulent, with the exception of small regions near the leading edge. Consequently, the skin friction on these surfaces is the higher, turbulent value. For the aerodynamicist, who is usually striving to reduce drag, this is unfortunate. Today, aerodynamicists are still struggling to find ways to preserve laminar flow over a body—the reduction in skin friction drag and the resulting savings in energy are well worth such efforts. These efforts can take the form of shaping the body in such a way to encourage laminar flow; such "laminar flow bodies" are designed

to produce long distances of decreasing pressure in the flow direction on the surface (favorable pressure gradients) because an initially laminar flow tends to remain laminar in such regions. Figure 5 indicates how this can be achieved. It shows two airfoils, the standard airfoil has a maximum thickness near the leading edge, whereas the laminar flow airfoil has its maximum thickness near the middle of the airfoil. The pressure distributions on the top surface on the airfoils are sketched above the airfoils in Figure 5. Note that for the standard airfoil, the minimum pressure occurs near the leading edge, and there is a long stretch of increasing pressure from this point to the trailing edge. Turbulent boundary layers are encouraged by such increasing pressure distributions. The standard airfoil is generally bathed in long regions of turbulent flow, with the attendant high skin friction drag. Note that for the laminar flow airfoil, the minimum pressure occurs near the trailing edge, and there is a long stretch of decreasing pressure from the leading edge to the point of minimum pressure. Laminar boundary layers are encouraged by such decreasing pressure distributions. The laminar flow airfoil can be bathed in long regions of laminar flow, thus benefiting from the reduced skin friction drag.

The North American P-51 Mustang, designed at the outset of World War II, was the first production aircraft to employ a laminar flow airfoil. However, laminar flow is a sensitive phenomenon; it readily gets unstable and tries to change to turbulent flow. For example, the slightest roughness of the airfoil surface caused by such real-life effects as protruding rivets, imperfections in machining, and bug spots can cause a premature transition to turbulent flow in advance of the design condition. Therefore, most laminar flow airfoils used on production aircraft do not yield the extensive regions of laminar flow that are obtained in controlled laboratory tests using airfoil models with highly polished, smooth surfaces. From this point of view, the early laminar flow airfoils were not successful. However, they were successful from an entirely different point of view; namely, they were found to have excellent high-speed properties, postponing to a higher flight Mach number the large drag rise due to shock waves and flow separation encountered near Mach 1. As a result, the early laminar flow airfoils were extensively used on jet-propelled airplanes during the 1950s and 1960s and are still employed today on some modern high-speed aircraft.

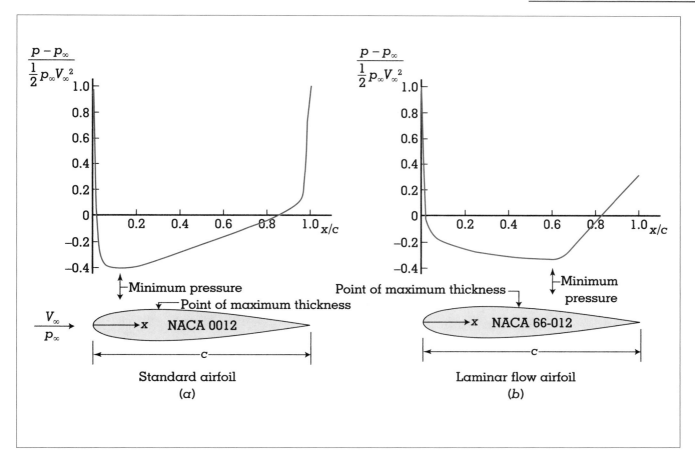

Figure 5.
Pressure distributions over (a) a standard airfoil shape and (b) a laminar low airfoil.

In reality, the boundary layer on a body always starts out from the leading edge as laminar. Then at some point downstream of the leading edge, the laminar boundary layer become unstable and small "bursts" of turbulent flow begin to grow in the flow. Finally, over a certain region called the transition region, the boundary layer becomes completely turbulent. For purposes of analysis, it is convenient to draw a picture, where transition is assumed to occur at a point located a distance x_{cr}, from the leading edge. The accurate knowledge of where transition occurs is vital to an accurate prediction of skin friction drag. Amazingly, after almost a century of research on turbulence and transition, these matters are still a source of great uncertainty in drag predictions today. Nature is still keeping some of her secrets from us.

Skin friction drag is by no means the whole story of aerodynamic drag. The pressure distribution integrated over the surface of a body has a component parallel to the flow velocity V_∞, called form drag, or more precisely pressure drag due to flow separation. In this type of drag, such as the flow over a sphere, the boundary layer does not totally close over the back surface, but rather separates from the surface at some point and then flows downstream. This creates a wake of low-energy separated flow at the back surface. The pressure on the back surface of the sphere in the separated wake is smaller than it would be if the flow were attached. This exacerbates the pressure difference between the higher pressure on the front surface and the lower pressure on the back surface, increasing the pressure drag. The bigger (fatter) the wake, the higher the form drag. Once again we see the different effects of laminar and turbulent flow. In the case where the boundary layer is laminar, the boundary layer separates near the top and bottom of the body, creating a large, fat wake, hence high pressure drag. In contrast, where the boundary layer is turbulent, it separates further around the back of the sphere, creating a thinner wake, thus lowering the pressure drag. Form drag, therefore, is larger for laminar flow than for turbulent flow. This is the exact opposite of the case for skin friction drag. To reduce form drag, you want a turbulent boundary layer.

For a blunt body, such as a sphere, almost all the drag is form drag. Skin friction drag is only a small percentage of the total drag. For blunt bodies a turbulent boundary layer is desirable. Indeed, this is the purpose of the dimples on the surface of a golf ball—to promote turbulent flow and reduce the aerodynamic drag on the ball in flight. The nose of an airplane is large compared to a golf ball. Hence, on the airplane nose, the boundary layer has already become a turbulent boundary layer, transitioning from laminar to turbulent in the first inch or two from the front of the nose. Therefore, dimples are not necessary on the nose of an airplane. In contract, the golf ball is small—the first inch or two is already too much, so dimples are placed on the golf ball to obtain turbulent flow right from the beginning.

For a body that is producing lift, there is yet another type of drag—induced drag due to lift. For example, consider an airplane wing, that produces lift by creating a higher pressure on the bottom surface and a lower pressure on the top surface. At the wing tips, this pressure difference causes the flow to try to curl around the tips from the bottom of the tip to the top of the tip. This curling motion, superimposed on the main freestream velocity, produces a vortex at each wing tip, that trails downstream of the wing. These wing tip vortices are like minitornadoes that reach out and alter the pressure distribution over the wing surface so as to increase its component in the drag direction. This increase in drag is called induced drag; it is simply another source of pressure drag on the body.

Finally, we note that if the body is moving at speeds near and above the speed of sound (transonic and supersonic speeds), shock waves will occur that increase the pressure on the front portions of the body, contributing an additional source of pressure drag called wave drag.

In summary, the principal sources of drag on a body moving through a fluid are skin friction drag,

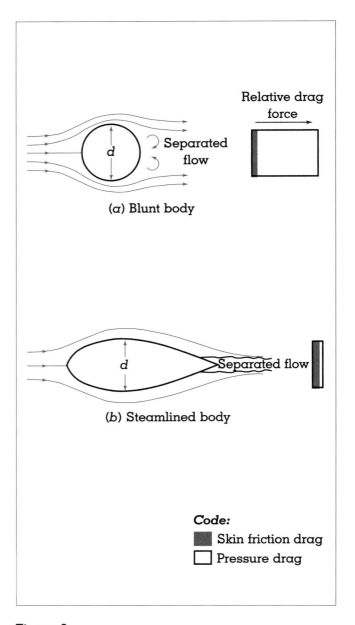

Figure 6.

Comparison of drag for a blunt body and a streamlined body.

form drag, induced drag, and wave drag. In terms of drag coefficients, we can write:

$$C_D = C_{D,f} + C_{D,p} + C_{D,i} + C_{D,w}$$

where C_D is the total drag coefficient, $C_{D,f}$ is the skin friction drag coefficient, $C_{D,p}$ is the form drag coefficient (pressure drag due to flow separation), $C_{D,i}$ is the induced drag coefficient, and $C_{D,w}$ is the wave drag coefficient.

STREAMLINING

The large pressure drag associated with blunt bodies such as the sphere, leads to the design concept of streamlining. Consider a body of cylindrical cross section of diameter d with the axis of the cylinder oriented perpendicular to the flow, as shown in Figure 6a. There will be separated flow on the back face of the cylinder, with a relatively fat wake and with the associated high pressure drag. The bar to the right of the cylinder denotes the total drag on the cylinder; the shaded portion of the bar represents skin friction drag, and the open portion represents the pressure drag. Note that for the case of a blunt body, the drag is relatively large, and most of this drag is due to pressure drag. However, look at what happens when we wrap a long, mildly tapered afterbody on the back of the cylinder, creating a teardrop-shaped body sketched in Figure 6b. This shape is a streamlined bod, of the same thickness d as the cylinder. Flow separation on the streamlined body will be delayed until much closer to the trailing edge, with an attendant, much smaller wake. As a result, the pressure drag of the streamlined body will be much smaller than that for the cylinder. Indeed, as shown by the bar to the right of Figure 6b, the total drag of the streamlined body will be almost a factor of 10 smaller than that of the cylinder of same thickness. The friction drag of the streamlined body will be larger due to its increased surface area, but the pressure drag is so much less that it dominates this comparison.

Streamlining has a major effect on the energy efficiency of bodies moving through a fluid. For example, a bicycle with its odd shaped surfaces, has a relatively large drag coefficient. In contrast, a streamlined outer shell used for recumbent bicycles reduces the drag and has allowed the speed record to reach 67 mph. Streamlining is a cardinal principle in airplane design, where drag reduction is so important.

Streamlining has a strong influence on the lift-to-drag ratio (L/D, or C_I/C_D) of a body. Lift-to-drag ratio is a measure of aerodynamic efficiency. For example, the Boeing 747 jumbo-jet has a lift-to-drag ratio of about 20. This means it can lift 20 lb at the cost of only 1 pound of drag—quite a leverage. In airplane design, an increase in L/D is usually achieved by a *decrease* in D rather than an increase in L. Vehicles that have a high L/D are that way because they are low-drag vehicles.

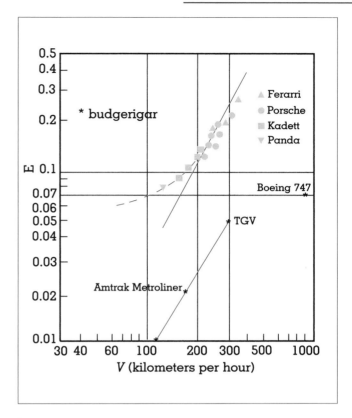

Figure 7.
Specific energy consumption versus velocity.

DRAG AND ENERGY

We now make the connection between aerodynamic drag and energy consumption, The drag of a moving vehicle must be overcome by the thrust from a propulsive mechanism in order to keep the vehicle in sustained motion. The time rate of energy consumption is defined as *power*, P. The power required to keep the vehicle moving at a given speed is the product of drag times velocity,

$$P = D V_\infty.$$

Because

$$D = \tfrac{1}{2}\rho_\infty V_\infty^2 S C_D,$$

we have

$$P = \tfrac{1}{2}\rho_\infty V_\infty^3 S C_D.$$

That is, the power required varies as the *cube* of the

13

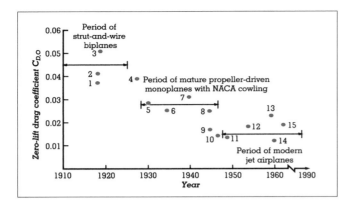

Figure 8.
Evolution of drag reduction for airplanes, drag coefficient versus years.

velocity, and directly as the drag coefficient. This clearly indicates why, as new vehicles are designed to move at higher velocities, every effort is made to reduce C_D. Otherwise, the velocity-cubed variation may dictate an amount of energy consumption that is prohibitive. Note that this is one of the realities facing civil transport airplanes designed to fly at supersonic speeds. No matter how you look at it, less drag means more energy efficiency.

The effect of aerodynamics on the energy consumption of transportation vehicles can be evaluated by using the dimensionless specific energy consumption, E, defined as $E = P/WV$, where P is the power required to move at velocity V and W is the weight of the vehicle, including its payload (baggage, passengers, etc.). Although power required increases as the cube of the velocity, keep in mind that the *time* required to go from point A to point B is inversely proportional to V, hence a faster vehicle operates for less time between two points. The quantity $E = P/WV$ is the total energy expended per unit distance per unit weight; the smaller the value of E, the smaller the amount of energy required to move 1 lb a distance of 1 ft (i.e., the more energy efficient the vehicle). Representative values of E for different classes of vehicles (trains, cars, airplanes) are given in Figure 7. Using E as a figure of merit, for a given long distance trip, trains such as the Amtrak Metroliner and the French high-speed TGV are most efficient, airplanes such as the Boeing 747, are next, and automobiles are the least efficient.

DRAG OF VARIOUS VEHICLES

Let us examine the drag of various representative vehicles. First, in regard to airplanes, the evolution of streamlining and drag reduction is clearly seen in Figure 8, which gives the values of drag coefficient based on wing planform area for a number of different aircraft, plotted versus years. We can identify three different periods of airplanes, each with a distinctly lowered drag coefficient: strut-and-wire biplanes, mature propellerdriven airplanes, and modem jet airplanes. Over the past century, we have seen a 70 percent reduction in airplane drag coefficient. Over the same period, a similar aerodynamic drag reduction in automobiles has occurred. By 1999, the drag coefficients for commercialized vehicles have been reduced to values as low as 0.25. There are experimental land vehicles with drag coefficients on par with jet fighters; for example, the vehicles built for the solar challenge races, and some developmental electric vehicles.

The generic effect of streamlining on train engines is similar, with the drag coefficient again based on frontal area. The high-speed train engines of today have drag coefficients as low as 0.2.

For motorcycles and bicycles, the drag coefficient is not easy to define because the proper reference area is ambiguous. Hence, the drag is quoted in terms of the "drag area" given by D/q, where q is the dynamic pressure; $q = \frac{1}{2}\rho_\infty V_\infty^2$. A typical drag area for a motorcycle and rider can be reduced by more than fifty percent by wrapping the motorcycle in a streamlined shell.

SUMMARY

Aerodynamics is one of the applied sciences that plays a role in the overall consideration of energy. We have explained some of the more important physical aspects of aerodynamics, and illustrated how aerodynamics has an impact on energy efficiency.

John D. Anderson, Jr.

See also: Automobile Performance; Tribology.

BIBLIOGRAPHY

Allen, J. E. (1982). *Aerodynamics: The Science of Air in Motion*, 2nd ed. New York: McGraw-Hill.

Anderson, J. D., Jr., (1991). *Fundamentals of Aerodynamics*, 2nd ed. New York: McGraw-Hill.

Anderson, J. D., Jr. (1997). *A History of Aerodynamics, and Its Impact on Flying Machines*. New York: Cambridge University Press.

Anderson, J. D., Jr. (2000). *Introduction to Flight,* 4th ed. Boston: McGraw-Hill.

Tennekes, H. (1992). *The Simple Science of Flight.* Cambridge, MA: MIT Press.

von Karman, T. (1963). *Aerodynamics.* New York: McGraw-Hill. (Originally published by Cornell University Press, 1954).

AGRICULTURE

The earliest human inhabitants of the earth were hunters and gatherers. Under favorable conditions they required at least 1.5 square kilometers to provide food for one person, and in harsher environments as much as 80 to 100 square kilometers. Population pressure eventually led humans to raise plants and animals.

With the introduction of agriculture, humans began to use energy to control the growth of plants and animals, to make more efficient use of the solar energy stored in plants by photosynthesis. However, for many thousands of years, the only energy used for this purpose was human energy. The energy drawn from the biosphere was limited to the dietary energy provided by plant food and meat, and to the use of wood and grasses as fuel for heating and cooking. Later, humans learned to use animal, water, and wind energy to obtain power for transport and for simple agricultural and industrial processes. As population grew, the use of energy increased steadily, but all of it came from renewable resources.

Shifting cultivation, one of the first agricultural practices developed, is still widely used. In the early 1970s, about 36 million square kilometers of land were farmed under this system, producing food for about 250 million people. Each hectare (2.47 acres or

A pair of oxen pull a wooden plow through a muddy paddy field in Sri Lanka. (Corbis-Bettmann)

10,000 square meters) under shifting cultivation can provide adequate food for one person. Degradation of soil and vegetation usually occurs when the population density exceeds one person per four hectares. At higher population densities, shorter fallow periods and eventually annual cropping become necessary.

Annual cropping by traditional methods requires more labor, and yields are generally lower. Although animal power can help reduce human labor and provide manure for fertilizer, draught animals must be either fed by some of the crop or pastured, thereby increasing the land area required per person unless yields per unit of land increase accordingly. Average livestock ate one-fourth of the yield in the early nineteenth century.

This land per person dynamic has changed in the past one hundred years as humans moved beyond the limitations of human and animal power by drawing upon nonrenewable sources of energy in the form of fossil fuels—coal, oil, and natural gas. Rapid improvements in technology made it possible to locate and extract increasing quantities of fossil fuels with little or no increase in cost. By the 1950s and 1960s the world had come to take advantage of the large supply of fossil fuels to dramatically and economically boost agricultural production.

Energy dense fossil fuels can be converted very efficiently to heat and/or power. Use of energy for power has spread rapidly in both the industrial and agricultural sectors of the developed countries. The developing countries have followed the developed countries along the same path but at a slower pace. Although agriculture uses only a small share of the world's total energy consumption, it is generally recognized that its needs are crucial, because the existing technologies for increasing production rely so heavily on energy-intensive inputs. Farm use of energy accounts for only 4 percent of total commercial energy use in developing countries and 3 percent in the developed countries. Commercial energy includes oil, natural gas, and coal. Noncommercial sources of energy include fuel-wood, plant and animal residues, and human and animal energy. Noncommercial energy uses are very important in the developing countries, especially in rural areas.

There is a two-way relationship between agriculture and energy because agriculture is both a consumer and a producer of energy. Directly or indirectly, the whole agricultural process involves the input of energy. The ability of plants to capture and store the sun's energy through photosynthesis can be increased through the input of energy in the form of fertilizer, pesticides, and fuel to drive machinery used in production. The production of nitrogen fertilizers is very energy intensive input because it requires hydrogen-dissociation from water or fossil fuels (natural gas) to fix nitrogen in synthetic fertilizers. Production of pesticides similar to nitrogen fertilizers requires large amounts of energy. The gasoline and diesel fuel used for agricultural machinery also consumes a lot of energy, but has dropped significantly since the early 1970s because of better technology.

The role of energy in agricultural production has become crucial as population and income growth put pressure on the demand for food. By discovering new technologies (hybrid seeds, drought and disease-resistant crops, as well as genetically modified crops), dramatic increases in crop yields per hectare have been achieved in the developed countries. The pace of growth, however, has been uneven among developing countries. Asia has shown significant increases in agricultural productivity, while many countries in Africa and Latin America have had modest productivity growth. Although many developing countries have reserves of unused but potentially productive land, most have to meet rising food demands by substantially raising yields on both used and new land. The growth in yields does not come automatically. An increase in agricultural research investment is essential both to increase the yield and to expand the area under cultivation. Clearing less accessible new land for cultivation, which is the goal of many environmentalist groups, will require greater investment in inputs of commercial energy and often the provision of drainage, irrigation, and soil conservation systems. Once the land is under production, additional energy will be needed for maintenance.

CLASSIFICATION OF AGRICULTURAL PRODUCTION

World production of food and fiber could be classified into three distinct groups: traditional, transition from traditional to modern, and modern farming methods.

Traditional Agriculture

Traditional farms rely on labor, draught animals, and hand tools. The only commercial energy input in traditional farming is that required to produce hand

English farmers use a motorized tractor to drive a plow through a field around 1905. At that time most farmers still used horses for this purpose. (Corbis-Bettmann)

tools and animal implements. No commercial energy inputs are used for irrigation; the land either gets rain or it's irrigated by traditional methods.

Traditional methods continue to be the mode of operation in many parts of the world. Currently agriculture production in many of the lowest income countries in Africa, Latin America, and Asia are based on traditional methods. Agriculture is largely operated by small family farms using human and animal power and organic fertilizer with little access to or knowledge of modern inputs such as chemicals, fertilizers, hybrid seeds, or mechanical drive. Low soil fertility and inadequate or irregular rainfall sharply limit the productivity of low-input farms in developing countries.

The Transition to Modern Farming

In the transition to modern farming methods, the use of commercial energy, especially machinery and fertilizer, increases sharply. Primary tillage is usually one of the first operations to be mechanized, requiring an increase in commercial energy use not only for the production of farm machinery, but also for its operation. Improved crop varieties are often introduced during the transitional phase, requiring commercial energy for their production and distribution. To help realize their yield potential entails the use of chemical fertilizers and pesticides. Both of these inputs require commercial energy for their manufacture. In addition to the traditional irrigation methods, supplementary irrigation with mechanically powered pumps is often introduced during the transitional phase. This process substantially increases commercial energy requirements but also increases yields. The growing needs for investment during transition influence farm size. To achieve economies of scale, the number of family farms is reduced and they are replaced with larger commercial farms.

This general trend toward larger farms, greater mechanization, and greater use of commercial inputs in most countries results in greater productivity but at the cost of greater direct and indirect energy use. The combination of increased irrigation, use of high yield variety crops, and new inputs has contributed to steady increases in both absolute and percapita agricultural production.

Modern Commercial Agriculture

To feed the growing population, agricultural productivity must increase. Modern inputs are needed to increase agricultural production. Commercial fuel inputs to agriculture include mechanized land preparation, mechanized irrigation, and synthetic fertilizer. Modern commercial agriculture is greatly dependent on high yielding varieties, and modern pesticides.

The degree of utilization of the above inputs varies widely, but generally increases with economic development. These modern technologies reduce the time and labor needed for land preparation, plowing, planting, and harvesting crops. In favorable areas, it also aids double cropping management.

In the developed countries, stages in the historical development of agricultural production have been characterized by differing combinations of inputs: differences in relative proportion of land, labor and capital, and in the composition of the capital inputs.

Such changes reflect primarily the changing structure of the economy and the successive advances that have been made in agricultural technology.

The well-documented case of the United States serves as an illustration. During the period 1870 to 1900 the farm population was increased through a rapid expansion of the agricultural area. The agricultural labor force increased by 60 percent, but there was a replacement of labor by nonland capital in the form of horses and mules. New and more efficient types of horse-drawn machinery including plows, cultivars, seed drills, grain harvesters, and mowers became available.

The following period, from 1900 to 1930, was a period of transition: the first half was the beginning of the end for traditional farming, based predominantly on large inputs of relatively unskilled labor, and the second half the beginning of commercial agriculture, technologically oriented and capital intensive. The crop-land use continued to increase until about 1920 but remained relatively stable thereafter, and the agricultural labor force continued to increase until about 1918. Equally significant, however, were the shifts that became evident in the composition of non-real-estate capital input: the replacement of horses and mules by tractors, automobiles and trucks; of manpower by motor-driven machinery and equipment; and the purchase of production inputs such as fertilizer, lime, seed, and feed required for the application of improved production techniques.

Between 1935 and 1960 outputs per man-hour of labor increased about 4.5 times, and crop production per hectare of crop-land almost doubled. Inputs of labor were decreased by 50 percent, inputs of land remained relatively stable, but inputs of non-real-estate capital inputs were nearly tripled. Among these capital inputs, those of seed, feed, and livestock purchased increased by about four times, and those of mechanical power and machinery by more than 2.5 times.

It has been estimated that of the total U.S. increase in farm output between 1940 and 1955, 43 percent is attributable to increased crop yields per hectare, 27 percent to increases in value added by livestock production, 23 percent to reduction in farm-produced power, and 7 percent to changes in the amount of capital used. While it is not possible to isolate the effect of a single input, it is estimated that increased use of fertilizer accounted for more than half of the increase in crop production per hectare. Other important causes of increased crop yields include the use of hybrid maize and other improved plant varieties, irrigation, better soil tillage practices, more timely planning, cultivation and harvesting operations, and better weed, insect, and disease control.

ENERGY USE IN TRADITIONAL AND MODERN AGRICULTURE

Commercial energy plays a major role in modern methods of production. Energy use even with modern methods varies by crop. Crops requiring irrigation use much more energy than rain-fed crops. About 42 percent of total energy is used for irrigation, 20 percent for the manufacture and operation of farm machinery, 18 percent for fertilizers, and 7 percent for drying. For modern rice production in the United States, the commercial energy input is more than 500 times that in traditional production and more than 10 times in transitional production in the Philippines. For modern corn production in the United States, total commercial energy use is only about half that in modern rice production, mainly because little or no irrigation is required. Regardless of the crop produced, energy use is much higher with modern production method. For example, commercial energy input in the United States is more than 174 times that in traditional production in the Mexico.

With modern production methods 1500 kilograms of petroleum per hectare are needed for rice and 700 kilograms for maize. However, with this commercial energy use, yields of 5.8 metric tons per hectare have been obtained for rice and 5 metric tons per hectare for maize—about five times those obtained with traditional methods, Thus, 20 to 25 people can be fed on an all-grain diet from a single hectare compared with 4 to 6 people by traditional methods.

Total energy used to produce a kilogram of rice on Philippine traditional farms requires three grams of petroleum and in transitional farms 55 grams. In the U.S. modern rice farming system the requirement increases to 258 grams of petroleum. Similarly, the ratio for corn production in Mexico for traditional farming is equal to four grams of petroleum, and for modern corn farming in the United States is equal to 137 grams of petroleum.

Despite higher dependence on energy in modern methods of production, agriculture is responsible for only a small part of total energy use. Total energy use as well as energy used in agricultural production in selected developing and developed countries and for the largest food producers in the world; the United States and China are compared in Table 1. In 1996, agricultural shares of energy use for all developed countries were 2 percent, whereas in the developing countries the proportion was slightly higher. Agricultural shares of total commercial energy use in the developing countries ranged from 0.67 percent in Egypt to 5.08 percent in Brazil. For the developed countries agricultural share of energy ranged from 1.09 percent in the United States to 3.39 percent for the European Community.

COMMERCIAL ENERGY USE IN AGRICULTURE

Agriculture uses a variety of commercial energy forms, directly for operating machinery and equipment on the farm and indirectly through the use of fertilizer and pesticides produced off the farm. In addition, commercial energy is used in manufacturing of farm machinery and farm equipment. During 1972 to 1973 farm machinery manufacture and operation, the largest user of commercial energy in agriculture, accounted for 51 percent of the world total and ranged from 8 percent in the Far East to 73 percent in Oceania. Chemical fertilizer was second with 45 percent of the world total and ranged from 26 percent in Oceania to 84 percent in the Far East. However, in the developing countries chemical fertilizer was first.

About 1.5 billion hectares of crop land are planted for annual crops and permanent crops. About one third of the arable and permenant crop-land is located in the United States, India, and China. Out of 1.5 billion hectares, 263 million hectares are irrigated. Irrigated lands in China and India account for the largest share—about 110 million hectares.

Reports by the International Energy Agency indicate that total renewable energy use in developing countries and in agriculture is high, but no data is available. The available data on selected countries shows that the use of renewable energy in developing countries varies from 5 percent in Egypt to 86 percent in Nigeria. Energy share of the renewable energy in the developed countries ranged from 1 percent in Japan to 6 percent in Australia.

To measure the efficiency of energy use in agri-

	1996 Total energy use				
	Fossil	Renewable	Energy used in agriculture	Energy share of Renewable	Agriculture
	Million tons of oil equivalent			percent	percent
Developing Countries:					
Bangladesh	22.32	15.76	0.48	71	
Brazil	138.23	33.57	7.02	24	5.08
China	865.86	206.08	26.52	24	3.06
Egypt	25.2	1.19	0.17	5	0.67
India	350.26	188.65	8.1	54	2.31
Indonesia	99.76	43.83	1.67	44	1.67
Nigeria	73.12	63		86	
Pakistan	46.82	21.09	0.84	45	1.79
Total developing	**3270.3**	**824.73**		**25**	
Developed Countries:					
Australia	66.11	3.99	1.5	6	2.27
Canada	181.92	8.58	4.13	5	2.27
EC(15) 1/	1010.54	26.37	23.15	3	2.29
Japan	337.08	3.42	11.44	1	3.39
USA	1443	30.09	15.8	2	1.09
Total developed	**3479.11**	**96.32**	**72.61**	**3**	**2.09**
World	**6749.41**	**921.05**		**14**	

Table 1.
Total fossil and renewable energy use, energy use in agriculture, and agriculture energy share for selected developed and developing countries, 1996.
Note: European Community (15) includes Austria, Belgium-Luxembourg, Denmark, Finland, France, Germany, Greece, Ireland, Italy, Netherlands, Portugal, Spain, Sweden, and United Kingdom.
SOURCE: *Energy Statistics and Balances of OECD and Non-OECD countries, 1995–96*, International Energy Agency. Paris, 1998.

culture, energy use in agriculture is compared with agriculture's share of gross domestic products (GDP). For many developing countries, the share of agricultural GDP is much larger than the share of energy used in agriculture, which suggests that much of the agricultural sector is still engaged in subsistence, low input, labor-intensive farming. For some middle income countries such as Brazil, Egypt, and Indonesia the two ratios are in better balance. Developed countries have a very close balance between agriculture's share of GDP and share of energy use in agriculture. Energy use share is lower than agricultural GDP share except for Japan and the European Union which both rely heavily on intensive production methods.

Energy use per hectare of arable and permanent crop-land can be used as an indicator of the efficiency of energy use in agriculture, although definite conclusions from this indicator are hard to draw unless countries with the same scale of operations are compared. Low energy use per hectare can reflect low-input, low-yield agriculture, which is the case for some developing countries such as Bangladesh, India, Indonesia, Nigeria, and Pakistan. On the other hand, low energy use (dry-land farming with less fertilizer and pesticide use) per hectare can also indicate the extensive farming practices of countries such as Australia and Canada where low yield relative to high-input use countries have not prevented these countries from being competitive agricultural exporters. Japan and the European Union stand out as being very high energy users relative to their land base. The United States has a rela-

Irrigation with mechanically powered pumps substantially increases commercial energy requirements but also increases yields. (Corbis-Bettmann)

tively low energy use per hectare, especially when compared to the European countries.

INPUT USE

Land Use and Irrigated Area

Using data from the United Nations Food and Agriculture Organization (FAO) for arable and permanent crop-land for 1966 to 1996, developing countries have experienced the most significant growth in land area. The highest growth was realized in Brazil, China, and India. In contrast, most developed countries have experienced a decline in arable and permanent crop-land, with Australia and Canada being the exceptions. For some developing countries, such as Egypt and India, increases in area as well as yields have been important in increasing output. For developed countries, increases in output have been a function of higher yields rather than increases in area. The U.S. agricultural area increased more than 12 million hectares during the 1970s and 1980s and declined in the 1990s, with an overall decline of one-half million hectares.

Irrigated areas increased in the last three decades in all countries except Japan. Growth has been rapid in Bangladesh and Brazil. Growth in irrigated areas in the United States has been among the slowest of the selected developed countries. The share of agricultural area under irrigation varies considerably, reflecting resource endowments and crop composition. In 1996, Egypt had 100 percent of its area under irrigation, followed by Pakistan with 81 percent and Japan with 63 percent. More than a third of land is irrigated in China, India, and Bangladesh. Canada, Australia, and the United States, with large-scale, extensive farming practices, have lower shares of irrigated areas.

Fertilizer Use

Fertilizer use (measured in tons of active ingredients) has increased most rapidly in developing countries, where yields generally lagged behind those in developed. Among developed countries, fertilizer use per hectare is very high in Japan and the European countries. Fertilizer use is also high in some of the developing countries, including Egypt, China, Bangladesh, and Pakistan. Fertilizer use in the United States is low compared to the European countries, as is true for other large crop producers such as Canada, Australia, and Brazil.

Pesticide Use

Data on pesticide use, including insecticides, herbicides, fungicides, and other chemicals is scarce. Similar to fertilizer use, pesticide use is the highest in the United States followed by the European countries, and Australia. Pesticide use per hectare is much higher than in other developed countries because farmers are trying to maximize the yield per hectare. Among developing countries pesticide use is also high in Egypt, Brazil, and Pakistan. The trend in developing countries is for more use of chemicals in agriculture to increase yield per hectare.

Machinery Use

Machinery use has followed a similar pattern to fertilizer use, with the largest increases occurring in developing countries and smaller increases or declines in developed countries. The data do not reflect changes in quality or complexity of agricultural machinery. The relative intensity of use is highest in Japan followed by the European countries. Machinery use in the United States is much lower than in Europe and has remained steady because the average farm is much larger in the United States, which translates into better utilization of farm equipment. Machinery use increased in all developing countries. Growth has been rapid in Pakistan, Indonesia, and India, followed by Japan. In the developing countries, machinery use per hectare is lowest in Bangladesh and highest in Egypt.

Roger Conway
Hosein Shapouri

See also: Biological Energy Use, Ecosystem Functioning of; Chemical Energy, Historical Evolution of the Use of; Nitrogen Cycle.

BIBLIOGRAPHY

Dovring, F. (1988). *Farming for Energy: The Political Economy of Energy Sources in the United States.* New York: Praeger Publishers.

Food and Agricultural Organization of the United Nations. (1977). *The State of Food and Agriculture, 1976*, pp. 79–110. Rome: United Nations.

Food and Agricultural Organization Homepage. (1999). *Data on Production, Trade, and Agricultural Inputs.* Rome: United Nations.

International Energy Agency. (1998). *Energy Statistics and Balances of OECD and Non-OECD Counties, 1995-96.* Paris: United Nations

Stanhill, G. (1984). *Energy and Agriculture.* Germany: Springer-Verlag.

Stout, B. A. (1979). *Energy for World Agriculture.* Rome: United Nations.

Stout, B. A. (1984). *Energy Use and Management in Agriculture.* North Scituate, MA: Breton Publishers.

U.S. Congress. (1991). *Energy in Developing Countries. Office of Technology Assessment.* Washington, DC: U.S. Government Printing Office.

U.S. Department of Agriculture. Economic Research Service. (1997). *Agricultural Resources and Environmental Indicators, 1996–97.* Washington, DC: U.S. Government Printing Office.

World Bank. (1998). *World Development Indicators, 1998.* Washington, DC: Author.

AIR CONDITIONING

Air conditioning is the treatment of air to control simultaneously its temperature, humidity, cleanliness, and distribution to meet the requirements of a conditioned space. Common use of the term "air conditioning" applies it to the cooling of air; however, true air conditioning treats all aspects of indoor atmospheric comfort.

An air conditioning system uses an assembly of equipment to treat air. Normally the assembly includes a heating system for modifying winter indoor temperature and humidity; a refrigeration system for modifying summer temperature and humidity, a means to maintain indoor air quality (i.e., air filters and fresh air intake); a method of distribution of conditioned air; and a control system, such as a thermostat, to maintain desired comfort conditions.

Air conditioning systems fall into two broad classes. Comfort air conditioning, which accounts for most applications, is used to modify and maintain the indoor environment for habitation. Process air conditioning is the modification of the indoor environment to enhance an industrial or a scientific process.

Most air conditioning systems utilize a vapor-compression refrigeration system (Figure 1) to transfer the indoor heat to a suitable heat sink such as the outdoors. Vapor-compression refrigeration systems employ a cycle in which a volatile liquid, the refrigerant, is vaporized, compressed, liquefied, and expanded continuously in an enclosed system. A compressor serves as a pump, pressurizing the refrigerant and circulating it through the system. Pressurized refrigerant is liquefied in a condenser, liberating heat. Liquid refrigerant passes through an expansion device into an evaporator where it boils and expands into a vapor, absorbing heat in the process.

Some air conditioning systems use an absorption refrigeration system (Figure 2). Absorption refrigeration systems work by evaporating refrigerant in an evaporator, with the refrigerant vapor then absorbed by an absorbent medium, from which it is subsequently expelled by heating in a generator and changed back into liquid in a condenser. Absorption systems may use a pump to help circulate the refrigerant. The most common absorption systems used for air conditioning use water as an absorbent and ammonia as a refrigerant or lithium bromide salt as an absorber and water as a refrigerant.

For heating purposes, most air conditioning systems use fossil-fueled furnaces to heat air, or boilers to heat water or produce steam. Forced air systems use a blower fan and ductwork to distribute conditioned air to points of use. Air quality is enhanced in forced-air systems through the use of filters. The filters are normally placed in the return air, just before the heating and cooling components. Provision may be made for fresh outdoor air to be added to recirculated room air. A hydronic heating system uses hot water to convey heat from a boiler to radiators, convectors, or wall and floor panel coils. Most steam heating systems use boiler-produced steam to heat buildings via radiators, convectors, or heating coils placed in ductwork. Heating systems may employ humidifiers in winter to counter the drying effect of heated air, particularly in forced-air systems.

Summer comfort cooling can increase electrical loads as much as 50 percent over average consump-

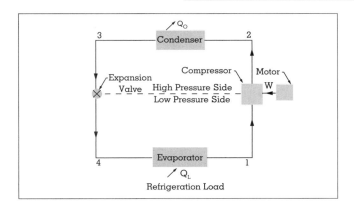

Figure 1.

Equipment diagram for refrigeration using a basic vapor compression cycle.

tion. The most uncomfortable summer days, those with both high temperature and high relative humidity, increase both sensible and latent system load. The sensible load increases because the difference between indoor and outdoor temperature is greater, requiring the cooling system to move heat to a higher level. The latent load increases as humidity rises, since the cooling system must extract more moisture from the air.

Many air conditioning systems use simple thermostats to cycle equipment; however, more sophisticated control systems employing electronics and microprocessors can reduce energy consumption.

Automotive air conditioning systems provide simple heating and cooling/dehumidification functions. There is no provision for filtration of the air. All current automotive air conditioning utilizes vapor-compression refrigeration systems coupled to the automobile's engine.

Air conditioning systems are categorized by the method used to control cooling or heating in a conditioned space. They are further described based on their terminal cooling or heating media. The most common type for cooling is the refrigerant based all-air system which uses a refrigeration cycle to transfer heat from indoor air to outdoor air with heat exchangers commonly called *coils*. Most heating systems in residences and small buildings are all-air-using fossil-fueled furnaces. Hydronic (hot water) or steam heating systems also are common, particularly in large buildings. Many systems employ unitary equipment—that is, they consist of

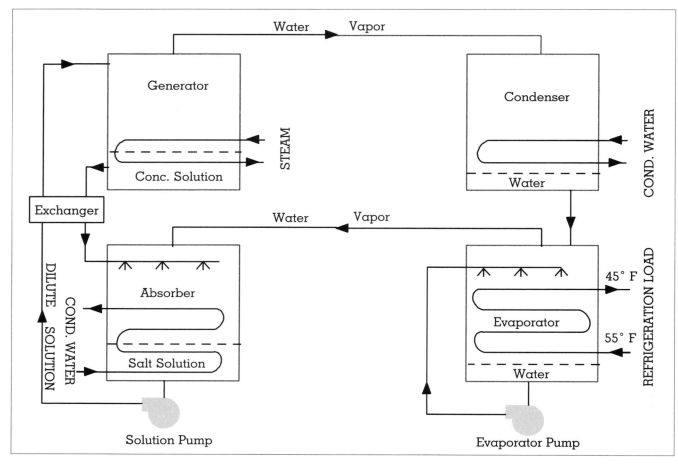

Figure 2.
Absorption type refrigeration using water as a refrigerant and lithium-bromide salts as an absorber.

one or more factory-built modules. Larger buildings may require built-up systems made of various system components that require engineering design tailored for the specific building. Sometimes multiple unitary units are used in large buildings for ease of zone control where the building is divided into smaller areas that have their own thermostats controlling that space.

Selection of air conditioning equipment depends on balancing various criteria, including performance, capacity, special requirements, initial and continuing operating costs; reliability; flexibility and maintainability.

Air conditioning systems have been traditionally compared and rated by cooling and/or heating capacity and, more recently, energy efficiency. Capacity is expressed in British thermal units per hour (Btuh) or in watts. Energy efficiency is expressed as the operat-

ing efficiency using the term "energy efficiency ratio" (EER). In the U.S. EER is expressed as British thermal units per hour per watt (Btuh/w). For window air conditioners, an energy usage label is required by the 1975 Energy Policy and Conservation Act. Unitary air conditioners are rated at operating conditions standardized by the Air Conditioning and Refrigeration Institute (ARI). Standard ratings provide the consumer comparisons between competing models. This information is published annually in ARI's *Unitary Directory*.

Higher EERs in air conditioning equipment are not the sole answer to reducing energy consumption. Proper sizing, reduction of building air leakage, increasing insulation, reducing unnecessary internal energy usage, proper conditioned air distribution, use of night temperature setback or increase, and proper maintenance of equipment can be greater contribute factors for reducing energy cost. For example, a properly sized low-efficiency system operating in a well-sealed and insulated building can consume less energy than a high-efficiency system in a poorly insulated building with lots of outside air infiltration from poorly fitting windows and doors.

HISTORY

The invention of air conditioning is actually a progression of the applied ideas of many individuals starting in the early nineteenth century and dramatically accelerating in the twentieth century. Most air conditioning developments occurred in the twentieth century.

Because of the limitations of energy sources and the technology to top that energy, mechanical air conditioning was not a practical possibility until the dawn of the Scientific Age and the Industrial Revolution. In 1813, Sir John Leslie of England made one of the earliest proposals to use artificial cooling for human comfort. Jean Frederic Marquis de Chabannes followed in 1815 with a British patent for use of a centrifugal fan to force heated or cooled air through ducts to rooms. The major breakthrough came in 1834 when Jacob Perkins invented the vapor compression refrigeration system, making it possible to cool air mechanically. David Boswell Reid designed the first building air conditioning system, using water sprays, for the British Houses of Parliament in about 1836.

In the United States, Dr. John Gorrie, unaware of the work of Perkins, proposed that mechanical refrigeration be used for comfort cooling, and he constructed mechanical systems for cooling his patients at his home in Florida in about 1842.

Although limited experiments, such as Gorrie's were being conducted, there was little understanding of the science involved in cooling and dehumidification. The first engineering textbook for heating and cooling, *Guide to Calculating and Design of Ventilating and Heating Installations* by Hermann Rietschel, was published in Germany in 1894. Rietschel's chapter on room cooling was the earliest comprehensive example of a real scientific approach to comfort cooling.

Rietschel's engineered approach was introduced in the United States by consulting engineer Alfred Wolff, who designed the first modern energy-saving air conditioning system for the New York Stock Exchange in 1901. Wolff's huge system used waste steam from the building power plant to power an absorption-type refrigeration system to provide comfort cooling. The fact that the system used steam that would have been thrown away meant that the energy needed for the cooling plant was free! The system operated successfully for twenty years. Wolff designed several other comfort cooling systems for large buildings before his death in 1909.

Textile engineer Stuart Cramer first published the term "air conditioning" in 1906, and G. B. Wilson defined air conditioning as the control of temperature, humidity, and air quality in 1908.

Process or industrial air conditioning was proposed, and a few examples installed, by the late nineteenth century. Willis Haviland Carrier devoted his engineering career to air conditioning, catering to industrial needs beginning in 1902. Carrier took an engineering, scientific, and business approach to air conditioning, becoming its greatest early-twentieth-century practitioner. Carrier patented dewpoint control, used for precise humidity control, in 1907. At about the same time, Carrier devised a psychrometric chart for calculating air conditions, which became an essential engineering tool in use to this day. He founded the first design and manufacturing company exclusively devoted to air conditioning. In the 1920s Carrier expanded his interest to comfort cooling.

Before the twentieth century, few homes or public venues experienced mechanical comfort cooling. A curious public was exposed enmasse to the pleasures of summer cooling at the Louisiana Purchase Exposition in St. Louis in 1904, where the Missouri State Building had an air conditioned amphitheater. A hospital in Boston had air conditioned wards in 1906. Some hotels installed cooling systems for lobbies, meeting halls, and restaurants after 1907. Motion picture theaters began to install mechanical comfort cooling systems after 1915.

Air conditioned theaters produced a two-pronged demand for comfort cooling. Consumers liked it, asked for it, and patronized those theaters offering it. Increased attendance at cooled theaters showed that the installation and operating costs were worthwhile expenditures, causing more theater owners to decide to purchase comfort cooling systems.

The aforementioned applications of air conditioning were possible because a commercial advantage was present. Limited refrigeration technology contributed to high installation costs. These costs could

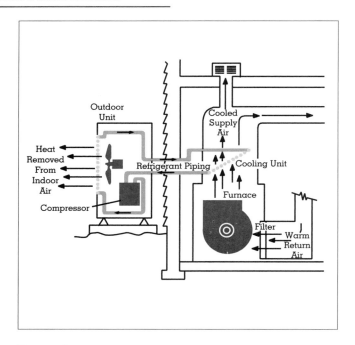

Figure 3.
Schematic drawing of a central air conditioning
system.

be reduced by using ice instead of mechanical refrigeration, however, ice-type systems did not dehumidify as well, and could present higher operation costs, dependent upon the cost of ice. Still, it often paid to incur the costs of air conditioning where a commercial advantage was present.

A commercial advantage was present in many industrial processes. Uncontrolled heat and humidity impacted some products such as chocolate, pasta, textiles, and tobacco. Air conditioning allowed for uniform and continuous production despite weather conditions, reduced spoilage and waste, and thus saved money—enough in many cases to easily justify the installed and operating costs.

Thus, most of the uses of air conditioning before the 1930s concentrated on applications that had a viable financial payback. There was no obvious financial advantage in air conditioning homes, and this branch of air conditioning developed later. Home air conditioning passed from luxury to necessity only when the installed cost decreased and when air conditioning systems became worry-free.

The perfection of the electric household mechanical refrigerator by the 1930s provided a technology applicable to home air conditioning. This technology

allowed mass production of lower cost, reliable package air conditioners for homes. It was no accident that the first room cooler was introduced by the major refrigerator manufacturer, Frigidaire, in 1929, followed swiftly by others such as General Electric, Kelvinator, and Westinghouse. These early package air conditioners were not central systems and thus were applicable to homes and small commercial establishments without any modification of an existing heating system. These simple cooling devices, forerunners of the window air conditioners introduced in the late 1930s, were the beginnings of affordable air conditioning.

Central air conditioning systems for homes were available in the early 1930s from several manufacturers (Figure 3). These were combined with automatic fossil-fueled heating systems, a new innovation of the time. Reliable, thermostatically controlled automatic oil, gas, or coal-firing systems had only become available after the late 1920s. Before that, most homes and buildings had used coal that was hand-fired in all but the largest installations.

The safety of refrigeration and air cooling systems had always been an issue due to the toxicity or flammability of most refrigerants. In fact, increasing prevalence of refrigeration caused accidents and deaths, and the bad publicity and restrictive legislation were becoming serious threats to the growth of refrigeration and air conditioning. Fortunately, a solution was found when Thomas Midgley, Albert Henne, and Robert McNary invented chlorofluorocarbon refrigerants (CFCs) for Frigidaire. They were introduced in 1930 and, with the realization of their overall importance to health, safety, and the future of refrigeration and air conditioning, CFCs were made available to the entire industry. The CFCs made it possible to engineer air conditioning systems for any application without fear of safety issues. All other refrigerants used for small refrigeration and most air conditioning systems were soon completely displaced. The CFC refrigerants were applied to air conditioning systems in the early 1930s. One early use was in air conditioning for passenger trains. By 1936 all long-distance dining and sleeping cars on U.S. railroads were air conditioned.

Most homes were not air conditioned by the 1940s, and the costs involved in installing air conditioning in existing homes were prohibitive until the window air conditioner was introduced. By the late 1940s the cost and reliability of a window air condi-

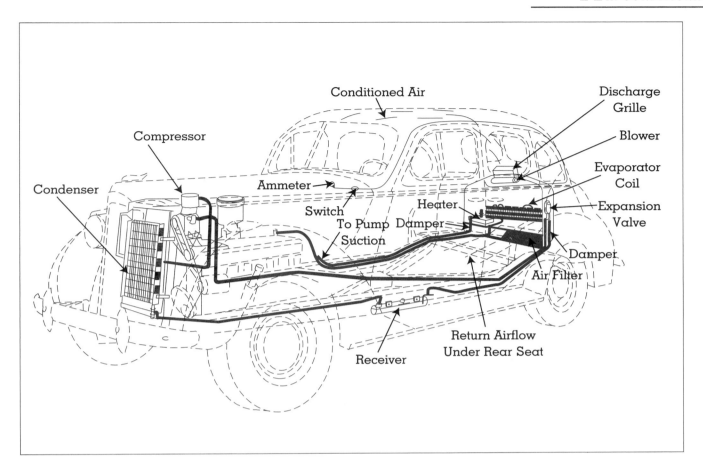

Figure 4.
The first automotive air conditioning system (1938).

tioner were such that middle-class homeowners could afford them. Sales rocketed from about 50,000 in 1946 to more than 1 million in 1953. The window air conditioner has become so relatively inexpensive that it ceased to be a luxury. Sixteen million window air conditioners were manufactured in 1992. The experience of window air conditioned homes no doubt played a part in later demand for centrally air conditioned homes as homeowners moved on to newer homes. The window air conditioner, combined with automatic central heating, now made it possible to live, work, and play indoors oblivious to the environment outside. In fact, the ability to control the indoor environment so effectively is credited with helping to reverse population migration out of the southern United States after 1960.

Automotive air conditioning systems using vapor compression refrigeration systems and CFC refriger-ants were proposed in 1932 and debuted in the late 1930s, but system problems retarded its popularity until the 1950s. Auto air conditioning soared in the 1960s; for example, American car installations tripled between 1961 and 1964.

By the 1960s, central residential air conditioning was becoming increasingly popular. Equipment had developed to the point that a number of manufacturers were producing and marketing unitary air conditioning equipment. Most new homes were designed to include central air conditioning. This was true of commercial buildings also. In fact, the use of air conditioning changed architectural design. By removing all environmental restraints, air conditioning made it possible to design almost any type of building. Hollow ventilation cores were no longer necessary, and the "windowless" building made its debut. Residential construction changed too. Deep porches for shading were no longer necessary, and a home site was no longer dependent on prevailing summer breezes.

During the 1960s, air conditioning system control began to shift from simple electrical or pneumatic control to electronic and rudimentary computer control. This new technology was applied to commercial, institutional and commercial buildings where the high cost and complicated nature justified its use.

The oil embargo of the early 1970s gave impetus to increasing energy efficiency. Manufacturers and designers responded by developing and using higher-efficiency compressors for cooling. Condenser size was increased to lower system pressures, resulting in energy savings. Heating boilers and furnaces were redesigned for higher efficiencies. Building construction methods were revised to include more insulation and to reduce outside air infiltration. Existing buildings were scrutinized for energy inefficiencies. Some building owners went to the extreme of closing up outside air intakes; the unintended effect being stale building air and occupant complaints.

Overall efficiency of air conditioning equipment steadily rose starting in the mid-1970s, attributed to consumer demand, government mandate, and incentive programs. For example, the average efficiency, as expressed in seasonal energy efficiency ratio, of new central air conditioners increased about 35 percent between 1976 and 1991. After national standards took effect in 1992, efficiency has increased as much as another 15 percent.

Increased system efficiency has resulted in higher equipment costs but lower operating costs. A higher initial cost for equipment can often be justified by the monetary savings from lower energy consumption over the life of the equipment.

CFC PHASEOUT

The air conditioning industry was challenged to reinvent one of its vital system components, refrigerants, when CFCs were targeted as a prime cause of high-level atmospheric ozone depletion. Ozone, an active form of oxygen, is present in the upper atmosphere. One of its functions is to filter out solar ultraviolet radiation, preventing dangerous levels from reaching the ground.

A hypothesis that halocarbons, including CFCs, diffused into the upper atmosphere, could break down, and an ozone-destroying catalytic reaction could result, was published in 1974. Computer modeling of the hypothesis showed that such destruction could happen. The resulting increase in UV radiation would have adverse health and biological system consequences. This scenario so alarmed nations that various measures to control CFC emissions were undertaken. Although there were scientific uncertainties concerning the depletion hypothesis, the United States became the first nation to ban nonessential CFC use in aerosol sprays, in 1978.

Other nations followed with various control measures, but CFCs were also widely used as refrigerants, fire-extinguishing agents, insulation components, and solvents. The UN Environment Program began working on a worldwide CFC control scenario in 1981, culminating in the Montreal Protocol of 1987 which called for a phaseout of CFC production over time. In the United States, an excise tax on CFCs was passed, steadily increasing year by year so as to make CFCs increasingly more costly to use. As a result of various measures, CFC consumption by 1999 had decreased more than 50 percent.

Phaseout of the CFC refrigerants posed a challenge for the refrigeration and air conditioning industries. If the primary refrigerants in use were to be replaced, what was to be used instead? Some previously used refrigerants, such as hydrocarbons and ammonia, were proposed; however, safety and litigation fears eliminated them from serious consideration. The producers of CFCs, pursuing their vested interest, conducted extensive research, resulting in a number of alternative refrigerants that are widely accepted today. These alternatives do not contain chlorine, the element responsible for the ozone-depleting reaction.

Replacement of halocarbon refrigerants has increased the cost of refrigeration and air conditioning systems since the new refrigerants are more costly, system components need to be redesigned for the new refrigerants, and service and installation are more complicated. For example, most refrigeration and air conditioning systems used one of four refrigerants before 1987, but now there are more than a dozen alternatives.

RECENT TRENDS IN AIR CONDITIONING

Energy efficiency had always been a goal of building owners simply because they were trying to reduce operating costs. Sometimes there was a trade-off in personal comfort. Today's emphasis on energy conservation also must consider a balance of comfort in all forms: temperature and humidity control, and indoor air quality. Air conditioning engineers now have awareness that energy must be saved, but at the same time reasonable comfort, and therefore productivity, must be maintained. Modern air conditioning system and equipment design coupled with responsible building architecture have resulted in indoor environments that minimize the trade-offs. The percentage of new homes built with central air conditioning increased from 45 percent in 1975 to 80 percent in 1995. As of 1997, 41 percent of all U.S. households used central air conditioning, and an additional 30 percent used room air conditioners. By 1990, 94 percent of new cars sold in the United States had air conditioning systems.

The trend has been toward progressively higher energy efficiencies. Per-capita end-use energy consumption began decreasing in the early 1970s and has trended downward ever since. National minimum efficiency requirements for room and central air conditioners were enacted through the National Appliance Energy Conservation Act (NEACA) in 1987. The requirements took effect in 1990 for room units and in 1992 for central air conditioners.

For air conditioning systems used in commercial buildings, minimal efficiency targets are published in the form of Standards by the American Society of Heating, Refrigerating and Air Conditioning Engineers (ASHRAE). These standards influenced building energy codes and legislated minimum standards included in the U.S. Energy Policy Act of 1992. ASHRAE, in addition to developing standards for design of equipment, publishes the most comprehensive air conditioning design and application information in their *Handbook*, published annually.

Several innovations have contributed to more efficient and comfortable air conditioning. In large buildings, indoor environmental conditions can be determined and even corrected from a central, computer-controlled console. Programmable, microprocessor control systems are seeing increasing use in large buildings, and their cost is decreasing, improving the probability that they will be used in residences. Microprocessors permit design of air conditioning systems that can respond to occupant and building needs at a scale only dreamed of before. Direct digital control of conditioned air outlets permits buildings to

be divided into mini zones. The result is customized energy use to match the zone's needs.

Motion detectors and daylight detectors reduce artificial lighting use, reducing summer cooling loads.

Variable-speed fan drives permit conditioned air distribution to be matched more closely to a building's needs. High-efficiency electric motors are used to drive the fans, saving as much as half the energy once used. Both variable-speed and high- efficiency motors are being applied even in residential air conditioning systems.

New compressor technology employing rotary scroll-type compressors is replacing previously used reciprocating technology. Scroll compressors operate at higher efficiencies over wider operating conditions. Electrical and electronic technology are making variable-capacity compressors cost-effective for the twenty-first century. Compressor performance can be optimized for a wider operation range, and matching compressor capacity to the actual demand for cooling increases the system efficiency, saving energy.

Thermal storage is being used to reduce energy needs. Cooling systems are being designed to "store" cooling at night, when energy cost and consumption may be lower, and release the stored cooling during the day. Building design itself has begun to change so the energy storage capability of the building mass itself can be used. Shading and "low e" glass is being used to reduce heat gains and losses through windows.

Energy recovery ventilators, which move energy between outgoing stale building air and incoming fresh air, are being used to improve indoor air quality without large increases in energy consumption.

Cooling systems that use refrigeration systems to cool and partially dehumidify air are being combined with desiccant dehumidification to reduce energy costs. Dessicant systems use a regenerable moisture absorbant to extract humidity from the air to be conditioned. Saturated absorbant is exposed to a higher air temperature, releasing the absorbed moisture. The absorbant is then recycled. Some systems use rotating wheels to continuously recycle the absorbant. Absorption-type refrigeration systems are used in these hybrid systems where the waste heat can be used to continuously regenerate the desiccant.

Research continues in both industry and government to find ways to innovate and improve air conditioning systems. For example, ASHRAE maintains an ongoing research program covering energy conservation, indoor air quality, refrigerants and environmentally safe materials. The society's journal maintains an HVAC&R search engine on the Internet covering more than 700 related web sites. In addition, the U.S. Department of Energy funds research and development on advanced air conditioning technologies such as innovative thermally activated heat pumps.

THE FUTURE OF AIR CONDITIONING

Precise and sophisticated control of air conditioning systems will be the trend of the near future. The explosion of innovation in computer and electronic technology will continue to impact the design and operation of air conditioning systems. Buildings will become "intelligent," their internal systems responding to changing environmental and occupancy conditions. "Cybernetic" building systems will communicate information and control functions simultaneously at multiple levels for various systems, including heating, cooling, and ventilation energy management, fire detection, security, transportation, and information systems. "Interoperability" in control systems will allow different controls to "talk to each other" in the same language. Wireless sensors will be used, allowing easy retrofit of older buildings.

The recent trend of integrating building design with the environment will result in energy savings as old concepts of natural ventilation, shading, and so on are reapplied. Technological innovation will permit increased use of solar technology as costs decrease. Hybrid cooling systems using both electricity and gas will be used in greater numbers.

The current debate over controversial global warming theories will continue. The impact of carbon dioxide levels in the atmosphere, whether they are increasing or not over time, and the effect on climate and economics will continue to be discussed. A solution, if it is needed, may evolve—or not.

Science, technology, and public need will continue to interact in ways that cannot be accurately predicted, each providing a catalyst for change at various times. However, the history of the development of air conditioning has shown that the trend has been beneficial. No doubt it will continue to be.

Bernard A. Nagengast

See also: Air Quality, Indoor; Building Design, Commercial; Building Design, Energy Codes and; Building Design, Residential; Energy Management Control Systems; Heat and Heating; Heat Pumps; Insulation; Refrigerators and Freezers; Water Heating.

BIBLIOGRAPHY

Air Conditioning and Refrigeration Institute. (1999). *ARI Unitary Directory*. <http://www.ari.org>.

American Society of Heating, Refrigerating and Air Conditioning Engineers. (1978). *ASHRAE Composite Index of Technical Articles, 1959–1976*. Atlanta: Author.

American Society of Heating, Refrigerating, and Air Conditioning Engineers. (1989). *CFCs: Time of Transition*. Atlanta: Author.

American Society of Heating, Refrigerating and Air Conditioning Engineers. (1989) *Standard 62-1989, Ventilation for Acceptable Indoor Air Quality*. Atlanta: Author.

American Society of Heating, Refrigerating, and Air Conditioning Engineers. (1989). *Standard 90.1-1989, Energy-Efficient Design of New Buildings Except Low-Rise Residential Buildings*. Atlanta: Author.

American Society of Heating, Refrigerating, and Air Conditioning Engineers. (1993) *Standard 90.2-1993, Energy-Efficient Design of New Low-Rise Residential Buildings*. Atlanta: Author.

American Society of Heating, Refrigerating, and Air Conditioning Engineers. (1995) *Standard 100-1995, Energy Conservation in Existing Buildings*. Atlanta: Author.

American Society of Heating, Refrigerating, and Air Conditioning Engineers. (1996). *ASHRAE Handbook: Heating, Ventilating, and Air Conditioning Systems and Equipment*. Atlanta: Author.

American Society of Heating, Refrigerating, and Air Conditioning Engineers. (1997). *Refrigerants for the 21st Century*. Atlanta: Author.

American Society of Heating, Refrigerating, and Air Conditioning Engineers. (1999). *ASHRAE Handbook: Heating, Ventilating, and Air Conditioning Applications*. Atlanta: Author.

Bhatti, M. S. (1999). "Riding in Comfort: Part II: Evolution of Automotive Air Conditioning." *ASHRAE Journal* 41 (Sept.):44–52.

Chartered Institution of Building Services Engineers. (1994). *The International Dictionary of Heating, Ventilating, and Air Conditioning*. London: Author.

Cooper, G. (1998). *Air Conditioning America: Engineers and the Controlled Environment, 1900–1960*. Baltimore: Johns Hopkins University Press.

Donaldson, B., and Nagengast, B. (1994). *Heat & Cold: Mastering the Great Indoors: A Selective History of Heating, Ventilation, Refrigeration, and Air Conditioning*. Atlanta: American Society of Heating, Refrigerating, and Air Conditioning Engineers.

Elliott, C. D. (1992). *Technics and Architecture.* Cambridge, MA: MIT Press.

Howell, R.; Sauer, H.; and Coad, W. (1997). *Principles of Heating, Ventilating, and Air Conditioning.* Atlanta: American Society of Heating, Refrigerating, and Air Conditioning Engineers.

Hunn, B. D. (1996). *Fundamentals of Building Energy Dynamics.* Cambridge, MA: MIT Press.

Lorsch, H. G. (1993). *Air Conditioning Design.* Atlanta: American Society of Heating, Refrigerating, and Air Conditioning Engineers.

AIRCRAFT

Today the airplane is part of everyday life, whether we see one gracefully winging overhead, fly in one, or receive someone or something (package, letter, etc.) that was delivered by one. The invention and development of the airplane is arguably one of the three most important technical developments of the twentieth century—the other two being the electronics revolution and the unleashing of the power of the atom.

The first *practical* airplane was invented by Orville and Wilbur Wright, two bicycle shop proprietors from Dayton, Ohio. On December 17, 1903, the Wright Flyer lifted from the sand of Kill Devil Hill near Kitty Hawk, North Carolina, and with Orville at the controls, flew a distance of 120 ft above the ground, staying in the air for 12 sec. It was the first successful, sustained flight of a heavier-than-air piloted airplane. The photograph of the Flyer as it is lifting off the ground, with Wilbur running alongside to keep the right wing tip from digging into the sand, is the most famous photograph in the annals of the history of aeronautics. There were three more flights that morning, the last one covering a distance of 852 ft above the ground, and remaining in the air for 59 sec. At that moment, the Wright brothers knew they

First flight of the Wright Flyer, December 17, 1903. (Library of Congress)

Otto Lilienthal, with his collapsible glider, near Rathenow, East Germany. (Corbis-Bettmann)

had accomplished something important—a feat aspired to by many before them, but heretofore never achieved. But they had no way of knowing the tremendous extent to which their invention was to dominate the course of events in the twentieth century—technically, socially, and politically.

The history of the technical development of the airplane can be divided into four eras: pre-Wright; the strut-and-wire biplane; the mature propeller-driven airplane, and the jet-propelled airplane. We will organize our discussion in this article around these four eras.

THE PRE-WRIGHT ERA

Before the Wright brothers' first successful flight, there were plenty of attempts by others. Indeed, the Wrights did *not* invent the first airplane. They inherited a bulk of aeronautical data and experience achieved by numerous would-be inventors of the airplanes over the previous centuries. In many respects, when the Wright brothers began to work on the

invention of the practical airplane, they were standing on the shoulders of giants before them.

From where and whom did the idea of the modern configuration airplane come? The modern configuration that we take for granted today is a flying machine with fixed wings, a fuselage, and a tail, with a separate mechanism for propulsion. This concept was first pioneered by Sir George Cayley in England in 1799. Cayley is responsible for conceiving and advancing the basic idea that the mechanisms for lift and thrust should be separated, with fixed wings moving at an angle of attack through the air to generate lift and a separate propulsive device to generate thrust. He recognized that the function of thrust was to overcome aerodynamic drag. In his own words, he stated that the basic aspect of a flying machine is "to make a surface support a given weight by the application of power to the resistance of air."

To key on Cayley's seminal ideas, the nineteenth century was full of abortive attempts to actually build and fly fixed-wing, powered, human-carrying flying machines. Cayley himself built several full-size aircraft over the span of his long life (he died in 1857 at the age of eighty-three), but was unsuccessful in achieving sustained flight. Some of the most important would-be inventors of the airplane were William Samuel Henson and John Stringfellow in England, Felix Du Temple in France, and Alexander Mozhaiski in Russian. They were all unsuccessful in achieving sustained flight. In regard to the nature of airplane performance and design, we note that these enthusiastic but unsuccessful inventors were obsessed with horsepower (or thrust). They were mainly concerned with equipping their aircraft with engines powerful enough to accelerate the machine to a velocity high enough that the aerodynamic lift of the wings would become large enough to raise the machine off the ground and into the air. Unfortunately, they all suffered from the same circular argument—the more powerful the engine, the more it weighs; the heavier the machine is, the faster it must move to produce enough lift to get off the ground; the faster the machine must move, the more powerful (and hence heavier) the engine must be—which is where we entered this circular argument. A way out of this quandary is to develop engines with more power without an increase in engine weight, or more precisely, to design engines with large horsepower-to-weight ratios. The thrust-to-weight ratio, T/W, for the entire aircraft, is a critical parameter in airplane performance and design. In the nineteenth century, inventors of flying machines functioned mainly on the basis of intuition, with little quantitative analysis to guide them. They knew that, to accelerate the aircraft, thrust had to be greater than the drag; that is, $T - D$ had to be a positive number. And the larger the thrust and the smaller the drag, the better things were. In essence, most of the nineteenth-century flying machine inventors were obsessed with brute force—given enough thrust (or horsepower) from the engine, the airplane could be wrestled into the air. The aviation historians call such people "chauffeurs." They were so busy trying to get the flying machine off the ground that they paid little attention to how the machine would be controlled once it got into the air; their idea was that somehow the machine could be chauffeured in the air much as a carriage driven on the ground. This philosophy led to failure in all such cases.

The antithesis of the chauffeur's philosophy was the "airman's" approach. In order to design a successful flying machine, it was necessary to first get up in the air and experience flight with a vehicle unencumbered by a power plant; that is, you should learn to fly before putting an engine on the aircraft. The person who introduced and pioneered the airman's philosophy was Otto Lilienthal, a German mechanical engineer, who designed and flew the first successful gliders in history. Lilienthal first carried out a long series of carefully organized aerodynamic experiments, covering a period of about twenty years, from which he clearly demonstrated the aerodynamic superiority of cambered (curved) airfoils in comparison to flat, straight surfaces. His experiments were extensive and meticulously carried out. They were published in 1890 in a book entitled "Der Vogelflug als Grundlage der Fliegekunst" ("Bird Flight as the Basis of Aviation"); this book was far and away the most important and definitive contribution to the budding science of aerodynamics to appear in the nineteenth century. It greatly influenced aeronautical design for the next fifteen years, and was the bible for the early work of the Wright brothers. Lilienthal's aerodynamic research led to a quantum jump in aerodynamics at the end of the nineteenth century.

The last, and perhaps the most dramatic, failure of the pre-Wright era was the attempt by Samuel P. Langley to build a flying machine for the U.S. government. Intensely interested in the physics and technology of powered flight, Langley began a series of aerodynamic experiments in 1887, using a whirling arm apparatus. At the time, he was the director of the Allegheny Observatory in Pittsburgh. Within a year he seized the opportunity to become the third Secretary of the Smithsonian Institution in Washington, D.C. Langley continued with his aeronautical experiments, including the building and flying of a number of elastic-powered models. The results of his whirling arm experiments were published in 1890 in his book *Experiments in Aerodynamic*. In 1896, Langley was successful in flying several small-scale, unmanned, powered aircraft, which he called aerodromes. These 14-ft-wingspan, steam-powered aerodromes were launched from the top of a small houseboat on the Potomac River, and they flew for about a minute, covering close to 1 mi over the river. These were the first steam-powered, heavier-than-air machines to successfully fly—a historic event in the history of aeronautics that is not always appreciated today.

This was to be the zenith of Langley's success. Spurred by the exigency of the Spanish-American War, Langley was given a $50,000 grant from the War

Department to construct and fly a full-scale, person-carrying aerodrome. He hired an assistant, Charles Manly, who had just graduated from the Sibley School of Mechanical Engineering at Cornell University. Together, they set out to build the required flying machine. The advent of the gasoline-powered internal-combustion engine in Europe convinced them that the aerodrome should be powered by a gasoline-fueled reciprocating engine turning a propeller.

By 1901 Manly had assembled a radically designed five-cylinder radial engine. It weighed 200 lb, produced a phenomenal 52.4 hp, and was the best airplane power plant designed until the beginning of World War I. The full-scale aerodrome, equipped with his engine, was ready in 1903. Manly attempted two flights; both resulted in the aerodrome's falling into the water moments after its launch by a catapult mounted on top of a new houseboat on the Potomac River.

Langley's aerodrome and the fate that befell it are an excellent study in the basic aspects of airplane design. Despite excellent propulsion and adequate aerodynamics, it was the poor structural design that resulted in failure of the whole system.

ERA OF STRUT-AND-WIRE BIPLANES

The 1903 Wright Flyer ushered in the era of successful strut-and-wire biplanes, an era that covers the period from 1903 to 1930. There is no doubt in this author's mind that Orville and Wilbur Wright were the first true aeronautical engineers in history. With the 1903 Wright Flyer, they had gotten it all right—the propulsion, aerodynamic, structural, and control aspects were carefully calculated and accounted for during its design. The Wright brothers were the first to fully understand the airplane as a whole and complete system, in which the individual components had to work in a complementary fashion so that the integrated system would perform as desired.

Let us dwell for a moment on the Wright Flyer as an airplane design. The Wright Flyer possessed all the elements of a successful flying machine. Propulsion was achieved by a four-cylinder in-line engine designed and built by Orville Wright with the help of their newly hired mechanic in the bicycle shop, Charlie Taylor. It produced close to 12 hp and weighed 140 lb, barely on the margin of what the Wrights had calculated as the minimum necessary to

get the flyer into the air. This engine drove two propellers via a bicycle-like chain loop. The propellers themselves were a masterpiece of aerodynamic design. Wilbur Wright was the first person in history to recognize the fundamental principle that a propeller is nothing more than a twisted wing oriented in a direction such that the aerodynamic force produced by the propeller was predominately in the thrust direction. Wilbur conceived the first viable propeller theory in the history of aeronautical engineering; vestiges of Wilbur's analyses carry through today in the standard "blade element" propeller theory. The Wrights had built a wind tunnel, and during the fall and winter of 1901 to 1902, they carried out tests on hundreds of different airfoil and wing shapes. Wilbur incorporated these experimental data in his propeller analyses; the result was a propeller with an efficiency that was close to 70 percent (propeller efficiency is the power output from the propeller compared to the power input to the propeller from the engine shaft). This represented a dramatic improvement of propeller performance over contemporary practice. For example, Langley reported a propeller efficiency of only 52 percent for his aerodromes. Today, a modern, variable-pitch propeller can achieve efficiencies as high as 85 to 90 percent. In 1903, the Wrights' propeller efficiency of 70 percent was simply phenomenal. It was one of the lesser-known but most compelling reasons for the success of the Wright Flyer. With their marginal engine linked to their highly efficient propellers, the Wrights had the propulsion aspect of airplane design well in hand.

The aerodynamic features of the Wright Flyer were predominately a result of their wind tunnel tests of numerous wing and airfoil shapes. The Wrights were well aware that the major measure of aerodynamic efficiency is the lift-to-drag ratio L/D. They knew that the lift of an aircraft must equal its weight in order to sustain the machine in the air, and that almost any configuration could produce enough lift if the angle of attack was sufficiently large. But the secret of "good aerodynamics" is to produce this lift with as small a drag as possible, that is, to design an aircraft with as large an L/D value as possible. To accomplish this, the Wrights did three things:

1. They chose an airfoil shape that, based on the collective data from their wind tunnel tests, would give a high L/D. The airfoil used on the

Wright Flyer was a thin, cambered shape, with a camber ratio (ratio of maximum camber to chord length) of 1/20, with the maximum camber near the quarter-chord location. (In contrast, Lilienthal favored airfoils that were circular arcs, i.e., with maximum camber at midchord.) It is interesting that the precise airfoil shape used for the Wright Flyer was never tested by the Wright brothers in their wind tunnel. By 1903, they had so much confidence in their understanding of airfoil and wing properties that, in spite of their characteristic conservative philosophy, they felt it unnecessary to test that specific shape.

2. They chose an aspect ratio of 6 for the wings. (Aspect ratio is defined as the square of the wing span divided by the wing area; for a rectangular wing, the aspect ratio is simply the ratio of the span to the chord length.) They had experimented with gliders at Kitty Hawk in the summers of 1900 and 1901, and they were quite disappointed in their aerodynamic performance. The wing aspect ratio of these early gliders was 3. However, their wind tunnel tests clearly indicated that higher-aspect-ratio wings produced higher values of L/D. (This was not a new discovery; the advantage of high-aspect-ratio wings had been first theorized by Francis Wenham in 1866. Langley's whirling arm data, published in 1890, proved conclusively that better performance was obtained with higher-aspect-ratio wings. Based on their own wind tunnel results, the Wrights immediately adopted an aspect ratio of 6 for their 1902 glider, and the following year for the 1903 flyer. At the time, the Wrights had no way of knowing about the existence of induced drag; this aerodynamic phenomenon was not understood until the work of Ludwig Prandtl in Germany fifteen years later. The Wrights did not know that, by increasing the aspect ratio from 3 to 6, they reduced the induced drag by a factor of 2. They only knew from their empirical results that the L/D ratio of the 6-aspect-ratio wing was much improved over their previous wing designs.

3. The Wrights were very conscious of the importance of parasite drag, which in their day was called head resistance. They used empirical formulas obtained from Octave Chanute to esti-mate the head resistance for their machines. (Octave Chanute was a well-known civil and railroad engineer who had become very interested in aeronautics. In 1893 he published an important survey of past aeronautical work from around the world in a book entitled "Progress in Flying Machines." It has become a classic; you can still buy reprinted copies today. From 1900, Octave Chanute was a close friend and confidant of the Wright brothers, giving them much encouragement during their intensive inventive work in 1900 to 1903.) The Wrights choice of lying prone while flying their machines, rather than sitting up, or even dangling underneath as Lilienthal had done, was a matter of decreasing head resistance. In early 1903, they even tested a series of wooden struts in an airstream in order to find the cross-sectional shape that gave minimum drag. Unfortunately, they did not appreciate the inordinately high drag produced by the supporting wires between the two wings.

The Wrights never quoted a value of L/D for their 1903 Wright Flyer. Modern wind tunnel tests of models of the Wright Flyer carried out in 1982 and 1983 as reported by Culick and Jex at Cal Tech indicate a maximum L/D of 6. This value is totally consistent with values of $(L/D)_{max}$ measured by Gustave Eiffel in 1910 in his large wind tunnel in Paris for models of a variety of aircraft of that time. It has been estimated that the Fokker E-111, an early World War I aircraft had an $(L/D)_{max}$ of 6.4. In 1903 the Wrights had achieved a value of $(L/D)_{max}$ with their flyer that was as high as that for aircraft designed 10 years later.

The control features of the Wright Flyer are also one of the basic reasons for its success. The Wright brothers were the first to recognize the importance of flight control around all three axes of the aircraft. Pitch control, obtained by a deflection of all or part of the horizontal tail (or the forward canard such as the Wright Flyer), and yaw control, obtained by deflection of the vertical rudder, were features recognized by investigators before the Wrights; for example, Langley's aerodrome had pitch and yaw controls. However, no one except the Wrights appreciated the value of roll control. Their novel idea of differentially warping the wing tips to control the rolling motion of the airplane, and to jointly control roll and yaw for coordinated turns, was one of their most important contributions to aeronautical engineering. Indeed, when Wilbur Wright finally carried out the first public demonstrations of their flying machines in LeMans, France, in August 1908, the two technical features of the Wright machines most appreciated and immediately copied by European aviators were their roll control and their efficient propeller design.

Finally, the structural features of the Wright Flyer were patterned partly after the work of Octave Chanute and partly after their own experience in designing bicycles. Chanute, inspired by the gliding flights of Lilienthal, carried out tests of gliders of his own design beginning in 1896. The most important technical feature of Chanute's gliders was the sturdy and lightweight Pratt-truss method of rigging a biplane structure. The Wright brothers adopted the Pratt-truss system for the Wright Flyer directly from Chanute's work. Other construction details of the Wright Flyer took advantage of the Wrights' experience in designing and building sturdy but lightweight bicycles. When it was finished, engine included, the empty weight of the Wright Flyer was 605 lb. With a 150-lb person on board, the empty weight-gross weight ratio was 0.8. By comparison, the empty weight of the Fokker E-111 designed 10 years later was 878 lb, and the empty weight-gross weight ratio was 0.65, not greatly different from that of the Wright Flyer. Considering that 10 years of progress in aircraft structural design had been made between the 1903 flyer and the Fokker E-111, the structural design of the 1903 Wright Flyer certainly seems technically advanced for its time. And the fact that the flyer was structurally sound was certainly well demonstrated on December 17, 1903.

In summary, the Wright brothers had gotten it right. All the components of their system worked properly and harmoniously—propulsion, aerodynamics, control, and structures. There were no fatal weak links. The reason for this was the natural inventiveness and engineering abilities of Orville and Wilbur Wright. The design of the Wright Flyer is a classic first study in good aeronautical engineering. There can be no doubt that the Wright brothers were the first true aeronautical engineers.

The Wright Flyer ushered in the era of strut-and-wire biplanes, and it basically set the pattern for subsequent airplane design during this era. The famous World War I fighter airplanes—such as the French Nieuport 17 and the SPAD XIII, the German Fokker

D. VII, and the British Sopwith Camel—were in many respects "souped-up" Wright flyers.

First, the wing warping method of roll control used by the Wrights was quickly supplanted by ailerons in most other aircraft. (The idea of flaplike surfaces at the trailing edges of airplane wings can be traced to two Englishmen: M. P. W. Boulton, who patented a concept for lateral control by ailerons in 1868; and Richard Harte, who also filed for a similar patent in 1870). Ailerons in the form of triangular "winglets" that projected beyond the usual wingtips were used in 1908 by Glenn Curtiss on his June Bug airplane; flying the June Bug, Curtiss won the Scientific American Prize on July 4, 1908, for the first public flight of 1,000 m or longer. By 1909, Curtiss had designed an improved airplane, the Gold Bug, with rectangular ailerons located midway between the upper and lower wings. Finally, in 1909 the Frenchman Henri Farman designed a biplane named the Henri Farman III, which included a flaplike aileron at the trailing edge of all four wingtips; this was the true ancestor of the conventional modern-day aileron. Farman's design was soon adopted by most designers, and wing warping quickly became passé. Only the Wright brothers clung to their old concept; a Wright airplane did not incorporate ailerons until 1915, six years after Farman's development.

Second, the open framework of the fuselage, such as seen in the Wright Flyer, was in later designs enclosed by fabric. The first airplane to have this feature was a Nieuport monoplane built in 1910. This was an attempt at "streamlining" the airplane, although at that time the concept of streamlining was only an intuitive process rather than the result of real technical knowledge and understanding about drag reduction.

Third, the demands for improved airplane performance during World War I gave a rebirth to the idea of "brute force" in airplane design. In relation to the thrust minus drag expression $T - D$, designers of World War I fighter airplanes, in their quest for faster speeds and higher rates of climb, increased the thrust rather than decreasing the drag. The focus was on more powerful engines. The SPAD XIII, one of the best and most famous aircraft from World War I, had a Hispano-Suiza engine that produced 220 hp—the most powerful engine used on a fighter aircraft at that time. Because of this raw power, the SPAD XIII had a maximum velocity of 134 mph which made it one of the fastest airplanes during the war. The SPAD XIII typifies the strut-and-wire biplane: the struts and wires produced large amounts of drag, although this was not fully understood by most airplane designers at that time. In fact, in the March 1924 issue of the Journal of the Royal Aeronautical Society, the noted British aeronautical engineer Sir Leonard Bairstow was prompted to say, "Our war experience showed that, whilst we went forward as regard to horsepower, we went backwards with regard to aerodynamic efficiency." Aircraft design during World War I was an intuitive "seat-of-the-pants" process. Some designs were almost literally marked off in chalk on the concrete floor of a factory, and the completed machines rolled out the front door two weeks later.

ERA OF THE MATURE, PROPELLER-DRIVEN AIRPLANE

The period from 1930 to 1950 can be classified as the era of the mature, propeller-driven airplane. During this time, airplane design matured, new technical features were incorporated, and the speed, altitude, efficiency, and safety of aircraft increased markedly. The 1930s are considered by many aviation historians as the "golden age of aviation" (indeed, there is currently a gallery at the National Air and Space Museum with this title). Similarly, the 1930s might be considered as a golden age for aeronautical engineering—a period when many improved design features, some gestating since the early 1920s, finally became accepted and incorporated on "standard" aircraft of the age.

The maturity of the propeller-driven airplane is due to nine major technical advances, all of which came to fruition during the 1930s.

Cantilevered-Wing Monoplane

Hugo Junkers in Germany during World War I and Anthony Fokker in Holland during the 1920s pioneered the use single-wing aircraft (monoplanes). This was made possible by the introduction of thick airfoils, which among other advantages allowed room for a large cantilevered spar that structurally supported the wing internally. This eliminated the need for the biplane box structure with its external supporting struts and wires. Consequently, the drag of monoplanes was less than that of comparable strut-andwire biplanes.

The All-Metal Airplane

The vast majority of airplanes before 1930 were constructed from wood and fabric, with some having a steel tube frame mechanism for the fuselage, over which fabric was stretched. Although Hugo Junkers designed and built the first all-metal airplane in 1915, this design feature was not adopted by others for many years. The case for the all-metal airplane was strengthened when the famous Notre Dame football coach Knute Rockne was killed on March 31, 1931, in the crash of a Fokker tri-motor transport. This shook the public's faith in the tri-motor design, and essentially led to its demise in the United States. Such concern was misdirected. Later investigation showed that the wooden wing spar (the entire wing of the Fokker tri-motor was made from wood) had rotted, and the crash was due to this structural failure. What better case could be made for all-metal construction?

Air-Cooled Engines and the NACA Cowling

Propeller-driven airplanes have two types of reciprocating engines—liquid cooled, or air-cooled engines. Since the early days of flight, liquid-cooled engines had the advantage of being longer and thinner, allowing them to be enclosed in relatively streamlined housings with less frontal drag. However, such engines were more vulnerable to damage during combat—a bullet through any part of the liquid cooling system would usually spell failure of the engine. Also, liquid-cooled engines were heavy due to all the machinery and cooling jackets that were associated with the liquid cooling mechanism. In contrast, air-cooled engines, where the cylinder heads are directly exposed to, and cooled by, the airstream over the airplane, are lighter. They require fewer moving parts, and therefore tend to be more reliable. The development of the powerful and reliable Pratt and Whitney Wasp series and the Curtiss-Wright Cycline series of air-cooled radial engines during the late 1920s and the 1930s resulted in the widespread adoption of these engines. But with the cylinders exposed directly to the airstream, the drag created by these was inordinately large.

This set the stage for a major technical development during this era, namely the National Advisory Committee for Aeronautics (NACA) cowling for radial piston engines. Such engines have their pistons arranged in a circular fashion about the crankshaft, and the cylinders themselves are cooled by airflow over the outer finned surfaces. Until 1927, these cylinders were usually directly exposed to the main airstream of the airplane, causing inordinately high

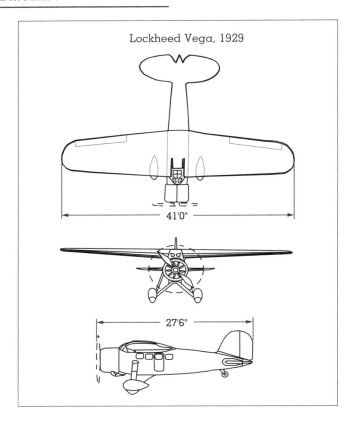

Lockheed Vega, 1929

41'0"

27'6"

Figure 1.

drag. Engineers recognized this problem, but early efforts to enclose the engines inside an aerodynamically streamlined shroud (a cowling) interfered with the cooling airflow, and the engines overheated. One of the earliest aeronautical engineers to deal with this problem was Colonel Virginius E. Clark (for whom the famous Clark-Y airfoil is named). Clark designed a primitive cowling in 1922 for the Dayton-Wright XPS1 airplane; it was marginal at best, and Clark had no proper aerodynamic explanation as to why a cowling worked. The first notable progress was made by H. L. Townend at the National Physical Laboratory in England. In 1927, Townend designed a ring of relative short length that wrapped around the outside of the cylinders. This resulted in a noticeable decrease in drag, and at least it did not interfere with engine cooling. Engine designers who were concerned with the adverse effect of a full cowling on engine cooling were more ready to accept a ring.

The greatest breakthrough in engine cowlings was due to the National Advisory Committee for Aeronautics in the United States. Beginning in 1927,

at the insistence of a group of U.S. aircraft manufacturers, the NACA Langley Memorial Laboratory at Hampton, Virginia, undertook a systematic series of wind tunnel tests with the objective of understanding the aerodynamics of engine cowlings and designing an effective shape for such cowlings. Under the direction of Fred E. Weick at Langley Laboratory, this work quickly resulted in success. Drag reduction larger than that with a Townend ring was obtained by the NACA cowling. In 1928, Weick published a report comparing the drag on a fuselage-engine combination with and without a cowling. Compared with the uncowled fuselage, a full cowling reduced the drag by a stunning 60 percent. By proper aerodynamic design of the cowling, the airflow between the engine and the inside of the cowling resulted in enhanced cooling of the engine. Hence, the NACA cowling was achieving the best of both worlds. One of the first airplanes to use the NACA cowling was the Lockheed Vega, shown in Figure 1. Early versions of the Vega without a cowling had a top speed of 135 mph; after the NACA cowling was added to later versions, the top speed increased to 155 mph. The Lockheed Vega went on to become one of the most successful airplanes of the 1930s. The Vega 5, equipped with the NACA cowling and a more powerful engine, had a top speed of 185 mph. It was used extensively in passenger and corporate service. In addition, Amelia Earhart and Wiley Post became two of the most famous aviators of the 1930s—both flying Lockheed Vegas. Not only is the Vega a classic example of the new era of mature propeller-driven airplanes, but its aesthetic beauty supported the popular adage "If an airplane looks beautiful, it will also fly beautifully."

Variable-Pitch and Constant-Speed Propellers

Before the 1930s, a weak link in all propeller-driven aircraft was the propeller itself. For a propeller of fixed orientation, the twist of the propeller is designed so that each airfoil section is at its optimum angle of attack to the relative airflow, usually that angle of attack that corresponds to the maximum lift-to-drag ratio of the airfoil. The relative airflow seen by each airfoil section is the vector sum of the forward motion of the airplane and the rotational motion of the propeller. When the forward velocity of the airplane is changed, the angle of attack of each airfoil section changes relative to the local flow direction. Thus a fixed-pitch propeller is operating at maximum efficiency only at its design speed; for all other speeds of the airplane, the propeller efficiency decreases.

The solution to this problem was to vary the pitch of the propeller during the flight so as to operate at near-optimum conditions over the flight range of the airplane—a mechanical task easier said than done. The aerodynamic advantage of varying the propeller pitch during flight was appreciated as long ago as World War I, and H. Hele-Shaw and T. E. Beacham patented such a device in England in 1924. The first practical and reliable mechanical device for varying propeller pitch was designed by Frank Caldwell of Hamilton Standard in the United States. The first production order for Caldwell's design was placed by Boeing in 1933 for use on the Boeing 247 transport. The 247 was originally designed in 1932 with fixed-pitch propellers. When it started flying in early 1933, Boeing found that the airplane had inadequate takeoff performance from some of the airports high in the Rocky Mountains. By equipping the 247 with variable-pitch propellers, this problem was solved. The new propellers increased its rate of climb by 22 percent and its cruising velocity by over 5 percent. Later in the 1930s, the variable-pitch propeller, which was controlled by the pilot, developed into the constant-speed propeller, where the pitch was automatically controlled so as to maintain constant rpm over the flight range of the airplane. Because the power output of the reciprocating engine varies with rotational speed, by having a propeller in which the pitch is continuously and automatically varied to constant engine speed, the net power output of the engine-propeller combination can be maintained at an optimum value.

High-Octane Aviation Fuel

Another important advance in the area of propulsion was the development of high-octane aviation fuel, although it was eclipsed by the more visibly obvious breakthroughs in the 1930s such as the NACA cowling, retractable landing gear, and the variable-pitch propeller. Engine "pinging," an audible local detonation in the engine cylinder caused by premature ignition, had been observed as long ago as 1911. An additive to the gasoline, tetraethyl lead, was found by C. F. Kettering of General Motors Delco to reduce this engine knocking. In turn, General Motors and Standard Oil formed a new company, Ethyl Gasoline Corporation, to produce "ethyl" gasoline with a lead additive. Later, the hydrocarbon compound of octane was also found to be effective in preventing engine

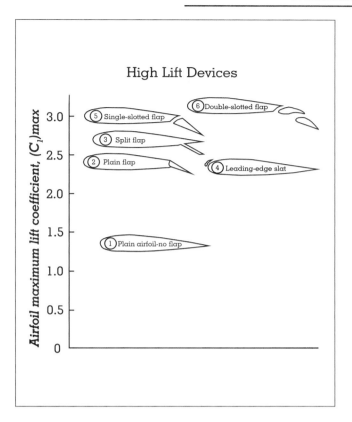

Figure 2.

knocking. In 1930, the Army Air Corps adopted 87-octane gasoline as its standard fuel; in 1935, this standard was increased to 100 octane. The introduction of 100-octane fuel allowed much higher compression ratios inside the cylinder, and hence more power for the engine. For example, the introduction of 100-octane fuel, as well as other technological improvements, allowed Curtiss-Wright Aeronautical Corporation to increase the power of its R-1820 Cycline engine from 500 to 1,200 hp in the 1930s.

High-Lift Devices

When a new airplane is designed, the choice of wing area is usually dictated by speed at takeoff or landing (or alternatively by the desired takeoff or landing distances along a runway). The wing area must be large enough to provide sufficient lift at takeoff or landing; this criterion dictates the ratio of airplane weight to wing area, that is, the wing loading W/S—one of the most important parameters in airplane performance and design. After the airplane has taken off and accelerated to a much higher cruising speed, the higher-velocity airflow over the wing creates a larger pressure difference between the upper and lower wing surfaces, and therefore the lift required to sustain the weight of the airplane can be created with a smaller wing area. From this point of view, the extra wing required for takeoff and landing is extra baggage at cruising conditions, resulting in higher structural weight and increased skin friction drag. The design of airplanes in the era of strut-and-wire biplanes constantly suffered from this compromise. A partial solution surfaced in the late 1920s and 1930s, namely, the development of high-lift devices such as flaps, slats, and slots. Figure 2 illustrates some of the standard high-lift devices employed on aircraft since the 1920s, along with a scale of lift coefficient indicating the relative increase in lift provided by each device. By employing such high-lift devices, sufficient lift can be obtained at takeoff and landing with wings of smaller area, allowing airplane designers the advantage of high wing loadings at cruise. High-lift devices were one of the important technical developments during the era of the mature propeller-driven airplane.

Pressurized Aircraft

Another technical development of the late 1930s is the advent of the pressurized airplane. Along with the decrease in atmospheric pressure with increasing altitude, there is the concurrent decrease in the volume of oxygen necessary for human breathing. The useful cruising altitude for airplanes was limited to about 18,000 ft or lower. Above this altitude for any reasonable length of time, a human being would soon lose consciousness due to lack of oxygen. The initial solution to the problem of sustained high-altitude flight was the pressure suit and the auxiliary oxygen supply breathed through an oxygen mask. The first pilot to use a pressure suit was Wiley Post. Looking like a deep-sea diver, Post set an altitude record of 55,000 ft in his Lockheed Vega in December 1934. This was not a practical solution for the average passenger on board an airliner. The answer was to pressurize the entire passenger cabin of the airplane, so as to provide a shirtsleeve environment for the flight crew and passengers. The first airplane to incorporate this feature was a specially modified and structurally strengthened Lockheed IOE Electra for the Army Air Corps in 1937. Designated the XC-35, this airplane had a service ceiling of 32,000 ft. It was the forerunner of all the modern pressurized airliners of today.

Superchargers for Engines

Along with pressurization for the occupants, high-altitude aircraft needed "pressurization" for the engine. Engine power is nearly proportional to the atmospheric density; without assistance, engine power dropped too low at high altitudes, and this was the major mechanical obstacle to high-altitude flight. Assistance came in the form of the supercharger, a mechanical pump that compressed the incoming air before it went into the engine manifold. Supercharger development was a high priority during the 1930s and 1940s; it was a major development program within NACA. All high-performance military aircraft during World War II were equipped with superchargers as a matter of necessity.

Streamlining

One of the most important developments in the era of the mature propeller-driven airplane was the appreciation of the need for streamlining the airframe. The rather box-like shape of the World War I vintage SPAD was characteristic of airplanes of that day. There was little if any attempt to shape the airplane into a streamlined configuration. The Douglas DC-3, however, was designed and began airline service in the mid-1930s. Here is streamlining personified. By comparison, the zero-lift-drag coefficient for the SPAD is 0.04, whereas that for the DC-3 is about 0.025, a considerable improvement. Part of the concept of streamlining was to retract the landing gear flush with the external airframe.

Summary

The Douglas DC-3 epitomizes the mature, propeller-driven aircraft of the 1930s. Here you see a cantilever wing monoplane powered by radial engines enclosed in NACA cowlings, and equipped with variable-pitch propellers. It is an all-metal airplane with retractable landing gear, and it uses flaps for high lift during takeoff and landing. For these reasons, the 1930s can indeed be called the golden age of aeronautical engineering.

ERA OF THE JET-PROPELLED AIRPLANE

The jet engine was invented independently by two people: Frank Whittle in England and Dr. Hans von Ohain in Germany. In 1928, as a student at the Royal Air Force technical college at Cranwell, Frank Whittle wrote a senior thesis entitled "Future

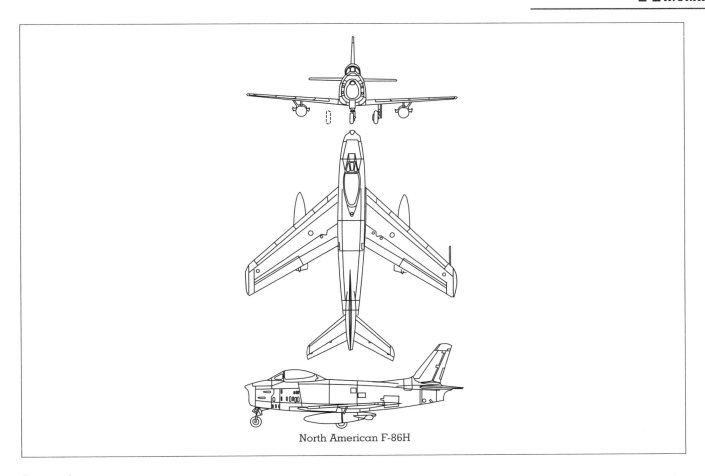

North American F-86H

Figure 3.

Developments in Aircraft Design" in which he expounded on the virtues of jet propulsion. It aroused little interest. Although Whittle patented his design for a gas-turbine aircraft engine in 1930, it was not until five years later that he formed, with the help of friends, a small company to work on jet engine development. Named Power Jets Ltd., this company was able to successfully bench-test a jet engine on April 12, 1937—the first jet engine in the world to successfully operate in a practical fashion. It was not the first to fly. Quite independently, and completely without the knowledge of Whittle's work, Dr. Hans von Ohain in Germany developed a similar gas-turbine engine. Working under the private support of the famous airplane designer Ernst Heinkel, von Ohain started his work in 1936. On August 27, 1939, a specially designed Heinkel airplane, the He 178, powered by von Ohain's jet engine, successfully flew; it was the first gas turbine-powered, jet-propelled air-

plane in history to fly. It was strictly an experimental airplane, but von Ohain's engine of 838 lb of thrust pushed the He 178 to a maximum speed of 360 mph. It was not until almost two years later that a British jet flew. On May 15, 1941, the specially designed Gloster E.28/39 airplane took off from Cranwell, powered by a Whittle jet engine. It was the first to fly with a Whittle engine. With these first flights in Germany and Britain, the jet age had begun.

The era of jet-propelled aircraft is characterized by a number of design features unique to airplanes intended to fly near, at, or beyond the speed of sound. One of the most pivotal of these design features was the advent of the swept wing. For a subsonic airplane, sweeping the wing increases the airplane's critical Mach number, allowing it to fly closer to the speed of sound before encountering the large drag rise caused by the generation of shock waves somewhere on the surface of the wing. For a

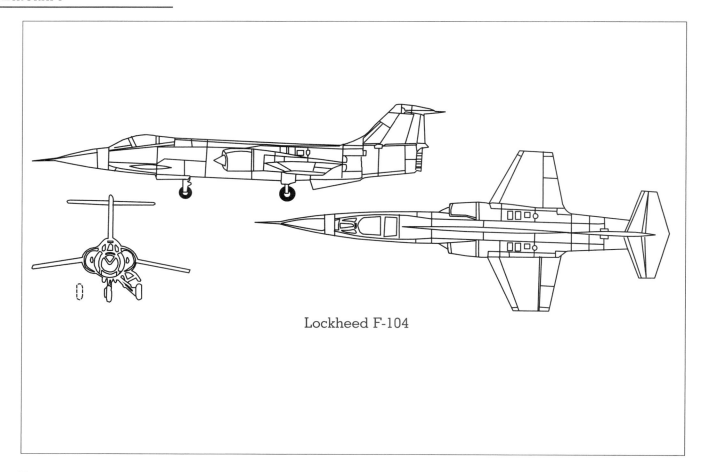

Lockheed F-104

Figure 4.

supersonic airplane, the wing sweep is designed such that the wing leading edge is inside the Mach cone from the nose of the fuselage; if this is the case, the component of airflow velocity perpendicular to the leading edge is subsonic (called a subsonic leading edge), and the resulting wave drag is not as severe as it would be if the wing were to lie outside the Mach cone. In the latter case, called the supersonic leading edge, the component of flow velocity perpendicular to the leading edge is supersonic, with an attendant shock wave generated at the leading edge. In either case, high subsonic or supersonic, an airplane with a swept wing will be able to fly faster than one with a straight wing, everything else being equal.

The concept of the swept wing for high-speed aircraft was first introduced in a public forum in 1935. At the fifth Volta Conference, convened on September 30, 1935, in Rome, Italy, the German aerodynamicist Adolf Busernann gave a paper in

which he discussed the technical reasons why swept wings would have less drag at high speeds than conventional straight wings. Although several Americans were present, such as Eastmann Jacobs from NACA and Theodore von Karman from Cal Tech, Busernann's idea went virtually unnoticed; it was not carried back to the United States with any sense of importance. Not so in Germany. One year after Busemann's presentation at the Volta Conference, the swept-wing concept was classified by the German Luftwaffe as a military secret. The Germans went on to produce a large bulk of swept-wing research, including extensive wind tunnel testing. They even designed a few prototype swept-wing jet aircraft. Many of these data were confiscated by the United States after World War II, and made available to U.S. aircraft companies and government laboratories. Meanwhile, quite independently of this German research, Robert T. Jones, a NACA aerodynamicist,

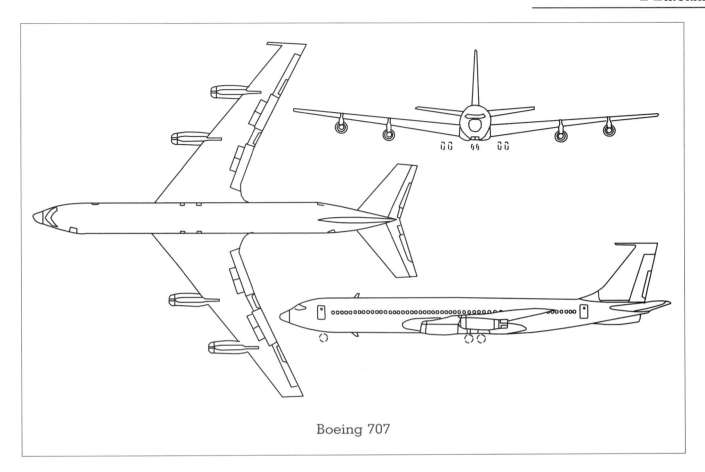

Boeing 707

Figure 5.

had worked out the elements of swept-wing theory toward the end of the war. Although not reinforced by definitive wind tunnel tests in the United States at that time, Jones's work served as a second source of information concerning the viability of swept wings.

In 1945, aeronautical engineers at North American Aircraft began the design of the XP-86 jet fighter; it had a straight wing. The XP-86 design was quickly changed to a swept-wing configuration when the German data, as well as some of the German engineers, became available after the war. The prototype XP-86 flew on October 1, 1947, and the first production P-86A flew with a 35° swept wing on May 18, 1948. Later designated the F-86, the swept-wing fighter had a top speed of 679 mph, essentially Mach 0.9—a stunning speed for that day. Shown in Figure 3, the North American F-86 Sabre was the world's first successful operational swept-wing aircraft.

By the time the F-86 was in operation, the sound

barrier had already been broken. On October 14, 1947, Charles (Chuck) Yeager became the first human being to fly faster than the speed of sound in the Bell X-1 rocket-powered airplane. In February 1954, the first fighter airplane capable of sustained flight at Mach 2, the Lockheed F-104 Starfighter, made its first appearance. The F-104, Figure 4, exhibited the best qualities of good supersonic aerodynamics—a sharp, pointed nose, slender fuselage, and extremely thin and sharp wings. The airfoil section on the F-104 is less than 4 percent thick (maximum thickness compared to the chord length). The wing leading edge is so sharp that protective measures must be taken by maintenance people working around the aircraft. The purpose of these features is to reduce the strength of shock waves at the nose and trailing edges, thus reducing supersonic wave drag. The F-104 also had a straight wing with a very low aspect ratio rather than a swept wing. This exhibits an

Aircraft	Payload Capacity	Top Speed	Cruising Speed	Fuel Consumption (1000 mile flight)	4D Ratio
737	128 passengers	M=0.89	M=0.83	1600 gal.	17
737 Stretch	188 passengers	0.89	0.83	1713	17
747-400	413 passengers	0.87	0.82	6584	21
SST Concorde	126 passengers	2.2	2.0	6400	8
Turboprop DHC-8	37 passengers	325 MPH	305 MPH	985	15
Fighter Plane F-15		M=2.5		750	6
Military Cargo Plane C-17	120,000 lbs.	0.80	0.76	5310	14

Table 1.
Comparison of Some Aircraft Figures
Payload capacity for civil transports is given in number of passengers; for the C-17 it is in pounds. Payloads and cruising speeds were unavailable for the fighter planes.

alternative to supersonic airplane designers; the wave drag on straight wings of low aspect ratio is comparable to that on swept wings with high aspect ratios. Of course, this low-aspect-ratio wing gives poor aerodynamic performance at subsonic speeds, but the F-104 was point-designed for maximum performance at Mach 2. With the F-104, supersonic flight became an almost everyday affair, not just the domain of research aircraft.

The delta wing concept was another innovation to come out of Germany during the 1930s and 1940s. In 1930, Alexander Lippisch designed a glider with a delta configuration; the leading edges were swept back by 20°. The idea had nothing to do with high-speed flight at that time; the delta configuration had some stability and control advantages associated with its favorable center-of-gravity location. When Busemann introduced his swept-wing ideas in 1935, Lippisch and his colleagues knew they had a potential high-speed wing in their delta configuration. Lippisch continued his research on delta wings during the war, using small models in German supersonic wind tunnels. By the end of the war, he was starting to design a delta wing ramjet-powered fighter. Along with the German swept-wing data, this delta wing technology was transferred to the United States after the war; it served as the basis for an extended wind tunnel test program on delta wings at NACA Langley Memorial Laboratory.

The first practical delta wing aircraft was the Convair F-102. The design of this aircraft is an inter-esting story in its own right—a story of the interplay between design and research, and between industry and NACA. The F-102 was designed as a supersonic airplane. Much to the embarrassment and frustration of the Convair engineers, the prototype F-102 being tested at Edwards Air Force Base during October 1953 and then again in January 1954 exhibited poor performance and was unable to go supersonic. At the same time, Richard Whitcomb at NACA Langley was conducting wind tunnel tests on his "area rule" concept, which called for the cross-sectional area of the fuselage to be reduced in the vicinity of the wing. By so doing, the transonic drag was substantially reduced. The Convair engineers quickly adopted this concept on a new prototype of the F-102, and it went supersonic on its second flight. Convair went on to produce 975 F-102s; the practical delta wing airplane was a finally a reality.

The area rule was one of the most important technical developments during the era of jet-propelled airplanes. Today, almost all transonic and supersonic aircraft incorporate some degree of area rule. For his work on the area rule, Whitcomb received the Collier Trophy, the highest award given in the field of aeronautics.

One of the most tragic stories in the annals of airplane design occurred in the early 1950s. Keying on England's early lead in jet propulsion, de Havilland Aircraft Company designed and flew the first commercial jet transport, the de Havilland Comet. Powered by four de Havilland Ghost jet engines, the

Comet carried 36 passengers for 2,000 mi at a speed of 460 mph, cruising at relatively high altitudes near or above 30,000 ft. The passenger cabin was pressurized; indeed, the Comet was the first pressurized airplane to fly for extended periods at such high altitudes. Inasmuch as good airplane design is an evolutionary process based on preceding aircraft, the de Havilland designers had little precedent on which to base the structural design of the pressurized fuselage. The Comet entered commercial service with BOAC (a forerunner of British Airways) in 1952. In 1954, three Comets disintegrated in flight, and the airplane was quickly withdrawn from service. The problem was later found to be structural failure of the fuselage while pressurized. De Havilland used countersunk rivets in the construction of the Comet; reaming the holes for the rivets produced sharp edges. After a number of pressurization cycles, cracks in the fuselage began to propagate from these sharp edges, leading eventually to catastrophic failure. At the time, de Havilland had a massive lead over all other aircraft companies in the design of commercial jet aircraft. While it was in service, the Comet was very popular with the flying public, and it was a moneymaker for BOAC. Had these failures not occurred, de Havilland and England might have become the world's supplier of commercial jet aircraft rather than Boeing and the United States.

In 1952, the same year as the ill-fated de Havilland Comet went into service, the directors of Boeing Company made a bold and risky decision to privately finance and build a commercial jet prototype. Designated the model 367-80, or simply called the Dash 80 by the Boeing people, the prototype first flew on July 15, 1954. It was a bold design that carried over to the commercial field Boeing's experience in building swept-wing jet bombers for the Air Force (the B-47 and later the B-52). Later renamed the Boeing 707, the first production series of aircraft were bought by Pan American Airlines and went into service in 1958. The Boeing 707 (Figure 5), with its swept wings and podded engines mounted on pylons below the wings, set the standard design pattern for all future large commercial jets. The design of the 707 was evolutionary because it stemmed from the earlier experience at Boeing with jet bombers. But it was almost revolutionary in the commercial field, because no other airliner had ever (not even the Comet) looked like that. Boeing's risky gamble paid

off, and it transformed a predominately military aircraft company into the world's leader in the design and manufacture of commercial jet transports.

Boeing made another bold move on April 15, 1966, when the decision was made to "go for the big one." Boeing had lost the Air Force's C-5 competition to Lockheed; the C-5 at the time was the largest transport airplane in the world. Taking their losing design a few steps further, Boeing engineers conceived of the 747, the first wide-body commercial jet transport. Bill Allen, president of Boeing at that time, and Juan Trippe, president of Pan American Airlines, shared the belief that the large, wide-body airplane offered economic advantages for the future airline passenger market, and they both jointly made the decision to pursue the project. It was an even bolder decision than that concerning the 707.

The gamble paid off. The Boeing 747 first flew in February 1969, and it entered service for the first time in January 1970 on Pan American's New York—London route. Boeing is still producing 747s.

What about commercial transportation at supersonic speeds? In the 1960s this question was addressed in Russia, the United States, England, and France. The Tupolev Design Bureau in Russia rushed a supersonic transport design into production and service. The Tu-144 supersonic transport first flew on December 31, 1968. More than a dozen of these aircraft were built, but none entered extended service, presumably due to unspecified problems. One Tu-144 was destroyed in a dramatic accident at the 1973 Paris Air Show. In the United States, the government orchestrated a design competition for a supersonic transport; the Boeing 2707 was the winner in December 1966. The design turned into a nightmare for Boeing. For two years, a variable-sweep wing supersonic transport (SST) configuration was pursued, and then the design was junked. Starting all over again in 1969, the design was caught up in an upward spiral of increased weight and development costs. When the predictions for final development costs hit about $5 billion, Congress stepped in and refused to appropriate any more funds. In May 1971, the SST development program in the United States was terminated. Only in England and France was the SST concept carried to fruition.

The first, and so far only, supersonic commercial transport to see long-term regular service was the Anglo-French Concorde (Figure 6). In 1960 both

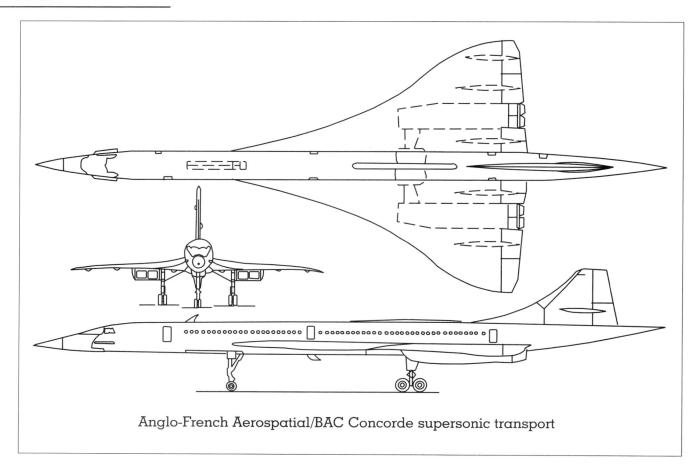

Anglo-French Aerospatial/BAC Concorde supersonic transport

Figure 6.

the British and French independently initiated design studies for a supersonic transport. It quickly became apparent that the technical complexities and financial costs were beyond the abilities of either country to shoulder alone. On November 29, 1962, England and France signed a formal agreement aimed at the design and construction of a supersonic transport. The product of this agreement was the Aerospatiale-British Aerospace Corporation's Concorde. Designed to cruise at Mach 2.2 and carry 125 passengers, the Concorde first flew on March 2, 1969. It first exceeded Mach 1 on October 1, 1969, and Mach 2 on November 4, 1970. Originally, orders for 74 Concordes were anticipated. When the airlines were expected to place orders in 1973, the world was deep in the energy crisis. The skyrocketing costs of aviation jet fuel wiped out any hope of an economic return from flying the Concorde, and no orders were placed. Only the national airlines of France and Britain, Air France and British Airways,

went ahead, each signing up for seven aircraft after considerable pressure from their respective governments. After a long development program, the Concorde went into service on January 21, 1976. In the final analysis, the Concorde was a technical, if not financial, success. It was in regular service from 1976 until the entire fleet was grounded in the wake of the first concorde crash in August 2000. It represents an almost revolutionary airplane design in that no such aircraft existed before it. The Concorde designers were not operating in a vacuum. Examining Figure 6, we see a supersonic configuration which incorporates good supersonic aerodynamics—a sharp-nosed slender fuselage and a cranked delta wing with a thin airfoil. The Concorde designers had at least fifteen years of military airplane design experience with such features to draw upon. Today, we know that any future second-generation SST will have to be economical in service and environmentally acceptable. The design of such a vehicle is one of the great challenges in aeronautics.

In summary, the types of aircraft in use today cut across the flight spectrum from low-speed, propeller-driven airplanes with reciprocating engines, moderate speed turboprop airplanes (propeller driven by gas turbine engines), and high-speed jet-propelled airplanes. For low-speed flight, below about 250 mph, the reciprocating engine/propeller combination has by far the best propulsive efficiency. For moderate speeds (250–400 mph) the turboprop is superior. This is why most high-performance commuter aircraft are powered by turboprops. For high speeds (at least above 500 mph) the jet engine is the only logical powerplant choice; a propeller rapidly looses efficiency at higher flight speeds. In short, a reciprocating engine/propeller combination is a high efficiency, but comparably low thrust powerplant, and a jet engine is a lower efficiency but higher thrust powerplant. The turboprop is a middle-ground compromise between thrust and efficiency. The wide variety of airplanes in use today draw on the technology developed in both the era of the mature, propeller-driven airplane and the era of the jet-propelled airplane. In the future, airplane design will continue to be influenced by the desire to fly faster and higher, but moderated by the need for environmental effectiveness, economic viability, and energy efficiency.

John D. Anderson

See also: Aerodynamics; Air Travel; Efficiency of Energy Use; Propellers.

BIBLIOGRAPHY

Anderson, J. D., Jr. (1997). *A History of Aerodynamics, and Its Impact on Flying Machines,* New York: Cambridge Universtiy Press.

Anderson, J. D., Jr. (1999). *Aircraft Performance and Design.* Boston: McGraw-Hill.

Anderson, J. D., Jr. (2000). *Introduction to Flight,* 4th ed. Boston: McGraw-Hill.

Loftin, L. K. (1985) *Quest for Performance: The Evolution of Modern* Aircraft, Washington, DC: NASA.

Greenwood, J. T. ed. (1989). *Milestones of Aviation.* Washington, DC: National Air and Space Museum, Smithsonian Institution.

AIR POLLUTION

The pernicious effects of air pollution were first documented long ago. As early as 61 C.E., Seneca, a Roman

Aerial view of Los Angeles obscured by smog. (Phototake)

philosopher and noted essayist, wrote, "As soon as I had gotten out of the heavy air of Rome, and from the stink of the chimneys thereof, which being stirred, poured forth whatever pestilential vapors and soot they had enclosed in them, I felt an alteration to my disposition" (Miller and Miller, 1993). With technology being as simple as it was in Rome, however, there was not much that could be done about the problem.

Thirteen hundred years later, controls on the use of coal in London were passed, marking the recorded start of air pollution control. But such controls were not enough to prevent the buildup of pollutants as by-products of industrialization; air pollution was common to all industrialized nations by 1925. Air pollution is still a significant problem in urban centers worldwide. In the United States, pollutant emissions and air pollution concentrations have been falling, for the most part, since the 1970s.

The major constituents of unpolluted air (not including water) at ground level are nitrogen (78.08%) and oxygen (20.95%). The next most abundant constituents are argon (at 0.934%) and carbon dioxide (about 0.034%), followed by the other noble gases:

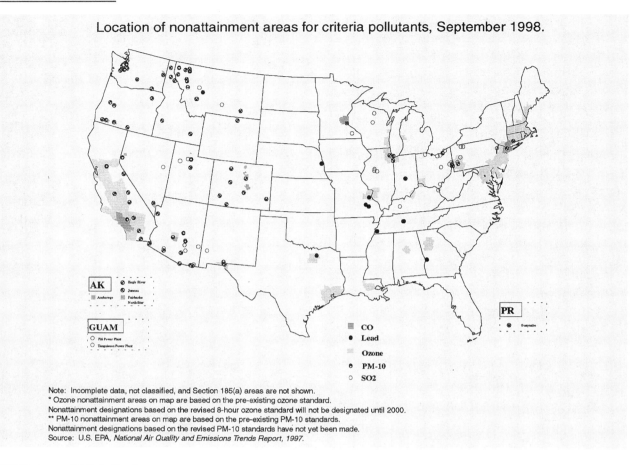

Location of nonattainment areas for criteria pollutants, September 1998.

AK
⊘ Eagle River
⊙ Juneau
▨ Anchorage
▨ Fairbanks North Star

GUAM
○ Piti Power Plant
○ Tanguisson Power Plant

PR
⊗ Guayabo

▨ CO
● Lead
▨ Ozone
◓ PM-10
○ SO2

Note: Incomplete data, not classified, and Section 185(a) areas are not shown.
* Ozone nonattainment areas on map are based on the pre-existing ozone standard.
Nonattainment designations based on the revised 8-hour ozone standard will not be designated until 2000.
** PM-10 nonattainment areas on map are based on the pre-existing PM-10 standards.
Nonattainment designations based on the revised PM-10 standards have not yet been made.
Source: U.S. EPA, National Air Quality and Emissions Trends Report, 1997.

Source: U.S. Environmental Protection Agency.

neon (0.002%), helium (0.0005%), krypton (0.0001%), and xenon (0.000009%). A variety of chemicals known as trace gases (some quite toxic) also are found in unpolluted air, but at very low concentrations.

When certain substances in the air rise to the level at which they can harm plant or animal life, damage property, or simply degrade visibility in an area more than it would be in the absence of human action, those substances are considered to be pollutants. Such pollutants enter the atmosphere via both natural processes and human actions.

The pollutants most strongly damaging to human, animal, and sometimes plant health include ozone, fine particulate matter, lead, nitrogen oxides (NO_x), sulfur oxides (SO_x), and carbon monoxide. Many other chemicals found in polluted air can cause lesser health impacts (such as eye irritation). VOC compounds comprise the bulk of such chemicals. Formaldehyde is one commonly mentioned pollutant of this sort, as is PAN (peroxyacyl nitrate). Such chemicals are common components of photochemical smog, a term that refers to the complex mixture of chemicals that forms when certain airborne chemicals given off by plant and human activity react with sunlight to produce a brownish mixture of thousands of different chemical species.

Finally, there are also pollutants that do not cause direct health impacts but that may have the potential to cause harm indirectly, through their actions on the overall ecology, or as they function as precursor chemicals that lead to the production of other harmful chemicals. The major indirect-action pollutants include volatile organic carbon (VOC) compounds that act as precursors to more harmful species; chemicals called halocarbons; and chemicals called greenhouse gases.

ENERGY USE AND POLLUTION

Energy use is the predominant source of global air pollution, though other human actions produce significant amounts of pollution as well. The burning of biomass for agriculture, such as the burning of crop stubble is one such nonenergy source, as is the use of controlled-burn management of forest fires. Natural events such as forest fires as well as volcanic outgassing and eruption also contribute to air pollution. Finally, even living things produce considerable quantities of emissions considered "pollutants." Animals and insects produce a large share of the world's methane emissions, while plants emit significant quantities of volatile organic carbon compounds—enough, in some areas to produce elevated ozone levels with no other pollutant emissions at all.

Ozone

Ozone, or O_3, a chemical consisting of three atoms of oxygen, is a colorless, odorless gas produced by a variety of chemical reactions involving hydrocarbons and nitrogen oxides in the presence of sunlight. Often there is confusion about ozone's effects because it is found at two different levels of the atmosphere, where it has two very different effects. At ground level, ozone is known to be a respiratory irritant, implicated in causing decreased lung function, respiratory problems, acute lung inflammation, and impairment of the lungs' defense mechanisms. Outdoor workers, elderly people with pre-existing lung diseases, and active children who spend significant amounts of time in areas with elevated ozone levels are thought to be particularly at risk.

Low-altitude ozone forms when sunlight reacts with "precursor" chemicals of both human and non-human origin. These precursors include volatile organic carbon compounds, carbon monoxide, and nitrogen oxides. Volatile organic carbon compounds are created both naturally (by plants) and through human activity such as fuel use, biomass burning, and other industrial activities such as painting and coating. Nitrogen oxides are generated by stationary sources of energy use such as power plants and factories, and mobile sources such as cars, trucks, motorcycles, bulldozers, and snowmobiles. Because of the diversity of ozone precursors, sources vary by region.

At high altitudes, ozone forms naturally from the interaction of high-energy solar radiation and normal diatomic oxygen, or O_2. High-altitude ozone is not considered a pollutant, but is actually beneficial to life on Earth, as it screens out solar radiation that can damage plants, or cause skin cancer and cataracts in animals. High-altitude ozone can be destroyed through

VOC Source Breakdown	Tg/yr
Human emissions sources	98
Biomass burning	51
Continental biogenic sources	500
Oceans	30-300
Total	**750**

Table 1.
VOC Amounts by Source
Note: One Tg (teragram) is equivalent to 1 million metric tons. Continental biogenic sources includes animal, microbial, and foliage emissions
SOURCE: Finlayson-Pitts & Pitts (2000)

the action of chemicals called halocarbons, which are commonly used in refrigeration and air conditioning. Ozone levels across the United States fell, on average, by four percent between 1989 and 1998.

Volatile Organic Carbon Compounds

There are many different chemical species in the VOC compounds group, including such commonly known chemicals as formaldehyde and acetone. The common feature that VOCs share is that they are ring-shaped (organic) molecules consisting principally of carbon, hydrogen, oxygen, and nitrogen. VOCs are released into the environment as a result of human activity, but also because of natural biological processes in plants and animals. Table 1 shows the percentage of global contribution to VOC concentrations from all sources.

VOCs react in the presence of sunlight to produce photochemical smog, a mixture of organic chemicals that can irritate the eyes and other mucous membranes. VOCs also constitute a major precursor chemical leading to ozone production. VOC levels across the United States fell, on average, by 20.4 percent between 1989 and 1998.

Particulate Matter

Particulate matter, generated through a range of natural and manmade processes including combustion and physical abrasion, has been implicated in increased mortality for the elderly, as well as those members of the population with damaged respiratory systems. Studies also have linked particulate matter with aggravation of preexisting respiratory and cardiovascular disease, resulting in more frequent and/or serious attacks of asthma in the elderly or in children.

The particulate matter of most concern consists of

National Ambient Air Quality Standards

National Ambient Air Quality Standards (NAAQS) have been established by the U.S. Environmental Protection Agency for the following six criteria air pollutants:

NATIONAL AMBIENT AIR QUALITY STANDARDS			
Pollutant	Averaging Time	Primary Standard	Secondary Standard
CO	8 Hours 1 Hour	9 ppm 35 ppm	None None
Lead (Pb)	Calendar Quarter	1.5 $\mu g/m^3$	Same as Primary
NO_2	Annual	0.053 ppm	Same as Primary
O_3	1 Hour	0.12 ppm	Same as Primary
PM_{10}	Annual 24 Hours	50 $\mu g/m^3$ 150 $\mu g/m^3$	Same as Primary Same as Primary
SO_2	Annual 24 Hours 3 Hours	0.03 ppm 0.14 ppm None	None None 0.5 ppm
The TSP NAAQS is no longer applicable. It was superseded by the PM_{10} NAAQS on 07/01/87. The old TSP NAAQS is provided for information only.			
TSP	Annual 24 Hours	75 $\mu g/m^3$ 260 $\mu g/m^3$	60 $\mu g/m^3$ 150 $\mu g/m^3$

The NAAQS are the allowable ambient (outdoor) concentrations that must be maintained in order to protect public health and welfare. Limits have been set for carbon monoxide (CO), lead (Pb), nitrogen dioxide (NO_2), ozone (O_3), sulfur dioxide (SO_2), and particulate matter (PM_{10}). EPA is currently reviewing the adequacy of the ozone and PM_{10} standards.

Table 2.
National Ambient Air Quality Standards. (NAAQS).

particles that are most likely to be trapped in tiny air sacs of the lung (alveoli) after inhalation. Studies suggest that such particles are in the microscopic range, from less than 1 micrometer in diameter to about 2.5 microns in average diameter.

Although it has been shown that exposure to certain air pollutants can aggravate preexisting lung ailments such as asthma, no causal link has been identified between exposure to low-level air pollution and asthma. In fact, while asthma levels have been rising, air pollution levels have been declining, suggesting that there is probably a different cause for the increase in asthma rates. While various indoor air pollutants have been suggested as the cause, no definitive cause for the increase in asthma rates has yet been found.

Particulate matter is generated by stationary sources such as power plants and factories and by mobile sources such as cars, trucks, motorcycles, bulldozers, and snowmobiles. Particulate matter lev-

els across the United States fell, on average, by 25 percent between 1989 and 1998.

Carbon Monoxide

Carbon monoxide, or CO, is a highly toxic chemical that chemically binds to hemoglobin, rendering it incapable of carrying oxygen to the tissues of the body. CO is produced by the incomplete combustion of fossil fuels. Carbon monoxide levels across the United States fell, on average, by 39 percent between 1989 and 1998.

Lead

Lead is an element used in many industrial processes and also has been used in fuels and coatings. Tetraethyl lead was added to gasoline to improve performance as a motor fuel, and elemental lead was extensively used in paints and coatings to improve coverage and durability until the 1970s, when phase-out efforts began to reduce lead emissions to the environment.

Long-term exposure to airborne lead was shown to

lead to a variety of health problems, primarily neurological. Atmospheric lead concentrations fell dramatically through the 1970s and continue to fall: Atmospheric lead concentrations fell, on average, by 56 percent between 1989 and 1998. The amount of lead found in the bloodstream of children growing up in urban environments also fell dramatically.

Sulfur Oxides

Sulfur oxide emissions enter the atmosphere from a variety of sources, some of human origin, others of natural origin. The main sulfur oxide is sulfur dioxide, or SO_2.

High concentrations of SO_2 can produce temporary breathing difficulties in asthmatic children and in adults who are active outdoors. Sulfur dioxide also can directly damage plants and has been shown to decrease crop yields. In addition, sulfur oxides can be converted to sulfuric acid and lead to acid rain. Acid rain can harm ecosystems by increasing the acidity of soils as well as surface waters such as rivers, lakes, and streams. Sulfur dioxide levels fell, on average, by 39 percent between 1989 and 1998.

Nitrogen Dioxide

Nitrogen dioxide (NO_2) is a reddish-brown gas that is formed through the oxidation of nitrogen oxide (NO). The term "nitrogen oxides," or NO_x is used to encompass NO_2 as well as NO and the other oxides of nitrogen that lead to NO_2 production.

Nitrogen oxides are generated by both human and nonhuman action, but the major sources of NO_x are high-temperature combustion processes such as those occurring in power plants and automobile engines. Natural sources of NO_x include lightning, chemical processes that occur in soil, and the metabolic activities of plants.

Short-term exposure to elevated levels of NO_x have been shown to cause changes in the function of human lungs, while chronic exposures have been linked to increased susceptibility to lung infections and to lasting changes in lung structure and function. Nitrogen oxides also are of concern because of their roles as ozone precursors, and through their contribution to acid rain in the form of nitric acid.

Nitrogen dioxide concentrations have changed little since 1989, and alterations in measuring techniques make it difficult to accurately assess the trend. Data suggest that NO_2 concentrations may have increased by 2 percent between 1989 and 1998.

Human industry is one of the leading causes of air pollution. (Corbis Corporation)

AIR POLLUTION REGULATION IN THE UNITED STATES

Air pollution in the United States is regulated at federal, state, and local levels. Allowable concentrations of the major air pollutants are set by the U.S. Environmental Protection Agency (EPA) under the auspices of the Clean Air Act. States and localities implement pollution control plans in accordance with the provisions of the Clean Air Act in regions where air pollutant concentrations exceed the federal standards. Some states and localities have air pollution standards of their own, and in the past, such standards have occasionally been more stringent than those of the EPA.

The EPA sets two kinds of national ambient air quality standards. The primary standard is set at a level intended to protect human health with an adequate margin of safety. The secondary standard, usually less stringent, is set based on protecting the public welfare, which can include factors other than health impacts, such as reduced visibility, and damage to crops.

Pollutant	Primary Standard (Health Related)		Secondary Standard (Welfare Related)
	Type of Average	Allowable Concentration	
CO	8-hour	9 parts per million	No secondary standard
	1-hour	35 parts per million	No secondary standard
Pb	Maximum quarterly average	1.5 micrograms/cubic-meter of air	Same as primary standard
NO_2	Annual arithmetic mean	0.053 parts per million	Same as primary standard
O_3	Maximum daily 1-hr average	0.12 parts per million	Same as primary standard
	4th Maximum daily 8-hr average	0.08 parts per million	Same as primary standard
PM_{10}	Annual arithmetic mean	50 micrograms/cubic-meter of air	Same as primary standard
	24-hour	150 micrograms/cubic-meter of air	Same as primary standard
$PM_{2.5}$	Annual arithmetic mean	15 micrograms/cubic-meter of air	Same as primary standard
	24-hour	65 micrograms/cubic-meter of air	Same as primary standard
SO_2	Annual arithmetic mean	0.03 parts per million	3-hour/0.50 parts per million
	24-hour	0.14 parts per million	

Table 3.
National Ambient Air Quality Standards in Effect as of December 1999
SOURCE: U.S. EPA National Air Quality and Emissions Trends Report, 1998, Table 2-1, p 9.

Table 3 shows the current health-related national ambient air quality standards set by the EPA as of 1999. Regions that violate these standards may be classified as "nonattainment" areas by the EPA, and can face sanctions if they do not promulgate pollution control plans that are acceptable to the agency.

Environmental regulations to curtail air pollution have had a major impact on energy producers, manufacturers of energy-using products, and energy consumers. Before the Clean Air Act of 1970, coal-fired steam turbines were the least expensive means available for utilities to generate electricity. Coal still remains the least expensive fossil fuel, yet the high capital costs of installing the technology to comply with ever more stringent environmental regulations have resulted in many utilities reconsidering their options for new electric-power-generating facilties. The combined cycle natural gas turbine is being favored not only because of the much easier compliance with current air quality regulations but also because of the likelihood of stricter future regulations. A typical coal power plant burns more than 70 lb of coal each second, and considering that these plants number in the hundreds, coal-burning emissions are likely to be a major future target of legislators and regulators eager to curtail emissions further.

Transportation is another sector that is a major contributor to air pollution. To improve air quality, particularly in urban areas, regulations require the use of reformulated gasoline and alternative fuels to reduce emissions of nitrogen oxides and carbon monoxide. Reflecting the higher cost of refining, these regulations have added at least 5 to 10 cents to the price of gasoline at the pump.

Besides cleaner fuels, vehicle makers have developed many emission-reducing technologies—both in "cleaner combustion" and in catalytic converter technologies—to comply with ever stricter tailpipe emission standards. The U.S. EPA stringent standards proposed in 1999 for model year 2004 vehicles will result in new vehicles emitting less than 1 percent of the VOC and NO_x emissions of their 1960s counterparts.

Kenneth Green

See also: Acid Rain; Air Quality, Indoor; Atmosphere; Automobile Performance; Climatic Effects; Emission Control, Vehicle; Emission Control, Power Plant; Environmental Economics; Environmental Problems and Energy Use; Gasoline and Additives; Transportation, Evolution of Energy Use and; Turbines, Gas.

BIBLIOGRAPHY

Finlayson-Pitts, B. J., and Pitts, J. N., Jr. (2000). *Chemistry of the Upper and Lower Atmosphere: Theory, Experiments, and Applications.* San Diego: Academic Press.

Klaassen, C. D. (1996) *Casarett & Doull's Toxicology: The Basic Science of Poisons,* 5th ed. New York: McGraw-Hill.

Miller, F. W., and Miller, R. M. (1993). *Environmental Hazards: Air Pollution—A Reference Handbook.* Santa Barbara, CA: ABC-CLIO.

U. S. Environmental Protection Agency. (1998). "National Air Quality and Emissions Trends Report, 1998." <http://www.epa.org/>.

AIR QUALITY, INDOOR

While most of us are aware of the dangers posed by outdoor air pollution, awareness of airborne chemical and biological pollutants present indoors and its implications for human health is more limited. Some indoor air gases and pollutants such as radon, asbestos, carbon monoxide, biological contaminants, and volatile organic compounds (VOCs) pose a serious threat to our health and well-being. Over the past several decades, our exposure to indoor air pollutants is believed to have increased due to a variety of factors, including the construction of more tightly sealed buildings; reduced ventilation rates to save energy; use of synthetic building materials and furnishings; and use of chemically formulated personal care products, pesticides, and household cleaners. Since an average person spends increasing amount of time indoors, it is important to understand the health risks posed by prolonged exposure to indoor pollutants and the energy and comfort implications of different methods to control and mitigate these pollutants in order to ensure acceptable indoor air quality.

Acceptable indoor air quality (IAQ) is defined as "air in which there are no known contaminants at harmful concentrations as determined by cognizant authorities and with which a substantial majority (80%) of the people exposed do not express dissatisfaction" (ASHRAE, 1989). Some of these indoor air contaminants are particulates, vapors, and gases that may be generated by occupants and their activities, building materials, furniture, equipment and appliances present in indoor space, operations and maintenance activities, or brought in from outside. Examples of indoor pollutants are certain gases (radon, carbon monoxide, and carbon dioxide), volatile organic compounds or VOCs (environmental tobacco smoke, formaldehyde, solvents, and fragrances, etc.), bioaerosols (mold spores, pollen, bacteria, animal dander, etc.), and particles from buildings, furnishings and occupants (fiberglass, paper dust, lint from clothing, carpet fibers, etc.).

As the science of indoor air quality has matured, indoor air professionals have realized that many indoor air contaminants and the associated health effects are linked to specific types of buildings and their characteristics. For example, radon is primarily an indoor air concern in homes because of the ease with which it can be transported inside residential construction from the soil beneath. On the other hand, Sick Building Syndrome (SBS) primarily afflicts office building occupants who experience acute health and comfort effects that appear to be linked to time spent in a specific building.

It has been estimated that hundreds of billions of dollars per year is lost due to decreased workplace productivity and increased health costs that can be saved by maintaining good indoor air quality in commercial buildings. The financial benefits of improving IAQ can accrue from reducing costs for health care and sick leave, as well as the costs of performance decrements at work caused by illness or adverse health symptoms and of responding to occupant complaints and costs of IAQ investigations.

Indoor air quality problems have grown with the increased use of heating, ventilating, and air-conditioning (HVAC) systems in commercial and residential buildings. Greater use of HVAC systems have also resulted in the closer examination of the energy impacts of maintaining good indoor air quality. The trade-off between reducing heating and cooling loads of the HVAC system by recirculating as much indoor air as possible and providing an optimum amount of fresh outdoor air forms the underpinnings of many IAQ standards and guidelines in climate-controlled commercial buildings.

HVAC SYSTEM

A wide variety of HVAC systems are used in residential and commercial buildings to thermally condition and ventilate the occupied spaces. While HVAC systems can provide enhanced levels of thermal comfort in very hot or very cold weather conditions, thermodynamic processes required to condition outdoor air and deliver conditioned air to occupied spaces to maintain indoor comfort conditions are fairly energy intensive.

A typical HVAC system uses a combination of heating, cooling, humidification (adding moisture) and dehumidification (removing moisture) processes to thermally condition air. This conditioned air, which is a mixture of outdoor air and recirculated indoor air, is known as supply air. The supply airstream typically passes through filters, heat exchangers that add or remove heat from the supply airstream, a supply fan, air ducts, dampers that are used to regulate the rate of airflow, and finally diffusers located either in the ceiling or floor to the occupied space. The return air is drawn from the occupied spaces and flows back to the mechanical rooms either through return air ducts or through the plenum between suspended ceiling and the floor of the next-higher story. A portion of the return air is exhausted to the outdoors, and the remainder is mixed with the fresh outdoor air and re-supplied to the space after filtering and thermal conditioning. In general, the supply air contains more recirculated air than fresh outdoor air to keep the energy cost of air conditioning down.

In an all-air system, the indoor temperature can be controlled either by a constant air volume (CAV) system, which varies the temperature of the air but keeps the volume constant, or by a variable air volume (VAV) system, which maintains a constant temperature and varies the volume of the air supplied to internal spaces.

To save energy, many HVAC systems employ a mechanism for regulating the flow of outdoor air called an economizer cycle. An economizer cycle takes advantage of milder outdoor conditions to increase the outside air intake and in the process reduces the cooling load on the system. Controlling the rate of flow of outdoor air appears simple, in theory, but often works poorly in practice. The small pressure drop required to control the flow rate of outdoor air is rarely controlled and monitored. Quite often, the damper system used to regulate the airflow is nonfunctional, disconnected from the damper actuators, or casually adjusted by building operators (Institute of Medicine, 2000).

Outdoor airflow concerns are some of the many reasons that make maintaining good indoor air quality in a controlled indoor environment a particularly challenging task. Maintaining a safe, comfortable indoor environment for workers and residents is challenging under the best of circumstances because apart from temperature control, HVAC systems are also responsible for moisture control and ventilation of buildings. Hot, humid climates, in particular, present some of the biggest challenges, and solutions that are tested and proven in temperate climates may actually worsen IAQ problems in hot, humid climates (Odom and DuBose, 1991).

INDOOR AIR POLLUTANTS: GENERATION, MITIGATION, AND EXPOSURE

To maintain acceptable indoor air quality, the concentration of pollutants known to degrade indoor air quality and affect human health must be controlled. If the origin of the contaminant is known, it is more effective to exercise source control over any mitigation strategy. If the origin of the contaminants is not known, building ventilation and air cleaning and filtration are the two most commonly used processes to dilute or remove all types of contaminants from the indoor air and maintain acceptable indoor environmental conditions.

Source Control

Source control is one of the most important methods to achieve healthy indoor air. Methods vary depending on the pollutants and can range from simple solutions (not using pressed wood furniture that use formaldehyde) to complex and costly solutions (identifying moisture infiltration through the building envelope). Source control may also require behavioral changes on the part of the affected population, a remedy that may be achieved either through environmental regulation or by raising public awareness. A case in point is the changing perception toward environmental tobacco smoke (ETS). Banning smoking in public spaces or discouraging smoking in homes and in front of children can reduce the risks of second hand smoke greatly. Other examples are selecting furniture that doesn't contain formaldehyde and using paints and carpets that don't emit chemicals that are known to degrade indoor air. If the air contaminants are being transported inside from outdoor sources, precautions should be taken to either plug the pathway to stop such transport (e.g., sealing the construction joints and cracks in the basement to reduce radon infiltration) or take steps to minimize the infiltration of contaminants (e.g., positioning the fresh air intake away from loading docks and parking spaces).

For radon, moisture, ETS, asbestos, lead-based paint, building materials and products that emit VOCs, pesticides, and household products, one can develop effective source control strategy as well. For example, assuring building envelope integrity at the time of construction of the building would help control and restrict the airflow into the building. The envelope (walls, roof, floor or slab system) should also control moisture infiltration by installing a continuous vapor barrier on the warm side of the insulation system. Operating the building at a slight positive pressure can also help in keeping the moisture outdoors. All these precautions can help control the growth of mold and mildew that thrive under moist conditions and can adversely affect the health of building occupants.

Ventilation

In addition to minimizing the emissions of pollutants from indoor sources, ventilation with outside air must be provided at an adequate rate to maintain acceptable IAQ. The ventilation rate—the rate of outside air supply—is usually defined per unit of floor area (liters per second per sq. meter), number of occupants (liters per person), or indoor air volume (air changes per hour). For indoor-generated particles, the effects of ventilation rate is highly dependent on particle size because the depositional losses of particles increases dramatically with particle size. The predicted change in pollutant concentrations with ventilation rate is greatest for an "ideal" gaseous pollutant that is not removed by deposition or sorption on surfaces (Institute of Medicine, 2000). Ventilation rate may have a very significant indirect impact on indoor concentrations of some pollutants because they affect indoor humidities, which in turn modify indoor pollutant sources.

Buildings are ventilated mechanically with the HVAC systems where it is a controlled process, as well as via air infiltration and through the openable windows and doors where it is largely an uncontrolled process. However, as discussed earlier, mechanical ventilation is one of the most energy-intensive methods of reducing indoor pollutant concentrations primarily because of the need to thermally condition air before it can be circulated inside the occupied spaces. It is estimated that the ventilation needs are responsible for consuming 10 percent of all the energy consumed in buildings in developed countries.

On average, buildings with air conditioning that have inadequate supply of fresh air are far more likely to suffer from poor indoor air quality than naturally ventilated buildings. On the other hand, one can find serious IAQ problems in homes and apartment buildings that are naturally ventilated as well.

There are two commonly used techniques to control odors and contaminants. Both depend on ventilation to achieve their goals. One of them relies on the concept of "ventilation effectiveness," which is defined as the ability of the ventilation system to distribute supply air and dilute internally generated pollutants by ensuring a consistent and appropriate flow of supply air that mixes effectively with room air. The second technique isolates odors and contaminants by maintaining proper pressure relationship between outdoors and indoors and between different indoor spaces. This is accomplished by adjusting the air quantities that are supplied to and removed from each room. In many large commercial buildings, particularly in warm humid climates, the design intent is to pressurize the building slightly with the mechanical ventilation system in order to prevent undesirable infiltration of unconditioned air, moisture, and outdoor air pollutants. On the other hand, smoking rooms, bathrooms, and laboratories are often depressurized so that pollutants generated within these rooms do not leak into the surrounding rooms.

Often, local dedicated exhaust ventilation is used in rooms with high pollutant or odor sources as it is more efficient in controlling indoor pollutant concentrations than general ventilation of the entire space (U.S. Environmental Protection Agency, 1991; U.S. Department of Energy, 1998). In practice, however, indoor-outdoor pressure differences are often poorly controlled, and many buildings are not pressurized (Persily and Norford, 1987). There is considerable uncertainty in predicting the rate of dilution of indoor contaminants in actual complex indoor environment, with rates of pollutant loss by deposition on indoor surfaces being one of the largest sources of uncertainty.

Air Cleaning and Filtration

Particle air cleaning is any process that is used intentionally to remove particles from the indoor air. Filtration and electronic air cleaning are the two most

Pollutants	Major Sources	Health Effects
Biological Contaminants		
1. **Infectious communicable bioaerosols** contain bacteria or virus within small droplet nuclei produced from the drying of larger liquid droplets and can transmit disease.	Human activity such as coughing and sneezing; wet or moist walls, ceilings, carpets, and furniture; poorly maintained humidifiers, dehumidifiers, and air conditioners; bedding; household pets.	Eye, nose, and throat irritation; dizziness; lethargy; fever. May act as asthma trigger; may transmit humidifier fever; influenza, common cold, tuberculosis and other infectious diseases.
2. **Infectious non-communicable bioaerosols** are airborne bacteria or fungi that can infect humans but that have a non-human source.	Cooling towers and other sources of standing water (e.g., humidifiers) are thought to be typical sources of Legionella in buildings.	The best known example is Legionella, a bacterium that causes Legionnaires Disease and Pontiac Fever.
3. **Non-infectious bioaerosols** include pollens, molds, bacteria, dust mite allergens, insect fragments, and animal dander.	The sources are outdoor air, indoor mold and bacteria growth, insects, and pets.	The health effects of non-infectious bioaerosols include allergy symptoms, asthma symptoms, and hypersensitivity pneumonitis.
Carbon Monoxide (CO) is a colorless and odorless gas that can prove fatal at high concentrations. High carbon monoxide concentration is more likely to occur in homes.	Unvented kerosene and gas space heaters; leaking chimneys and furnaces; back-drafting from furnaces, gas water heaters, woodstoves, and fireplaces; automobile exhaust from attached garages; environmental tobacco smoke.	At low concentrations, fatigue in healthy people and chest pain in people with heart disease. At higher concentrations, impaired vision and coordination; headaches; dizziness; nausea. Fatal at very high concentrations.
Carbon dioxide (CO_2) is one of the gaseous human bioeffluents in exhaled air. Indoor concentrations are usually in the range of 500 ppm to a few thousand ppm.	Humans are normally the main indoor source of carbon dioxide. Unvented or imperfectly vented combustion appliances can also increase indoor CO_2 concentrations.	At typical indoor concentrations, CO_2 is not thought to be a direct cause of adverse health effects; however, CO_2 is an easily-measured surrogate for other occupant-generated pollutants.
Environmental tobacco smoke (ETS) is the diluted mixture of pollutants caused by smoking of tobacco and emitted into the indoor air by a smoker. Constituents of ETS include submicron-size particles composed of a large number of chemicals, plus a large number of gaseous pollutants.	Cigarette, pipe, and cigar smoking.	Eye, nose, and throat irritation; headaches; lung cancer; may contribute to heart disease; build-up of fluid in the middle ear; increased severity and frequency of asthma episodes; decreased lung function. ETS is also a source of odor and irritation complaints.
Fibers in indoor air include those of asbestos, and man-made mineral fibers such as fiberglass, and glass wool.	Deteriorating, damaged, or disturbed insulation, fireproofing, acoustical materials, and floor tiles	No immediate symptoms, but long-term risk of chest and abdominal cancers and lung diseases.

Table 1.
Reference Guide to Major Indoor Air Pollutants

Pollutants	Major Sources	Health Effects
Moisture is not a pollutant but it has a strong influence on indoor air quality. In some situations, high relative humidity may contribute to growth of fungi and bacteria that can adversely affect health.	Water vapor is generated indoors due to human metabolism, cooking and taking showers, unvented combustion activities and by humidifiers; water and moisture leaks through roof or building envelope; improperly maintained HVAC equipment	Condensation of water on cool indoor surfaces (e.g., windows) may damage materials and promote the growth of microorganisms. The presence of humidifiers in commercial building HVAC systems has been associated with an increase in various respiratory health symptoms.
Particles are present in outdoor air and are also generated indoors from a large number of sources including tobacco smoking and other combustion processes. Particle size, generally expressed in microns (10-6 m) is important because it influences the location where particles deposit in the respiratory system (U.S. Environmental Protection Agency 1995), the efficiency of particle removal by air filters, and the rate of particle removal from indoor air by deposition on surfaces.	Some particles and fibers may be generated by indoor equipment (e.g. copy machines and printers). Mechanical abrasion and air motion may cause particle release from indoor materials. Particles are also produced by people, e.g., skin flakes are shed and droplet nuclei are generated from sneezing and coughing. Some particles may contain toxic chemicals.	Increased morbidity and mortality is associated with increases in outdoor particle concentrations (U.S. Environmental Protection Agency 1995). Of particular concern are the particles smaller than 2.5 micrometers in diameter, which are more likely to deposit deep inside the lungs (U.S. Environmental Protection Agency 1995). Some particles, biological in origin, may cause allergic or inflammatory reactions or be a source of infectious disease.
Volatile Organic Compounds (VOCs): VOCs are a class of gaseous pollutants containing carbon. The indoor air typically contains dozens of VOCs at concentrations that are measurable.	VOCs are emitted indoors by building materials (e.g., paints, pressed wood products, adhesives, etc.), equipment (photocopying machines, printers, etc.), cleaning products, stored fuels and automotive products, hobby supplies, and combustion activities (cooking, unvented space heating, tobacco smoking, indoor vehicle use).	Eye, nose, and throat irritation; headaches, nausea. Some VOCs are suspected or known carcinogens or causes of adverse reproductive effects. Some VOCs also have unpleasant odors or are irritants. VOCs are thought to be a cause of non-specific health symptoms.
Radon (Rn) is a naturally occurring radioactive gas. Radon enters buildings from underlying soil and rocks as soil gas is drawn into buildings.	The primary source of radon in most buildings is the surrounding soil and rock, well water, earth-based building materials.	No immediate symptoms but estimated to contribute to between 7,000 and 30,000 lung cancer deaths each year. Smokers are at higher risk of developing radon-induced lung cancer.

Table 1 (cont.).
Reference Guide to Major Indoor Air Pollutants
SOURCE: Excerpted from U.S. Environmental Protection Agency (1995) and Indoor Environmental Quality Appendix to International Performance Measurement & Verification Protocol (U.S. Department of Energy, 1998)

common examples. Typically, portable air cleaning devices are used in rooms, while in typical commercial HVAC systems the filter is placed upstream of many of the HVAC components in the path of supply airstream to filter particles. Two facts that are applicable to both air cleaners and filters are

- Efficiency of any air cleaner or filter is a function of the particle size present in the indoor air and the velocity and volume of air flowing through the device.
- Pressure drop is a concern wherever filters and, to a lesser extent, air cleaners are employed in the path of normal forced ventilation system.

The technologies for removing particles include mechanical filters; electrostatic precipitators, which charge particles and then collect them onto a surface with the opposite charge; and ion generators, which charge particles and thereby facilitate their deposition. Among mechanical filters, high efficiency particulate air (HEPA) filters are highly efficient in removing particles of a wide range of sizes. However, there is little evidence of either direct health benefits or reduced concentration of pollutants resulting from air cleaning or filtration applications (American Thoracic Society, 1997). New stricter standards that also allow filter selection based on offending contaminants and their particle sizes found in buildings are more likely to show direct health benefits.

Table 1 describes some of the common indoor air pollutants found in buildings, their sources, and their adverse health effects on human beings.

INDOOR AIR QUALITY AND ENERGY

Energy conservation measures instituted following the energy crisis of 1974 resulted in the elimination of openable windows and in the recycling of as much air as possible to avoid heating or cooling outside air. The amount of outdoor air considered adequate for proper ventilation has varied substantially over time. The current guideline, widely followed in the United States, was issued by ASHRAE in 1989. To achieve good IAQ in all-air systems, large but finite amount of fresh air needs to be brought in, heated or cooled depending on the climate and season, and distributed to various parts of the building. The energy implications are obviously huge because the temperature and humidity of the supply air stream must be maintained within a very narrow range to satisfy the ther-

mal comfort requirements of the building's occupants. Furthermore, temperature and humidity are among the many factors that affect indoor contaminant levels. Quality design, installation, and testing and balancing with pressure relationship checks are critically important for the proper operation of all types of HVAC systems and for maintaining good IAQ (U.S. Environmental Protection Agency, 1991; ASHRAE, 1989).

Energy professionals and equipment manufacturers are more cognizant of the indoor air problems and are coming out with new products and strategies that reduce energy use without degrading indoor air quality. According to the U.S. Department of Energy (1998), some of these energy efficient technologies can be used to improve the existing IAQ inside buildings:

Using outdoor air economizer for free cooling—An air "economizer" brings in outside air for air conditioning a commercial building. This strategy is particularly effective in mild weather where the temperature and humidity content of outside air is suitable for increasing the rate of outside air supply above the minimum setpoint. Generally, IAQ will improve due to the increase in average ventilation rate. Care must be exercised in using the economizer cycle in regions where outdoor air is of suspect quality. Also, in very humid regions, one must employ enthalpy-based control systems to take advantages of free cooling with the economizer cycle without encountering mold and mildew problems.

Heat recovery from exhaust air—If a heat recovery system allows an increase in the rate of outside air supply, IAQ will usually be improved. Proper precautions must be taken to ensure that moisture and contaminants from the exhaust air stream are not transferred to the incoming air stream. An innovative way of recovering heat and reducing the dehumidification cost is to use the waste heat to recharge the desiccant wheels that are then used to remove moisture from the supply air. In this method, the energy savings have to be substantial to offset the high cost of the desiccant wheels.

Nighttime pre-cooling using outdoor air—Nighttime ventilation may result in decreased indoor concentrations of indoor-generated pollutants when occupants arrive at work. Once again, proper precautions must be taken to ensure that outdoor air with the right level of moisture content is used for this purpose, otherwise condensation on heating, ventilation, and air conditioning equipment or building components may result, increasing the risk of growth of microorganisms.

Using radiant heating/cooling systems—Because of the higher thermal capacity of water compared to air, water is a better heat transfer medium. In hydronic climate conditioning systems, heat exchangers transmit heat from water to indoor environment, or vice-versa. These heat exchangers can either be convectors or radiators depending on the primary heat transfer process. The decoupling of ventilation from heating and cooling can save energy and improve IAQ. However, one must take appropriate measures to avoid condensation problems.

CONCLUSIONS

Research done by experts in the field as well as in laboratories has helped them understand the relationship between IAQ, ventilation, and energy. More research is needed to link specific health symptoms with exposure to specific or a group of pollutants. The policy challenge will be to raise awareness of indoor air quality so that healthy, comfortable environments can be provided by energy efficient technology.

Satish Kumar

See also: Air Conditioning; Building Design, Commercial; Building Design, Residential; Furnances and Boilers.

BIBLIOGRAPHY

American Society of Heating, Refrigerating, and Air Conditioning Engineers (ASHRAE). (1989). *Ventilation for Acceptable Indoor Air Quality*. Standard 62. Atlanta: Author.

American Society of Heating, Refrigerating, and Air Conditioning Engineers (ASHRAE). (1996). *1996 ASHRAE Handbook: HVAC Systems and Equipment*. Atlanta: Author.

American Society of Heating, Refrigerating, and Air Conditioning Engineers (ASHRAE). (1999). *Method of Testing General Ventilation Air Cleaning Devices for Removal Efficiency by Particle Size*. Standard 52.2. Atlanta: Author.

American Thoracic Society. (1997). "Achieving Healthy Indoor Air." *American Journal of Respiratory and Critical Care Medicine* 156(Suppl. 3):534–564.

Blank, D. M. (1998). "Earning It; What's in the Office Air? Workers Smell Trouble." *New York Times*, February 22, Sunday Section: Money and Business/Financial Desk.

Barbosa, D. (2000). "3M Says It Will Stop Making Scotchgard." *New York Times*, May 17.

Dorgan, C. E., and Dorgan, C. B. (1999). "Developing Standards of Performance (SOP) for IAQ in Building." In *Proceedings of the 8th International Conference on Indoor Air Quality and Climate*. Edinburgh, Scotland.

Fisk, W. J., and Rosenfeld, A. H. (1998). "Potential Nationwide Improvements in Productivity and Health from Better Indoor Environments." In *Proceedings of the ACEEE 1998 Summer Study of Energy Efficiency in Buildings*. Washington, DC: American Council for an Energy-Efficient Economy.

Institute of Medicine. (2000). "Clearing the Air: Asthma and Indoor Air Exposures." Washington, DC: National Academy Press.

"Is Your Office Killing You?" (2000). *Business Week,* no. 3684, p. 114.

Liddament, M. W. (1999). "A review of Ventilation and the Quality of Ventilation Air." In *Proceedings of the 8th International Conference on Indoor Air Quality and Climate*. Edinburgh, Scotland.

Mendell, M. J. (1993). "Non-specific Health Symptoms in Office Workers: a Review and Summary of the Epidemiologic Literature." *Indoor Air* 3(4):227–236.

National Academy of Science. (1999). *Health Effects of Exposure to Radon: BEIR VI, Committee on Health Risks of Exposure to Radon*. Washington, DC: National Academy Press.

Odom, D., and DuBose, G. H. (1991). "Preventing Indoor Air Quality Problems in Hot, Humid Climates: Problem Avoidance Guidelines." Denver, CO: CH2M Hill, in cooperation with the Disney Development Company.

Persily, A., and Norford, L. (1987). "Simultaneous Measurements of Infiltration and Intake in an Office Building." *ASHRAE Transactions* 93(2):42–56.

Roulet, C.; Rossy, J.; and Roulet, Y. (1999). "Using Large Radiant Panels for Indoor Climate Conditioning." *Energy and Buildings* 30:121–126.

Seppanen, O. (1999). "Estimated Cost of Indoor Climate in Finnish Buildings." In *Proceedings of the 8th International Conference on Indoor Air Quality and Climate*. Edinburgh, Scotland.

Smith, K. R. (1999). "The National Burden if Disease from Indoor Air Pollution in India." In *Proceedings of the 8th International Conference on Indoor Air Quality and Climate*. Edinburgh, Scotland.

U.S. Department of Energy. (1998). *Indoor Air Quality Appendix to International Performance Measurement & Verification Protocol*. Washington, DC: Author.

U.S. Environmental Protection Agency. (1991). *Building Air Quality—A Guide for Building Owners and Facility Managers*. Washington, DC: Author.

U.S. Environmental Protection Agency. (1992). *A Citizen's*

Guide to Radon, The Guide to Protecting Yourself and Your Family From Radon, 2nd ed. Washington, DC: Author.

U.S. Environmental Protection Agency. (1995). *The Inside Story: A Guide to Indoor Air Quality.* Washington, DC: Author.

U.S. Environmental Protection Agency. (2000). "EPA and EM." Headquarters Press Release, Washington, DC, May 16.

AIR RESISTANCE

See: Aerodynamics

AIR TRAVEL

Since its inception in the 1920s, commercial air travel has steadily grown. At first available only for the discretionary exclusive use of the wealthy, air travel is now an indispensable tool of business and the world economy. Today, with the competitive marketplace dictating routes and fares, it is now possible for almost anyone in the developed world to fly almost anywhere. It is critical to remember that the airline industry is just that—a business—and the key to business success is efficiency. No matter how romantic the image of air travel may be, airlines exist only as long as they are able to provide a reliable service for a reasonably profitable price. Starting in the United States in 1978 and gradually spreading around the globe, the economic deregulation of the airline industry has profoundly changed the face of this critical enterprise by greatly expanding the market for air travel while exposing it to the vicissitudes of the free market.

In 1999 an estimated 1.5 billion people from around the world chose to fly on commercial carriers. Within the United States alone, 530 million passengers chose to fly. During that year airliners took off and landed 7.3 million times on domestic flights alone. These aircraft consumed over 13 billion gallons of jet fuel. According to the Air Transport Association the U.S. airline industry spent more than $10 billion for fuel for its members' domestic and international flights.

ORIGINS

With more than $100 billion in assets in U.S. carriers alone, the airlines have constantly sought ways to improve their ability to make money. They have searched continuously for better, more efficient aircraft in the endless search for greater yield and productivity. In Europe the airline industry was nationalized for most of its history. Consequently, the emphasis was more on service and less on profits. In the United States the experience was different. Although the early commercial airlines were completely dependent on government subsidies administered as air mail contracts, federal policy encouraged the airlines to develop passenger and freight service to offset the subsidies. In the early 1930s the airlines were given bonuses for carrying passengers and for developing faster and safer aircraft. The government's successful plan was to wean the airlines off the subsidies gradually by encouraging the development of larger, faster, and more efficient aircraft. The immediate result of this far-sighted policy was the creation of the Boeing 247, the world's first modern airliner, and the subsequent Douglas series of commercial aircraft. In fact, according to former American Airlines president C. R. Smith, the 21-passenger Douglas DC-3 was the first aircraft capable of making a profit just by carrying passengers. By 1952 the air mail subsidy was withdrawn, as it was no longer needed.

THE JET AGE

During the 1940s and 1950s, the airline industry developed a new generation of large four-engine aircraft that expanded air travel around the globe. Douglas DC-4s, 6s, and 7s, along the Boeing Stratoliners and the graceful Lockheed Constellations, became the industry standard for speed, comfort, and profitability. While these piston-engine aircraft dominated the world's air routes, a new generation of much faster aircraft powered by jet engines was poised to revolutionize air travel. First developed by de Havilland in Great Britain, the sleek Comet entered service in 1952 but experienced serious technical problems that were not overcome until a new series of aircraft built in the United States seized the market. The Boeing 707 and the Douglas DC-8 entered service in the late 1950s and found immediate success throughout the world. These aircraft offered high subsonic speeds significantly faster than their piston-engine counterparts and much larger passenger-carrying capability. The combination of the greater reliability of the jet and the use of much cheaper kerosene-based fuel instead of expensive high-octane gasoline greatly increased efficiency and productivity.

By the end of the 1960s jets had virtually replaced all piston-engine airliners on most of the world's air routes. Newer, more powerful and more efficient turbofan engines replaced the first generation of thirsty and polluting turbojet engines. In 1970 the arrival of the Boeing 747 dramatically increased aircraft productivity by greatly decreasing seat-mile costs. Equipped with four new-generation high-bypass turbofans that were much quieter, more powerful, and more efficient, the 747 and the wave of wide-bodied aircraft that followed carried twice the number of passengers previously possible, with lower operating costs.

THE DEREGULATION REVOLUTION

By the late 1970s, economists and members of the U.S. Congress realized that despite the great efficiencies now possible because of technological advancements, air fares were not decreasing. In a bipartisan effort, fifty years of regulation dating back to the days of the Post Office contract were ended with the deregulation of the U.S. airline industry in October 1978. While the federal government continued to regulate safety matters, the airlines were now free to charge whatever fares the market would allow and were now free to enter or exit any market they wished. The immediate result was a flood of discount fares as millions of people, many of whom had never flown before, took to the sky. New airlines emerged almost overnight, while numerous local and regional carriers sought to compete directly with the major airlines.

These were heady days for the consumer but chaos for the airlines. Passengers could now compare prices as the airlines drastically cut fares to attract business. Despite a national economy in recession, ridership increased from 254 million to 293 million between 1978 and 1979. But problems loomed. The airline industry is capital-intensive, with high barriers to entry, high operating costs, and thin profit margins. Ruthless competition did not produce long-term stability as most of the new and many of the old airlines fell by the wayside, unable to pay for expensive aircraft and unable to fight against dominant carriers with experience and large cash reserves and access to credit. No longer protected by the government, airlines were free to succeed or fail against the invisible hand of market forces. Many failed. Many more were consumed during a merger mania in the 1980s that sought to cut overhead costs by consolidating companies. Aviation enthusiasts lamented this logical business step as many grand old names such as Pan American, Eastern, Braniff, Piedmont, Frontier, and others disappeared. An oligopoly of larger, more efficient companies survived, ironically little different from the oligopoly of companies that dominated before 1978.

Efficiency became the watchword after deregulation. With competition opened up to all comers, the airlines fought hard to make money by cutting costs and increasing productivity. Coupled with an unexpected crisis in the supply of fuel in 1979, there was an industrywide effort to acquire the latest, most efficient equipment using the latest engine technology.

Following the success of the large high-bypass turbofan engines built for the massive wide-bodied airliners, engine manufacturers turned their attention to building a new generation of smaller engines using the same advanced technology. The result was a new series of engines that produced 20,000 to 40,000 pounds of thrust while using much less fuel than earlier engines of comparable power.

These new engines, from Pratt & Whitney, Rolls-Royce, and the international consortium of General Electric and Snecma, found wide acceptance on new and old narrow-bodied airframes. Boeing produced its 757 in 1982, while McDonnell Douglas updated its proven DC-9 series with the improved MD-80 line. First flown in 1965, the Boeing 737 had been a slow seller. With deregulation and the replacement of its low-bypass engines with two GE/Snecma CFM56 power plants, the 737 quickly became a sales leader because of its high serviceability, low operating costs, and high productivity. It became the aircraft of choice for a host of new entrant and expanding airlines, such as Southwest. The 737 in its many forms is today the most successful jet airliner in history, with well over 4,500 sold.

In Europe, the international consortium of Airbus Industrie took notice of the postderegulation environment in the United States and the changing environment in Europe as well and produced its own efficient narrow-bodied airliner, the A320, in 1988. Incorporating the latest technology, the A320 is the first commercial airliner equipped with a fully computerized "glass cockpit" and digital fly-by-wire con-

Airline industry deregulation in 1978 led to upheaval and change for Delta Airlines and many other passenger carriers. (Delta Airlines)

trol system that cuts weight, enhances safety, and maximizes efficiency. So successful is this revolutionary aircraft that Airbus has sold more than 2,000 A320s in just over ten years.

AIR TRAVEL TODAY

The sudden rush of growing traffic because of the resulting lower fares taxed the aviation infrastructure to its limits. Aircraft utilization soared from six hours per day to twelve hours a day as load factors increased. This led immediately to much congestion at overcrowded airports and severe air traffic problems. Airlines sought to maximize productivity by funneling passengers though centralized airports, the "hub and spokes," where traffic could be efficiently concentrated and dispersed throughout the airlines' route networks. While maximizing productivity for the airlines, the hub and spokes system strained the facilities of these airports during peak hours of very

high traffic often causing serious delays and inconveniencing passengers. The sheer volume of aircraft all trying to converge on the same airport at the same time often overwhelmed the air traffic control system. The installation of flow control whereby the movement of aircraft is controlled throughout the system has alleviated some of the worst problems. By holding aircraft at the gate of the departing airport until the route and the gate of the receiving airport will be clear has cut fuel consumption by preventing the massive stacking of aircraft that used to occur at overwhelmed airports. It has also improved the flow of traffic around bad weather and helped to rationalize the system. Problems with aging equipment, the difficult introduction and integration of new technologies, and the issues of an overworked and understaffed labor force still hamper the effectiveness of the nation's air traffic control.

By the early 1990s some semblance of order had returned to the industry. After the trying period of consolidation in the late 1980s, a solid core of well-managed, stable airlines emerged despite the cyclical nature of the business caused by variations in the economy and unpredictable fluctuations in the price of fuel. A solid oligopoly of large airlines now dominates the market and, while problems remain, much improvement has been made. Airports are gradually catching up to the demand for their services by spending billions of dollars to upgrade their facilities, while gradual improvements in air traffic control and more rational scheduling are reducing the worst congestion problems. In the late 1990s the airlines spent heavily in acquiring new, efficient aircraft of all types. Boeing and Airbus, the two largest manufacturers of airliners, are building hundreds of aircraft each year to satisfy the continuous demand for ever-better transports.

FUTURE DEVELOPMENTS

As the new century unfolds, Airbus and Boeing are heavily engaged in market studies to determine the scale and scope of the next generation of airliners. Air travel is steadily growing domestically and internationally, though the rate of growth is slowing. Through 2020, the U.S. Department of Energy projects that the growth in jet fuel consumption should outpace that of all other liquid fuels. The *Annual Energy Outlook* for 1999 foresees that by 2020 the consumption of gasoline will drop from 65 to 61 percent, diesel from 18 to 14 percent, while jet fuel use will rise from 13 to 17 percent.

It is estimated that more than 80 percent of the American population has flown at least once; if that is true, it is effectively a fully mature market. Outside the United States, the demand for air travel is expanding as well, although it has been tempered by economic downturns, particularly in the Asian market. These factors are leading Airbus and Boeing to the conclusion that there will exist a need for an aircraft larger than any now flying. They differ, however, on the question of when that aircraft will be needed and what form it should take.

Airbus Industries is staking its plans on the assumption that the market now exists for 1,300 aircraft that can each seat 550 to 650 passengers. Their marketing experts point to the heavily traveled North Atlantic route, particularly that between New York and London, as well as several high-density routes in the Asia-Pacific region as evidence for this new market and are assuming a steady 5 percent growth in

traffic. This aircraft will be needed to handle the increased traffic expected between major population centers and will be necessary to help overburdened airports and air traffic control systems to handle more people, but with a manageable number of aircraft. The large airliner, bigger than the current 747-400 series that carries 416 and 524 passengers, will also enjoy the lowest seat-mile costs of any aircraft, thereby addressing the airlines' ever-present need to improve revenue yield.

Similarly, Boeing predicts a 4.7 percent growth in between 2000 and 2020 but anticipates market fragmentation instead of consolidation as airlines choose to expand service away from the traditional population centers and begin serving many more city pairs not previously served. The competitive marketplace will dictate an increased number of flights from many more cities; therefore the demand will be for more frequency of smaller aircraft. Boeing argues that this is already happening over the North Atlantic, where in 1980 the large four-engine 747 dominated travel. Today smaller, twin-aisle aircraft such as the 767, 777, and A330, dominate as airlines seek to bypass congested hubs. Boeing believes that governments will keep pace with the rise in traffic by building new airports, improving existing facilities, and investing in improved air traffic control.

Boeing does concede that the market for a very large aircraft will exist, particularly along the Pacific Rim, but that that will not occur until the second decade of the twenty-first century, ten years after Airbus's forecast. Boeing believes that there will be a market for 1,030 large aircraft between 2000 and 2020. Of this number, only 425 will be for aircraft with more than 500 seats, one-third that of Airbus's prediction. Boeing argues that in the future, market fragmentation will occur in the United States as well because of increased competition and the availability of existing long-range, large-capacity airliners such as the 777, which can seat up to 394 passengers.

Only time and the marketplace will determine which manufacturer is correct. One thing that both manufacturers agree on is that the airlines will always demand more efficient aircraft in their constant struggle to maintain profitability. The competition between the two manufacturers will help ensure that the best products will be available to the customer and the passengers.

F. Robert van der Linden

See also: Aircraft; Aviation Fuel; Efficiency of Energy Use, Economic Concerns and; Engines; Kerosene; Subsidies and Energy Costs; Supply and Demand and Energy Prices; Transportation, Evolution of Energy Use and.

BIBLIOGRAPHY

Air Transport Association Home Page. <http://www.air-transport.org/default.htm>.

Airbus Industrie Home Page. <http://www.airbus.com>.

Davies, R.E.G. (1982). *Airlines of the United States Since 1914*. Washington, DC: Smithsonian Institution Press.

Dienel, H.-L., and Lyth, P., eds. (1998). *Flying the Flag: European Commercial Air Transport Since 1945*. New York: St. Martin's Press.

Energy Information Administration Home Page. Department of Energy. <www.eia.doe.gov>.

FAA Statistical Handbook of Aviation. Department of Transportation, Federal Aviation Administration. <http://www.bts.gov/NTL/data/tables>.

Heppenheimer, T. A. (1995). *Turbulent Skies: The History of Commercial Aviation*. New York: John Wiley and Sons.

Lynn, M. (1997). *Birds of Prey: Boeing vs Airbus—A Battle for the Skies*. New York: Four Walls Eight Windows.

Morrison, S., and Winston, C. (1995). *The Evolution of the Airline Industry*. Washington, DC: The Brookings Institution.

Ott, J., and Neidl, R. E. (1995). *Airline Odyssey: The Airline Industry's Turbulent Flight into the Future*. New York: McGraw-Hill.

ALCOHOL

See: Biofuels

ALEXANDERSON, ERNST FREDERIK WERNER (1878–1975)

Ernst F. W. Alexanderson was a Swedish American engineer and inventor who is best remembered for his pioneering work on the high frequency alternator that made long-distance radio communication possible. He was born on January 25, 1878, in Uppsala, Sweden. His father, Aron M. Alexanderson, taught classical languages at the University of Uppsala and was later chair of classical languages at the University of Lund. Alexanderson's mother was the former Amelie von Heidenstam. The young Alexanderson was educated at Lund High School and then at the University of Lund (between 1896 and 1897). He continued his studies at the Royal Institute of Technology in Stockholm, and later, in Berlin, he studied under the instruction of Adolf K. H. Slaby, the inventor of a primitive form of radio communication.

Alexanderson was anxious to put his knowledge to practical use. America, which seemed at that time to be the fountainhead of many important technological advancements, beckoned. Arriving in New York in 1901, he immediately went to work as a draftsman for the C. & C. Electrical Company in New Jersey. Alexanderson sought out and was quickly befriended by the esteemed inventors Thomas Edison and Charles Steinmetz. In 1904, after passing General Electric's engineering exams, he became a member of that company's engineering staff.

Alexanderson's big break came when he was commissioned by Reginald Fessenden, a pioneering wireless operator, to build a generator that could produce alternating, high frequency currents. These currents would be used to generate a continuous, dependable wave for radio transmission and thus enable a broadcast of more complexity. On Christmas Eve of 1906, Alexanderson's invention was used to broadcast the first radio show that featured singing and conversation.

Guglielmo Marconi, the Italian engineer and inventor, visited Alexanderson in 1915 and bought one of his 50-foot alternators for the transatlantic Marconi Center in New Jersey. Within a few years, Alexanderson's alternators were to be found in numerous countries. Using a 200-foot Alexanderson alternator, Marconi broadcast radio transmissions during World War I that were heard all over Europe.

In 1916, Alexanderson made another important contribution to radio broadcasting when he unveiled his tuned radio receiver, which allowed for selective tuning. It quickly became an integral part of radio broadcasting.

Alexanderson's alternator played an important part in history when President Woodrow Wilson used it to broadcast his 1918 ultimatum to Germany, ending the war. Afterwards, the Marconi company sought to buy exclusive world rights to the alternator, but was rebuffed by the U.S. government. Wishing to keep control of the invention within American hands, the government set up the Radio Corporation of America (RCA) in 1919, with Alexanderson as its chief head engineer. Concurrently, Alexanderson continued to work for General Electric, an association that lasted forty-six years.

In 1919, he made history with yet another of his inventions when his multiple-tuned antenna, antistatic receiver, and magnetic amplifier were used to transmit the first two-way radio conversation. This great event took place 900 miles out to sea, between the Trans-Atlantic Marconi Company station at New Brunswick and the steamship *George Washington,* with President Woodrow Wilson on board as a witness.

The magnetic amplifier was outmoded by another Alexanderson invention, the electronic modulator, which used vacuum tubes to help generate high frequency transmitters of great power. Alexanderson also helped to create the amplidyne, a direct current generator. By the use of compensating coils and a short circuit across two of its brushes, the amplidyne uses a small power input to precisely control a large power output. Its system of amplification and control originally was designed for use in steel mills, but later hundreds of other applications, including an adaption to fire antiaircraft guns during World War II. Alexanderson also held patents for his inventions of telephone relays, radiant energy guided systems for aircraft, electric ship propulsion, automatic steering, motors and power transmission systems, railway electrification systems, as well as inventions in the fields of radio and television.

In 1924 Alexanderson began his television research. By 1927 his group was able to broadcast mechanical television into the home and in 1930 General Electric gave the first large screen demonstration of television in a theatre in Schenectady. His team then transferred to RCA where they helped develop our modern system of television.

Alexanderson retired from General Electric in January 1948, although he remained a consultant engineer to the company. In 1952 Alexanderson was consulted for RCA. In all, Alexanderson held 322 patents for his inventions.

Ernst Alexanderson's enormous contribution to technology was acknowledged more than once. Some of the honors and awards he received during his long life are: The Gold Medal of the Institute of Radio Engineers in 1919, the Order of the Polonia Restituta in 1924, the John Ericcson Medal in 1928, the Edison Medal of the American Institute of Electrical Engineers in 1944, the Cedergren Medal of the Royal Institute of Technology of Sweden in 1945, and the Valdemar Poulsen Gold Medal and the Royal Danish Medal, both in 1946. He received honorary degrees from Union College, Schenectady, New York in 1926, and the University of Uppsala in 1938. He was a member, fellow, and later president of the Institute of Radio Engineers, a member and president of the Institute of Radio Engineers, and a member of the Swedish Royal Academy and Sigma Xi. In his spare time and retirement, Alexanderson enjoyed sailing, and was elected the first Commodore of the Lake George Yacht Club in New York. Alexanderson died at the age of ninety-seven on May 14, 1975.

Albert Abramson

BIBLIOGRAPHY

Brittain, J. E. (1992). "Alexanderson: Pioneer in American Electrical Engineering." Baltimore: Johns Hopkins Press.
IEEE. (1975). "Ernst Alexanderson: Radio and TV Pioneer at 97." *Spectrum* 12(August):109.
Krebs, A. (1975). "Dr. Ernst Alexanderson, Radio Pioneer, Dies at 97." *New York Times Biography Series* 6:547–548.
"Last of the Pioneers Is Gone—Alexanderson Dies at Age 97." (1975). *Radio-Electronics,* July, p. 6.
"Our Respects to Ernest Frederik Werner Alexanderson." (1945). *Broadcasting-Telecasting,* December, pp. 48–50.

ALTERNATING CURRENT

See: Electricity

ALTERNATING CURRENT MOTOR

See: Electric Motor Systems

ALTERNATIVE FUELS AND VEHICLES

The term alternative fuels first appeared in the energy literature in the late 1970s as a way to refer to nonconventional fuels—fuels that are not gasoline, diesel or aviation fuel. Alternative fuels excludes all fuels refined from petroleum that are normally liquid at ambient conditions, such as gasoline through heavy fuel oil. It does include the highly volatile fractions: liquefied petroleum gas (propane), liquefied natural gas, and compressed natural gas. The category also comprises all fuels made from other fossil fuels, such as coal and oil shale, biofuels originating from plant material, and chemically derived fuels such as methanol and hydrogen. The nonfossil plant-derived fuels such as ethanol and bio-diesel (from vegetable oils), and hydrogen made from water via solar powered electrolysis, are the only renewable energy alternative fuels. Electric vehicles are considered alternative-fuel vehicles since only about 3 percent of electricity comes from burning petroleum.

Alternative fuels should not be confused with alternative energy, which is another term whose origins date back to the 1970s. Alternative energy and alternative fuel both exclude petroleum energy, yet alternative energy goes further by excluding all fossil fuel sources and nuclear. However, sometimes energy sources such as hydroelectric, which accounts for about 10 percent of U.S. electricity production, may be considered alternative even though it has been a major energy source for centuries.

Of all the fossil fuels, petroleum is considered the least sustainable fuel option (most finite and fastest rate of depletion), which is the main reason for the development of alternative fuels. Resources of natural gas and coal are far greater, and depletion slower, which makes alternative fuels developed from these sources more sustainable. The nonfossil plant-derived fuels of ethanol and bio-diesel (from vegetable oils) are the only renewable energy alternative fuels and, in theory, are considered the most sustainable. However, there is not enough agricultural land for biofuels alone to ever become a total replacement for petroleum at the year 2000 world petroleum consumption rate of about 30 million barrels a day

In the United States, the leading use of alternative fuels is not as standalone fuels, but as additives to petroleum-based gasoline and diesel fuel. For example, gasoline sold in much of the United States is 10 percent ethanol or 10 percent methyl tertiary butyl ether (MTBE).

RECENT HISTORY

The first major government investment in an alternative fuel was for the purpose of energy security and oil import independence. Beginning in the late 1970s, billions of dolars were spent on synthetic fuels (converting coal and oil shale into gasoline and diesel). When oil prices began to fall in the early 1980s, and it became apparent that the costs of producing synthetic fuels would remain well above that of petroleum fuels, the program was abandoned.

Interest in alternative fuels grew again in the late 1980s in response to urban air quality problems. Early in the 1990s, environmental regulations calling for oxygenated fuels (nonpetroleum fuels containing oxygen blended with gasoline) to cut carbon monoxide emissions went into effect that significantly increased the sales of MTBE and ethanol. In 1995, alternative fuels comprised about 3 percent of all fuels consumed in the United States (4.4 billion gallons versus 142 billion for gasoline and diesel). MTBE and ethanol consumption is greatest since these fuels are blended with gasoline as required by environmental regulations for carbon monoxide reduction. However, when the 1999 National Research Council study found that there was no statistically significant reduction in ozone and smog based on the data available, the continued requirement of using MTBE was questioned, especially since MTBE leaking from storage tanks can quickly contaminate groundwater.

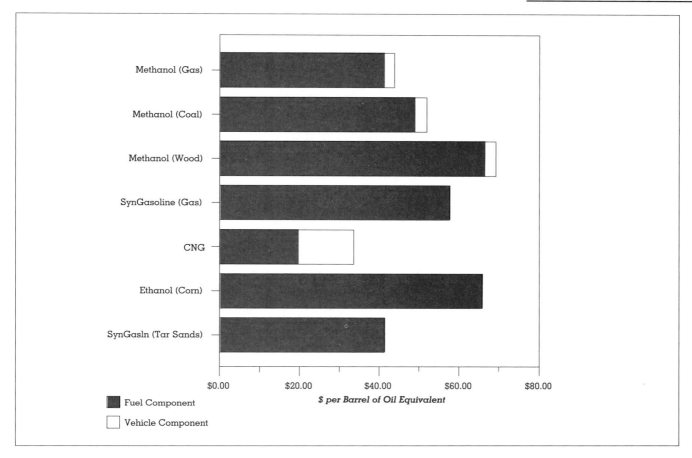

Figure 1.
NRC estimated costs of alternative fuels, including incremental vehicle costs, compared to oil at $20 per barrel.

SOURCE: National Research Council (1990), Table D-4.

MARKET BARRIERS

Alternative fuels are advocated as a way to improve the environment, enhance energy security, and replace dwindling petroleum reserves. Thus, the federal government continues to generously fund research and development for alternative fuels either as a replacement for, or for blending with, conventional fuel. Among the federal subsidies and regulations to promote alternative fuel use are the Energy Policy Act of 1992 (requiring alternative fuel vehicles in fleets and providing tax breaks for people who buy these vehicles), the Intermodal Surface Transportation Efficiency Act of 1991 (providing grants for purchasing alternative-fuel vehicles and for building refueling stations), and the Alternative Motor Fuels Act of 1988 (allowing automakers to sell more large, higher-profit conventional cars with poorer fuel economy if they also sell alternative fuel vehicles). However, even with subsidies and favorable regulations, alternative fuels face significant hurdles before becoming practical replacements for conventional fuels. Foremost is cost, followed by safety, practicality and reliability, and finally, the development of infrastructure (production, distribution and retailing availability).

Cost-Competitive

Natural-gas-derived fuels are the most cost-competitive because natural gas does not need to be refined like gasoline and diesel fuel from petroleum (Figure 1). Ethanol, a heavily subsidized alternative fuel, is not as cost-competitive as natural-gas-derived fuels. If not for the subsidies and environmental regulations requiring oxygenates, ethanol would not be used at all. The chances of ethanol ever becoming cost-competitive in the free market are slim since extensive land is needed to raise high-energy-yield plants for fuel, and the energy that must be expended to raise, harvest and dry the plants for the fermentation alcohol results in a low net energy yield.

When gasoline-powered automobiles are modified to burn a fuel such as ethanol alone, they are known as dedicated ethanol vehicles—risky investments for buyers who have concerns about future availability. For example, Brazil's Proalcool program promoted and heavily subsidized ethanol, and thus dedicated ethanol vehicles, from 1975 to 1988. Once the subsidies were curtailed and then eliminated (estimates of the costs of the subsidy to the government range from $7 to $10 billion), shortages resulted. Many of the owners of ethanol-dedicated vehicles either had to junk or retrofit the vehicles to run on gasoline, and the sales of ethanol-dedicated vehicles went from 50 percent of the market in 1988 to 4 percent by mid-1990.

Aside from the difference in fuel costs, the cost of redesigning and equipping vehicles engines and fuel tanks to run on alternative fuels has to be considered. Responding to the desire to switch fuels for cost reasons, or refueling security when the alternative fuel is not readily available, several auto makers offer flexible fuel vehicles that run primarily on compressed natural gas, but also gasoline when compressed natural gas is not available. These vehicles are sold at a premium and have shown little success in attracting buyers since low fuel prices ensure their return on investment will be poor in comparison to standard gasoline vehicles.

Practicality and Reliability

When the energy density of the alternative fuels is considerably less than gasoline and diesel fuel, it greatly impacts the practicality of the fuel for transportation. Most of the alternative fuels have a much lower energy density (Table 1). For vehicles, much more storage space is required to accommodate much larger fuel tanks to achieve comparable range or, for gaseous fuels, storage tanks that can withstand greater compression. Moreover, it will always take longer to refuel a vehicle using a lower-energy-density liquid fuel or a gaseous fuel.

The lower energy density of alternative fuels is even more problematic for aircraft. Methanol has been sug-

Energy Density	lb./ gal. @ 60 degrees F.
No. 2 Diesel Fuel	6.7 - 7.4
Gasoline	6.0 - 6.5
Methanol	6.63
Ethanol	6.61
MTBE	6.19
Propane (LPG)	4.22
Methane	1.07

Table 1.
Energy Densities of Fuels

gested as a jet fuel replacement. But using methanol would seriously curtail range and payload since the plane's weight is a principle determinaent of how much fuel is needed. A typical four-engine commercial jet will carry 775,000 pounds of aviation fuel to maximize range; to achieve the same range with methanol would require one million more pounds of fuel.

Since most of the alternative fuel vehicles burn cleaner, experience has found that this reliability is equal or better than that of comparable gasoline or diesel fuel vehicles.

Infrastructure Needs

A vast petroleum production, refining, distribution and retailing operation exists to deliver gasoline and diesel fuel. The major oil companies have invented billion of dollars in the setup and delivery of liquid fuels that can be stored in underground tanks; thus, any alternative fuel that requires massive new investments in infrastructure will face considerable market resistance. Moreover, there are personal investments in over 200 million vehicles on the road that are designed to consume either gasoline or diesel fuel, and a dauntingly immense and specialized infrastructure of industry building these vehicles and small businesses maintaining them. Since so much of the economy has a vested interest in the internal combustion engine burning gasoline or diesel fuel, a market transition to alternative fuels and vehicles is likely to be gradual.

In the free market, as long as petroleum supplies are plentiful, there is little incentive for oil companies to transition to any of the alternative fuels, which is a major reason that the U.S. Department of Energy projects petroleum consumption will rise from 18.6 million barrels per day in 1997 to 22.5-26.8 million barrels by 2020. As the crude oil reserves dwindle, the marketplace will either transition to the electrification the transportation system (electric and fuel cell vehicles and electric railways), or see the development of alternative fuels. Any short-term transition to an alternative fuel is likely to meet environmental air quality regulations. Beyond 2020, the transition is likely to occur due to the depletion of oil reserves resulting in steeply rising gasoline and diesel prices, or from advances in technologies that make alternative fuels and alternative transportation more attractive.

John Zumerchik

See also: Biofuels; Capital Investment Decisions; Hydrogen; Kinetic Energy, Historical Evolution of the Use of; Methanol; Natural Gas, Processing and Conversion of.

BIBLIOGRAPHY

Greene, D. L. (1996). *Transportation and Energy.* Landsdowne, VA: Eno Transportation Foundation, Inc.

Hadaller, O. J., and Momenthy, A. M. (1993). *"Characteristics of Future Aviation Fuels."* In *Transportation and Global Climate Change*, edited by D. L. Greene and D. J. Santini. Washington DC: American Council for an Energy Efficient Economy.

Howes, R., and Fainberg, A. (1991*). The Energy Sourcebook: A Guide to Technology, Resources and Policy.* New York: American Institute of Physics.

Lorenzetti, M. S. (1995). *Alternative Motor Fuels: A Nontechnical Guide.* Tulsa, OK: Pennwell Press, Inc.

National Research Council, Energy Engineering Board. (1990). *Fuels to Drive Our Future.* Washington DC: National Academy Press.

ALTERNATOR

See: Electric Motor Systems

AMPERE

See: Units of Energy

AMPÈRE, ANDRÉ-MARIE (1775–1836)

André-Marie Ampère was born in Lyons, France, the son of a wealthy merchant. Ampère's education was determined by his father, Jean-Jacques, who followed Jean Jacques Rousseau's theories of education. Ampère was left to educate himself, as his inclinations dictated, among the books of his father's extensive library. At an early age Ampère discovered a talent for mathematics, working out the early books of Euclid by himself. On finding that some of the books he wished to consult in the library in Lyons were in Latin, he taught himself the language. Ampère's mother was a devout Catholic, who ensured he was thoroughly instructed in the faith.

Ampère's domestic life was beset with tragedy. In 1787 the French Revolution began; Jean-Jacques assumed the post of *Juge de Paix,* a role with considerable police powers. When Lyons fell to the troops of the Republic in 1793, Jean-Jacques was tried and guillotined. In 1799 Ampère married, supporting his wife by teaching mathematics in Lyons, where their son was born the next year. Weakened by childbirth, his wife died in 1803. Ampère moved to Paris and took a post in mathematics at the École Polytechnique. In 1806 he remarried, but this union was ill-advised. After the birth of a daughter, his wife and mother-in-law made life for Ampère so unbearable that he was forced to seek a divorce. Ampère persuaded his mother and sister to come to Paris and take charge of his household. Ampère's expectations of his children were never realized, and in domestic life he faced constant money problems. In 1808 Ampère was named Inspector General of the newly formed university system, a post he held until his death. In 1824 Ampère was elected to the Chair of Experimental Physics at the Collège de France.

As a deeply religious person whose personal life was beset by a series of calamities, Ampère searched in the field of science for certainty. He constructed a philosophy that enabled him to retain a belief in the existence of both God and an objective natural world. Ampère's philosophy contained two levels of knowl-

André-Marie Ampère. (Public domain)

edge of the external world. There are phenomena witnessed through the senses, and the objective causes of phenomena—*noumena*—that can only be apprehended through intellectual intuition. Although Ampère's philosophical system was the one continuing intellectual passion of his life, he also devoted himself to other fields of scientific research. From 1800 to around 1814, mathematics was Ampère's primary interest, with his spare time spent in chemical investigations. From 1820 to 1827 he carried out the scientific work for which he is best known, pioneering the science of electrodynamics.

In 1820 Hans Christian Oersted discovered electromagnetism. A report of Oersted's work was delivered before a sceptical meeting of the Académie des Sciences held on September 4, 1820. Oersted's work was contrary to established ideas, based on Coulomb's work of the 1780s, that there could not be any interaction between electricity and magnetism. Ampère however, immediately accepted Oersted's discovery, and set to work, reading his first paper on the subject to the Académie on September 18, 1820.

Oersted's discovery suggested to Ampère, that two wires, each conducting current, might effect one another. Deducing the pattern of magnetic force around a current carrying wire to be circular, Ampère went on to visualize the resultant force if the wire were coiled into a helix. One week later Ampère announced to the Académie, his discovery of the mutual attraction and repulsion of two helices. In doing so Ampère presented a new theory of magnetism as electricity in motion.

Ampère's researches followed his own philosophy on the nature of science and scientific explanation. The phenomenon of electromagnetism had been discovered by Oersted, and the relationship between two current-carrying wires by Ampère; what remained was the discovery of the noumenal causes of the phenomenon. In his first memoir on electrodyamics Ampère investigated the phenomenon and provided factual evidence to show that magnetism was electricity in motion. He concluded that two electric currents attract one another when moving parallel to one another in the same direction; they repel each other when they are parallel but in opposite directions. Ampère felt that electrical phenomena could be explained in terms of two fluids, a positive one flowing in one direction and a negative fluid going in the other.

In 1820 Ampère described the magnetism of a needle placed within a helical coil. With the assistance of Augustin Fresnel, Ampère unsuccessfully attempted to reverse the procedure by wrapping a coil around a permanent magnet. They did not investigate the effect of moving the magnet within the coil. If magnetism is only electricity in motion, then, Ampère argued, there must be currents of electricity flowing through ordinary bar magnets. It was Fresnel who pointed out the flaw in Ampère's noumenal explanation. Since iron was not a good conductor of electricity, there should be some heat generated, but magnets are not noticeably hot. In a letter to Ampère, Fresnel wrote, since nothing was known about the physics of molecules, why not assume that currents of electricity move around each molecule. Ampère immediately accepted the suggestion, assuming each molecule to be traversed by a closed electric current free to move around its center. The coercive force, which is low in soft iron, but considerable in steel, opposes this motion and holds them in any position in which they happen to be. Magnetism consists in giving these molecular currents a parallel direction;

the more parallel the direction, the stronger the magnet. Ampère did not say why molecules should act this way, but it was sufficient that this electrodynamic model provided a noumenal foundation for electrodynamic phenomena.

In 1821 Michael Faraday sent Ampère details of his memoir on rotary effects, provoking Ampère to consider why linear conductors tended to follow circular paths. Ampère built a device where a conductor rotated around a permanent magnet, and in 1822 used electric currents to make a bar magnet spin. Ampère spent the years from 1821 to 1825 investigating the relationship between the phenomena and devising a mathematical model, publishing his results in 1827. Ampère described the laws of action of electric currents and presented a mathematical formula for the force between two currents. However, not everyone accepted the electrodynamic molecule theory for the electrodynamic molecule. Faraday felt there was no evidence for Ampère's assumptions and even in France the electrodynamic molecule was viewed with skepticism. It was accepted, however, by Wilhelm Weber and became the basis of his theory of electromagnetism.

After 1827 Ampère's scientific work declined sharply; with failing health and family concerns, he turned to completing his essay on the philosophy of science. In 1836 Ampère died alone in Marseilles during a tour of inspection.

Robert Sier

See also: Faraday, Michael; Oersted, Hans Christian.

BIBLIOGRAPHY

Hofmann, J. R. (1995). *André-Marie Ampère.* Oxford: Blackwell.

Williams, L. P. (1970). "Andre-Marie Ampere." In *Dictionary of Scientific Biography,* edited by C. C. Gillispie. New York: Scribners.

ANIMAL AND HUMAN ENERGY

Culture is the primary mechanism of human behavior and adaptation. Cultures are passed on socially from generation to generation. Tools and the ways they are made and used both shape and are shaped by the social organizations of which they are a part. In his classic study, *The Science of Culture*, Leslie A. White suggested that cultures grow in relation to the degree of efficiency of their tools in liberating energy from their natural environments. A working hypothesis in present day anthropology suggests that, in general, as tools became more efficient and numerous, populations increased in size and density, and new forms of social organization had to develop to cope with these population changes. It would be useful to consider three very general evolutionary stages in the development of our topic: hunting-gathering societies, early agricultural societies, and state agricultural societies.

HUNTING-GATHERING SOCIETIES

There is archaeological evidence that the earliest stone tools were used in Africa more than two million years ago. These "pebble tools" were often made from flat, ovoid river stones that fit in the palm of one's hand. Another stone was used to chip off a few adjacent flakes to form a crude edge. However crude the edge, it could cut through the thick hide of a hunted or scavenged mammal when fingernails and teeth could not, and thus provide the group with several dozen to several hundred pounds of meat—a caloric and protein windfall. It is a common misconception that stone tools are crude and dull. Human ancestors learned to chip tools from silicon-based stones such as flint and obsidian that have a glass-like fracture. Tools made from such stones are harder and sharper than finely-honed knives of tool steel. As early as 500,000 years ago, well-made hand axes were being used in Africa, Europe and Asia. There is evidence that people were using fire as early as one million years ago. With the advent of the use of fire, it is believed that the amount of energy used by early man doubled from 2,000 kilocalories per person per day (energy from food) to about 4,000 kilocalories per person per day (Figure 1). Thus, fire and eventually the apparatus employed to make it were important energy liberating tools. Cooking food allowed humans to expand their diet. Controlled fire could also be used to scare game into the open to be hunted. Present-day hunters and gatherers burn berry patches in a controlled way, to encourage new plants and more berries. Fire also subsidized body heat, allowing people to colonize colder regions.

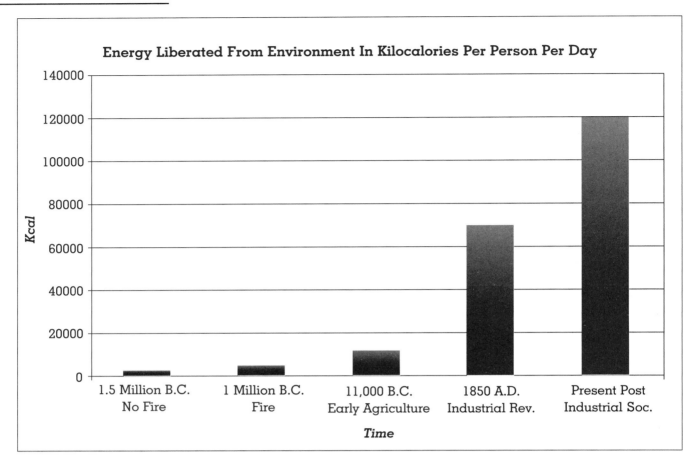

Figure 1.

SOURCE: Based on data from Earl Cook's *Man, Energy and Society* (1976, pp. 166–167).

The Neanderthal people who lived from about 130,000 to 35,000 years ago invented gracile and efficient "blade tools" that allowed them to make lighter and sharper spearheads. The Upper Paleolithic hunters who succeeded the Neanderthals made many smaller, lighter, and super-efficient blade tools called *burins*, each of which could be used for a special purpose. These burins could be transported easily from campsite, thus saving energy. Upper Paleolithic hunters also made many small and efficient tools from bone, ivory, shell and wood. The Upper Paleolithic invention of the *atlatl* or "throwing stick" allowed hunters to throw spears and darts at animals over longer distances and with greater force and accuracy than comparable missiles thrown by hand.

Just at the end of the Upper Paleolithic a Mesolithic cultural period emerged and with it several new innovations made hunting even more energy efficient. The bow could launch very lightweight arrows at game, and poison for arrows decreased the time spent stalking wounded prey. Finally, the domestication of the dog, itself a master hunter, lightened the hunter's task.

Anthropological research with modern hunter-gatherers suggests an ideal type or model for this kind of society. They were nomadic and exhibited low population size and density—on the order of thirty people per thousand square miles. Paramount in maintaining this low size and density was an imperative common to all hunter-gatherer women. A nomad woman had to move herself, all that her family owned (which was very little), and her children at a moments notice. Modern hunters and gatherers often have to walk twenty miles a day, so mothers cannot carry more than one small child. Faced with this restriction, women are careful to space their children so that the two or rarely three children they have will not burden them when they must move. There is every reason to believe that this pattern also existed in remote times.

There was little social differentiation. No one was particularly rich or poor. Leadership was informal and transitory and based on cooperation and sharing, especially with regard to meat. Although homicide was known, intragroup conflict was usually resolved by breaking the group into two or more subgroups that avoided each other until the conflict was resolved or forgotten.

Warfare was not a major cultural focus of hunter-gatherer communities and most anthropologists are of the opinion that when it did occur it was due to population pressure exerted by agricultural and/or state societies. Although hunter-gatherer social structure was informal and free-form, it was efficient with regard to hunting big animals. Studies of modern hunter-gatherer societies indicate that very little work is done in many such groups. With some exceptions, hunter-gatherers worked approximately four hours per day per adult.

EARLY AGRICULTURAL SOCIETIES

Beginning about eleven thousand years ago in the Middle East and somewhat later in other parts of the world, people began to experiment with the more sedentary subsistence patterns of growing crops and animal husbandry, and thus the Neolithic revolution began. Scholars cite two reasons for this change. First, the last glacial period began to wane as did many of the big mammals associated with it. Second, hunter-gatherer tools and organization became so efficient that human population numbers became relatively large and depleted the supply of game and wild food plants. Population expansion resulted in competition for space and directed humans toward the invention of more intensive modes of food production.

Early Neolithic peoples domesticated the more productive local plants, cared for them in densely planted plots, protected them from animals and other plants (weeds) and harvested the results. Likewise they tamed, bred and cared for local animals and ate them as they deemed fit. In the cases of cattle, horses, sheep and goats, milk and its products became staple foods. In some places larger domestic animals became beasts of burden. For very sound ecological reasons, agriculture allowed even early farmers to lib-erate much more caloric energy from plants than could be liberated from hunting and gathering and, thus, many more humans could be supported per square mile. In ecosystems, the most numerous organisms (measured in mass) are plants. Animals that eat plants (herbivores) are less abundant, while animals that hunt and eat these herbivores are least numerous. The reason for this decrease in "biomass" as one proceeds up the food chain lies in the fact that there is a considerable loss of energy due to inefficiency as animals search out, eat, digest and change the plants they consume into heat, growth and kinetic energy. The same can be said with regard to the killing and utilization of herbivores by hunting carnivores. Thus when people began cultivating plants, in effect, they stopped being carnivores and became herbivores and their population size and density increased. Since agricultural peoples were sedentary, they lived in more or less permanent settlements and their women could and did have more children. In early agricultural societies the amount of energy liberated per person per day rose to about 11,000 kilocalories.

In woodland areas early farmers developed *swidden* agriculture. They cut down forest plots using new and more efficient axes that were ground and polished from hard and dense stone. The plots were left to dry and then were burned, allowing nutrients to return to the soil. One to three plantings could be grown on these plots every year. Other plots were then prepared and cropped until the original plot had been overgrown and the process was repeated. Swidden cycles lasted from about thirty to one hundred years.

In general there is a correlation between human population size and complexity of social organization. In the 1960s, Robert Carneiro showed that there is a rough positive correlation between the number of organizational traits (N) in single community societies and their population (P), expressed by the equation . Neolithic populations increased and there was an organizational reaction. Society became stratified based on access to the extra calories of agricultural surplus. It was now possible for some people to "own" more land and animals than other people. With sedentary lifeways it was possible to accrue and own duplicate things and heavy things. Coercive institutions like war evolved to protect what one had and as the means to get more. Although more work was done in early agricultural societies, the people in these societies

This ancient Egyptian fresco evinces the long-standing relationship between humans and other animals for agricultural purposes. (Corbis Corporation)

probably had, on the whole, more leisure time than people today. The cultural florescence of the Iroquois Indians of New York State is a case in point. Around 1000 C.E. a Proto-Iroquoian culture became evident in the archaeological record. This *Owasco* culture was characterized primarily by swidden agriculture of maize, beans and squash supplemented by hunting and gathering. Over the next six hundred years village size increased as did size of dwellings. Villages became palisaded and cannibalism became evident, indicating that warfare was an institution of growing importance. When Europeans began to establish contact with the Iroquois in the 1600s, they found a people with sufficient leisure to engage in elaborate warfare patterns that were connected with a rich and complex ritual life. At first contact the Iroquois were building advanced political and governmental institutions and establishing larger orbits of political influence based on the collection of tribute.

STATE AGRICULTURAL SOCIETIES

Late Neolithic times saw the evolution of a technical innovation that fostered the growth of societies that were monumental both in population size and organization. This innovation was irrigation. Hence, historians such as Karl Wittfogel speak of the "hydraulic" theory of the state. In Old World societies like Egypt, Mesopotamia and Northern China, farmers began to grow crops on the flood plains of great river systems, taking advantage of the water and nutrients that these rivers deposited on an annual basis. Irrigation works and the subsequent population increase they stimulated required more irrigation and an evolving bureaucratic organization to manage workers, works and increasing surpluses. The animal-drawn plow became very important at this juncture and thus there was a dramatic rise in the calories that farmers could wrest from their environments. A man pushing a

APPLIANCES

Modern State Societies depend on irrigation farming as well as dry field agriculture based on mechanized energy subsidies. At first steam power accounted for these subsidies. By 1800 Watt's steam engine could generate 40 horsepower. Later the internal combustion engine replaced steam. At the height of the Industrial Revolution (c.1850) each person used about 70,000 kilocalories per day. Today we have entered into a new phase of social ecology subsidized by an ever-increasing use of fossil fuels. As a result people in developed countries now use about 140,000 kilocalories per person per day.

David Mulcahy

lever could generate about 1/20 horsepower, while an ox pulling a load (or plow) could generate 1/2 horsepower. This new technology could support still larger populations than could simple agriculture. The large-scale civilizations which evolved in all three areas supported hundreds of people per square mile. Similar state societies based on irrigation evolved independently in the New World even though Conquest States like the Aztecs and the Incas of Peru lacked both the plow and the wheel and had no beasts of burden except the dog and the llama.

BIBLIOGRAPHY

Carneiro, R. L. (1967). "On the Relationship between Size of Population and Complexity of Social Organization." *Southwestern Journal of Anthropology* 23:234-243.

Cohen, M. (1977). *The Food Crisis in Prehistory*. New Haven: Yale University Press.

Cook, E. (1976). *Man, Energy and Society*. San Francisco: W. H. Freeman and Company.

Funk, R. E. (1983). "The Northeastern United States." In *Ancient North Americans*, Chapter 8, ed. Jesse D. Jennings. San Francisco: W. H. Freeman and Company.

Kottak, C. P. (1997). *Anthropology: The Exploration of Human Diversity*, 7th ed. New York: McGraw-Hill.

Jones, S., et al., eds. (1992). *The Cambridge Encyclopedia of Human Evolution*. Cambridge, Eng.: Cambridge University Press.

Robbins, R. H. (1997). *Cultural Anthropology: A Problem-Based Approach*, 2nd ed. Itasca, IL: F. E. Peacock Publishers, Inc.

Sahlins, M. (1972). *Stone Age Economics*. Chicago: Aldine Atherton Press.

Stein, P. L., and Rowe, B. M. (1996). *Physical Anthropology*, 6th ed. New York: McGraw-Hill.

White, L. A. (1949). *The Science of Culture: A Study of Man and Civilization*. New York: Grove Press.

Wittfogel, K. A. (1970). *Oriental Despotism, A Comparative Study of Total Power*. New Haven: Yale University Press.

INTRODUCTION

Appliances refers to a vast array of devices that account for about 35 percent of overall energy consumption in the United States. As was pointed out in an early California Energy Commission hearing determining the scope of the law that delegated to that Commission authority to regulate the efficiency of appliances, *Webster's Unabridged Dictionary* (2nd ed., 1979) defines *appliance* as "something applied to a particular use; a device or machine...." This broad definition allows products that use energy directly or indirectly for virtually any purpose to be considered as appliances.

This review adopts the broad perspective, recognizing that the primary policy mechanisms applied to improve energy efficiency—minimum efficiency standards, incentive programs, normative and informative labeling programs, and technology-driven market forces—can address a very wide variety of products.

Examples of products considered appliances under U.S. federal law, most recently modified by the Energy Policy Act of 1992, include residential products such as refrigerators, freezers, clothes washers, dishwashers, water heaters, heating and cooling equipment, televisions, computers and their power supplies, showerheads, toilets, plumbing fittings, and cooking products. Also considered appliances under U.S. federal law commercial and industrial appliances such as air conditioning and heating systems and their components, water heaters and storage tanks, boilers, lighting system components such as fluorescent lamps, ballasts and fixtures, in addition, incandescent lamps, and motors. These products are involved in virtually all energy use in the residential and commercial sectors, which exceeds 35 percent of the U.S. total, including the upstream effect of electricity consumption. Excluding motors and lighting, appliance energy use is about one–fifth of the U.S. total.

Each of these products is different; each uses a different technology, is subject to different market forces, different production characteristics, distribution practices, and pricing systems. Yet, the dominant trends in energy efficiency, and in the policies used to affect it, show remarkable similarity across end uses.

For most appliances, substantial efficiency improvements have taken place over the last thirty years. Most of these improvements have been responsive to policy initiatives at the state or federal level, although in a few cases technology improvements that were aimed at some other goal also included substantial energy efficiency improvements as a by-product. The range of efficiency improvement has varied dramatically, from reductions of over 75 percent in energy use of refrigerators that comply with the 2001 Department of Energy standard compared with similiar (but somewhat smaller) products in 1972, to water heaters, where current products appear to be no higher in efficiency than they were in the 1940s.

Energy policies and the progress of technology are deeply intertwined for appliances. Appliances are sold to provide a specific energy service, and that service is the focus of their marketing and promotion as well as of consumer acceptance. Energy efficiency is not widely perceived by the marketplace to be important. Even when the payback period of energy efficiency investment by the consumer is short, surveys have shown that energy efficiency is never the first, or even second or third most important feature determining consumer choice.

This trend is self-reinforcing in the broader market and creates a vicious circle. Manufacturers recognize that the consumer will not accept a product that costs more to achieve greater energy efficiency, even if the payback period is as short as three years. Because of this observation, manufacturers do not produce high– efficiency options for the consumer marketplace in sufficient quantities to make their prices competitive. The consumer, therefore, does not perceive efficiency as a product differentiation feature. This reinforces an indifference to energy efficiency. In recognition of this indifference, policy interventions—in the form of mandatory efficiency standards, incentive programs such as rebates for products meeting specified efficiency levels or competitive acquisition of products that achieve the highest cost/efficiency results, bulk purchases of efficient products, and normative labeling—have led to the introduction of vastly improved technologies.

Indeed, before widespread government policy interventions into appliance efficiency, trends in efficiency were as likely to represent decreases in efficiency as increases. The same market barriers that recently have impeded the introduction of new technologies for efficiency have, in the past, caused products to be redesigned with new technologies for lower energy efficiency in order to cut first costs.

Refrigerators declined in efficiency following World War II due, apparently, to the substitution of lower-cost, lower-efficiency motors with improved thermal insulation to protect them from waste heat generation, and there is some evidence that water heater efficiency declined between the end of World War II and 1975, and that industrial dry-type transformers declined in efficiency between 1970 and 1993.

Energy consumption also can increase through the introduction of new product features, a process that should be distinguished from reductions in efficiency. The move from black and white to color television, and the introduction of automatic defrost on refrigerators, are examples of additions of features that compromise energy conservation. These must be balanced against other features or technologies, such as electric components for television and micro-channel heat exchangers for air conditioners, that improve product performance while also improving efficiency.

The effects of trends toward lower efficiency and higher energy-consuming features and the countervailing force of policy interventions is illustrated in Figure 1, which displays the history of refrigerator energy consumption.

During the period before 1972, when energy policy ignored efficiency, energy consumption grew at an annual rate of about 6 percent. Some of this trend was due to the introduction of new features such as automatic defrosting and through-the-door ice service, some of it was due to increases in product size and in freezer compartment size, but two percentage points of the annual growth rate in energy was due to actual decreases in efficiency.

Following multiple energy policy initiatives of the post-1973 era, this trend was arrested and energy consumption began to decline, even as size and features continued to increase. The effect of the 1979 California efficiency standard is clearly apparent in the figure. Further inflection points in the curve appear when California implemented its 1987 standards, when the 1990 federal standards went into effect, and particularly when the 1993 amendments went into effect. In addition, a gradual slope toward decreasing energy consumption coincided with utility-sponsored incentive programs whose largest effects were felt in the mid-1980s and early 1990s.

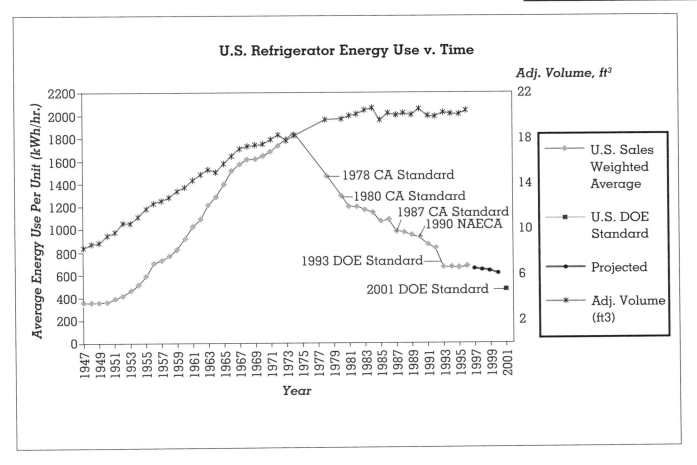

Figure 1.

SOURCE: S. M. Berman, et al. (1976). "Electrical Energy Consumption in California: Data Collection and Analysis." Lawrence Berkeley Laboratory, UCID 3847 (for 1947–1975 data). Association of Home Appliance Manufacturers (for 1972 and 1978–1995 data).

In sum, for appliances, energy efficiency technology improvements are not always adopted even after they become well understood and technology changes do not necessarily lead to improved energy efficiency.

HISTORY OF APPLIANCE EFFICIENCY IN THE UNITED STATES

During the era from the end of World War II to 1973, energy consumption for appliances increased dramatically. Overall, residential electricity consumption was increasing at an annual rate of 9 percent in the United States, while residential energy consumption was increasing by about 3 percent. These increases were driven by a number of factors that all compounded the growth: increasing population, increasing number of households due to declining household size, increasing saturation of products, increasing levels of energy service being provided by many of the products, and, in several cases, declining energy efficiency. Some, new products were introduced that had not existed previously, but if we define end uses broadly, this has not been a major effect.

Increasing saturation was a particularly important trend. While the vast majority of families owned refrigerators after World War II, virtually everyone owned a refrigerator in 1973, and over 10 percent of households had more than one. While televisions were relatively unknown in 1947, radios provided a similar service and consumed noticeable amounts of energy. Some other trends in saturation are noted in Table 1.

Appliances began to attract serious analytic interest toward the end of this era. The Northeast blackout of

Product	1970	1982	1990	1996
Washers	71.1%	73.6%	92.6%	95.3%
Dryers	41.6%	65.3%	80.6%	82.4%
Dishwashers	18.9%	44.5%	53.9%	56.9%
Microwave Ovens	small	25.6%	82.7%	90.5%

Table 1.
Trends in Appliance Saturation

1965 inspired engineers to take a second look at the consequences of continued exponential growth in electricity demand. On the West Coast, environmental concerns over the siting of power plants led to early analysis of the possibilities of policy interventions such as appliance efficiency standards.

The energy crisis of 1973 greatly accelerated efforts at analyzing the impact of appliances on regional and national energy consumption, and stimulating policy responses. On the West Coast, a major analysis of power plant siting options and efficiency and renewable energy alternatives led to the passage of the Warren-Alquist Act of 1974. This act established a California Energy Commission with the authority to forecast the needs for power plants under different scenarios of energy efficiency, analyze efficiency options, and propose efficiency standards. California adopted its first appliance standard in 1976.

In the east, a New York Public Service Commissioner testified before Congress in mid-1973 in support of appliance efficiency standards, even before the energy crisis, and New York State began to adopt its own appliance efficiency standards in 1976.

At the national level, the Ford Administration in response to the embargo began to take active steps to develop energy policy for presentation in the 1975 State of the Union Address. The Federal Energy Administration staff worked with state officials to provide a framework for a broad national energy policy. This policy, announced by President Gerald Ford, led to an executive order and ultimately, the Energy Policy and Conservation Act of 1975.

This act selected some dozen key residential appliances and proposed industry-wide voluntary targets for energy efficiency improvement. If industry could not meet these voluntary targets, which averaged a 20 percent reduction in energy use compared to then-current figures, mandatory standards would be estab-

lished. The legislation also required the establishment of nationally uniform test procedures for appliances, and mandatory labeling of the energy performance results obtained from the tests.

But before the effectiveness of the voluntary program could be determined other important events intervened. First, states began to adopt efficiency regulations of their own. The state proceedings generated considerable controversy, with manufacturers uniformly opposing the energy standards.

The prospect of a patchwork of state standards became a cause of great concern to manufacturers. When President Jimmy Carter was elected, he proposed that mandatory standards be set by the Department of Energy (DOE) to replace the voluntary efficiency targets. While opposing mandatory standards at the federal level, manufacturers acquiesced to the National Energy Conservation and Policy Act (NECPA) of 1978, which required DOE to set appliance efficiency standards for residential products. In return, manufacturers were able to obtain a requirement for DOE to evaluate the impacts on manufacturers that standards would impose and to consider them in setting the standards.

In addition, manufacturers obtained language allowing federal standards to preempt state efforts in most cases. Pursuant to this legislation, the Carter Administration proposed appliance standards in mid-1980, but was unable to issue a final rule before the Reagan Administration took over.

By the end of the 1970s, important new trends began to be manifest. First, one appliance manufacturer, Carrier, changed its initial advocacy position from opposition to appliance standards to support of those standards, because it found that the higher efficiency and higher profitability products that it would have preferred to sell were not doing well in states without standards, but were selling well in states with standards.

In addition, public interest organizations began to be active participants in the efficiency standards debates. Organizations such as the Natural Resources Defense Council and the American Council for an Energy-Efficient Economy were participating in regulatory and legislative proceedings on appliance efficiency standards and policies, supporting stronger standards than those proposed by state or federal officials.

Another important trend that had become established by 1980 was governmental support, primarily at the federal level but also in several states, of research and development efforts to improve end use energy efficiency. These efforts, funded primarily by the Department of Energy at the federal level, and the New York State Energy Research and Development Authority at the state level, but also by other organizations, focused both on expanding the technological opportunities for efficiency and on analyzing the markets for energy–efficient products and services, and market barriers that were impeding progress. Already by 1980, the intellectual basis of technologies that would prove very important commercially over the next twenty years had been initiated. Products such as low-emissivity coatings for windows, compact fluorescent lamps, electronic ballasts for fluorescent lamps, and high-efficiency compressors for refrigerators, are some of the noteworthy examples.

While initial policy efforts had focused at direct users of energy in the residential sector, increasing analysis identified other opportunities for efficiency, which began to be the target of standards efforts and later utility programs. State standards began to be established for shower heads and plumbing fittings, as well as for ballasts powering fluorescent lamps.

By the early 1980's the federal labeling requirements had taken effect and all residential appliances were labeled with yellow "Energy Guide" stickers. Initial studies, however, suggested that theses stickers were "not particularly effective in specific purchase decisions." More recent analysis has found that despite high levels of awareness of the label, significant comprehension problems exist with it.

The movement toward federal appliance efficiency standards stalled in the 1980s as the Reagan Administration, which opposed standards from an ideological perspective, began. That administration's approach was made evident by its refusal to finalize the DOE's 1980 standards proposal, and in 1983, by the issuance of a federal rule that determined that no standards were necessary. Both the delay and the "no standard" determination were challenged by NRDC, with the support of several large states, through the courts.

While these disputes were being settled, other activities were taking place. Utilities began to offer rebates to encourage the purchase of more efficient products, focusing first on refrigerators and air conditioners. State energy offices began considering adopting their own appliance efficiency standards. In California, the Energy Commission in 1983 adopted stringent standards for refrigerators, freezers, central air conditioners, and heat pumps. Following California's lead, several other states became interested in adopting appliance efficiency standards, since there was now a state model on which they could draw. By the end of 1986, six states had adopted new standards for one or more products.

The proliferation of state standards—and the 1985 court decision overturning DOE's "no-standard" stance—led manufacturers to accept federal standards. In 1986, the appliance industry offered to negotiate with NRDC, seeking a compromise that would adopt national appliance efficiency standards, but provide enhanced federal preemption of state efforts. NRDC, working directly with state energy offices, utilities, and other environmental and consumer organizations, reached an agreement with manufacturers over legislation that would adopt specific efficiency regulations explicitly in the law, and provide a schedule according to which DOE would consider increasingly stringent regulation in the future. The legislation passed Congress rapidly and overwhelmingly. It is known as the National Appliances Energy Conservation Act (NAECA) of 1987.

States also began to look at new products. Massachusetts promulgated legislation requiring its state energy office to set standards for fluorescent and incandescent lamps, and introduced legislation requiring standards for electric motors. Transformers were later added to the Massachusetts list.

Another forum for advancing the efficiency of appliances used in the commercial sector was the American Society of Heating, Refrigerating, and Air Conditioning Engineers (ASHRAE), which had first issued voluntary standards for energy efficiency of commercial buildings, including the efficiency of equipment installed in those buildings, in 1975, responding (albeit slowly) to the Northeast blackout. The ASHRAE/EIS standards, issued in 1989, became the basis for negotiations on national standards for commercial-sized heating, cooling, and water heating equipment, which were eventually incorporated into national law in the Energy Policy Act of 1992 (EPAct).

EPAct also nationalized state efforts that had gotten underway to regulate water consumption of toilets, plumbing fittings, and showerheads. The faucet and showerhead standards reduce hot water use, thereby saving energy for consumers. The toilet efficiency standards do not save energy in the home, but they do reduce municipal expenditures for water supply, disposal, and clean-up.

The late-1980s also saw the initiation of more widespread programs operated by utilities to promote energy–efficient appliances in the marketplace. Many of these focused on reducing peak loads by offering rebates for more efficient air conditioners and chillers in both residential and commercial applications.

As the 1990s began, a new concept in voluntary programs, market transformations, was developed and implemented for an increasing number of product. This approach invokes a strategic intervention in the marketplace, intended to produce significant improvements in energy efficiency features, generally through the introduction of new technologies, structured in such a way as to promote long-lasting effects.

Some of the first market transformation programs in the United States included the Energy Star program for computers and related equipment, established by the Environmental Protection Agency, which set a voluntary specification for efficiency based on the technologies used in laptop computers. In a laptop, the hard drive and display screens power down after brief periods of inactivity to low stand-by levels in order to conserve battery life. These same features were to be incorporated in desktop computers to qualify for an Energy Star label. The cost of the improvements was very low, and the Energy Star specification gained very high market penetration relatively quickly.

At the same time, a consortium of utilities working with public interest organizations, state energy offices, and the federal government, organized the Super Efficiency Refrigerator Program, a competitive challenge to refrigerator manufacturers to produce a product that saved at least 25 percent in energy use while eliminating chlorofluorocarbons. The winning manufacturer—the one that bid the lowest life cycle cost product—could receive up to $30 million obtained from utility subscriptions.

The progress of these programs, and encouraging results from similar programs in Sweden, led to an increasing number of market transformation programs during the 1990s.

Market transformation involves coordination between large numbers of market players. It recognizes that because of returns to scale in manufacturing, uniform specifications are necessary to create the climate necessary to make investment in energy efficiency improvements. This coordination was devel-

oped at the regional level through informal collaboration in California; the Northwest Energy Efficiency Alliance in Oregon, Washington, Montana, and Idaho; the Northeast Energy Efficiency Partnerships in the New England area; by a joint agreement between EPA and DOE to promote the Energy Star program at the national governmental level; and the nationwide Consortium for Energy Efficiency (CEE), an organization consisting of utilities, public interest organizations, and state energy offices.

Currently, CEE offers programs for air conditioners, heat pumps, clothes washers, dishwashers, refrigerators, industrial motors and drives, gas furnaces and boilers, lamps, light fixtures, and transformers.

In the early part of the 1990s deliberations on appliance efficiency standards appeared to be heading toward greater consensus. Manufacturers, efficiency advocates, and states joined together to discuss a negotiated joint proposal for the second DOE revision under NAECA of refrigerator standards, which was to be issued in 1995. All major parties submitted a joint proposal to DOE in late 1994.

But an ideological shift in Congress disrupted this process. In the 104th Congress, industrial opponents of appliance efficiency standards found sympathetic support, and passed a one-year moratorium on appliance efficiency standards in 1995. The moratorium held back DOE efforts on appliance standards for nearly two years. The refrigerator standard that was to be issued early in 1995 was delayed until 1997, and the effectiveness date set back three years until 2001. Progress toward new standards on ballasts, water heaters, air conditioners, clothes washers, and other products was delayed.

As of mid-2000, DOE was actively pursuing revised standards on clothes washers, water heaters, and residential central air conditioners, in addition to fluorescent lighting ballasts. These choices were the result of a formal prioritization proceeding, in which DOE, with stakeholder input, decided to concentrate its budgetary resources on the products that had the potential for the largest energy savings and economic value. Standards on fluorescent lamp ballasts and clothes washers had become a foregone conclusion as energy advocates and industry representatives reached negotiated agreements on joint support of new standards that would be promulgated by DOE in 2000.

The benefits from appliance efficiency improvements have been substantial to date. As of mid-2000,

Appliance	1970-1975 Energy Use		2000-2001 Energy Use	
	Average	Best	Average	Best
Refrigerator	1,725 kWh/yr	1,325 kWh/yr	Standard = 475 kWh/yr	
Clothes Washer	3.81 kWh/cycle		~2 kWh/cycle	0.7 kWh/cycle
Central Air Conditioners	7 EER	9.5 EER	9.8 EER	13EER
Dishwashers	4.2 kWh/cycle		~2 kWh/cycle	~1 kWh/cycle

Table 2.
Examples of Efficiency Changes

total projected energy savings from current standards by the year 2015, when most of the stock of appliances will have turned over, is 3.5 quads (a quad is a unit of thermal energy equal to one quadrillion Btu's) of primary energy per year, almost 4 percent of total U.S. energy used for all purposes. The net economic savings from these standards exceeds $175 billion.

The overwhelming bulk of these savings is due to appliance and equipment efficiency standards, although the synergistic relationship between standards and voluntary programs makes any assignment of credit somewhat arbitrary. Savings could be considered significantly larger if the base case against which it is being compared allows efficiencies to declines. Although this phenomenon has occurred several times, it is virtually never present in the economic models used to evaluate standards.

CURRENT STATUS

Despite considerable gains in the energy efficiency of most appliances over the past thirty years, substantial additional savings remain feasible and cost effective. A study by ACEEE and NRDC estimated that an aggressive set of new appliance efficiency standards could save one and one-half quads of energy, or thirty metric tons carbon equivalent (MTCE)of greenhouse pollution by 2010; appliance turnover would raise the savings from the same standards to over three quads and almost sixty MTCE by the year 2020. An unpredictable fraction of this potential can or will be achieved by other efficiency programs.

Additional savings from next generation standards or from additional products are likely. For virtually all products that have had efficiency regulations, new technologies became available following the implementation of the standards that were not available before, and in some cases were not even foreseeable.

Some of this progress is illustrated in Table 2.

These savings can be complemented by additional efficiency improvements brought forth by market transformation programs. For products such as room air conditioners, dishwashers, and residential lighting systems, the federal government, through the Energy Star program, and the Consortium for Energy Efficiency have issued specifications for higher levels of efficiency than required by standards. These specifications can be promoted in the market through utility-paid incentives, tax credits (such as those currently available in the state of Oregon), and the provision of information and marketing support.

Such programs are underway at the regional, national level, or international level, developed by government agencies, by utilities, or by non-governmental organization.

For distribution transformers, a voluntary standard assembled by the trade association NEMA (National Electrical Manufacturers Association) achieves substantial savings, particularly in standby energy, with a payback period typically of three years. The loss rate from the transformers is small, but the throughput of electricity accounts for a large fraction of total energy use in buildings, particularly when utility-owned transformers are considered.

A number of new technology promotion options are being explored in the lighting area. Work is underway on market transformation programs based on bulk procurement for improved efficacy incandescent light bulbs and for compact fluorescent lamps and fixtures. For incandescent lamps, adaptations of the infrared reflective lamp coating that is already in use on reflector bulbs are encouraged by the EPAct requirement.

The Department of Energy is issuing an Energy Star specification for compact fluorescent screw-in lamps, and EPA is revising its specification for energy efficiency (effectively compact fluorescent). These Energy Star specifications can serve as the base for marketing efforts and utility incentive programs. EPA recently issues an Energy Star specification for electronic equipment calling for greatly reduced standby losses in the small AC/DC transformers; Lawrence Berkeley National Laboratory has worked on developing a one-watt standby loss specification for such transformers worldwide.

Given the rapid growth of low voltage portable electronic equipment, including cellular phones, tape recorders and CD players, VCRs, televisions and associated equipment, and other products using rechargeable batteries. Considering the ever growing number of products that operate on standby, the potential savings from such a standard and other products using rechargeable batteries, the potential savings from such a standard could be quite large.

While tremendous progress has been made towards improving the efficiency of home refrigerators, a large number of commercial refrigeration units, both in the form of refrigerated beverage vending machines and in the form of supermarket-style refrigerators, use technologies essentially unchanged in decades. These products consume several times the energy use per unit volume of refrigerated storage compared to residential units. Work is underway for an Energy Star specification for these products as well as buyer group-based programs that will encourage the highest currently feasible efficiencies.

An additional potential for significant and growing energy savings is in the consumer electronics area. The explosive growth of laptop computers has demonstrated the feasibility of products that use about an order of magnitude less energy than their desktop counterparts, even when they are fully on. The rapid proliferation of video display screens in both home and office environments makes this a particularly attractive area for technology development and for policy encouragement of technological development aimed at energy efficiency.

In summary, there is no evidence to suggest that we have reached the point of diminishing returns in improving appliance efficiency in the United States. Instead, each step forward for a given product tends to reveal additional measures not previously analyzed that could reduce energy use even further, often while improving the quality of the energy service provided. The primary constraint on progress appears to be the intellectual effort of identifying the opportunities and developing programs directed at pursuing them.

David Goldstein

BIBLIOGRAPHY

Blumstein, C.; Kreig, B.; Schipper, L.; and York, C. (1980). "Overcoming Social and Institutional Barriers to Energy Efficiency." *Energy* 5(4):355–72.

Blumstein, C., and Harris, J. (1993). "The Cost of Energy Efficiency." Letter to the Editor. *Science*. 261.

Geller, H. S., and Goldstein, D. B. (1999). "Equipment Efficiency Standards: Mitigating Global Climate Change at a Profit." *Physics & Society*. 28(2).

Greening, L. A.; Sanstad, A. H.; and McMahon, J. E. (1997). "Effects of Appliance Standards on Product Price and Attributes: An Hedonic Pricing Model." *Journal of Regulatory Economics*, 11(2):181–194.

Greening, L. A., et al. (1996). "Retrospective Analysis of National Energy-Efficiency Standards for Refrigerators." *Proceedings of the 1996 ACEEE Summer Study on Energy Efficiency in Buildings*. Washington, DC: American Council for an Energy-Efficient Economy.

Goldstein, D. B. (1983). "Refrigerator Reform: Guidelines for Energy Gluttons." *Technology Review*. February/March.

Golove, W. H., and Eto, J. H. (1996). *Market Barriers to Energy Efficiency: A Critical Reappraisal of the Rationale for Public Policies to Promote Energy Efficiency*. Lawrence Berkeley National Laboratory. LBNL-38059.

Howarth, R., and Andersson B. (1993). "Market Barriers to Energy Efficiency." *Energy Economics* 15(4).

Hinnels, M., and McMahon, J. E. (1997) "Stakeholders and Market Transformation: An Integrated Analysis of Costs and Benefits." *Proceeding of the 1997 ACEEE Summer Study on Energy Efficiency in Buildings*. Washington, DC: American Council for an Energy-Efficient Economy.

Jaffe, A., and Stavins, R. (1994). "The Energy Efficiency Gap: What Does it Mean?" *Energy Policy* 22(10):804–810.

Koomey, J. G.; Mahler, S. A.; Webber C. A.; and McMahon, J. E. (1997). *Projected Regional Impacts of Appliance Efficiency Standards for the U.S. Residential Sector*." Lawrence Berkeley National Laboratory. LBNL-39511.

Levine, M. D.; Hirst, E.; Koomey, J. G.; McMahon, J. E.; and Sanstad, A. H. (1994) *Energy Efficiency, Market Failures, and Government Policy*. Lawrence Berkeley National Laboratory. LBL-35376.

Lovins, A. (1976). "Energy Strategy: The Road Not Taken?" *Foreign Affairs*. 55(1):65–96

Lovins, A. (1992). *Energy-Efficient Buildings: Institutional Barriers and Opportunities*. Boulder, CO: E-Source, Inc.

McMahon, J. E.; Pickle, S.; Turiel, I.; Chan, P.; Chan, T.; and Webber, C. (1996) "Assessing Federal Appliance and Lighting Performance Standards." *Proceedings of the 1996 ACEEE Summer Study on Energy Efficiency in Buildings*, 9:159–165. Washington, DC: American Council for an Energy-Efficient Economy.

Meier, A., ed. (1997). *Energy and Buildings: A Special Issue devoted to Energy Efficiency Standards for Appliances* 26.

Nadel, S., and Goldstein, D. (1996). "Appliance and Equipment Efficiency Standards: History, Impacts, Current Status and Future Directions." *Proceedings of the 1996 ACEEE Summer Study on Energy Efficiency in Buildings,* 2.163–2.172. Washington, DC: American Council for an Energy-Efficient Economy.

Sanstad, A., and Howarth, R. (1994). "'Normal Markets, Market Imperfections, and Energy Efficiency." *Energy Policy* 22(10):811–818.

Sutherland, R. (1991). "Market Barriers to Energy-Efficiency Investments." *Energy Journal* 12(3):15–34.

U.S. Department of Energy. (1995). *Technical Support Document: Energy Efficiency Standards for Consumer Products: Refrigerators, Refrigerator-Freezers, & Freezers.* DOE/EE-0064. Washington DC: Department of Energy.

ARC LAMP

See: Lighting

ARCHIMEDES (287 B.C.E.–212 B.C.E.)

Archimedes was a native of Syracuse, Sicily, the son of the astronomer Pheidias. The many achievements accredited to him include: showing that the value of π lies between the values 3 10/71 and 3 1/7 (this he obtained by circumscribing and inscribing a circle with regular polygons having 96 sides); showing that the problem of squaring the circle and rectifying its circumference were equivalent; developing a number system based on powers of myriad (10,000) to deal with large numbers; and establishing methods for finding the area under a parabola, a result that needed the integral calculus of Gottfried von Leibnitz and Isaac Newton by 2,000 years. His name is also attached to many fundamental ideas in hydrostatics and the use of levers.

Little is known of his early life other than that he studied in Alexandria and became friends with Conon, with whom he corresponded for many years.

This correspondence is the source of much that is known of Archimedes mathematics. A good deal of his work survived only in Arabic translations of the Greek originals, and was not translated into Latin until 1543. Perhaps due to the high regard contemporaries had for his geometrical work, much of it survived. It was standard reading for scholars into the late seventeenth century, and would have been read by Leibnitz and Newton.

It is thought that Archimedes had a lower regard for his mechanical work; however, this is difficult to validate because few writings about his mechanical devices remain. Archimedes used mechanics as a tool to think about abstract problems, rather than as a field of study itself. Contemporaries such as Plato frowned upon such a link between geometry and mechanics; they considered it as a corruption of the purity of geometry.

Despite his preference for pure geometry, Archimedes was not adverse to dramatic demonstrations of his discoveries of force-enhancing devices such as levers. Reports tell that he was able to manipulate a fully laden ship single-handed, using a series of levers and pulleys, after which he is said to have exclaimed, "Give me a place to stand on and I will move the earth." Applications of these ideas were exploited by Hieron II in the Punic wars when Marcellus, a Roman General, attacked Syracuse in 214 B.C.E. Marcellus's pride and joy was a primitive siege engine mounted on eight galleys lashed together, but Archimedes built a variety of far more advanced machines to defeat him. These included catapults that could launch massive stones to crash down on the fleet and sink Marcellus's galleys, and other devices, using systems of levers and counterweights, capable of lifting an entire galley until it was upright on its stern, and then plunging it to the bottom.

Archimedes' mechanical skill, together with his theoretical knowledge, enabled him to construct many ingenious machines. During his time in Egypt, he invented a hand-cranked manual pump, known as Archimedes' screw, that is still used in many parts of the world. Its open structure is capable of lifting fluids even if they contain large amounts of debris.

Archimedes' fascination with geometry was beautifully described by Plutarch:

> Oftimes Archimedes' servants got him against his will to the baths, to wash and anoint him; and yet

being there, he would ever be drawing out of the geometrical figures, even in the very embers of the chimney. And while they were anointing of him with oils and sweet savours, with his fingers he drew lines upon his naked body; so far was he taken from himself, and brought into ecstasy or trance, with the delight he had in the study of geometry. Archimedes discovered fundamental theorems concerning the center of gravity of plane figures and solids. His most famous theorem, called Archimedes' Principle, gives the weight of a body immersed in a liquid.

The reference to Archimedes' Principle is in connection with another problem posed by Hieron II. The story tells of how Hieron, suspecting that a disreputable jeweler had adulterated a gold crown with silver, asked Archimedes to determine whether the crown was pure gold or not. Legend has it that Archimedes discovered a solution while in his bath, yelled "Eureka-I've found it" and ran off to the palace, neglecting to dress first! It is not know whether the goldsmith was guilty, but for the sake of the story it is usually assumed that he was.

The result that Archimedes discovered was the first law of hydrostatics, better known as Archimedes' Principle. Archimedes studied fluids at rest, *hydrostatics*, and it was nearly 2,000 years before Daniel Bernoulli took the next step when he combined Archimedes' idea of pressure with Newton's laws of motion to develop the subject of fluid dynamics.

As enigmatics Archimedes was in life, he is perhaps better remembered for his death. An account is given by Livy (59 B.C.E.–17 C.E.) *History of Rome from its Foundation, Book XXV*. It tells how Archimedes, while intent on figures that he had traced in the dust, and regardless of the hideous uproar of an army let loose to ravage and despoil a captured city, was killed by a soldier who did not know who he was. Another version, by Plutarch, recounts that Archimedes was intent on working out some problem by a diagram, and having fixed both his mind and eyes upon the subject of his speculation, noticed neither the entry of the Romans nor that the city was taken. A soldier unexpectedly came up to him and commanded that Archimedes accompany him. When he declined to do this before he had finished his problem, the enraged soldier drew his sword and ran him through. Yet a third account by John Tzetzes in the twelfth century *Book of Histories (Chiliades), Book II*, tells a similar story with a slight twist. It says that when Archimedes

refused to stand clear of one of his diagrams when disturbed by a Roman soldier Archimedes cries out "Somebody give me one of my engines." The Roman, knowing that Archimedes' engines had defeated Marcellus's fleet, became frightened and slew him.

Douglas Quinney

See also: Bernoulli, Daniel.

BIBLIOGRAPHY

Bell, E. T. (1965). *Men of Mathematics*. London: Penguin

Dijksterhuis, E. J. (1987). *Archimedes*. Princeton, NJ.; Princeton University Press.

Fauvel, J., and Gray, J. (1987). *The History of Mathematics*. London: Macmillan.

Hollingdale, S. (1983). *Makers of Mathematics*. London: Pelican.

ARCHITECTURE

See: Building Design, Commercial; Building Design, Residential

ARGONNE NATIONAL LABORATORY

See: National Energy Laboratories

ATMOSPHERE

Man is able to directly utilize only a small portion of the energy of the Earth's atmosphere. Indeed, excessive concentrated energy in the atmosphere—hurricanes and tornadoes—represents risks to mankind. Most human demands created by atmospheric conditions involve consumption of energy to maintain comfort. The ambient surface air temperature, for instance, determines how much energy is needed for heating or cooling demands and the level of outside ambient illumination determines the need for artificial lighting. Electricity derived from fossil fuels powers the industrialized world Petroleum products directly power most forms of transportation. Pollution emitted by massive fossil fuel consumption affects man's well-being and quality of life on both immediate and long-term time scales, and mitigation this anthropogenic (manmade) pollution using emission-control devices requires even greater energy consumption.

ATMOSPHERIC COMPOSITION

Unpolluted air contains about 78 percent molecular nitrogen, 21 percent molecular oxygen, 1 percent argon, up to 3 percent water vapor, and a host of trace gases, including carbon dioxide, carbon monoxide, methane, nitrous oxide, helium, krypton, and radon. Oxygen is constantly released to the atmosphere by green plants during photosynthesis. Plants and animals excrete carbon dioxide during respiration. Water evaporates from the surface of the Earth and travels as a vapor through the atmosphere, eventually condensing and falling as precipitation. The atmosphere-ocean-geosphere and biosphere have maintained a natural chemical balance over many millennia, although steadily increasing anthropogenic trace gas emissions may have the potential to change this natural balance in the future.

Air pollution is produced by various natural and anthropogenic sources. Natural sources inject large amounts of particles into the atmosphere, including inorganic minerals, pollen, small seeds, bacteria, fungi, and effluvia from animals, especially insect parts. These natural particles usually have diameters greater than 10^{-5} cm. Many anthropogenic particles are continuously injected into the atmosphere, including latex and soot. Particles produced from combustion generally have diameters smaller than 10^{-5} cm. Tiny hygroscopic particulates from both natural and anthropogenic sources play an important role in the atmosphere, serving as condensation nuclei for water droplet and ice crystal formation. The period of time that particles remain in the atmosphere is influenced by their height and weight. Small particles in the stratosphere remain aloft much longer than small particles in the lower troposphere.

Polluted air often contains carbon monoxide and volatile organic carbon (VOC) gases, including ketones and aldehydes, as well as oxides of sulfur and oxides of nitrogen. Anthropogenic emissions of these gases arise from incomplete combustion and subsequent photochemical alterations in the atmosphere. Anthropogenic emissions also inject a number of relatively inert gases into the troposphere, including chlorofluorocarbons, sulfur hexafluoride, and carbon tetrachloride. Trees have been found to be a major natural source for VOCs.

Pollutants have various atmospheric residence times, with reactive gases and large aerosols being rapidly removed from air. In the London air pollution episode of December 1952, the residence time for sulfur dioxide was estimated to be five hours; daily emissions of an estimated 2,000 tons of sulfur dioxide were balanced by scavenging by fog droplets, which were rapidly deposited. Most relatively inert gases remain in the atmosphere for extended periods. Sulfur hexafluoride, used extensively in the electric power industry as an insulator in power breakers because of its inertness, has an estimated atmospheric lifetime of 3,200 years.

Emissions from fossil fuel combustion have caused increasing air pollution problems. Four major types of problems have been recognized: acid deposition and acid rain, air pollution episodes involving sulfur-rich smog from coal burning, photochemical smogs from gasoline-powered vehicles, and the threat of global warming as a result of increasing levels of carbon dioxide (a "greenhouse gas") in the atmosphere.

North American and Western European countries responded to acid rain, acid deposition, and acidic sulfurous smog episodes (which caused excess mortality and morbidity as well as greatly decreased visibility) by passing emission control laws. Sulfurous smogs are now rare in North America and Western Europe. Despite emission controls, acid rain and acid deposition are believed by many scientists to be the cause of forest decline, known as "neuartige Waldschäden" in Europe. This forest decline has been detected throughout Central Europe at all elevations and on all soil types. Evidence suggests that nitrates in acid rain play an important role in Waldschäden. Asian countries continue to burn fuel with few emission controls. A few tragic consequences include erosion of the Taj Mahal by acidic air pollutants, and occasional snoot-laden smog blanketing the Indian Ocean north of the equator.

Photochemical smogs arise worldwide because of the action of sunlight on emissions from gasoline-powered vehicles. Decreased visibility, increased morbidity, and crop damage as a result of photochemical smogs led to introduction of the catalytic converter on automobiles in the United States. This has had only a small impact on the occurrence of photochemical smogs in the United States.

Global warming has attracted growing worldwide concern, leading to the Kyoto Accords of 1997, which agreed that rich industrial nations would reduce greenhouse gas emissions. Legally binding reductions for each major greenhouse gas were set, with emphasis on reducing carbon dioxide emissions by 2008–2012. The Kyoto Accords were signed by President Clinton in 1998, with a carbon dioxide emissions reduction objective of 7 percent for the United States, although the U.S. Senate failed to ratify them. China, Brazil, India, and Mexico were among nations exempted from the Kyoto Accords. Canada (with an emissions reduction objective of 6%) and the European Union (with an emissions reduction objective of 8%) have developed (and implemented) some strategies to reduce carbon dioxide emissions. Norway has begun a program to sequester carbon in the ocean.

The concentration of chlorofluorocarbons in the atmosphere has been steadily increasing since they began being manufactured. It has been discovered that chlorofluorocarbons are slowly destroyed by chemical reactions in the stratosphere, especially heterogeneous reactions in polar stratospheric clouds above Antarctica. The chlorine released during these reactions in turn destroys stratospheric ozone, the most prominent result being the creation of the infamous "ozone hole," a zone with greatly diminished stratospheric ozone centered over Antarctica during winter. This ozone depletion occurs in the winter and early spring—when the sun's radiation strikes the Antarctic stratosphere. Ozone levels recover

Recognition of the threat of stratospheric ozone depletion posed by chlorofluorocarbons and chlorofluorohydrocarbons led 131 countries to sign the Montreal Protocol in 1987. Production of chlorofluorocarbons was banned as of January 1, 1996, because of their potential to further deplete stratospheric ozone. Chlorofluorohydrocarbons will be phased out of production by 2030; HCFC-22 will be phased out by 2020. However, large amounts of chlorofluorocarbon refrigerants produced over many decades remain in use worldwide, awaiting future release.

EARTH'S RADIATION BALANCE

Solar radiation continually reaches the earth, warming the atmosphere, ocean, and land surfaces on the sunlit portions of the planet. Although the sun emits a continuous spectrum of electromagnetic energy, its peak emissions are in the visible wavelengths, with a maximum at 500 nm wavelength. The average amount of solar energy received globally at the top of the atmosphere is relatively constant, about 1,353 W/m^2. Clouds and particles reflect some incident solar radiation back into space. Some large volcanic eruptions inject copious numbers of particles, which attenuate solar radiation reaching Earth's surface. When a volcanic eruption injects large amounts of sulfur into the stratosphere, sulfuric acid aerosols slowly form in the stratosphere, where they remain for months; these aerosols also reflect incident solar radiation.

About 51 percent of solar energy incident at the top of the atmosphere reaches Earth's surface. Energetic solar ultraviolet radiation affects the chemistry of the atmosphere, especially the stratosphere where, through a series of photochemical reactions, it is responsible for the creation of ozone (O_3). Ozone in the stratosphere absorbs most of the short-wave solar ultraviolet (UV) radiation, and some long-wave infrared radiation. Water vapor and carbon dioxide in the troposphere also absorb infrared radiation.

Considerable energy is radiated back from Earth's surface into space as long-wave infrared radiation. The atmosphere absorbs some of this infrared radiation, preventing its loss to space. This trapping is sometimes referred to as "the Greenhouse Effect."

THE HYDROLOGICAL CYCLE

Water is constantly evaporated from rivers, lakes, and oceans, and released from vegetation through evapotranspiration. Water vapor travels through the atmosphere, eventually forming small droplets or ice crystals in clouds. Some particles grow sufficiently

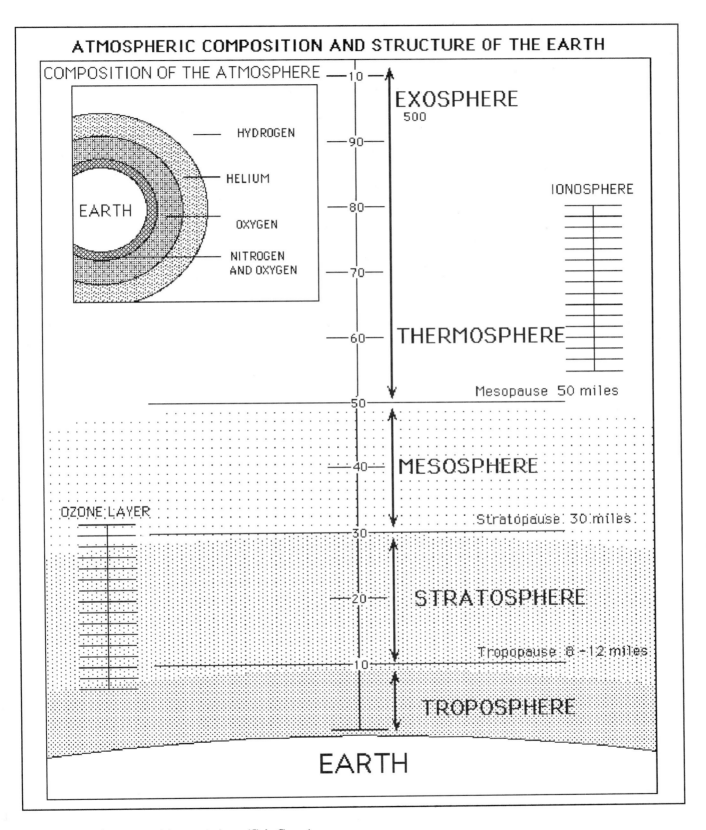

Composition and structure of the atmosphere. (Gale Group)

Earth's atmosphere as seen from outer space. (National Aeronautics and Space Administration)

large, and fall as rain or snow. Most precipitation occurs over the world's oceans. Much of the rain and snow falling over continental areas rapidly runs off into major river channels, returning water to the oceans. Some snow is deposited in glaciated areas, including high mountain peaks on the continents and on the Greenland and Antarctic ice sheets, where it may remain for millennia. About 75 percent of the Earth's fresh water is currently stored in glaciers ice sheets. Calving of icebergs from the ice sheets and periodic glacial retreat in major mountain ranges return some of this long-frozen water to the ocean. During the summer of 1999, no icebergs were seen to enter North Atlantic shipping lanes, possibly because of warmer than usual ocean temperatures.

The balance between evaporation, precipitation, glaciers, and oceans, known as the hydrological cycle, is usually considered to be in rough equilibrium over the Earth, although there is evidence that the Greenland ice sheet shrank substantially during the mid-1990s. There is also evidence that the West Antarctic ice sheet thinned during the same period.

During times of global cooling and glacial advance, ocean water levels dropped as increasing amounts of water were stored in ice sheets, as was the case during the Pleistocene glaciation of Eurasia and America. At warmer times in the geological record, glaciers have had dramatic retreats, resulting in a worldwide rise in ocean levels. The change from a cold period to a warm period may occur rapidly over the course of a century. Since 1960, mid-latitude glaciers have receded dramatically. Glaciers in the Caucasus are estimated to have lost about half their mass, while the Tien Shan Mountains are estimated to have lost about 22 percent of their glacial ice between 1960 and 2000. The decade of the 1990s is believed by most scientists to have been the warmest in many millennia. It has been postulated that these changes foreshadow a more prolonged global warming period that may be partially attributable to anthropogenic alterations in atmospheric composition.

TROPICAL STORMS

Solar heating of tropical oceans warms the surface water, promoting evaporation. Where the equatorial surface waters are warmest and the northeast and southeast trade winds meet, a band of cirrostratus and cirrus clouds spreads out from convective precipitation regions. This area is known as the Intertropical Convergence Zone.

When tropical ocean surface water temperatures exceed 26°C near the edge of the Intertropical Convergence Zone, and air aloft is warm and moist, conditions are favorable for the development of large tropical cyclones. These storms begin as weak depressions or disturbances, most of which fail to develop into organized systems. When conditions favor storm development, pressures drop in the center and winds increase in a tight 30–60 km band around a central eye. Large storms are powered by the latent heat of condensation released as clouds form from the moisture-laden air.

When a tropical storm has winds in excess of 120 km/hr, it is officially classed as a hurricane. Large hurricanes have a highly organized rotary structure, with a central eye surrounded by tightly curving bands of clouds extending up to 2,000 km in diameter, although most important activity occurs within 100 km of the eye. Large hurricanes draw enormous amounts of moisture from the Intertropical Convergence Zone.

The most powerful hurricanes (called "Category 5") have sustained winds exceeding 248 km/hr. In general, hurricanes move slowly with the average wind speed of the troposphere. When these hurricanes strike land, they bring a devastating combination of high winds, torrential rain, and a storm surge. The storm surge is an uplifting of the water level resulting from an air pressure drop and wind-driven water; the most powerful hurricanes have a storm surge exceeding 18 feet (5.5 m). Hurricane Gilbert, a massive Category 5 hurricane in 1988, dominated about 20 percent of the entire global Intertropical Convergence Zone, causing the cloudiness in the zone outside the storm to dissipate. Hurricane Andrew, which devastated South Florida in 1992, was also a Category 5 hurricane.

As a hurricane travels over warm ocean water, it lowers the sea surface temperature by about 3°C in a 100 km swath. When a hurricane is stationary, this surface ocean cooling weakens the storm intensity. Hurricanes also rapidly lose strength when they move over cold water or land.

How is energy utilized in a hurricane? Hurricanes derive energy mainly from the release of latent heat, and lose energy, in part, through precipitation and frictional loss of the wind. For an average hurricane, the rate of release of latent heat has been estimated at 10^{14} watts. This is equivalent to the energy output of a 20-megaton bomb every fifteen minutes. An average hurricane with maximum winds of 50 m/s and a radius of 30 km dissipates wind energy at a rate of 3×10^{12} watts. Thus, it takes only about 3 percent of the input energy to maintain the destructive winds of an average hurricane.

THUNDERSTORMS

Thunderstorms (cumulonimbus clouds) come in many sizes and shapes, ranging from small "air-mass" thunderstorms to large "supercells." Thunderstorms are influenced by the surrounding atmosphere and nearby convective activity. Sometimes a thunderstorm is composed of a single, isolated cumulonimbus cloud. At other times, cumulonimbus clouds are so numerous that they form a continuous sheet, losing any separate identity.

The air-mass thunderstorm is the least severe of all thunderstorms. In its simplest form, an air-mass thunderstorm grows as a single cell when solar radiation heats the surface air in an unstable atmosphere. Its life cycle lasts around 30 minutes. Towering cumulus clouds are formed as in-cloud updrafts push moisture upward. The tower may reach a height about five times the diameter of the cloud base in the growth phase.

When water vapor is deep enough for continued convective activity, the thunderstorm reaches an active phase, in which the top of the cloud glaciates, often forming a distinctive anvil. Strong updraft and downdraft regions form within the cumulonimbus cloud. The change from a towering cumulus cloud to a cumulonimbus cloud is usually quite rapid as the top turns to ice, and lightning and heavy rain begin.

The final stage in the life of a cumulonimbus cloud is marked by dissipation. The lower regions of the cloud break up, while the upper anvil spreads out.

Mixing with the environment lowers vertical wind velocities by reducing the in-cloud temperatures through evaporation and mechanical mixing with the cooler surrounding air.

Most air mass thunderstorms form in groups, facilitating growth by the reduction of environmental mixing. These multicellular storms may occur as compact clusters of cells or, if there is some external organization, laterally aligned in squall lines.

Supercells have greater size, organization, and duration than air mass thunderstorms. A supercell rotates, with persistent updrafts and downdrafts, and lasts for many hours. Updrafts in supercells may exceed 140 km/hr. Supercells develop when there are large changes in wind velocity with height. Moist, warm air entering from the front side is lifted at the cold air gust front until condensation occurs, releasing latent energy. This air parcel then moves rapidly upward and, usually, out ahead of the storm at upper levels in the atmosphere. Dry air moves in from the back side of the supercell, is cooled by rain falling out of the rising air, and then descends as a downdraft. Several different arrangements of this flow are possible. Supercells frequently move slower than the mean winds aloft. These storms are notorious for their ability to spawn tornadoes; they may show a tornadic "hook echo" on radar displays. Supercells and regular cells can combine in a multicellular complex, which then exhibits some characteristics of both types of storms.

Thunderstorms arise from convective activity, driven by energy derived from the latent heat of condensation and sublimation of water vapor within cumuliform clouds. Buoyant air movements caused by surface heating, by orographic (relating to mountains) forcing, and by lifting of warm moist surface air along frontal zones, are some of the important mechanisms for initiating the upward transfer of energy.

TORNADOES

Calculations of tornado energy are difficult to make—the aftermath of a large destructive tornado sometimes resembles carpet bombing in a war situation, with buildings ripped off foundations, large numbers of trees uprooted, and asphalt stripped from roadways. Several reports describe the derailment of up to five train cars as tornadoes have apparently lifted cars off the tracks. Large building debris has been found at a distance of 20 km from its original location.

Meteorologists categorize tornadoes by their wind speeds as deduced subjectively from severity of the damage. Each tornado is given a Fujita F-scale class: F0 (light damage), 40–72 mph; F1 (moderate damage), 73–112 mph; F2 (considerable damage), 113–157 mph; F3 (severe damage), 158–206 mph; F4 (devastating damage), 207–260 mph; F5 (incredible damage), 261–318 mph. The highest reported toronto wind speed (reported from Doppler reader) was 318 mph in the F5 Oklahoma City tornado of May 3, 1999.

The F-scale classification is only a first approximation to tornado damage. Some buildings are wind-sensitive while others are wind-resistant. The lower pressure of the tornado core also weakens the integrity of the building. Mobile homes, wood-frame houses, buildings with sheet metal roofs, and those with unreinforced masonry walls are particularly sensitive, often damaged by winds less than 100 mph. In rural counties without building codes, wood-frame houses using nails to anchor walls to foundations can be blown off and destroyed by 80-mph winds. Structurally engineered buildings are seldom destroyed or even severely damaged. People are most often injured and killed by falling building materials and by projectiles in the debris suspended in the tornado.

Tornadoes form in several ways. The most common tornadoes form at the edge of thunderstorm cold air outflow, and they are called gustnadoes. Gustnadoes fall into the F0 or F1 class and only rarely inflict intense damage along a short, narrow path. Waterspouts and landspouts form in areas where a pre-existing surface circulation becomes entrained, stretched, and intensified as a thunderstorm updraft passes over. Waterspouts and landspouts may attain in F2 class, and several have been reported to inflict moderate damage to marinas or poorly constructed buildings. The least frequent and most severe tornadoes form and descend from supercell thunderstorms, which may persist for many hours and spawn multiple tornadoes. However, the most severe tornadoes, although less frequent, are those that descend from supercells. Supercells may persist for many hours. A single supercell, moving over several hundred kilometers, has been observed to spawn a series or "family" of up to eight tornadoes along its route. These tornadoes are associated with rotating circulations called mesocyclones. The mesocyclone is 10 to 20 km in diameter, much bigger than a single tornado. It can sometimes be seen as a larger-scale rotation of the clouds at the bottom of the supercell. Rotation begins at an altitude between 4 km and 8 km and builds downward. Sometimes a mesocyclone produces more than one tornado at a time.

An average of 800 tornadoes is reported within the United States yearly, with possibly 2,000 small tornadoes going unreported. Tornadoes have been reported in every state, including Hawaii and Alaska. The Great Plains has the highest occurrence of damaging tornadoes. Occasionally, tornadoes occur in outbreaks. The super-outbreak of April 2 and 4, 1974, had 148 reported tornadoes in 13 states. Hurricanes can spawn tornadoes; in 1967, Hurricane Beulah generated 115 reported tornadoes. The majority of tornadoes occur in late afternoon and evening. However, tornadoes can form at any time of day or night. Nocturnal tornadoes are relatively common on the U.S. Gulf Coast.

Television news reporting gives the impression that people incur substantial risk in tornado-prone regions, but the likelihood of any particular building being hit is on the order of once every million years.

LIGHTNING

Within cumulonimbus clouds, precipitation processes and ambient physical conditions interact to produce regions of high electrical charge. The mechanisms by which charge separation occurs in cumulonimbus clouds are poorly understood by cloud physicists. Some researchers believe that electrical charges build in strength when ice pellets fall through a region of ice crystals and water droplets.

Lightning is the visible manifestation of a plasma channel. The plasma is very hot, with peak temperatures greater than 30,000°C, compared to 6,000°C for the sun. Although the peak current in a lightning stroke may be as high as 100 kiloamperes, charge transfer is limited by the brief duration of the flash. Movement within the plasma is limited; a typical electron in the lightning channel may move only two meters. Most of the charge transfer occurs by way of a continuing current between the strokes comprising the flash, and by relatively low amplitude currents following strokes. Usually lightning transfers negative charge to the ground. However, positively charged cloud-to-ground lightning also occurs.

After electrical potentials on the order of 300 to 400 kV/m are produced in discrete regions within the cloud, streamers extend their way forward along the cloud's charge gradient in a tree-like structure. When electrical potentials on the order of 1,000 kV/m develop, streamers become self-propagating. A plasma channel then moves toward regions of opposite charge within the cloud, neutralizing much of the electric charge within the cloud as it travels through diffusely charged regions. As the channel tip advances, it may branch in several directions simultaneously. If it penetrates into highly charged a region, a recoil streamer may flow along the channel to the initiating region.

About 80 percent of lightning channels begin and end in the cloud. The remaining 20 percent of streamers extend horizontally into the clear air outside the cloud. They propagate in a stepwise fashion called step leaders. Discharges ending in the clear air are usually highly branched, and generally quite weak. When a step leader approaches the earth's surface, an upward streamer propagates from the ground toward the channel tip aloft. These plasma channels intersect at an altitude of about 100 m above the ground. Completion of the circuit causes an upward rush of electrons called a return stroke, substantially increasing the brightness of the luminous plasma channel. Frequently, a second pulse of energy, the dart leader, moves smoothly down from the cloud, following the same path to the ground. Return strokes may follow the dart leader. Typically, cloud-to-ground flashes have four or more separate strokes.

Capturing electricity from a stray lightning flash is an intriguing but impractical idea. Presumably, Benjamin Franklin, in his famous kite experiment, transferred energy from a lightning flash to a Leyden jar, a primitive type of battery. A typical lightning flash has 25 coulombs of charge and 30,000 instantaneous amps. However, the stroke is very brief; 0.01 seconds. This is only enough energy to power one 100-watt bulb for a few months. One hundred thousand 1,000-ft towers would be needed to capture lightning energy equivalent to the output of a typical small power station.

Very powerful lightning discharges, known as superbolts, are about 100 times more powerful than the typical lightning stroke. Superbolts are most common in the wintertime off the coasts of Japan and the eastern United States. The radius of a superbolt channel is estimated to be 20 cm, compared to 2 cm of a typical lightning stroke. Because superbolts are thought to be rare over land, tapping energy from them is even more problematic than obtaining energy from a regular lightning strike.

High-altitude discharges above active thunderstorms have been studied with cameras that sense very low light levels. Several distinct, differently colored phenomena have been identified. Unusual cloud-to-air discharges from the anvil top upward to heights of 35 to 50 km (into the stratosphere) are called "blue jets." Blue jets propagate upward at about 100 km/sec in a narrow conical or trumpet shape. Red discharges extending 50 to 95 km upward above thunderstorm anvils are called "sprites." Sprites have widths ranging from 10 km to 50 km, and have been observed only above large (30,000 km²) multicellular thunderstorms. "Elves" are very brief (1 msec) red halos that form at altitudes of 60 to 100 km.

WINDS

Winds arise through a complex interplay of forces. As Earth rotates around its axis every 24 hours, the atmosphere moves along with the earth. In the troposphere, large-scale weather systems, covering regions of around two million square kilometers, form an interlocking grid pattern over the globe. The growth and decay of these large systems produces day-to-day changes in weather conditions around the world. Large-scale weather systems develop quickly; they may double in intensity in a period of 12 to 48 hours. Once formed, these systems decay slowly, generally halving in intensity in four days.

Temperature swings following frontal passages are common in North America and Eurasia, but are rare in the tropics, where differences in cloudiness and precipitation arise from seasonal variability in thermal forcing. Poleward from the tropical regions, extratropical cyclones transform latitudinal temperature gradients into kinetic energy.

In coastal areas, temperature differences between the land and the water produce air pressure variations, creating sea and lake breezes that are superimposed on the normal winds. These winds vary diurnally and as a function of cloudiness. During the daytime, winds blow from the cool sea toward the warm land, while at night the land becomes cooler than the sea surface, and the winds blow from land to sea.

On a larger scale, continents produce flows, known as "monsoon winds," over wide areas between the surrounding seas and lands. These winds respond to seasonal forcing. The best example is the Indian monsoon. During the summer monsoon, from June through September, moist winds blow northward off the Indian Ocean. Convergence of this moist air with other air masses produces intense precipitation. The monsoon slowly moves northward in spring and summer, traveling about 5 km/day. From December through February, the Siberian high dominates Eurasian air circulation, and the general flow of air is reversed, with cold, dry air traveling from the continental land masses southward over the warmer surface waters of the Indian Ocean.

Topography can substantially change air flow. Local mountain winds form when surface heating causes winds to flow up the sides of the mountain: technically known as "anabatic flow." Anabatic winds are generally strongest in early afternoon. At night, winds flow down off hills or mountains, technically known as "katabatic flow." In hilly terrain, with slopes of about two degrees, winds on the order of 3 km/hr descend as the ground surface cools.

Mountains modify the velocity and direction of wind. The coastal mountains along western North, Central, and South America play a major role in determining regional winds on the eastern rim of the Pacific Ocean. Dynamically induced winds may attain substantial speeds in mountainous regions, sometimes exceeding 100 km/hr. Some orographic winds have been given names associated with a specific region, such as the "Santa Ana" winds that occur as dry continental air descends from the Sierra Nevada Mountains to Southern California coastal areas during spring and autumn. A strong, warm wind on the leeward side of a mountain range is called a chinook (North America) or föhn (Europe). Strong chinooks, with damaging winds reaching 160 km/hr, occur several times each winter along the Front Range of the Rocky Mountains.

WIND ENERGY EXPLOITATION

Wind turbines produce power by converting the force of the wind into torque. The power produced is a function of the wind energy flux (power), which, in turn, is a function of the air density multiplied by the wind velocity raised to the third power. Changes of air density with time at a particular site are negligible compared to the fluctuations in wind velocity. Meteorologists usually report wind speed as an average. To get the potential wind power, the average

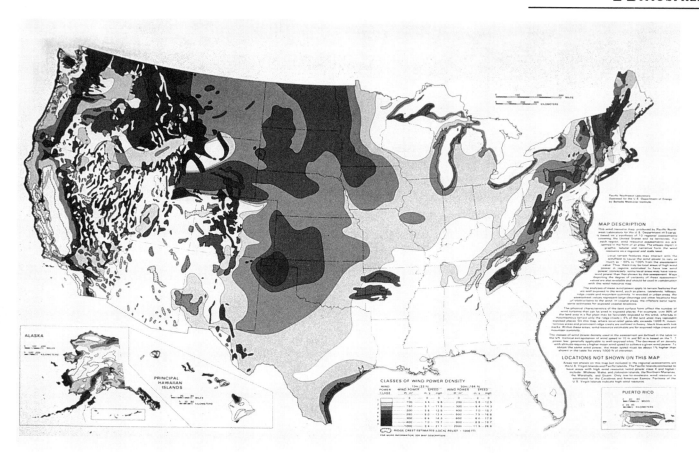

Areas with annual average wind speeds of 13 mph or greater are found throughout the United States. Regions with Class 4 winds are considered attractive for wind turbine siting. (Developed by Battelle Pacific Northwest Laboratories for the U.S. Department of Energy)

wind speed is raised to the third power and then adjusted using a Weibull statistical distribution too account for the natural instantaneous wind variability.

Wind speed, and thus available wind power, at any given location is a function of several factors: global variations; local variations, especially around coast lines with sea or lake breezes and topography; and diurnal variations of wind speed from differences in the stability of the air next to the ground. Turbulence associated with unstable air during the afternoon, or on cloudy days, mixes higher velocity winds aloft with the winds slowed by friction at the surface. On clear nights the air is stable, and there is little transport of the high winds aloft to the ground. Thus, wind speeds near the ground are normally higher during the daytime than at night, with the highest wind speeds occurring in the afternoon, and minimum wind speeds around dusk and dawn. In general, gusts are greatest in the afternoon.

Wind speed varies with height above the ground. Because surface wind speeds are routinely measured at 10 m, winds turbine heights (usually higher than 10 m) must be estimated. The turbulence level of wind also varies. Forests, buildings, and other obstacles slow wind down, and increase turbulence levels. Long grass, shrubs, and crops can slow the wind considerably. Such variations can be corrected by use of "roughness classes" or "roughness lengths." The sea is assigned a roughness class of zero, while a landscape with many trees and buildings has a class of three or four. (Sheep can keep the roughness down through grazing.) When no data are available for a site, a wind rose from the nearest observations may provide a rough estimate of wind speed. However, data availability frequently is sparse in areas with substantial wind generation potential.

During the late 1970s and early 1980s, there was considerable interest in harnessing wind energy in the

United States. During this time, efforts were made to determine the national wind energy potential. Maps were drawn using the "Batelle Wind Power Classes," ranking nominal wind energy at 10 m, 30 m, and 50 m elevations. These classes, which remain standard in mapping wind energy, are shown on the map. In general, Class 4 and higher winds are considered favorable for wind energy exploitation in the United States.

Because of its large population and the tacit assumption that its varied topography would be ideal for wind power exploitation, California conducted its own program to determine wind energy potential. This study demonstrated the meteorological difficulties in characterizing wind speeds in hilly terrain. Some wind turbines were constructed in areas thought to be ideal, but which proved to be quite marginal. Three California passes were identified as among the best wind energy sites in the world, with average wind speeds in excess of 8 m/s. Tehachapi and San Gorgonio have proven successful, and the Altamount Pass wind farm has over 7,500 wind turbines in operation.

Within the United States, some areas are especially suited for wind power generation, including North and South Dakota, Minnesota, Montana, Wyoming, the Front Range of the Rocky Mountains, the Cascade Mountains, the Great Lakes shoreline, and the ridge crests and peaks of the Appalachians. Close examination of specific geographical and topographical features may help wind power planners identify suitable sites. This has proven to be the case for Buffalo Ridge, a 100-km long ridge stretching from Spirit Lake, Iowa, through southwestern Minnesota north through Lake Benton to Sica Hollow, South Dakota. It has the potential to yield 3 Terawatt hours yearly.

Dennis G. Baker
Anita Baker-Blocker

See also: Acid Rain; Climatic Effects; Turbines, Wind.

BIBLIOGRAPHY

Church, C.; Burgess, D.; Doswell, C.; and Davies-Jones, R. (1993). *The Tornado: Its Structure, Dynamics, Prediction, and Hazards*. Washington, DC: American Geophysical Union.

DeHarpporte, D. (1983). *Northeast and Great Lakes Wind Atlas*. New York: Van Nostrand Reinhold.

DeHarpporte, D. (1984). *South and Southeast Wind Atlas*. New York: Van Nostrand Reinhold.

Eagleman, J. R. (1983). *Severe and Unusual Weather*. New York: Van Nostrand.

Eagleman, J. R.; Muirhead, V. U.; and Willems, N. (1975). *Thunderstorms, Tornadoes, and Building Damage*. Toronto: Lexington Books.

Emmanuel, K. A. (1999). *Weather*, Vol. 54: *The Power of a Hurricane*. Reading, UK: Royal Meteorological Society.

Gedzelman, S. D. (1980). *The Science and Wonders of the Atmosphere*. New York: John Wiley & Sons.

Gipe, P. (1995). *Wind Energy Comes of Age*. New York: John Wiley & Sons.

Jursa, A. S. (1985). *Handbook of Geophysics and the Space Environment*. Springfield VA: National Technical Information Service.

Kessler, E. (1986). *Thunderstorm Morphology and Dynamics*, 2nd ed. Norman, OK: University of Oklahoma Press.

Lane, F. W. (1965). *The Elements Rage*. Philadelphia: Chilton Books.

Linacre, E., and Geerts, B. (1997). *Climate and Weather Explained*. London: Routledge.

Lovelock, J. E. (1991). *GAIA, a New Look at Life on Earth*. New York: Oxford University Press.

Ludlum, F. H. (1980). *Clouds and Storms*. University Park: Pennsylvania State University.

Miller, K. B.; Matchett, J. M.; Purcell, C. W.; and Turner, M. H. (1995). *Strategies for Managing Ozone-Depleting Refrigerants, Confronting the Future*. Columbus, OH: Battelle Press.

Newton, C. W., and Holopainen, E. O. (1990). *Extratropical Cyclones*. Boston: American Meteorological Society.

Pacific Northwest Laboratories. (1991). *Wind Energy Resource Atlas of the United States*. Richland, WA: Author.

Pielke, R. A., Jr., and Pielke, R. A., Sr. (1997). *Hurricanes: Their Nature and Impacts on Society*. Chichester, England: John Wiley & Sons.

Uman, M A. (1984). *Lightning*. Minneola, NY: Dover.

Uman, M. A. (1986). *All About Lightning*. Minneola, NY: Dover.

Uman, M. A. (1987). *The Lightning Discharge*. Orlando, FL: Academic Press.

Viemeister, P. E. (1972). *The Nature of Lightning and How to Protect Yourself from It*. Cambridge, MA: MIT Press.

Williamson, S. J. (1973). *Fundamentals of Air Pollution*. Reading, MA: Addison-Wesley.

Yoshino, M. M. (1975). *Climate in a Small Area*. Tokyo: University of Tokyo Press.

ATOMIC BOMB

See: Nuclear Fission

AUDITING OF ENERGY USE

Energy conservation has always proceeded along two main avenues. One involves new technology—by continually improving the efficiency of appliances or the mileage of automobiles, the overall energy intensity (energy use per person) of society decreases. Both the government and private industry have spent large sums of research-and-development dollars on countless products that has lead to great improvements in the energy efficiency of products. The second major avenue is improvement of actual practice, whether at home, in a commercial building or on the factory floor. It is based on the assumption that through ignorance, poor operation, insufficient maintenance, priority conflicts or in some cases simply sloth, that energy is used less efficiently than the current state of technology allows.

One way of closing the gap between the current state of operations and what would be considered "best practice" is to formally examine energy use through energy audit. Using the term "audit" literally, one "counts" the energy consumed (and paid for) and matches that to necessary energy uses and practices that result in energy waste. Remedial actions are then planned to minimize the energy waste and save money. In reality, a one-for-one accounting of energy in versus energy out is rarely done. The term audit is often avoided because of negative connotations. Why are these negative connotations? Other terms used include "energy survey," "energy assessment," or "energy use analysis." All do essentially the same thing, namely examine how energy is consumed and try to identify areas where energy and money can be saved.

HOW THEY WORK

Energy audits are classified according to the client served, falling generally under the categories of residential, industrial, and commercial. Commercial audits include public and semipublic buildings like schools and hospitals and are sometimes referred to as institutional audits.

Audits are done by a variety of groups and agencies, again depending on the type of audit. Most utilities have residential auditing programs. State and community agencies run a number of auditing programs for institutions and low-income housing. The people actually doing the work are either employees of the funding organizations (from nonprofits and universities), or work at "for profit" energy service companies (ESCOs), which either do contract work for the funding sources or work directly for the client.

The scope of the audit also varies considerably. It can consist of anything from a brief walkthrough by an auditor who notes possible areas for improvement to a several month forty-person study at a major manufacturing operation. Nearly any type of audit is of some benefit. The simpler walkthrough type audits can be automated to point where a computer-printed report can be handed to the homeowner at the end of the audit. (Consequently, the costs are quite modest). In larger auditing efforts significant engineering analysis often is required to generate customized recommendations that quantify both the costs and benefits of a particular project.

A common phrase, which has been attributed to many different people, is that "you can't control what you can't measure." This is particularly true of energy use and energy waste. One way energy audits provide information to the client or end-user is by making measurements that show energy waste and allow its magnitude to be calculated. Therefore the toolbox the auditor carries is almost as important as the auditor him. The most important parameter to measure is temperature. A thermocouple can easily measure inside temperatures of the air and hot water systems, but often more is needed. One common tool is an infrared camera that is used to "map" the temperatures of walls and ceilings. This allows hot spots to be found where insulation is not functioning properly and excessive heat is escaping. Ceiling temperatures can reveal problems or suggest the installation of destratification fans to mix the air. Another important tool is a combustion analyzer. Most furnaces need to be tuned periodically to maximize performance. Typically furnace maintenance people adjust a flame by "eye," which cannot match the accuracy of measuring the composition of the flue gas with a combustion analyzer. Other important tools are electric power meters (which can measure for low power factor and line imbalance), flow meters (for fan and pump sizing), light meters and ultrasonic sound sensors (for picking up leaks in gas systems). For best results, measurements

A Southern California Edison employee logs test results at the company's electric meter site. (Corbis-Bettmann)

should be made over a significant period of time. To do this, data-loggers are available that provide inexpensive long-term measurements. Basically, the more sophisticated the audit, the greater the emphasis on measurements.

For some more advance audits, total building modeling can be done using any of several good software packages. One area where this is very important is in studies of ventilation. Excess ventilation wastes energy and some audits in the past involved measuring total infiltration into a building by putting the building under suction. Recently many of the old "rules of thumb" have come under scrutiny because insufficient ventilation can result in air quality problems. Changes in ventilation rates has led to more reliance on modeling.

An audit should also catalog the hardware used at a site. Older hardware can be operating well within its expected efficiency range and still be wasteful. The auditor needs to know about old devices and their performance specifications, as well as what is newly available. It is also essential that the auditor understand *why* the newer device works better. Many products that come on the market do not live up to their marketing hype. The auditor must filter through these and determine which are clearly indicated for their clients.

FINANCING

While some energy audits are paid for directly, most are either subsidized or leveraged in some way. Utilities in the United States are required to provide assistance in energy conservation through their demand-side management programs. Even with the deregulation of electricity, many states are requiring that all energy providers pay into a "public benefits" pool, and money from this pool be used in part to support energy auditing. Another way of leveraging the costs of audits is through "performance contracting." This process normally involves an ESCO that will provide energy audits and system upgrades for no direct costs to the client. The client agrees to pay some fraction of the savings from the system improvements to the ESCO for a contract period (normally 5-10 years). This type of contracting has worked best when energy improvements are clearly measurable, such as lighting upgrades. Other types of projects normally present a larger risk and are therefore less common.

Energy audits are also leveraged with a system commonly called "over the fence" energy. In this scheme, the end-user does not buy equipment, but purchases a commodity such as steam, heat, or compressed air from a third party. The third party owns, leases, or operates the equipment used to provide the energy service. It is then in the interest of that third party to ensure that energy audits are routinely carried out and unnecessary energy use is kept to a minimum.

During the 1990s, system commissioning started to become popular. Historically, builders, architects and owners have agreed on hardware. The owners are guaranteed that the building will be warm in the winter and cool in the summer. Owners now often ask for an additional guarantee—a guarantee of performance in terms of energy costs. A commissioning audit does not recommend improvements, but acts as a measuring device to ensure compliance, often of new buildings before owners take possession. In a commissioning audit, the auditor ensures that heating and air conditioning systems, energy management and control systems, and other complex systems are installed and operating properly.

TRENDS

The interest in energy audits by end-users has historically tracked with the price of energy. An additional motivation appeared in the 1990s with the concern about global warming. Energy use (the majority of which comes from the burning of fossil fuels) directly correlates with the emission of greenhouse gases. Saving energy now also reduces CO_2 emissions and the buildup of greenhouse gases in the atmosphere. For many who wish to be environmentally sensitive, auditing is a proactive step. Another change impacting energy auditing is energy deregulation. Deregulation is leading to open competition for both electricity and gas. For commercial and industrial customers, electricity prices are going down, probably decreasing interest in energy conservation. However, advances in electrical generation technolo-

gies are increasing the likelihood that businesses and institutions will be generating at least part of their own electricity needs. There have also been recent advances in engine-driven technologies to replace the electric motors in chillers, air compressors and pumps. Therefore the energy systems within a building or plant are becoming more complex. The way in which energy is purchased is also becoming more complicated. Consumers are being approached by different energy providers, some of whom are packaging technical services such as audits with their energy. Some fuel supplies are provided with "interruptible service" requiring backup fuels. Some energy is provided with "time-of-use" charges and "ratchets" that can result in twelve months of surcharges for the ill-advised use of energy at an inopportune time. This added complexity makes energy auditing an essential periodic check on operations.

Michael Muller

BIBLIOGRAPHY

Felters, J. L. (1998). *The Handbook of Lighting Surveys and Audits.* Boca Raton, FL: CRC Press.

Schueman, D., ed. (1992). *The Residential Engery Audit Manual,* 2nd ed. Lilburn, GA: Fairmont Press.

Thumann, A. (1998). *Handbook of Energy Audits,* 5th ed. Lilburn, GA : Fairmont Press.

AUTOMOBILE

See: Transportation, Evolution of Energy Use and

AUTOMOBILE PERFORMANCE

At the beginning of the twentieth century, the automobile was still a novelty. In the United States more cars were then powered by steam engines and battery-electric systems than by internal-combustion engines. By the end of the twentieth century, the automobile had become an integral part of the American lifestyle, with approximately one privately owned passenger vehicle in operation for every two people. Practically all of these vehicles were powered by internal-combustion engines burning a fuel derived from petroleum.

New cars are purchased on the basis of such qualities as performance, fuel economy, reliability, durability, ride quality, noise and vibration, comfort, convenience, maintenance requirements, styling, safety, environmental qualities, price, and resale value. Many of these attributes conflict. The conflict between performance and fuel economy serves as an example. The typical driver wants a vehicle with sufficient power to merge safely into high-speed freeway traffic, or to pass a slowly moving highway truck on an upgrade. To maintain performance when the vehicle is carrying a heavy load, or pulling a trailer, or operating at high altitude can call for even more installed engine power. That desire for performance potential persists with the typical consumer even though it is used only very occasionally in normal driving.

Installing a more powerful engine to meet performance expectations typically penalizes average fuel economy. To accommodate a range of expectations, manufacturers frequently offer a given car model with a choice of more than one engine. In a recent example, one car model is offered with its base engine and two engine options that exceed the base engine in rated power by 13 percent and 26 percent respectively. While offering greater performance, those optional engines decrease fuel economy in typical driving by 5 percent and 10 percent respectively. In striking a balance between performance and fuel economy, the typical United States consumer leans more heavily toward performance than his overseas counterpart because the pump price of gasoline in the United States is only about a third of what it is in many overseas nations. Moreover, tax policies in some overseas countries, discourage the use of large and powerful engines.

The petroleum from which gasoline is derived is a depletable resource. Passenger cars and light-duty trucks account for about 40 percent of national petroleum consumption. In recent years, imported petroleum has supplied an increasing proportion of U.S. needs. This adversely affects the U.S. balance of trade. Currently about half of the oil consumed is imported. Nearly a quarter of U.S. fuel consumed comes from Organization for Petroleum Exporting Countries (OPEC), a group organized to control the price of oil. About 10 percent of consumption is imported from the Persian Gulf segment of OPEC, where the potential for political instability is a national-security issue. Gasoline combustion also accounts for about 20 percent of the carbon dioxide generated in the combustion of fossil fuels in the United States. The growing concentration of carbon dioxide in the atmosphere threatens to increase world average temperature. For reasons such as these, the federal government has established minimum fuel-economy standards for light-duty vehicles.

Several different metrics are used to assess vehicle performance. Included are the time required to travel a specified distance from a standing start, the time required to accelerate from rest or from some initial speed to a specified final speed, and the speed that can be maintained on a specified upgrade without downshifting the transmission from its highest gear. In the United States, the most frequently used metric of this nature is the time required to reach 60 mph from a standing start.

This is not to imply that a full-throttle acceleration from rest to that speed is a maneuver frequently executed by the typical driver. The time to 60 mph is rather an easily measured parameter that serves as a surrogate for other performance metrics. A car that is slow from 0 to 60 mph will likely have slow response from 40 to 60 mph for freeway merging, or prove lethargic when climbing hills. Reflecting the market preference of the typical new-car buyer, for the average new U.S. passenger car, the acceleration time from 0 to 60 mph has decreased from about 14 seconds in 1975 to fewer than 11 seconds in 1995.

This gain has not come entirely at the expense of fuel economy, however. Over that same twenty-year span, the fuel economy of the average new U.S. car increased by 80 percent. These advances are attributable to lower vehicle weight, improved tires, reduced aerodynamic drag, improved transmissions, and gains in engine efficiency.

ACCELERATION PERFORMANCE

When a driver commands an increase in vehicle velocity, that vehicle obeys Newton's first law of motion, which states that when a force (F) acts on a body of mass (M) and initially at rest, that body will experience an acceleration (a). For an automobile, typical units for acceleration, which is the rate of change of velocity, would be miles per hour per second. Mass is further defined as the weight of the body (W) divided by the acceleration of gravity (g).

The product of force F and the rolling radius (R) of the tires on the drive wheels is the wheel torque (T). Power depends on both torque and rotational speed (N). By definition, power is given by $P = 2\pi NFR = 2\pi NT$. When driving at constant speed, the driver adjusts the accelerator pedal so the drive-wheel power exactly matches the *power required* (P_r) to overcome the resistance of the vehicle (discussed later in this article). To accelerate the vehicle, the driver further depresses the accelerator pedal so that the *power available* at the drive wheels (P_a) exceeds P_r.

When applying Newton's law to a moving automobile, acceleration depends on the excess of power over that required for constant-speed driving, namely $P_a - P_r$. From this it follows that the instantaneous acceleration (a) of the vehicle at a given road speed (V) is

$$a = g (P_a - P_r) / WV \qquad (1)$$

For maximum vehicle acceleration, the driver depresses the accelerator pedal to the floorboard and the engine operates with a wide-open throttle. The power required curve traces the power needed by the car as a function of vehicle velocity when it is operated at constant speed in still air on a level road. At any given speed, the difference between these curves, Pa–Pr in Equation 1, is available for accelerating and hill climbing.

In the above expression for vehicle acceleration, the proper weight to use is the effective vehicle weight, which is the actual weight plus an additional increment that accounts for the rotating inertia of the engine, drivetrain, and wheels. For a passenger car driven in high gear, the ratio used for normal driving, this increment amounts to about 10 percent of the vehicle weight. It is substantially higher in the lower gears, however, because at a given road speed, the rotational speed of the engine and part of the transmission is multiplied by the transmission gear ratio. This increase in rotational speed magnifies their influence on the effective weight of the vehicle. For illustrative purposes, however, this adjustment to vehicle weight for rotational inertia is ignored here.

POWER REQUIRED

The force required to move the vehicle forward is the sum of four components: rolling resistance, aerody-

namic drag, acceleration force, and grade requirement. This required force is converted into power required by multiplying by the forward velocity of the vehicle with respect to the road.

Rolling resistance stems from the energy expended in deforming the tire and the surface of the road at the contact patch between tire and roadbed. The power required to overcome rolling resistance (P_r) depends on the rolling resistance coefficient (C_r), and the vehicle weight (W) and velocity (V_v). It is given by

$$P_r = C_r\, W\, V_v \qquad (2)$$

C_r tends to increase slightly with speed but is often considered constant over the normal speed range of an automobile.

The rolling resistance coefficient of a tire depends on the construction of the tire carcass, the elastic characteristics of the tire material, the tread design, and characteristics of the roadbed. It increases with decreasing wheel diameter, tire underinflation, and roadbed compliance. It decreases as the operating temperature of the tire rises.

Before the 1960s, the bias-ply tire exemplified standard construction. It had a typical rolling resistance coefficient of 0.015 on hard pavement. Since then, the radial-ply tire has emerged, offering a coefficient closer to 0.010. Coefficients as low as 0.008 to 0.009 have been claimed in tires suitable for use on passenger cars. Cutting the coefficient from 0.015 to 0.008 offers the opportunity for about a 10 percent reduction in fuel consumption.

Other qualities sought in a tire include ride quality, cornering ability, traction characteristics on both dry and slippery roads, tire noise, life, and cost. Addressing these qualities often opposes the objective of lower rolling resistance.

The force of aerodynamic drag opposing forward motion of the vehicle depends on its drag coefficient (C_d), its frontal area (A_f), the air density (ρ), and the velocity of the wind with respect to the vehicle. In still air, this velocity is simply the vehicle velocity (V_v). If driving into a headwind of velocity V_w, however, the wind velocity with respect to the vehicle is the sum of these two. Multiplying the aerodynamic drag force by vehicle velocity provides the aerodynamic power requirement (P_a).

$$P_a = \tfrac{1}{2}\, \rho A_f V_v (V_v + V_w)^2 \qquad (3)$$

The drag coefficient for an automobile body is typically estimated from wind-tunnel tests. In the wind tunnel, the drag force acting on a stationary model of the vehicle, or the vehicle itself, is measured as a stream of air is blown over it at the simulated vehicle speed. Drag coefficient depends primarily on the shape of the body, but in an actual vehicle is also influenced by other factors not always simulated in a test model.

For example, the windage loss associated with the rotating wheels increases drag. Covering the wheel wells can reduce this adverse effect. Although suitable for rear wheels, a body-mounted cover over the front wheel wells interferes with steering because during sharp turns, the front wheels extend beyond the plan-view profile of the car body.

A smooth underbody also would improve fuel economy by reducing drag but is rarely used because of more pressing demands. For example, a continuous smooth underbody interferes with engine-compartment ventilation, blocking the normal exit route for engine cooling airflow that has passed through the radiator. It also interferes with accessibility for routine engine servicing. Farther to the rear, it is desirable to expose the exterior surfaces of the hot running exhaust system, which includes a catalytic converter and muffler, to the flowing airstream for cooling. Therefore, covering them with an underbody panel creates problems.

Another source of increased drag involves the external-surface details. In this regard, stand-alone bumpers, externally protruding door handles, and running boards alongside the passenger compartment to facilitate entry, all of which were once universal, have disappeared. Flush rather than recessed side windows have come into recent use. On the other hand, external rear view mirrors, which increase drag, have been added as a safety measure.

Over the years, the flat vertical windshield of 1920 has given way to an increasingly raked windshield, first curved in two dimensions, but now in three. The once ubiquitous vertical flat radiator that fronted the engine compartment has disappeared into a compartment covered by a streamlined front body. Fenders and headlights, both of which were once free standing, are now incorporated into that body.

A major fraction of the aerodynamic drag in the modern streamlined car is caused by flow separation at the rear of the body. Alleviating that separation calls for a long afterbody that tapers to a point.

Depending on driving conditions, this change could decrease fuel consumption by about 15 percent, with even greater improvement on the highway but less in the city. However, such a sharp tailpiece is useless for carrying passengers, of minimal utility in storing luggage, prohibitively dangerous for storing fuel, and impairs vehicle handling and parking.

The equation given above for the power required to overcome aerodynamic drag (P_a) is expressed for a vehicle driving into a headwind. If, instead, the wind direction is from the rear, the sign on the wind velocity changes from positive to negative, and the aerodynamic drag is reduced. In practice, however, the wind almost never blows directly from either the front or the rear. When the wind approaches from an angle oblique to the direction of travel, the wind velocity relative to the vehicle is the vector sum of the travel velocity and the wind velocity. The vehicle cross-sectional area encountered by this relative wind is greater than the frontal area A_f, and the vehicle shape was not designed for that oblique wind direction. As a result, the drag coefficient can increase as much as 50 percent above its value when the direction of the wind is aligned with the vehicle centerline.

The power expended in accelerating the vehicle represents an investment in kinetic energy that is stored in the vehicle by virtue of its motion. It can be partially recovered during vehicle coasting, but most of it is usually dissipated as heat in the brakes when the vehicle is decelerated or brought to a stop. The acceleration power (P_k) depends on vehicle weight (W), velocity (V_v), and the rate of change of velocity with respect to time, which is the instantaneous acceleration (a). The power for acceleration is given by

$$P_k = (W \, a \, V_v) / g \qquad (4)$$

where g is the gravitational constant. If the car is being powered in a decelerating mode, the sign on this acceleration term becomes negative.

The final term in the equation for required power is that which accounts for driving on a grade (P_h). The severity of a grade is normally defined as the ratio of its vertical rise (h) to its horizontal run (L), expressed as a percentage. On U.S. interstate highways, the grade is usually limited to 4 percent (h/L = 0.04). On public roads, grades as high as 12 percent may be encountered. For grades no steeper than that, P_h is closely approximated by

$$P_h = W V_v (h/L) \qquad (5)$$

If the vehicle is on a downgrade rather than an upgrade, the sign on this term becomes negative.

POWER AVAILABLE

The full-throttle capability of a typical gasoline engine, as delivered to the wheels by the drivetrain, was shown as a function of car speed for four transmission gear-ratios. In any given gear, engine full-throttle power rises with increasing engine speed, leveling off at maximum power and then falling again as engine speed is further increased. This characteristic is dominated by the influence of engine speed on mass airflow rate and on engine friction.

Mass airflow rate is the principal determinant of the maximum power that can be developed within the engine cylinders. As speed increases, the cylinders ingest more air. Therefore, one might expect the power developed in the cylinders to increase in proportion to engine speed. However, the aerodynamic losses in the air passing through the cylinders rise at an increasing rate as engine speed is increased. This causes the mass airflow rate, and with it the power developed in the cylinders, to reach a maximum at some high engine speed that typically lies beyond the normal operating range of the engine.

The power delivered by the engine crankshaft is less than that developed within the engine cylinders by the power expended in overcoming engine friction. Friction power also increases with speed at an increasing rate. As a consequence, the power output delivered at the engine crankshaft peaks at some speed less than that at which the cylinders achieve their maximum power.

The power developed on the crankshaft is further depreciated by the requirements of such accessories as the electric alternator, power steering pump, and air conditioner. The drivetrain that connects the crankshaft output to the vehicle drive wheels causes a further loss in power. Drivetrain efficiency generally falls within 80 to 95 percent. The remaining useful propulsive power varies with speed.

Design features in the engine and transmission can effect seemingly subtle changes to the shape of the power available curve that are important to vehicle drivability. Automotive engineers normally address this issue in terms of available torque, rather than power, as a function of speed. Torque, of course, is

determined by dividing power by rotational speed. An objective is to produce high torque at low speeds in order to minimize the need for transmission shifting.

Before the 1980s, engine air was normally inducted through an air cleaner mounted directly on the carburetor. Replacing the carburetor with electronically controlled fuel injection just ahead of the intake valves has minimized large areas of intake-manifold walls wetted with liquid fuel, which impaired engine response to quick depression of the accelerator pedal. This virtual elimination of wetted walls in the air intake system has facilitated use of a carefully designed intake system that includes a plenum and long air pipes. The components of this system are carefully proportioned to enhance torque at low engine speeds.

Another method of advantageously reshaping the torque variation with engine speed is the incorporation of variable valve actuation. This includes the options of variable valve timing; variable valve lift; and in engines with two intake valves per cylinder, the deactivation of one of them at low engine speeds. Variable valve actuation may be used not only to enhance low-speed torque but also to improve other operating aspects of the engine.

In automatic transmissions, a torque converter has long been incorporated to increase delivered torque at low vehicle speeds, as during acceleration from a standing start. In the torque converter, the engine shaft is fastened to a centrifugal pump that converts the torque delivered by the engine into flow energy in a hydraulic fluid. An adjacent hydraulic turbine converts this flow energy back into shaft torque for delivery to the input side of the transmission gearbox. Because there is no mechanical connection between pump and turbine, turbine speed is independent of pump speed and allows the vehicle wheels to be stationary while the pump rotates at engine speed. Through the action of a row of stationary vanes located between the pump inlet and the turbine discharge, the torque converter is able to multiply engine output torque at the expense of gearbox input speed. This ability of the torque converter to multiply torque typically allows the gearbox of the automatic transmission to meet the needs of the vehicle with one less gear step than would be needed in an equivalent manual transmission.

When the vehicle is at a standstill, the torque converter delivers approximately twice the engine torque to the gearbox. This torque multiplication falls toward unity as the vehicle accelerates from rest and as the turbine speed catches up with the pump speed. At the coupling point, turbine speed nearly equals pump speed; torque multiplication ceases; and the stator is unlocked, allowing it to spin freely. The small speed difference, or slip, between pump and turbine then existing causes a minor loss in transmitted power. The 1970s witnessed increasing use of a torque-converter clutch that reduces this slip loss by joining the pump to the converter when the coupling point has been reached. (Actually, a slight slip in this clutch may be employed to minimize the transmission of engine vibrations to the driver.)

Because of the importance of mass airflow rate in establishing engine output power, power available is sensitive to ambient conditions. Full-throttle engine power varies approximately inversely with inlet-air absolute temperature, but more significantly, approximately directly with ambient pressure. Mountain passes exist on public roads in the United States. that have altitudes of over 12,000 ft. The normal atmospheric pressure at such altitudes results in a one-third loss in power capability in the typical passenger-car engine.

POWER RESERVE

Power available at full throttle in top gear is again plotted against vehicle speed in Figure 1, along with power required curves for three different conditions. Figure 1a is for driving on a level road in still air, Figure 1b for driving on a level road into a 25 mp/h headwind, and Figure 1c for driving up a 6 percent grade in still air. In each case, the difference between power available and power required that is available for acceleration, termed the power reserve, is shown shaded. Clearly, the power reserve in top gear is diminished significantly by driving into a headwind or by hill climbing.

Vehicle maximum speed is indicated in Figure 1 by the intersections of the power available and power required curves. It is seen to fall from more than 120 mph to 90 mph in going from conditions of Figure 1a to Figure 1c. In the early days of the automobile, top speed was of greater importance than today. The Panhard Levassor of 1886 was capable of only 12 mph. In about 1900, cars with a top speed of about 40 mph had become available, which may have been adequate for existing roads. When the first concrete road appeared in 1909, the Olds Limited could reach

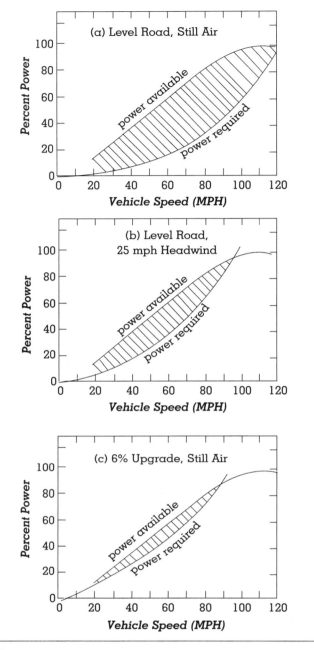

Figure 1.

Power available at wide-open throttle and power required on a level road in still air for car with a four-speed manual transmission.

65 mph. With its 108 horsepower engine, the 1935 Cadillac attained 85 mph. An additional 25 hp enabled the 1951 Cadillac to top 98 mph.

That a modern automobile may be able to exceed 100 mph is of no direct consequence in U.S. driving

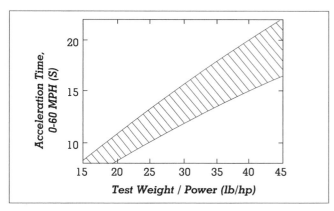

Figure 2.
Effect of vehicle weight/power ratio on acceleration time.

because these speeds are illegally high in nearly all locations. What accompanies the deterioration in top speed illustrated in Figure 1, however, is the obvious loss in power reserve at more commonly driven speeds. This reserve is so small—for example at 40 mph on a 6 percent grade—that in real driving, frequent gear shifting between fourth and third gear could be rendered necessary by minor disturbances such as wind gusts or on-off cycling of the air conditioner. Such shifting requirements, whether imposed on the driver by a manual transmission or done for the driver in an automatic transmission, are sources of driver dissatisfaction. In addition, fuel economy is poorer in the downshifted gear because the engine is forced to operate at a higher speed with lower thermal efficiency.

A corresponding situation occurs at high altitude, where one-third of the sea-level power available has been lost due to low atmospheric pressure. This low air density also reduces aerodynamic drag, but rolling resistance is unaffected by altitude. As a result, power reserve is seen to suffer. In fact, at this altitude, the power available in fourth gear is insufficient to operate the vehicle on a 6 percent grade at any speed without downshifting.

Past studies of passenger-car performance have shown that the acceleration time from 0 to 60 mph correlates well with vehicle weight/power ratio. Although a small difference was found between automatic and manual transmissions, most measured data falls within the band shown in Figure 2, where test weight corresponds to vehicle curb weight plus two 150-lb passengers or their weight equivalent. This

suggests that performance can be improved either by increasing installed engine power, which tends to depreciate fuel economy, or by decreasing vehicle weight, which tends to improve fuel economy.

FUEL ECONOMY

Vehicle fuel economy is normally measured in miles per gallon. At any given instant, it depends on the energy content of a gallon of fuel (Q_f), the vehicle velocity (V_v) and power required (P_{req}), the thermal efficiency with which the engine converts fuel energy into useful output work (η_e), and the mechanical efficiency with which the driveline delivers that work to the vehicle wheels (η_d). Specifically,

$$\text{Miles/Gallon} = Q_f V_v \eta_e \eta_d / P_{req} \qquad (6)$$

On a transient driving schedule, this instantaneous fuel economy must be averaged over the distance driven.

In earlier times it was customary to eliminate vehicle velocity as a variable by measuring fuel economy while driving at a constant speed, typically 40 mph. Fuel economy at this speed generally falls within 10 percent of its maximum level-road value. Fuel economy falls off at both much lower and much higher speeds.

The shortcoming of a constant-speed test is that in traffic, nobody drives at constant speed. Consequently, with the onset of the federal fuel-economy standards that took effect in the United States in 1978, the Environmental Protection Agency (EPA) prescribed transient driving schedules deemed representative of both urban and highway driving.

The EPA urban driving schedule is diagramed in Figure 3a. Its 18 start-and-stop cycles cover 7.5 miles in 1,372 seconds, with an average speed of 19.5 mph and a peak speed of 56.7 mph. The Federal Test Procedure (FTP) used for measuring exhaust emissions is based on this schedule. It involves operating a car, which has been initially stabilized at room temperature, for the prescribed 7.5 miles, shutting off the engine for 10 minutes, then restarting and repeating the first five cycles. The FTP is performed on a chassis dynamometer, with the vehicle stationary and the drive wheels turning a roller in the floor that is connected to an electric generator. That generator is loaded to simulate the rolling resistance and aerodynamic drag that would be encountered when actually driving the vehicle on the road. The urban fuel economy (MPG_u) is determined from this test.

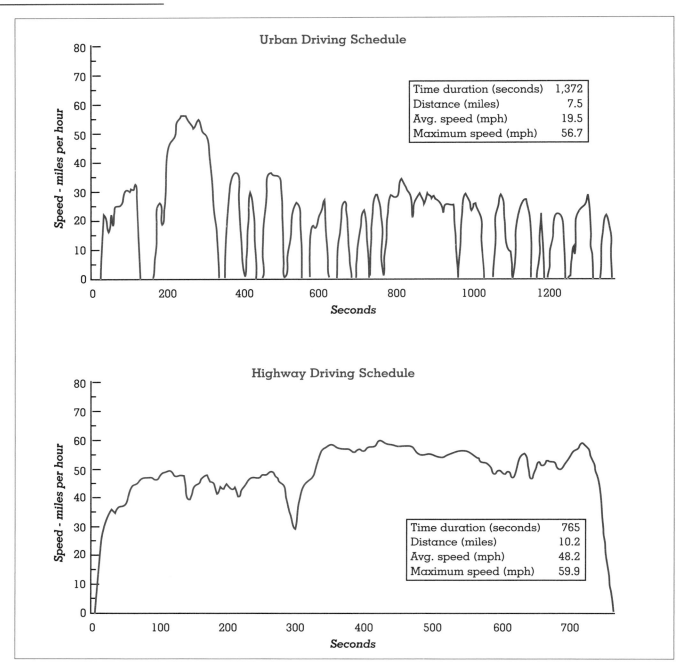

Figure 3.

Standard U.S. driving schedules: (a) urban and (b) highway.

SOURCE: *Code of Federal Regulations*, "Subpart B-Fuel Economy Regulations for 1978 and Later Model Year Automobiles-Test Procedures," July 1, 1988 ed., p. 676.

The schedule used to measure highway fuel economy (MPG$_h$), also driven on a chassis dynamometer, is diagramed in Figure 3b. It covers 10.2 miles in 765 seconds at an average speed of 48.2 mph and a peak speed of 59.9 mph. In contrast to the urban schedule, there are no stops and starts during the highway schedule.

For regulatory purposes the fuel economy assigned to the vehicle from these tests is based on the premise that the vehicle will accumulate 55 percent of its mileage on the urban schedule and 45 percent on the highway. The resulting composite fuel economy (MPG$_c$) is thus calculated according to:

$$1/MPG_c = 1/(0.55/MPG_u + 0.45/MPG_h) \qquad (7)$$

For a typical passenger car of the 1990s, the effect on fuel consumed over the urban and highway schedules, and also on composite fuel consumption, is illustrated in Figure 4 for independent 10 percent changes in vehicle weight, aerodynamic drag, and rolling resistance coefficient. On the urban schedule, vehicle weight has the greatest significance because of the dominant effect of acceleration power associated with the many start-and-stop cycles. This power, invested in kinetic energy during accelerations, is not recouped because the brakes are applied during the succeeding decelerations. On the highway schedule, which lacks frequent starts and stops, the high average speed makes aerodynamic drag the dominant factor. This reflects the exponential dependence of drag on speed.

At the bottom of Figure 4, the effect of fuel consumed during unpowered engine operation, which occurs during closed-throttle decelerations and engine idling at zero vehicle speed, also is shown. The fuel expended during unpowered operation is greatest on the urban schedule because of the time spent braking and standing during that schedule.

It is clear from Figure 4 that weight is the vehicle parameter of greatest significance to composite fuel economy. The power increments required to overcome rolling resistance, provide acceleration performance, and climb hills all vary directly with vehicle weight. The power required to overcome aerodynamic drag relates indirectly because vehicle frontal area, a factor in determining interior roominess, tends to correlate with vehicle weight. The importance of weight to performance was established in Figure 2.

Over the years, consumer desire for more interior roominess and greater luggage capacity has combined with mandated weight-increasing safety and emissions devices to complicate weight reduction. Despite this, the curb weight of the average new U.S. passenger car decreased 21 percent from 1975 to 1995, in which year it was 2,970 lb. Over those same years, however, the curb weight of the average light-duty truck increased 4 percent, to 3,920 lb in 1995. During those two decades, the category of light-duty trucks has been swollen by the popularity of light-duty vans and sports-utility vehicles included in this

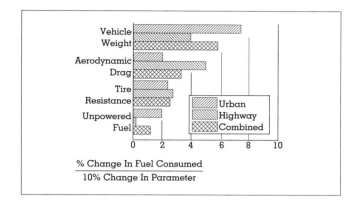

Figure 4.

Individual effects of design characteristics on fuel economy of a typical passenger car.

category. The sales ratio of these light-duty trucks to the lighter passenger cars climbed from about 20 percent in 1975 to near 40 percent in 1995. This increase in the proportion of truck sales meant less reduction in the weight of the fleet-average vehicle than might otherwise be expected.

Much of the passenger-car weight reduction has resulted from basic design changes, but some came from materials substitution. From 1978 to 1997, the portion of domestic-car weight attributable to ferrous materials has dropped from 74 to 67 percent. Concurrently, lighter aluminum climbed from 3.2 to 6.3 percent of car weight, and plastics and composites from 5.0 to 7.5 percent. Magnesium, which is lighter than aluminum, is finding limited application. Research is intense on composites, which combine glass or carbon fibers with a plastic binding material. Such composites are light, strong, and can be formed into complex shapes, but their high cost and long manufacturing cycle times have impeded their acceptance in high-production automobiles.

HISTORICAL PERSPECTIVE

U.S. automotive history reveals a fairly continuous improvement in both performance and fuel economy, but the relative interest in each is influenced by externalities. When gasoline is plentiful and inexpensive, the consumer is more interested in performance and/or larger vehicles, both of which tend to decrease fuel economy. During the Arab oil shocks of the 1970s, when the gasoline supply was stifled and driv-

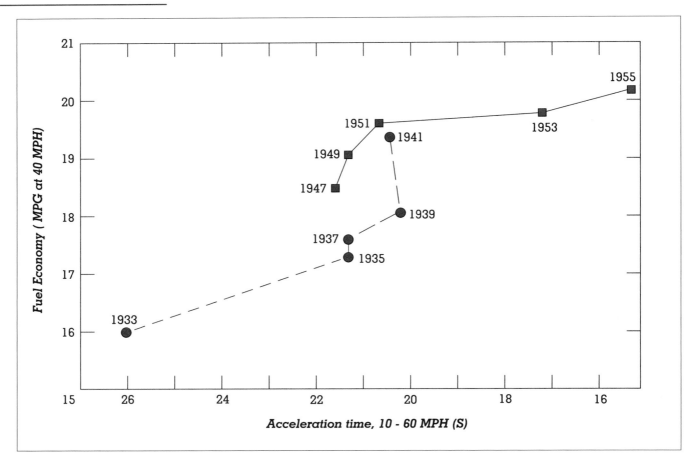

Figure 5.
Historical trend in trade-off between performance and fuel economy (average of thirty-eight car models, 1975–1995).

ers sometimes waited in line to fill their tanks, fuel economy became more important. Likewise, fuel economy received special emphasis from 1975 to 1980 as manufacturers scrambled to meet the newly imposed and rapidly rising fuel-economy standards.

In earlier days, performance was often measured in terms of acceleration time to 60 mph from an initial speed of 10 mph rather than from a standstill. Fuel economy at a steady 40 mh is plotted against that acceleration time in Figure 5. The acceleration scale is reversed so that the best-performing cars are on the right. The data are averaged over a fleet of thirty-eight models from eighteen U.S. manufacturers. The two branches of the plot represent fleets from 1933 to World War II, during which civilian car production was suspended, and from the postwar years until 1955. Over this twenty-two year span, fuel economy increased 25 percent and acceleration time was

improved 40 percent.

More recent data, from 1975 through 1995, are presented in Figure 6. The EPA classifies passenger-cars according to interior roominess. The four classes represented in Figure 6—sub compact, compact, mid size sedan, and large sedan—accounted for from 74 percent of the new passenger-car fleet in 1975 to 94 percent in 1995. The average composite fuel economy at five-year intervals is plotted for each car class against the corresponding average acceleration performance, as estimated by the EPA from vehicle installed power and test weight. In these estimates, each car is assumed to be carrying a 300-lb load. Curves on the plot indicate trends for each of the five years sampled.

The trend shown in a given year has been to offer somewhat better performance in larger cars than in smaller ones. However, the large sedan is able to carry a heavier load than a sub compact. Loading the

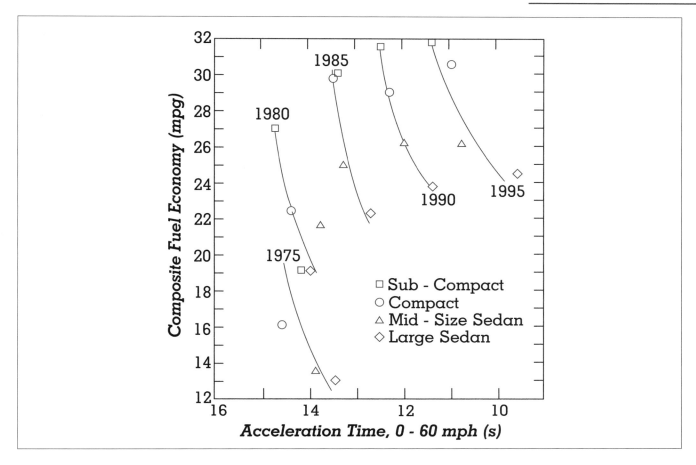

Figure 6.
Historical trend in trade-off between performance and fuel economy, segregated by car size (average for U.S. new-car fleet, 1975–1995).

large sedan more heavily, rather than with the same 300 lb as in the sub compact, would erode the apparent performance advantage of the large car.

It is seen from Figure 6 that in any given year, a fuel-economy spread of 6 to 8 mpg exists between the sub compact and the large sedan. A major contributor to this difference is the increased weight of the larger car. The average large sedan was 60 percent heavier than the average sub compact in 1955, with that spread decreasing to 30 percent in 1995.

Reduced weight has been a major contributor to the improvements in performance and fuel economy over the twenty-year span of Figure 6. From 1975 to 1980 the test weight of the average new passenger car fell nearly 25 percent. It has crept up slowly in the following fifteen years, but in 1995 was still 20 percent less than in 1975. The trend from rear-wheel to front-wheel drive aided in reducing weight. In 1975, front-wheel drive appeared in fewer than 10 percent of the new-car fleet, but its fleet penetration rose to 80 percent in 1995.

Although the average passenger-car engine produced about 10 percent more maximum power in 1995 than in 1975, it did so with approximately 40 percent less piston displacement as a result of advances in engine technology. The average 1995 engine also was lightened through greater use of non ferrous materials in its construction.

Typical passenger-car fuel economy was boosted slightly around 1980 with the entry of the light-duty diesel. A 25 percent increase in miles per gallon was claimed for that engine, about half of which resulted from the higher energy content of a gallon of diesel fuel relative to gasoline. Diesel-engine penetration of the new-passenger-car fleet peaked in 1981 at 5.9 percent, and a year later in the light-duty truck fleet at 9.3 percent. By 1985 the popularity of the diesel had waned, with penetration of the combined fleet being less than 1 percent as of the turn of the century.

In the drivetrain arena, the torque-converter automatic transmission has long dominated the U.S. passenger-car fleet, appearing in about 80 percent of new passenger cars in both 1975 and 1995. Its penetration in light-duty trucks has risen from 63 percent in 1975 to nearly 80 percent in 1995. However, the automatic transmission of 1995 was much improved over the 1975 model. In 1975, three-speed automatics and four-speed manual transmissions were the norm. By 1995 the more efficient four-speed automatic with torque-converter clutch had taken over, while manual-transmission vehicles enjoyed enhanced fuel economy through the addition of a fifth gear.

Looking back a century in the United States, when draft horses outnumbered automobiles and police enforced a speed limit of 10 mph on cars in some locales, the individual improvements that have since accrued to the automotive vehicle have transformed it into a necessity in the lives of many Americans. Gains in both performance and fuel economy have contributed importantly to its popularity. In the last quarter of the twentieth century, during which fuel economy regulation began in the United States, average new-car fuel economy has increased 80 percent at the same time that performance capability, as measured by acceleration time to 60 mph, has improved 25 percent. Both have benefited most significantly from a reduction in the weight of the average car, but over this time period the summation of effects from myriad other individual improvements has contributed even more to fuel economy. These individual improvements can be grouped into factors influencing engine efficiency, the efficiency of the drivetrain connecting the engine to the drive wheels of the vehicle, the rolling resistance of the tires, vehicle aerodynamic drag, and the power consumed by accessories either required for engine operation or desired for passenger comfort and convenience. Engine improvements can be further classified as to whether they improve the quality and control of the mixture inducted into the cylinders, increase the air capacity of the engine, improve the efficiency of the energy conversion process associated with combustion, or reduce the

parasitic losses associated with engine friction and the pumping of gas through the engine cylinder. With multiple technological improvements possible in each of these classifications, their additive effects can become significant even though the improvement attached to each individually is small. Sometimes gains are realized through interaction among separate components. For example, an improved transmission may enable the engine to run at a more efficient point in its operating range, even though the engine itself experiences no change in its overall efficiency characteristics.

Charles A. Amann

BIBLIOGRAPHY

Robert Bosch GmbH. (1986). *Automotive Handbook*, 2nd ed. Stuttgart, Germany: Author.

Burke, C. E.; Nagler, L. H.; Campbell, E. C.; Zierer, W. E.; Welch, H. L.; Lundsrom, L. C.; Kosier, T. D.; and McConnell, W. A. (1957). "Where Does All the Power Go?" *SAE Transactions* 65:713-737.

Caris, D. F.; Mitchell, B. J.; McDuffie, A. D.; and Wyczalek, F. A. (1956). "Mechanical Octanes for Higher Efficiency." *SAE Transactions* 64:77-96.

Gillespie, T. D. (1992). *Fundamentals of Vehicle Dynamics*. Warrendale, Pa: Society of Automotive Engineers.

Heavenrich, R. M., and Hellman, K. H. (1996). "Light-Duty Automotive Technology and Fuel Economy Trends Through 1996." Technical Report EPA/AA/TDSG/96-01. Ann Arbor, MI: U.S. Environmental Protection Agency.

Hoerner, S. (1965). *Fluid-Dynamic Drag*. Midland Park, NJ: Author.

Malliaris, A. C.; Hsia, H.; and Gould, H. (1976). "Concise Description of Auto Fuel Economy and Performance in Recent Model Years." SAE Paper 760045. Warrendale, PA: Society of Automotive Engineering.

McCuen, C. L. (1952). "Economic Relationship of Engine-Fuel Research." *SAE Transactions* 6:291-303.

Sovran, G.; Morel, T.; and Mason, W. T., Jr. (1978). *Aerodynamic Drag Mechanisms of Bluff Bodies and Road Vehicles*. New York: Plenum Press.

AVAILABLE ENERGY

See: Reserves and Resources

AVIATION FUEL

Aviation fuel is the fuel used to power aircraft in flight. It must satisfy the unique requirements of both the engine and the airframe of the aircraft. Currently the great majority (more than 99%) of aviation fuel used in both civil and military aircraft is jet fuel. A small quantity of aviation gasoline is still used in small aircraft. Early aircraft used motor gasoline to power their spark ignition engines because the aviation and auto worlds shared the same early engines. In recognition of aviation's more stringent requirements compared to ground transportation, separate specifications for aviation gasoline were developed after World War I. Subsequent aircraft spark ignition engine developments as World War II approached identified the need for high octane in aviation fuel for improved performance. This resulted in the development of 100-octane aviation gasoline and the unique refinery processes necessary to produce it. Beginning in the mid-1930s, research was initiated in both Great Britain and Germany on the development of a gas turbine aircraft engine, which was radically different from the spark-ignition, reciprocating engines used since the days of the Wright brothers. The new jet engine was capable of markedly improved high-speed performance. During this development, illuminating kerosene used as a fuel for lamps, was chosen as the liquid fuel for the jet engine because it did not conflict with the very strong military demand for high-octane aviation gasoline. This use for jet engines of distillate-based fuels different in composition from high-octane gasoline has continued to this day. The first operational use of jet-engine-powered aircraft occurred in a military aircraft (the German Me-262) late in World War II, and its performance proved so superior to propeller-powered, piston-engine aircraft that subsequently all air forces changed to the use of jet aircraft. The development and rapid growth of higher-speed commercial transport aircraft using jet engines began in the late 1950s. As a result of the switch of both military and commercial aircraft to jet engines from spark ignition engines, jet fuel demand rose rapidly, and jet fuel over time displaced aviation gasoline as the dominant fuel for aviation use.

FUEL TYPES

Jet fuels in use today are essentially all kerosene-based but differ somewhat in their compositions. For civil fuels, Jet A is used primarily in the United States and Jet A-1 throughout most of the rest of the world. Jet A and Jet A-1 differ principally in their freezing point, which is the temperature at which solid wax crystals form in the liquid fuel as it cools. Commercial aircraft store their fuel primarily in wing tanks, and there is a concern that during long international flights through cold-weather conditions the formation of wax could interfere with the flow of fuel from the wing tanks into the engines. Thus all jet fuels specify a freezing point suitable for its intended flight use. The military fuel used by both the U.S. Air Force and NATO air forces is JP-8, which is similar in composition to commercial Jet A-1, but employs military-designated additives. The U.S. Navy uses JP-5, a jet fuel with a higher flash point (a measure of the fire hazard associated with the fuel) than Jet A, Jet A-1, or JP-8 because of concern about fire safety aboard aircraft carriers, particularly in combat operations. In the past the U.S. Air Force used a very low flash point fuel called JP-4, composed of a mixture of kerosene and lighter-boiling refinery streams, but switched to the higher-flash-point kerosene-based JP-8 fuel to reduce combat losses and post-crash fire and handling incidents. A commercial low-flash-point fuel designated Jet B, similar to military JP-4, is used only in very cold Arctic areas because of difficulties in starting engines with the more viscous kerosene-type fuels.

SPECIFICATIONS

Jet fuel requirements are defined by engine and airframe technical needs, which are balanced against the need for a widely available and low-cost fuel. These technical and economic requirements are translated into fuel specifications that define physical properties and chemical compositional limits and that also require the fuel to pass a number of unique performance tests designed to predict satisfactory use. Jet fuel is a tightly specified, high-technology commodity. A number of commercial and military jet fuel specifications are used throughout the world. Commercial specifications include ASTM D 1655, which is an industry consensus specification; Defense Standard 91/91, issued by U.K. Ministry of Defense for their Civil Aviation Authority; and the

International Air Transport Association (IATA) Guidance Material. At major airports, where fueling systems are operated by a number of different companies rather than a single company, a combination of the most stringent requirements of ASTM D 1655, Defense Standard 91/91, and the IATA Guidance Material called the "Check List" is often used. Attempts are under way to harmonize the major commercial Western specifications and to get non-Western countries to join in using common worldwide specification and test methods. In addition, the U.S. military as well as other governments write specifications for jet fuel.

PROPERTY REQUIREMENTS

Because the jet engine was free of the demanding need for high-octane fuel, in the early days of the jet-engine development it was thought that it could use practically any liquid fuel. However, subsequent experience proved this to be untrue, as a number of potential problem areas indicated that control of fuel properties, reflecting both bulk and trace components, were important for satisfactory use. Over the years these important property requirements were translated into specification requirements that put restrictions on what is acceptable as jet fuel.

During the early jet-engine development work, it was recognized that the combustion system of the engine would be a critical component. Fuel-combustion-related properties are controlled via limits on the total concentration of aromatic-type compounds as well as the concentration of condensed ring aromatic compounds (i.e., naphthalenes). Since these types of hydrocarbon compounds tend to burn with higher levels of flame radiation than do other hydrocarbon compound types, it makes combustor wall cooling more difficult. In addition, specific combustion performance prediction tests such, as the smoke point test and the luminometer number test, were developed and added to specification requirements. Other concerns are energy content, density, and volatility. The minimum energy content of the fuel is specified for range considerations. The density of the fuel controls the weight of fuel that can be carried in a given volume. The volatility or boiling range of the fuel is controlled because it impacts on a number of properties. Lower boiling fuels are easier to use, either when starting a cold engine or when attempting to relight an engine at altitude, but early attempts to use very light

Net Heat of Combustion, MJ/kg	42.8 Min
Boiling Range, 10% Recovered, °C	205 Max
Boiling Range, Final Boiling Point, °C	300 Max
Flash Point, °C	38 Min
Density, kg/m3	775 to 840
Freezing (wax appearance) Point, °C	-40 Max

Table 1.
Important Jet Fuel Properties

fuels encountered problems with fuel boiling off from the vented wing tanks at the low pressure at higher altitudes. Subsequent experience also demonstrated the increased safety risks inherent in low-flash-point fuels in either civil or combat military use. Specifications include controls on the boiling range of the fuel as well as a flash point test measurement. The potential for harmful wax crystal formation in aircraft wing tanks at low temperatures in flight is controlled via the inclusion of a freeze point test requirement in all specifications. The viscosity of the fuel at low temperatures is limited to ensure the proper operation of fuel injection nozzles during low-temperature start-up. Problems with the stability of the fuel in storage leading to unwanted gums and deposit formation were anticipated in early fuel development work and led to restrictions on the olefin (unsaturated hydrocarbon) content of the fuel in specifications. Subsequent operational experience also discovered stability problems in flight caused by the exposure of the jet fuel to hot metal surfaces where reactions of the fuel with the dissolved oxygen in the fuel led to deposit formation in critical components such as the fuel nozzle, heat exchanger surfaces, and narrow-tolerance moving components in fuel control units. These high-temperature-thermal stability problems led to the development of tests designed to simulate the high-temperature exposure of the fuel, and all specifications require the fuel to pass such a test.

The use of high-sulfur-content fuels could enhance undesirable carbon-forming tendencies in the engine combustion chamber as well as result in higher amounts of corrosive sulfur oxides in the combustion gases. Mercaptans (a type of sulfur compound) cause odor problems and can attack some fuel system elastomers. Both the concentration of total sulfur compounds as well as the concentration of mercaptan sulfur compounds are controlled in

A Twin Otter turboprop airplane refuels at a cache before heading for the North Pole. (Corbis-Bettmann)

specifications. The corrosivity of the fuel toward metals caused by the presence of elemental sulfur or hydrogen sulfide is controlled by the use of tests such as the copper strip corrosive test. Acidic compounds present in the fuel, such as organic acids or phenols, are controlled by a total acidity test.

Another area of importance is contamination. Jet fuels are tested for the presence of heavier fuel contamination by use of an existent gum test, which detects the presence of heavier hydrocarbons from other products. Testing also is carried out to detect the presence of excessive levels of undissolved water and solids, as well as for surfactants that can adversely affect the ability of filters and coalescers to remove dirt and water from the fuel.

Additives also are used to enhance jet fuel quality in a manner similar to that of gasoline, but unlike gasoline, are tightly controlled. Only additives specifically cited in a specification can be used within allowed limits. The mandated or permitted use of additives varies somewhat in different specifications, with military fuels tending to the greater use of additives compared to civil fuels. A static dissipater additive is used in many fuels to enhance the rapid dissipation of any electrostatic charge in the fuel created by the microfiltration used for dirt removal. To prevent the formation of deleterious hydroperoxides during prolonged fuel storage, many specifications require that an antioxidant (a compound that slows down or prevents oxidation) additive be added to fuels that have been hydrotreated. This must be done because the natural antioxidants present in the fuel that were unavoidably removed.

Another additive used is a metal deactivator to chemically deactivate any catalytic metals such as copper accidentally dissolved in the fuel from metal surfaces. Uless they are chemically deactivated, dissolved metals cause the loss of good stability quality.

Corrosion inhibitor/lubricity improvement additives are used particularly in military fuel for the dual purpose of passivating metal surfaces and improving the lubricating properties of the fuel in equipment such as fuel pumps. The military also specifies the use of a fuel system icing inhibitor as an additive to prevent filter blocking by ice crystal formation, because military aircraft tend not to use fuel line filter heaters, which are standard equipment on civil aircraft.

COMPOSITION AND MANUFACTURING

Aviation turbine fuels are produced in refineries primarily using petroleum crude oil as the sole starting material. The exceptions are Canada, which uses some liquids produced from tar sands, and South Africa, which uses some liquids produced from coal. The overwhelming percentage of chemical compounds—present in jet fuel are hydrocarbon compounds, that is, compounds composed of carbon and hydrogen. These hydrocarbon compounds include branched and normal paraffins; single-ring and multiring cycloparaffins, which are also called naphthenes; and single-ring and multiring aromatics, hydroaromatics, and olefins. The distribution of hydrocarbon compound types varies considerably, primarily depending on crude source. Heteroatom compounds, which are hydrocarbon compounds that also contain sulfur, nitrogen, or oxygen, are present at trace levels, and are important because heteroatoms can have a disproportionate effect on fuel properties.

Much jet fuel is produced by simply distilling a kerosene fraction from the crude oil followed by some form of additional processing. The initial boiling points for the distillation are generally set to produce a jet fuel that meets the flash-point requirement, and the final boiling points are set to meet requirements such as freeze-point, smoke-point or naphthalene content. Jet fuel often is blended from a number of streams. In addition to simply distilled kerosene fractions, blend stocks are produced from heavier crude oil fractions or refinery product streams by breaking them down into lower-boiling fractions. The processing steps used to prepare blend stocks after distillation vary considerably, depending on factors such as crude oil type, refinery capabilities, and specification requirements. Crude oils, whose kerosene fractions are low in total sulfur content, can be chemically processed to reduce mercaptan sulfur or organic acid content. For example, a kerosene with a high organic acid content but a low total and mercaptan sulfur content can be simply treated with caustic (sodium hydroxide) to lower the acid level. Similarly, a jet fuel blend stock low in total sulfur but too high in mercaptan sulfur can be chemically treated in a so-called sweetening process, which converts the odorous mercaptan sulfur compounds into odor-free disulfide compounds. Chemical treatment is often followed by passage through both a salt drier to lower water levels, and a clay adsorption bed to remove any trace impurities still present.

Another type of processing employed is treatment in a catalytic unit with hydrogen at elevated temperatures and pressures. Catalytic processing is used, for example, when higher total sulfur levels require total sulfur removal, which cannot be achieved by chemical treatment. In addition, catalytic treatment with hydrogen, depending on the process conditions, can be used to break down heavier fractions into the kerosene range. More severe processing conditions such as higher pressures used to affect a boiling point reduction, will also generally extensively remove heteroatoms and markedly lower the level of olefins. Jet fuel normally sells at a premium compared to other distillates, and reflects the cost of crude oil.

Jet fuel is shipped in a highly complex system designed to prevent or eliminate excess water, particulates such as dirt and rust, microbial growths, and contamination from other products in the fuel being delivered into aircraft. Transportation may involve shipment in pipelines, railcars, barges, tankers, and/or trucks. Techniques employed include dedicated storage and transportation, the use of filters to remove particulates, and the use of coalescers and water-adsorbing media to remove water. The elimination of water will prevent microbial growth.

ENVIRONMENTAL ISSUES

As for all hydrocarbon fuels, the combustion of jet fuel produces carbon dioxide and water. Turbine engines are designed to be highly efficient and to produce low levels of unburned hydrocarbons and carbon monoxide. Nitrogen oxides and sulfur oxides also are emitted from the turbine engine. There is little organic nitrogen in the fuel, and oxides of nitrogen are produced from the nitrogen and oxygen in the air during the combustion process. As a result, control of nitrogen oxides emissions is essentially an engine combustor design issue. Sulfur oxides are produced from the low levels of sulfur compounds present in jet fuel during the combustion process, and thus control of sulfur oxides is essentially a fuel-related issue. Only a small fraction of all sulfur oxides emissions are produced by the combustion of jet fuel because of the relatively low use of jet fuel compared to total fossil-fuel combustion. However, jet-fuel-produced sulfur oxides emissions are unique because aircraft engines are the only source emitting these species directly into the upper troposphere and lower stratosphere an issue of growing interest to atmospheric and climate change researchers.

William F. Taylor

See also: Aircraft; Air Pollution; Air Travel; Climatic Effects; Gasoline and Additives; Gasoline Engines; Kerosene; Military Energy Use, Modern Aspects of; Transportation, Evaluation of Energy Use and.

BIBLIOGRAPHY

ASTM. (1998). "D1655 Standard Specification for Aviation Turbine Fuels." In *1998 Annual Book of ASTM Standards.* West Conshohocken, PA: Author.

Coordinating Research Council. (1983). *Handbook of Aviation Fuel Properties.* Atlanta, GA: Author.

Dukek, W. G. (1969). "Milestones in Aviation Fuels." AIAA Paper 69-779, July 14.

Dyroff, G. V., ed. (1993). "Aviation Fuels." In *Manual on Significance of Tests for Petroleum Products,* 6th ed. Philadelphia ASTM.

Smith, M. (1970). *Aviation Fuels.* Henley-on-Thames, Oxfordshire, Eng.: G. T. Foulis.

Taylor, W. F. (1997). "Jet Fuel Chemistry and Formulation." In *Aviation Fuels with Improved Fire Safety—National Research Council Proceedings.* Washington, DC: National Academy Press.

Waite, R., ed. (1995). *Manual of Aviation Fuel Quality Control Procedures.* West Conshohocken, PA: ASTM.

BATTERIES

A battery is a series of electrochemical cells. Electrochemical cells are devices that, whenever in use, can continuously and directly convert chemical energy into electrical energy.

The demand for energy from batteries is enormous. In the United States alone, battery sales exceeded $4.3 billion in 1997. Batteries have become an ubiquitous aspect of modern life, providing energy for almost every type of technology imaginable, from small-capacity devices such as watches, electronic toys, pocket radios, portable computers, cardiac pacemakers, hearing aids, and smart cards to the large-capacity applications such as motor vehicles, satellites, remote off-grid equipment, and uninterruptible backup power systems for hospitals.

Because of the ever-growing demand for a wide array of cordless electrical technology, and the markedly increased use of electronic devices and information technologies using integrated circuits, battery technology advanced remarkably in the latter half of the twentieth century. In particular, major advances in construction technology and new materials applications have resulted in smaller yet more powerful batteries. It has largely been the result of new combinations of more refined electrochemical materials that have made batteries lighter, thinner, and more efficient. Probably the most noteworthy advances have been alkaline batteries for consumer products, lithium rechargeable batteries for electronic technology, and nickel-hydrogen batteries for use in spacecraft.

These advances are likely to continue as industry and government liberally invest in battery research and development. Spending priorities of industry and government are far different, however. Whereas the majority of spending by industry is focused on improving battery technology for communication and computer technology, the majority of government-funded research is for military, spacecraft, transportation, and distributed energy applications.

BATTERY BASICS

All flashlight batteries, button batteries, compact rechargeable batteries and vehicle storage batteries operate under the same basic principles. An electrochemical cell is constructed of two chemicals with different electron-attracting capabilities. Called an electrochemical couple, these two chemicals, immersed in an electrolyte (material that carries the flow of energy between electrodes), are connected to each other through an external circuit.

The chemical process that produces an electrical current from chemical energy is called an oxidation-reduction reaction. The oxidation-reduction reaction in a battery involves the loss of electrons by one compound (oxidation) and the gain of electrons (reduction) by another compound. Electrons are released from one part of the battery and the external circuit allows the electrons to flow from that part to another part of the battery. In any battery, current flows from the anode to the cathode. The anode is the electrode where positive current enters the device, which means it releases electrons to the external circuit. The cathode, or "positive" terminal of the battery, is where positive current leaves the device, which means this is where external electrons are taken from the external circuit.

To generate electrical energy from chemical energy, the reactant molecules must be separated into oppositely charged ions (electrically charged atoms or groups of atoms) and the exothermic reaction (gives off heat) between the two reactants requires the proper conditions and hardware to drive an electric current through an external circuit. An electrolyte, which can be a free liquid, a paste, or even a rigid solid, is the ionic conductor serving as the charge transfer medium.

The ions move between electrodes in the electrolyte due to voltage potential gradients. The velocity of these chemical currents increases with temperature. Hence, electrolytic conductivity increases as temperature goes up. This is the opposite of electrical currents in metallic conductors, which increase as the temperature goes down.

The external circuit provides a path for the excess electrons at the negative electrode to move toward the positive electrode. Although the flow of electrons would seem to cancel out the two charges, chemical processes in the battery build the charges up as fast as the depletion rate. When the electrodes can no longer pass electrons between them, the battery "dies."

During charging, energy from electric current is stored on the plates of the cell. Substances on these plates are converted to new substances taking on a higher energy. Then when the battery is connected to any load with electrical resistance, the substances on the battery plates begin to retransform themselves back to the earlier state, producing electricity as a result of the electrochemical reactions.

The quantity of electric charge is measured in coulombs, and the unit of electric current—the number of coulombs per second that go past any point—is the ampere (A), named after French physicist André Marie Ampère:

$$1 \text{ ampere} = 1 \text{ coulomb/second}$$

Batteries are rated by ampere-hours, which is the total amount of charge that can be delivered. More specifically, it is a measure of the number of electrons that can be released at the anode and accepted at the cathode. It is also related to the energy content or capacity of the battery. If a battery has a capacity of 90 A-h, it can supply a current of 90 A for one hour, 30 A for three hours, or 1 A for 90 hours. The greater the current, the shorter the period the battery can supply it; likewise, the less the current, the longer the period the battery can supply it.

A battery produces a potential difference between its two electrodes that is brought about by a decrease in potential energy as one coulomb moves from the negative to positive electrode. Measured in volts, it is equal to the energy difference per unit charge. Voltage measures the "push" behind a current and is determined by the oxidation-reduction reactions that take place inside each cell. The better the oxidation-reduction characteristics, the higher the potential difference and the greater the current.

There is a huge range of cell capacities. At one extreme are miniature batteries providing milliamp-hours; at the other extreme are huge submarine batteries that can provide five million watt-hours. Battery cell voltages for any single cells fall within in a narrow range of one to two volts; cells are stacked in series to get higher voltages. The connection is always negative electrode to positive electrode so that each battery supplies more push to the electrons flowing through them. A typical golf cart uses three or four lead-acid batteries in series, the 1995 General Motors Impact electric car uses thirty-two lead-acid batteries in series, and the largest utility battery storage system, California Edison's substation at Chino, California connects hundreds of lead-acid batteries in series. In all, this substation houses 8,000 batteries capable of deep discharge (near-complete discharge without hampering the battery's capacity to be recharged), providing 40,000 kilowatt-hours of storage.

Another important concept for batteries is resistance. Expressed in ohms, resistance is what limits current. For example, if a 12V car battery were connected to a circuit with 4 ohms resistance, the current would be 3 A. If the battery had a capacity of 90 A-h, it would supply 3 A for 30 hours. (Resistance of 4 ohms here refers to external resistance of the battery and the headlight circuit.) The greater the resistance of the circuit and cell itself, the less the current for a given applied voltage.

The current from a cell is determined by the voltage level of the cell and the resistance of the total circuit, including that of the cell itself. If large amounts of current are needed to flow in the circuit, a low-resistance cell and circuit are required. The total area of the electrodes determine the maximum current. The amount of voltage that a battery can actually deliver is always less than the battery's advertised maximum deliverable voltage. That is because once current begins, cell voltage decreases because of the slowness of the chemical process at the electrodes and the voltage drop across the internal resistance of the cell.

Ohm's law, named after German physicist Georg Simon Ohm, explains the relationship between cur-

HOW BAD BATTERIES LEAD TO BETTER EFFICIENCY

The pace of innovations in batteries has not matched the remarkably fast evolution in the development of portable electronic technology. Batteries are widely regarded as the weak point slowing the rapid innovations in portable technology. Chief among consumer complaints are that batteries do not last long enough and that they do not last as long as the manufacturer claims they will.

Portable electronic manufacturers had the foresight to realize that battery technology was not keeping up with innovation in electronic technology, and it probably never would. To satisfy consumer demand for smaller and more powerful electronic devices that could go a longer time between charges, improvements in the energy efficiency of the devices themselves were required.

Consumer preference to "unplug" from the wall has been responsible for making all electronic equipment more energy-stingy. The growing demand for portable energy for ever-smaller phones, laptop computers, and DVD players has been the driving force behind the tremendous energy efficiency improvements in electronic equipment. Although the market for laptops, notebooks and palm-size computers is only a small fraction of what it is for desktops, much of the smart energy-saving electronics developed for these devices have been incorporated in desktop computers. In the United States, companies and consumers are saving millions of dollars in energy costs since new computers, mostly replacement systems, are being plugged in at a rate of over thirty million a year. Just as importantly, society receives the environmental benefit of reducing the need for additional electric power plants.

This rapid pace of innovation in electronics will continue. With each advance, electronic circuits become smaller, more sophisticated, more reliable, and ever more energy-stingy. In 2000, Transmeta Corporation introduced the Crusoe microprocessor that is specifically designed as a low-power option for notebooks and Internet appliances. Transmeta claims that a notebook equipped with a 500 to 700 MHz clock speed Crusoe chip requires slightly more than 1 watt of power on average—far less power than the average microprocessor's 6 to 10 watts. The chip garners additional savings since it emits less heat, eliminating the need for a cooling fan.

The Crusoe microprocessor and similar microprocessors coming from other chip manufacturers mark a revolutionary advance: all-day computing with full PC capabilities from a single smaller and lighter battery. As these more energy-stingy microprocessors migrate to the "plugged in" office world, it will help slow the rate of electricity demand for office equipment.

rent, potential difference, and resistance in a given metallic conductor: voltage = current × resistance (when voltage and current are constant). In other words, current depends on resistance and voltage applied. By simply altering the resistance of a battery, it can be designed for either fast current drain (low resistance) or slow current drain (high resistance) operation.

To ensure the safety and reliability of battery-using technology, the American National Standards Institute (ANSI), the International Electrotechnical Commission (IEC) and the International Standards Organization (ISO) have developed standards for battery sizes, voltages, and amperages. These standards are now accepted on a world-wide basis.

TYPES OF BATTERIES

There are two major types of electrochemical cells: primary batteries and secondary, or storage, batteries. Primary battery construction allows for only one continuous or intermittent discharge; secondary battery construction, on the other hand, allows for recharging as well. Since the charging process is the reverse of the discharge process, the electrode reactions in secondary batteries must be reversible.

Primary batteries include dry, wet, and solid electrolyte. The term "dry cell" is actually a misnomer, for dry cells contain a liquid electrolyte, yet the electrolyte is sealed in such a way that no free liquid is present. These cells are sealed to prevent damage to persons or materials that would result from seepage of the electrolyte or reaction products. Dry cells are typically found in products like flashlights, radios, and cameras. Wet cells use a free-flowing aqueous electrolyte. Because of the large capacity and moderately high currents of wet cells, the primary uses are for signal systems for highway, railway and marine applications. Solid electrolyte batteries use electrolytes of crystalline salts. Designed for very slow drain long-term operations, they are popular for certain kinds of electronic devices.

Secondary batteries consist of a series of electrochemical cells. The most popular types are the lead-acid type used for starting, lighting, and electrical systems in motor vehicles and the small rechargeable batteries used in laptops, camcorders, digital phones, and portable electronic appliances.

Because of the great diversity of applications requiring batteries, and the very distinct requirements for those applications, there is an array of batteries on the market in a wide variety of materials, sizes and capacities. Some estimates put the number of battery varieties at over 7,000. Decisions about batteries are not simple for battery manufacturers, designers of technology, or consumers. Many factors must be taken into account in choosing a battery. For any given application, important criteria may include longer life, high voltage, fast-draining, durability, lower weight, salvageability, recharging speed and control, number of charge-and-discharge cycles, and lower cost (material and construction costs).

Manufacturers often feel compelled to be very vague about how well a battery will perform because of the uncertainty about how the battery will be used. Technology manufacturers invest heavily in marketing research before deciding what type of battery to provide with a new product. Sometimes it is a clear-cut decision; other times there is no one best choice for all potential users. Perhaps the most important issue is whether a product will be used continuously or intermittently. For instance, selecting a battery for a DVD player is much easier than a cellular phone. The DVD player is likely to be used continuously by the vast majority of buyers; a cellular phone will be used very differently in a wide mix of continuous and intermittent uses.

The life span of a battery is also dependent on the power needs of a product. Some batteries produce a lot of power for a relatively short time before fully discharging, while others are designed to provide less peak power but more sustained power for a very low drain rate. Other important variables affecting battery lifespan are the design and efficiency of the device being used and the conditions of use. In particular, exposing a battery to excessive heat can dramatically curtail the length of a battery's life.

Zinc-Manganese Dioxide Primary Dry Cells

Zinc-manganese dioxide cells are the most prominent commercial battery. Over ten billion are manufactured each year for electronic equipment, camera flashes, flashlights, and toys. Initial voltage for such a cell starts in the range of 1.58 to 1.7 volts, and its useful life ends after voltage declines to about 0.8 volt. The three types are the zinc-carbon cell, the zinc chloride cell and the alkaline cell. The zinc-carbon cell was invented by the French engineer Georges Leclanché in 1866, and thus also referred to as the Leclanché cell. Because zinc and manganese were readily available at a low cost, the Leclanché cell immediately became a commercial success. It remains the least expensive general-purpose dry cell and still enjoys worldwide use.

Figure 1 shows a cross section of a Leclanché cell battery. The outer casing, or "cup," is the anode (negative electrode). It is primarily made of zinc, but also may include small amounts of lead, cadmium, and mercury in the alloy. Because of the environmental hazards of mercury, manufacturers in the late twentieth century dramatically reduced or eliminated mercury in the zinc alloy used in disposable batteries by developing better zinc alloys, reducing impurities in the electrolyte, and adding corrosion inhibitors. The moist paste electrolyte consists of ammonium chloride, zinc chloride, and manganese dioxide. Formed around the porous carbon cathode (positive electrode) rod is a pasty mixture of manganese dioxide and carbon black.

The zinc chloride cell, which was first patented in 1899, is actually an adaptation of the Leclanché cell. The major innovation was the development of plastic seals that permitted the replacement of ammonium chloride in the electrolyte.

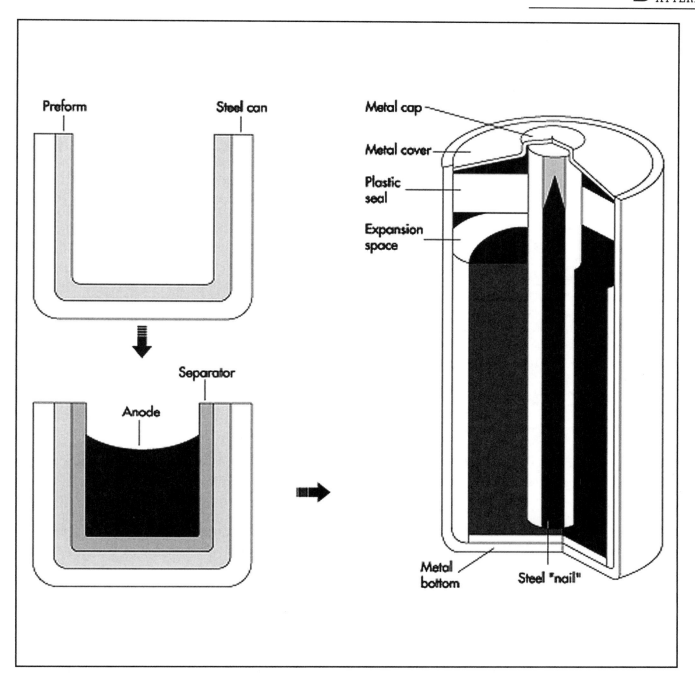

Figure 1.
Cross-section of a Leclanché cell battery. (Gale Group, Inc.)

The alkaline cell appeared commercially in the 1950s. It differs from the other zinc-manganese dioxide cells in that the electrolyte is alkaline, which primarily reduces corrosion and allows for a much longer battery life. Although these batteries look quite simple, manufacturing high-performance alkaline batteries is tedious. Many contain over twenty-

five different components and materials and require up to forty manufacturing steps, which makes them more expensive than standard batteries.

Alkaline cells offer the highest energy density (more energy per given volume) of any zinc-manganese dioxide cell, and the manufacturers continue to improve on performance. In 1998, Duracell introduced the Duracell Ultra, which they claimed will last 50 percent longer than the standard Duracell. And Eveready claims the Energizer Advanced Formula of 1998 provides 100 percent better performance than the standard Energizer of 1996.

Compact Alkaline Rechargeable Batteries

A variety of different rechargeable batteries provides the power for camcorders, laptops, digital phones and other portable electronic equipment. The following are some of the most common batteries used for these applications.

Nickel-Cadmium Cells. Nickel (hydroxide)-cadmium, or Nicad, is the most common rechargeable battery. These batteries are very durable and work well in extreme temperatures. Because the sealed cells use "jelly roll" electrodes that minimize resistance, high current can be delivered efficiently, which allows for quick recharging and works well for technology requiring high current. Other advantages of Nicad batteries are the better tolerance to overcharging and overdischarging and the ability to withstand 700 to 750 charge-and-discharge cycles.

However, Nicad batteries have some major drawbacks. In comparison to primary batteries and lead-acid batteries, Nicad batteries are heavier and have a relatively low energy density. They also require special attention from users to ensure full discharge each cycle before recharging to prevent the capacity-lowering memory effect (hysteresis) that also shortens battery life. When recharging a Nicad battery that is only partially discharged, chemical compounds around the cadmium anode are created that interfere with recharging. If a three-hour Nicad battery is repeatedly recharged after operating only one hour, the memory effect will set in and the battery will last for only one hour before it needs to be recharged again. Sometimes the memory effect can be erased by fully discharging the battery before recharging. The best way to prevent the memory effect is to closely monitor battery use and carry a backup battery. This makes it possible to fully drain each Nicad battery before recharging, and operate the laptop or camcorder for extended periods.

Nickel-metal hydride is a popular alternative to Nicad batteries since they are capable of operating 75 percent longer after each charge, are less likely to suffer memory effects, and pose less of an environmental disposal problem. The difference between nickel-metal and Nicad batteries is that the negative plate in sealed Nicad cells is replaced in the nickel-metal battery with hydrogen absorbed in a metal alloy. Hydrogen-absorbing alloys, which provide a reversible hydrogen sink that can operate at ordinary temperatures and pressures, were first used in the late 1960s in hydrogen-nickel oxide secondary cells for communication equipment and satellites. Nickel-metal hydride batteries are generally more expensive and offer a more uncertain life expectancy of between 500 and 1,000 charge-and-discharge cycles.

Lithium-Ion Cells. Lithium-ion cells and the newer alternative, lithium-ion-polymer, can usually run much longer on a charge than comparable-size Nicad and nickel-metal hydride batteries. "Usually" is the key word here since it depends on the battery's application. If the product using the battery requires low levels of sustained current, the lithium battery will perform very well; however, for high-power technology, lithium cells do not perform as well as Nicad or nickel-metal hydride batteries.

Because of lithium's low density and high standard potential difference (good oxidation reduction characteristics), cells using lithium at the anode have a very high energy density relative to lead, nickel and even zinc. Its high cost limits use to the more sophisticated and expensive electronic equipment.

The lithium-ion-polymer battery, which uses a cathode that contains lithium instead of cobalt, is likely to eventually replace lithium-ion. Lithium-ion-polymer batteries boast a longer life expectancy (over 500 charge-and-discharge cycles as opposed to around 400), much more versatility (they are flat and flexible and can be cut to fit almost any shape), and better safety (far less likely to vent flames while recharging).

For "smart" cards, micro-robots and small precision instruments, thin laminated micro-cells are being developed. Some of these developmental thin-film devices—using an electrolyte of lithium, a copper cathode, and lithium again for the electrode—can charge and discharge up to 3 volts, and can be expected to tolerate up to 1,000 charge-and-discharge cycles.

Rechargeable Alkaline Manganese Cells. Rechargeable alkaline manganese cells began to reach the market in the 1990s. Compared to Nicad batteries, these batteries are less expensive, weigh less (because no heavy metals are used), and boast higher energy density. The major disadvantage is its more limited life, particularly if deeply discharged.

The major innovation that led to alkaline manganese rechargeable batteries was the ability to prevent discharge of manganese dioxide beyond the 1 electron level. This has been accomplished by either limiting discharge to 0.9 volt or by using a zinc-limited anode. Other innovations that made alkaline rechargeables possible were the development of a special porous zinc anode and the development of a laminated separator that cannot be penetrated by zinc.

Button Primary Cells

The button cells that provide the energy for watches, electronic calculators, hearing aids, and pacemakers are commonly alkaline systems of the silver oxide-zinc or mercuric oxide-zinc variety. These alkaline systems provide a very high energy density, approximately four times greater than that of the alkaline zinc-manganese dioxide battery.

Lithium batteries began to replace mercury-zinc batteries in the late 1970s since lithium is the lightest of all the metals and loses electrons easily. Before the introduction of lithium batteries, pacemakers usually failed within one to two years. Since the pulse generator could not be sealed tightly enough, body fluids leaking in caused an early short circuit, significantly shortening the life expectancy of five years. Small and lightweight lithium-iodine batteries, which remain most common, dramatically cut the weight of the pulse generator to about 1 ounce for battery and circuitry combined. This extended the generator's life expectancy to almost ten years.

Lithium primary cells have also been introduced up to the AA-size. These "voltage compatible" cells have a working voltage of 1.5 V, and can deliver up to four times the energy of a comparable alkaline cell.

Lead-Acid Secondary Cells

Lead-acid batteries have been a workhorse battery for nearly a century. Over 300 million units are manufactured each year. One of the early uses of lead-acid batteries was for electric cars early in the twentieth century. The development of the electric starter made it possible for battery power to replace hand crank power for starting automobiles in the 1920s. First-generation portable radios with vacuum tubes and discrete transistors, which were critical for military operations, used very large lead acid batteries. The development of very large-scale integrated circuits made it possible to vastly reduce the size of these large and bulky portable radios and use much smaller batteries, since energy needs also were significantly reduced.

SMART BATTERY TECHNOLOGY

Smart battery technology started appearing on the market in the mid-1990s, mostly in high-end electronic products such as laptop computers and camcorders. Sensors send temperature, current output, and battery voltage data to microprocessors, which prevent improper charging. These features are helpful to users who do not care to take the time to regularly monitor their batteries; for those who do, they offer a way to eliminate the guesswork from recharging. It also reduces anxiety for the user and lessens the need for emergency backup batteries. For those relying on off-grid solar or wind energy systems, smart battery technology also provides critical monitoring for the times when the owner of the equipment is away from the site.

Smart batteries have been aided by tremendous advances in smart electronic circuitry. Found in everything from laptops to pacemakers, this circuitry has improved the operating time of everything from pacemakers to laptops. In pacemakers, this circuitry not only senses a patient's own heartbeat, and fires only when necessary, but also senses other body signals such as rises in body temperature and respiratory rate from exercise and respond by increasing the heart rate to meet the body's changing needs. In laptop computers, smart electronic circuitry entails power saving functions that shut off the monitor and put the computer processor in a sleep mode when not being used. Eventually most electronic technology probably will come with specialized computer chips devoted to power management, monitoring charge levels and optimizing the charging process.

Lead-acid batteries remain popular because of their capability to service high and low current demand, produce high voltage, provide capacity up to 100 A-h, and recharge well. Moreover, the lead-acid battery has important material and construction advantages, such as simple fabrication of lead components, the low cost of materials (lead is abundant and much less expensive than nickel or lithium), and excellent salvageability because of lead's low melting point.

The lead-acid battery is constructed of lead plates or grids. The negative electrode grid is coated in lead and other additives such as calcium lignosulfate, and the positive electrode grid is coated in lead oxide. Between them is the electrolyte of sulfuric acid. During discharge, the lead dioxide at the positive electrode combines with the electrolyte to produce lead sulfate and water. Meanwhile, the lead at the negative electrode combines with the sulfuric acid to produce lead sulfate and hydrogen, which replaces the hydrogen consumed at the positive electrode. The reversible reaction:

$$Pb + PbO_2 + 2\,H_2SO_4 \rightleftarrows$$
$$2\,PbSO_4 + 2\,H_2O + \text{electrical energy}$$

When the above reaction is read from right to left, it reflects what happens when the cell is discharging; read from left to right, it represents what happens when the cell is charging.

A 12-V storage battery, the type in automobiles, consists of six 2-V cells connected in series. Although the theoretical potential of a standard lead-acid cell is 1.92 volts, the actual potential can be raised to 2 V by properly adjusting the concentrations of the reactants. These batteries provide high current, and are designed to tolerate thousands of shallow-depth discharges over a period of several years. Traditionally, for automobiles and trucks, three or six cells have been connected in series to make a 6- or 12-volt battery, respectively. Manufacturers are likely to begin equipping automobiles and trucks with higher voltage batteries to accommodate increased electricity requirements and to improve the efficiency of air conditioning systems.

Unlike the automobile-type battery that is quite portable, the stationary lead-acid batteries that provide uninterruptible power to hospitals and other important facilities are not. Some may weigh over several tons because of the much heavier grid structure and other features to extend life expectancy and improve deep discharge capabilities.

Weight is usually inconsequential for stationary applications but of foremost concern for electric vehicles. Vehicle weight is detrimental in transportation because the greater the vehicle weight, the greater the energy consumed overcoming resistance. In fact, reductions in vehicle weight was one of the major rea-

sons for the doubling of the fuel efficiency for the U.S. vehicle fleet from 1973 to 1993. Electric vehicles also present other concerns for battery makers such as charge time and vehicle range on a single charge.

ADVANCED BATTERIES FOR ELECTRIC VEHICLES

To hasten development of batteries for electric vehicles, Chrysler, Ford, and General Motors formed the U.S. Advanced Battery Consortium (USABC). In 1991 USABC, battery manufacturers, the Electric Power Research Institute (EPRI), and the U.S. Department of Energy (DOE) launched a joint research effort to identify, develop and license promising battery technology for electric vehicles—vehicles with the range, performance and similar costs of gasoline-powered vehicles.

Whatever the battery construction and mix of materials being considered, the technical challenges facing developers are

1. Improve electrolyte material, using better conductors that are still chemically compatible with the electrodes.
2. Reduce the amount of change in the electrolyte and electrodes per charge-and-discharge cycle to extend life expectancy.
3. Improve the interface of the electrodes with the electrolyte by enlarging the effective surface area of the electrodes.
4. Identify electrode and electrolyte material that is inexpensive and readily available in order to achieve a low-cost battery.
5. Improve the energy density in terms of unit mass, unit volume, or both.

Advanced lead-acid batteries have resulted in higher energy density, longer life expectancy, and weight reductions. Energy density has improved by making use of the "starved" electrolyte concept in which a fiberglass and polyethylene separator immobilizes the electrolyte, allowing for the passage of oxygen from the anode to the cathode where it recombines with hydrogen to form water. Besides improving energy density, this concept also eliminated the need to add water and vent hydrogen. Weight reduction has been achieved by replacing the electrode grids of lead-antimony and lead-arsenic with lead-cadmium. Lead-cadmium also has extend the life expectancy.

To further reduce weight and improve energy density, several companies are developing thin lead film electrodes in a spiral-wound construction with glass fiber separators. Already on the market for cordless electric tools, this battery technology may eventually be used in electric vehicles.

However, since the chemical reactions for lead-acid batteries cannot sustain high current or voltage continuously during charge, fast charging remains a major problem. Charging has to be deliberate because the lead grids heat up quickly and cool slowly. When excessive heating or "gassing" (liquids inside the battery becoming a gas) occurs during charging, hydrogen is released from the battery's vent cap, which reduces battery performance and life.

Other Advanced Batteries

Despite the many virtues of lead-acid batteries, there are major limitations: range, long recharge times, and more frequent replacement than most potential electric vehicle buyers are likely to accept. Much research is taking place to develop alternatives that store more energy per given volume and mass, cost less to manufacture and operate, weigh less, and last longer. Some of the most promising materials include sodium-sulfur, zinc-bromine, nickel, and lithium.

Sodium-Sulfur Batteries. The sodium-sulfur battery consists of molten sodium at the anode, molten sulfur at the cathode, and a solid electrolyte of a material that allows for the passage of sodium only. For the solid electrolyte to be sufficiently conductive and to keep the sodium and sulfur in a liquid state, sodium-sulfur cells must operate at 300°C to 350°C (570°F to 660°F). There has been great interest in this technology because sodium and sulfur are widely available and inexpensive, and each cell can deliver up to 2.3 volts.

Though sodium-sulfur batteries have been under development for many years, major problems still exists with material stability. It is likely that the first commercial uses of this battery will not be for electric vehicles. Sodium-sulfur storage batteries may be more well-suited for hybrid electric vehicles or as part of a distributed energy resources system to provide power in remote areas or to help meet municipal peak power requirements.

Zinc-Bromide. Unlike sodium-sulfur batteries, zinc-bromide batteries operate at ordinary temperatures. Although they use low-cost, readily available

components, these batteries are much more complex than lead-acid and sodium-sulfur batteries because they include pumps and reservoirs to store and circulate the zinc-bromide electrolyte within the cell.

Nickel-Hydrogen, Nickel-Iron, and Nickel-Metal Hydride. First developed for communication satellites in the early 1970s, nickel-hydrogen batteries are durable, require low maintenance, and have a long life expectancy. The major disadvantage is the high initial cost. For these batteries to be a viable option for electric vehicles, mass production techniques will have to be developed to reduce the cost.

A more appropriate battery for transportation applications is probably a nickel-iron or nickel-metal hydride battery. These batteries are not as susceptible to heat and gassing as lead-acid batteries, so they can better withstand high current or high voltage charges that can dramatically shorten charging time.

In the mid-1990s, Chrysler and EPRI developed engineering prototypes of an electric minivan called the TEVan using a nickel-iron battery. Compared to a lead-acid battery, this battery was lighter, stored more energy, and lasted twice as long, resulting in a vehicle range of about 60 to 100 miles and a top speed of 65 mph. However, nickel-iron are much more expensive than lead-acid batteries and about half as energy efficient due to higher internal resistance. (This is not just the battery, but the battery-charger-vehicle combination according to Model Year 1999 EPA Fuel Economy Guide for vehicles offering both battery types.) In warmer climates, nickel-metal hydride energy efficiency drops even farther since there is a need to expend battery capacity cooling the battery.

In the late 1990s, Toyota began leasing the RAV4-EV sports utility vehicle using twenty-four nickel metal hydride batteries. Rated at 288 volts, these vehicles achieved a top speed of 78 mph, and a combined city/highway driving range of around 125 miles. The electric RAV4 weighed over 500 pounds more than the internal combustion engine version because the batteries weighed over 900 pounds. Nickel-metal hydride batteries are also proving to be the technology of choice for hybrid vehicles such as the Honda Insight because high current and voltage can be delivered very quickly and efficiently for the initial drive-away and during acceleration.

Lithium-Ion and Lithium Polymer. Major

research efforts are taking place in the United States, Japan, and Europe to develop lithium-ion technology for potential use in electric vehicles. Sony Corporation, which was the first electronics manufacturer to offer lithium batteries for top-end video cameras, teamed up with Nissan in the late 1990s to develop the lithium-ion batteries for the Altra electric vehicle. Lithium-ion batteries in electric vehicles offer a higher power-to-weight ratio for acceleration and a much better energy-to-weight ratio for extending range. Nissan claims these batteries have an energy density three times that of conventional lead-acid batteries. The Altra can attain a top speed of 75 mph and a between-charges range of 120 miles in combined city and highway driving. In the future, it is likely that the conventional non-liquid electrolyte will be replaced with polymer electrolytes to permit the fabrication of solid-state cells.

John Zumerchik

See also: Electric Vehicles; Fuel Cells; Fuel Cell Vehicles; Office Equipment; Storage; Storage Technology.

BIBLIOGRAPHY

Gray, F. (1991). *Solid Polymer Electrolytes: Fundamentals and Technological Applications.* Cambridge, MA: VCH.

Ingersoll, J. (1991). "Energy Storage Systems." In *The Energy Sourcebook,* ed. R. Howes and A. Fainberg. New York: AIP.

Kassakian, J. G.; Wolf, H.-C.; Miller, J. M.; and Hurton, C. (1996). "Automotive Electrical Systems Circa 2005." *IEEE Spectrum* 33 (8) :22–27.

Linden, D. (1995). *Handbook of Batteries,* 2nd ed. New York: McGraw-Hill.

Pistoia, G. (1994). *Lithium Batteries.* Amsterdam: Elsevier.

Scrosati, B., and Vincent, C. (1997). *Modern Batteries,* 2nd ed. London: Arnold.

BEARINGS

A world without bearings would look far different. Although rarely noticed, bearings are ubiquitous: computers have a few, electric appliances contain several, and an automobile has hundreds. Without bearings, much of the motion that is taken for granted would not be possible.

Since the first wheel was invented, people understood that it takes less effort to move an object on rollers than to simply slide it over a surface. People later discovered that lubrication also reduces the effort to side objects. Bearings combine these two basic findings to provide rolling motion necessary for things as simple as a tiny in-line skate wheel and as complex and large as a steam turbine. Bearings save energy, which is otherwise required to counteract friction arising from any elements related rotation, the better the bearing, the greater the energy savings.

TYPES OF BEARINGS

Bearings can be divided into rolling element bearings and sleeve (or plain) bearings. Sleeve (or plain) bearings consist of many sizes, shapes, and types, each of which functions essentially as a band (or sleeve) of close fitting material that encloses and supports a moving member. The sleeve is usually stationary and is called the bearing. The moving member is generally called the journal. Rolling element bearings are generally constructed with one or two rows of steel balls or rollers positioned between inner and outer rings. A retainer is used to equally space these rolling elements. Grooves, or raceways, are cut into the inner and outer rings to guide the rolling elements.

Rolling bearings have a lot of advantages compared with the sleeve bearings. Just to name a few: lower energy consumption, lower starting moment, lower friction at all speeds, and higher reliability. This is why there is a larger variety of rolling bearings than sleeve bearings. The variety of applications calls for the multitude of the rolling bearing types and designs. There are simple applications such as bicycles, in-line skates and electric motors. There are also complex applications such as aircraft gas turbines, rolling mills, dental drill assemblies, gyroscopes and power transmission products. Better automobile transmissions, containing hundred of bearings and delivering more mechanical energy to the wheels, have resulted in dramatic improvement in fuel economy and performance from 1975 to 2000. In delivering more mechanical energy to the wheels, and much of that improvement can be attributed to better bearings.

Rolling bearings can be further classified into ball and roller bearings. The following are some of the very common types of rolling bearings.

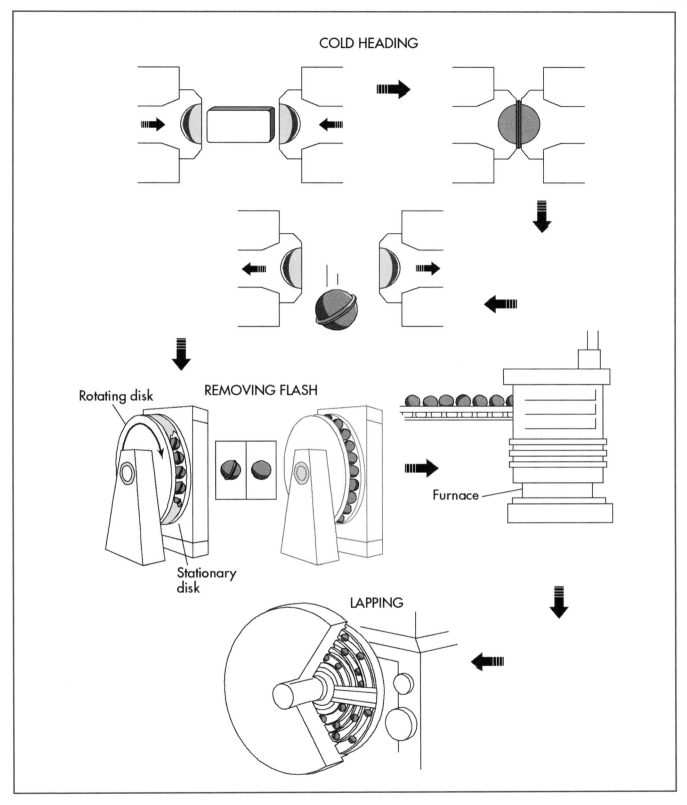

Components: (1) inner race, outer race, cage, and ball; (2) completed ball bearing. Manufacturing process: (1) cold heading—blank steel compressing between mold which forms ball with flash; (2) removing flash—balls are in machine with rotating disk and stationary disk, then balls enter furnace; (3) lapping—balls get polished. (Gale Group)

Bearing type	Advantages	Cost/ Performance	Applications
Deep grove ball bearing	high speed and high precision, average radial and thrust load	excellent	automobiles, cutting tools, water pumps, agricultural machinery
Self-aligning ball bearings	supporting radial and thrust load where shaft and housing are subject to misalignment	excellent	rubber mixers, oil field drilling rigs, vertical pumps
Angular contact ball bearings	average speed and support of radial and thrust load	fair	orbital sanders, food-processing machinery
Thrust ball bearings	support only thrust load	good	automble clutches, gauges and instruments
Cylindrical roller bearings	low speed and heavy load, but only support radial load	excellent	tractors, machine tools, mid- and small-size motors
Needle roller bearings	support of radial load where radial dimension is limited	good	oil pumps, harvester combines
Spherical roller bearings	support of radial and thrust load, expecially when the shaft is long	excellent	paper mill machinery, air compressors, speed reducers, cranes

Table 1.
Comparison of Some Ball Bearing Types

Deep Groove Ball Bearings

Deep groove ball bearings, widely found in automobile applications, are the most popular of all rolling bearings. They are available in single and double row designs. The single row bearings are also available in a sealed version. They are simple in construction as well as easy to operate and maintain. They can run at high speeds and can support both radial and thrust loads imposed by rotating shaft and other moving objects. They are versatile, quiet, lubricated-for-life, and maintenance-free. The bearing cost/performance ratio for deep groove ball bearings is excellent. They are widely found in automobile applications.

Self-Aligning Ball Bearings

Commonly found in vertical pumps, self-aligning ball bearings have two rows of balls with a common spherical outer ring raceway. This feature gives the bearings their self-aligning property, permitting angular misalignment of the shaft with respect to the housing. Self-aligning ball bearings show very low vibration and noise level owing to the high accuracy of form and smoothness of the raceways.

Angular Contact Ball Bearings

Angular contact ball bearings are available in single and double row designs as well as four-point contact ball bearings. They are designed for a combined load and provide stiff bearing arrangements. Angular contact ball bearings have raceways in the inner and outer rings, which are displaced with respect to each other in the direction of the bearing axis. This means that they are particularly suitable for the accommodation of combined loads (i.e., simultaneously acting radial and thrust loads such as for orbital sanders). The benefits are high-load carrying capacity and speed capability, low operating temperatures, long relubrication intervals and quiet operation.

Thrust Ball Bearings

Thrust ball bearings are manufactured in single direction and double direction designs. They are only able to accept thrust loads but can be operated at relatively high speeds. Mounting is simple because the various bearing components (shaft washer, housing washer, ball and cage thrust assembly) can be installed separately. The benefits of using thrust ball bearings derive from their high running accuracy and high load carrying capacity, which is why they are used in automobile clutches and speed reducers.

Cylindrical Roller Bearings

Cylindrical roller bearings, found in tractor and machine tools, can carry heavy radial loads at high speeds because the rollers and raceway are in linear contact. Single row bearings have optimized internal geometry that increases their radial and thrust load carrying capacity, reduces their sensitivity to misalignment, and facilitates their lubrication. Full complement bearings incorporate the maximum number of rollers and have no cage, and are intended for very heavy loads and moderate speeds.

Needle Roller Bearings

Needle roller bearings can support heavy radial load such as clutches. A wide variety of designs, including bearings for combined radial and thrust loads, provide simple, compact and economic bearing arrangements. Their small sectional height makes them suitable for limited radial space of the housing.

Spherical Roller Bearings

Spherical roller bearings are robust, self-aligning bearings that are insensitive to angular misalignment. They offer high reliability and long life even under difficult operating conditions. They are mounted on an adapter assembly or withdrawal sleeve and housed in plummer blocks. They are also available with seals for maintenance-free operation.

Tapered Roller Bearings

Tapered roller bearings are designed for heavy combined loads or impact loading such as freight train locomotives and rail cars. Composed of a cup, a cone and a number of rollers, tapered roller bearings can do a much better job of withstanding sideward forces. These three components have tapered surfaces whose apexes converge at a common point on the bearing axis. Their excellent load carrying capacity/cross section ratios provide an economic bearing arrangement.

APPLICATIONS

High-carbon chrome bearing steel specified in SAE 52100 is used as a general material in bearing rings and rolling elements. The cages can be made of various materials, such as steel sheet, steel, copper alloy and synthetic resins. Once relegated to high-end applications, such as aircraft wing-flap actuators and precision instruments, hybrid bearings with ceramic balls are moving into the mainstream. Hybrid ceramic bearings offer many new options for demanding applications. Benefits include high speed, corrosion resistance, durability, reduced vibration, ability to operate with less lubricant, and electrical insulation.

Today's in-line skate market takes the advantage of the newly available technology in bearing design. For example, the ABEC (Annular Bearing Engineers' Committee) scale classifies different accuracy and tolerance ranges for bearings. A good in-line skate bearing typically adopts a high ABEC rating bearing, chromium steel for rings and balls, 100 percent synthetic speed oil for skate lube, ultrafast self-lubricating synthetic resin retainers, and unique and attractive packaging.

Jiang Long

See also: Materials; Tribology.

BIBLIOGRAPHY

Harris, T. A. (2000). *Rolling Bearing Analysis*, 4th ed. New York: John Wiley.

BECQUEREL, ALEXANDRE-EDMOND (1820–1891)

Edmond Becquerel was one of a family of scientists. His father, Antoine-César, was professor of physics at the Muséum d'Histoire Naturelle, and his son, [Antoine-]Henri Becquerel, also a physicist, discovered the phenomenon of radioactivity (for which he received the Nobel Prize in 1903).

The scientific work of Edmond began in 1838, at the very early age of eighteen. When the Chair of

Professor of Physics Applied to Natural History was created for his father at the Muséum d'Histoire Naturelle, Edmund had the dilemma of choosing to attend l'Ecole Normale, l'Ecole Polytechnique, or become an assistant to his father for the course that went with the professorship. He chose to assist his father, and their collaboration continued for decades. Thus, his title on the title page of the book published in 1855-1856 with his father is given as: "Professeur au Conservatoire impérial des Arts et Métiers, Aide-naturaliste au Muséum d'Histoire Naturelle, etc." After a short period as assistant at la Sorbonne, and then as Professor at the Institut Agronomique de Versailles, he became Professor at the Conservatoire des Arts et Métiers in 1852, where he worked for almost forty years. When his father died in 1878, Edmond succeeded him as director of the Muséum in addition to his professorship. He received a degree as Doctor of Science from the University of Paris in 1840, and was elected a member of l'Académie des Sciences in 1863.

He published a great number of scientific articles and a number of books: the three volume *Traité d'électricité et de magnétisme, et des applications de ces sciences á la chimie, á la physiologie et aux arts.* (1855-1856 with his father); *Recherches sur divers effets lumineux qui résultent de l'action de la lumiére sur les corps* (1859); and *La lumiére, ses causes et ses effets* 1867, in two volumes.

Electricity, magnetism, and light were the main subjects of his work. At the time, these subjects were "hot" topics. Hans Christian Ørsted had made his discovery that an electric current had an effect on a magnet in the year Edmond was born. Michael Faraday had just (in 1831) discovered the effect of induction, and Louis Daguerre invented the photographic plate in 1837, the year before Edmund began his scientific work. Edmond set out to study the chemical effect of light, and in 1839 he discovered a remarkable effect: electricity was emitted following the chemical actions due to the light—the photoelectric effect. He was thus led to the construction of the "actinometer," which allows the measurement of light intensities by measuring the electric current generated by the light. Using photographic plates, he examined the sunlight spectrum and discovered that the dark lines, observed by Fraunhofer in the visible part, continue into the violet and ultraviolet region, and that the plates, when exposed briefly to ultraviolet radiation, become sensitive to the red part as well,

and can actually acquire an image without development of the plates.

Parallel to these investigations, he continued (with his father) to study electricity. He used the method of compensation to measure the resistivity of a large number of materials, including liquid solutions. The effect of the electrodes was, in the latter case, taken into account by using tubes in which it was possible to change the distance between the electrodes. Electrochemical effects and their practical applications were also a main concern. The second volume of the "*Traité ...*" is mainly concerned with the feasibility of extracting silver from minerals in Mexico by electrochemical methods as opposed to the methods then in use involving either charcoal or mercury. Mercury was expensive and charcoal was becoming increasingly expensive due to the shortage of wood. In the introduction to the "*Traité ...*" he mentions that if the consumption of wood in Mexico continued at the ten current rate, it would have severe effects, and that the Mexican government should be concerned. (It is at this point worth noting that Mexico had obtained independence in 1821, and the subject of French intervention in the internal wars that followed was a major political issue.) He studied extensively the electromotive force and internal resistance of a large number of batteries. He and his father used the thermoelectric effect to construct thermometers that could measure temperatures that were otherwise difficult to measure and at places that were hard to access by other means. The temperature in the ground was, for instance, measured throughout the year.

Another study was begun, in 1839, with Jean-Baptiste Biot (who had measured quantitatively the force that an electric current produces on a magnet, the effect that Ørsted had discovered qualitatively), namely, on phosphorescence, fluorescence and luminescence. To study the phenomenon that certain substances emit light after having been exposed to light, Becquerel devised an ingenious apparatus. The main idea was to have two discs with holes in them rotating about an axis parallel to the beam of light illuminating the sample placed between the discs. The sample receives light only when a hole passes in front of it; otherwise, the disc blocks the path of light. Likewise, the emitted light is observed only when a hole in the other disc passes the sample and can be examined at varying times after the exposure by changing either the relative positions of the holes or by changing the speed of rotation. Furthermore, a prism could be inserted in the path of the emitted light and spectral analysis performed. With this simple apparatus he was able to reduce to 1/40,000 of a second the time separating the luminous excitation and the observation. A number of important results were obtained; for instance, that fluorescence differed from phosphorescence only by its very short duration, and that the spectrum of the fluorescent light is characteristic for each substance (one of the first instances of nondestructive testing).

The discovery and detailed investigations of the phenomenon of fluorescence is generally considered the main contribution of Edmond Becquerel. It had the further impact of leading later to the discovery of radioactivity by his son Henri, as Henri continued these studies, including among the substances examined salts of uranium.

Edmond Becquerel was interested in and dedicated to science in general. He was a very careful and imaginative experimenter with an acute sense of the practical aspects of science. He put great effort and insight into exploring the practical uses of physics, especially the new phenomena of electricity and magnetism or, when combined, electromagnetism.

Stig Steenstrup

BIBLIOGRAPHY

Becquerel, H. (1892). "La chaire de physique du Muséum." *Revue Scientifique* 49:674-678.

Harvey, E. N. (1957). *A History of Luminescence from the Earliest times Until 1900.* Philadelphia: American Philosophical Society.

Violle, J. (1892). "L'œuvre scientifique de M. Edmond Becquerel." *Revue Scientifique* 49:353-360.

BEHAVIOR

Americans for years have been embracing new technology to make life more convenient, and at the same time grew accustomed to cheap, abundant energy to power that technology. Each year more than $500 billion is spent in the United States for energy to perform work and provide heat, and the energy needed for creating and building businesses, home-making, getting around, purchasing goods, and seeking pleasure. The decision-making exhibited while engaging in these daily activities varies widely, which results in diverse energy use behavior.

THE LONG-TERM IMPACTS OF DECISIONS

Energy-use behavior cannot be looked at just in the here and now or in isolation. Americans developed a greater reliance on heat, light, and power than any other nation, primarily due to decisions made by industry, government, and individuals of earlier generations that largely established the patterns for behavior today. Collective choices made long ago have behavioral consequences today, just as collective choices made today will have behavioral consequences for many more years to come.

The high-energy-consumption culture evolved primarily as the automobile became affordable in the 1920s and 1930s, offering tremendous mobility and the possibility of distancing home from work, school, and pleasure. It was then accelerated by the post-World War II federal policies of funding a vast network of highways and offering subsidies (home mortgage and property tax deductions) to make home ownership more affordable. Government policymakers at the time never realized the energy consequence of these nonenergy policies. Highways to everywhere, and home ownership subsidies promoted flight from city centers to suburbia and a high-energy-consumption economy largely based on the automobile. With these policies in place, the energy use patterns established a trend of greater and greater consumption. Moreover, the many benefactors of these policies amassed significant power and influence in the political system, discouraging any reversal.

Another factor contributing to high-consumption behavior is the ever-growing affluence of the population. In the 1950s, U.S. consumers spent around 25 percent of disposable income for food, which fell to less than 10 percent by the 1990s. For many, this translates into less than three hours of wages to cover the family's weekly food bill. (For much of the developing world, it can take seven or eight hours of wages to feed a family of four for one day.) Moreover, six or seven hours of wages in the United States can cover the cost of a month's worth of gasoline for the cars, and electricity and natural gas for the home.

This growing affluence allows a greater percentage of disposable income to go toward purchases of energy-using technology and ever-more-elective applications, rather than the cost of energy to power the technology. For example, while energy costs had remained flat or risen only slightly from 1980 through 2000, the average cost of an automobile tripled, going from around $7,000 to $21,000. Furthermore, as the United States became more affluent, the distinction between want and need—elective and essential—has blurred. Technology that was rare or nonexistent in the 1960s became a want in the 1970s and 1980s, and evolved into what many consider a need or necessity in the 1990s. Most households are now dependent on all sorts of energy-hungry technology that did not even exist in the 1960s—appliances such as lawn trimmers, personal computers, and microwave ovens, to name a few. Greater affluence, and an economy geared toward high consumption, has resulted in U.S. per capita energy use growing to more than thirty times that of developing world nations, and more than twice that of Western Europe or Japan.

One of the biggest fears with massive consumption and the high energy lifestyle is the high level of immigration that has accompanied it. Although the U.S. birthrate has been slightly more than self-sustaining since the 1970s, a liberal immigration policy had added more than 30 million people to the population by 1995, and is likely to push the population to more than 400 million by 2050. Millions of people from around the world try to immigrate, legally or illegally, to enjoy the freedom and the opportunities of the United States. Because most immigrants, regardless of culture and country of origin, tend to quickly assume the U.S. norm of high-energy-use behavior that increases carbon emissions, exacerbates air pollution (through a need for more power plants and automobile use), and worsens the dependence on imported crude oil, overpopulation and poverty is as much a United States problem as a world problem.

LIFESTYLES AND ENERGY USE

Energy is the ability to do work. The work can be done by man or machine; the fuel to do the work can come from food or fuel. Humans often have a choice of whether or not to continue to develop technology that replaces the work of man as well as do other types of work that man is not capable of doing. Historically, the choice has been almost always to embrace technology, driving the steady increase in demand for energy.

There is often a mental disconnect between technology and the energy needed to make technology work. Except for a small percentage of the population, people are not much interested in energy itself, but only what it can do for them. That is because energy is largely invisible. The units of energy—a kilowatt hour of electricity or a cubic foot of natural gas—cannot be seen. Thus, energy is primarily thought of in financial terms: the dollar amount of the electric or natural gas bill, or the cost of refueling the car.

Energy behavior can be categorized loosely as conserving (frugal), efficient, or high-energy (Table 1). It is not always fair to make such clear-cut distinctions because many behavior patterns exist within each income level that are beyond easy classification. For example, a few cross-classifications could be rich and frugal, rich and efficient, poor and frugal, and poor and wasteful. Ironically, the poor are often the most wasteful and inefficient because they own older cars and live in leaky homes.

The distinction between high-use versus conserving or efficient is well-known, but the difference between energy-conserving behavior and energy-efficient behavior is a more subtle distinction. Energy is used to perform important tasks such as getting to work. A vehicle will use a given amount of British thermal units (Btus) per mile. Multiplying the Btus per mile by the number of miles traveled, results in the total Btus used. The goal of energy conservation is to reduce the number of Btus used. Energy-conserving behavior is not strictly about forgoing consumption. The conserving person still needs to get to work; he just wants to get there by using less energy, say, by substituting the train, bus or carpool for driving alone. On a per passenger basis, the difference between the two could be 100 to 160 miles per gallon (mpg) for carpooling or mass transit, compared to 15 to 40 mpg for driving alone. Remembering to turn off the lights when leaving a room, or turning down the thermostat to 60°F, or shoveling the walk instead of buying a snow blower, are other examples of energy-conserving behavior.

Energy-efficient behavior usually involves embracing all technology to do work, but picking the most energy-efficient products to do that work. The energy-efficient choice is to stay in the single passenger automobile, but reduce the Btus per mile by buying an efficient model that gets 40 mpg rather than

	Conserving (Frugal) Lifestyle	Efficiency Lifestyle	High Use Lifestyle
Food	Eating a healthy diet and only the required 2,500 to 3,000 calories.	Prone to overeating, but tries to keep a healthy diet.	Overeating and an overabundance of meat and empty calories, highly processed junk foods.
Exercise	Commuting by walking or riding a bike.	Membership in fitness club to overeat yet remain thin.	Very little or none. Likely to become obese.
Recreational Activities	Walking and bicycling.	Running, swimming and aerobics.	Golfing with a cart, personal watercraft, all-terrain vehicles, and snow-mobiling.
Hobbies	Participatory activities.	Participatory activities.	Watching television and attending sporting events.
Home	1,500 sq. ft. apartment or condominium in a city high-rise.	5,000 sq. ft. energy-efficient home in the suburbs.	5,000 sq. ft., nonenergy-efficient home in the suburbs.
Heating	High-efficiency furnace and water heater. Bundles up with a sweater to keep thermostat at 65 degrees.	High-efficiency furnace and water heater. Keeps thermostat at 72 degrees, but programs it to 60 degrees for overnight hours.	Low-efficiency furnace and water heater. Keeps thermostat at 78 degrees at all times.
Air Conditioning	Only a few rooms, and only when occupied.	Central air that is programmed to come on an hour before arriving home.	Central air all the time, regardless if anybody is home.
Appliances	With limited space, fewer and smaller; only the necessities. The necessities must be energy-efficient. Buys food supplies daily on the way home from work; no need for a large refrigerator and freezer.	Desires all technology, but wants the most energy-efficient products.	Desires all technology and does not put a priority on energy efficiency in making buying decisions. Huge refrigerator and freezer for the convenience of limiting food shopping to once a week.
Vehicles	None	Two energy-efficient models.	Three or four automobiles and a recreational vehicle.
Commute	3 miles by walking, bicycling, or mass transit.	40 miles in an energy-efficient automobile. If there are high-occupancy vehicle lanes, will make effort to carpool.	60 miles in a 14 mpg sports utility vehicle, and a disdain for mass transit, carpooling and high-occupancy vehicle lanes.
Schools	Children walk to nearby school.	Bus to local school.	Drives children to school.
Children's activities	Nearby so that they can walk or take public transportation.	Effort to carpool and make it convenient for drop-off and pick-up on the way to and from work.	Little regard for proximity. The baby sitter will do driving.
Shopping	The majority is done locally and daily; accessible by walking or mass transit.	Drives to nearby shopping. Realizes that the additional gasoline cost of driving negates much of the discount achieved by giant retailer shopping.	Willing to drive many additional miles to get to malls and giant retailers.
Recycling	Yes	Yes, if there is a financial incentive.	No, unless mandated. Throw-away mentality.

Table 1.
Lifestyles for Three Affluent Four-Person Families
Note: This is an example for comparison only, to show how a high-energy-use lifestyle can result in energy consumption two or three times the level of an energy-efficient lifestyle, and over ten times that of an energy-conserving lifestyle. In reality, rarely do individuals exhibit behavior that is solely conserving, efficient or high-use.

one that gets 15 mpg, or replace a 75-watt incandescent light bulb with a 25 watt fluorescent tube. The amount of illumination is the same, but the amount of energy used is one-third less. In terms of total energy savings to a nation, energy conservation is more important than energy efficiency since there is less energy used when no light is on than when an efficient one is on.

A person exhibiting energy-conserving behavior is likely to exhibit energy-efficient behavior as well, but someone who exhibits energy-efficient behavior is not necessarily going to be into energy conservation. Energy conservation usually indicates a sacrifice in something, be it safety, aesthetics, comfort, convenience, or performance. To many energy-efficient people, this is an unacceptable compromise. Energy-efficient homeowners will take energy efficiency into account in making decisions about home lighting; yet, if they have a strong preference for art gallery quality lighting, they do not hesitate to spend more for electricity than a conserving neighbor whose only concern is a good reading light. Whereas energy-efficient behavior primarily entails an approach to making technology-buying decisions energy-conservation habits require a full-time, conscious effort to reduce the amount of technology used.

DECISION-MAKING THEORY

The reasons for the range of energy-use lifestyles, and the motivation behind energy decision-making, are varied and complex. Prior to the 1970s, there was little study of energy-use behavior. But energy scarcities and the growth of the environmental movement made the energy problem a social problem that economists, psychologists, sociologists, and anthropologists all began to address.

Economists primarily look at energy decision-making, like all other decision-making, as a function of price and utility: The individual is a rational utility maximizer who gathers and weighs all the relevant information to make cost-benefit evaluations to arrive at decisions. If an energy conservation or efficiency product proves beneficial, the purchase will be made. It is uncertain how much of the population realizes that often the greater initial out-of-pocket cost for energy-conservation measures and energy-efficient products will be made up by greater savings through the lifetime of the product. And even when aware of the life-cycle savings, the purchase of more

energy-efficient products may be rejected because of an unacceptable compromise in safety, aesthetics, comfort, convenience, quality, or performance. For example, the five-passenger Mercedes might be preferred over the five-passenger Hyundai since the Hyundai's acceleration, reliability, and luxury shortcomings more than outweigh the benefits of its lower price and better fuel economy.

The current price of energy is a primary factor entering into decision-making, yet the expected future price of energy can be even more important since the product life cycles of most technology is over ten years (autos and appliances 14 or more years, real estate 30 or more years). After the Energy Crisis of 1973, people adopted energy-conserving and efficient behavior more because of the fear of where energy prices were headed than the price at that time. Almost all energy forecasters in the 1970s and early 1980s predicted that energy prices would soar and oil resources would soon disappear. *National Geographic* in 1981 predicted an increase to $80 per barrel by 1985—a quadrupling of gasoline prices. By the 1990s, when these forecasts turned out to be wildly erroneous, most people reverted back to their old lifestyles, and became much more skeptical of any proclaimed impending crisis.

Whereas the focus of economists is on price and utility, the focus of psychologists has been on attitudes and social norms. For example, the nonprice reasons for choosing an energy-conserving or efficient lifestyle have been for social conformity and compliance—a sense of patriotism, good citizenship, or because it was the environmentally friendly thing to do. The main theory is that attitudes determine behavior, and if attitudes can be changed, it is more likely that behavior will change too. However, there is also the opposite theory: Behavior causes attitudes to change. If persons assume an energy-conserving lifestyle, they will assume attitudes to support that action to avoid internal conflict or hypocrisy (Stern, 1992).

Another problem in studying attitudes and behavior toward energy is the pace of change. Even in times of energy shortage, attitudes and behavior tend to change slowly and are influenced as much by cultural and geographical factors as price. A family living in a 5,000 sq. ft. home can turn down the thermostat, yet still needs to heat a significant amount of space. And if the family lives 75 miles from work, they can switch to a more-fuel-efficient car, yet still must burn considerable gasoline commuting. Of course, this family can make a drastic lifestyle change and consume a fraction of the energy by moving into a 1,200 sq. ft. condominium a few blocks from work and school, but this choice is unlikely for many who have grown accustomed to the comfort, lifestyle and perceived safety of suburban life and feel it is well worth the inconvenience and higher energy costs.

THE COMMERCIAL SECTOR

Energy is a capital investment decision that is often neglected by corporations. When a corporation decides to build or rent office space, the energy needed to heat, air-condition, and illuminate the facility usually is not a top priority. In the 1990s, layout, ergonomics, aesthetics, corporate image, proximity of highways and mass transportation, and other productivity factors ranked much higher. Moreover, there remains skepticism about the ability of energy-efficient technologies and design strategies to simultaneously save energy and improve labor productivity. If the price of energy skyrocketed, lowering energy costs would quickly become a priority again. Many office facilities, not competitive on an energy-efficiency basis, could become obsolete well before their expected lifetime of 50 to 100 years.

THE INDUSTRIAL SECTOR

Lean production—the improvement of processes and operations—is something that all industrial executives encourage, and is largely why the U.S. industrial sector energy consumption dropped about 20 percent from 1973 to 1983. With much of energy efficiency gains already achieved, company executives of the 1990s found the payback from other investments more economically efficient, and consequently have put the emphasis elsewhere. A top priority in the 1990s was faster and cleaner processes that reduced manufacturing costs and reduced waste. If these processes also reduced energy use, it was an added bonus.

Besides operational energy use, corporate decisions have an impact throughout society. One reason that per capita energy consumption in North America is much higher than in the rest of the world was the decision of corporate leaders to expand and relocate away from city centers and to major beltway loops in the suburbs. This necessitated more trucking and a workforce reliant on the private automobile instead of mass transit. In an era of tremendous job insecurity, even the most energy-conserving person is hesitant to reside near work, or live without the mobility of the personal vehicle.

THE TRANSPORTATION SECTOR

Driving by personal vehicle is the most popular mode of transportation. And although there is a desire for a fuel-efficient automobile, fuel efficiency is a consideration well behind style, performance, comfort, durability, reliability, status, and safety. The weak demand for a 40 mpg automobile occurs for several reasons: It is not a status symbol (not stylistic), accelerates too slowly (smaller engine), cramps the driver and occupants (smaller interior), and often offers inadequate protection (too light) in case of an accident.

Another reason for weak demand is that fuel cost in 1999 were a much smaller fraction of the cost of owning and operating a vehicle than it was in 1975. While the average cost of owning and operating an automobile more than tripled from 1975 to 1999, the price of fuel increased only marginally, and average fuel economy improved from 15 to 27 miles per gallon. As fuel cost declines relative to all other costs of operating an automobile, the purchase of a fuel-efficient automobile increasingly becomes a more secondary financial consideration. Most motorists express a preference for greater size, luxury, and performance from the automobiles and trucks purchased, knowing that such attributes usually are detrimental to fuel economy.

The Private Auto Versus Public Transit

The freedom of private transportation is something most Americans have taken for granted and consider a necessity. Only a generation ago, an automobile was considered a luxury, and a generation before that, a rarity. The ascension of the automobile coincided with the decline in mass transit in the 1920s. This trend accelerated after World War II when American society decided to build a vast interstate highway system, emphasizing private automobiles at the expense of mass transit. It is not easy to undo the technological impetus of large infrastructure changes of this nature. According to the U.S. Census Bureau, by 1990, 73 percent of the population drove alone to work, up from 64 percent in 1980. In percentage terms, walking (3.9%), bicycling (0.4%), public transit (5.3%), and carpooling (13.4%) all declined between 1980 and 1990. For other travel (shopping, family business, and leisure), which was responsible for two-

Traffic crawls past the skyline of downtown Seattle on Interstate 5. (Corbis-Bettmann)

thirds of all travel, the share of collective modes and nonmotorized modes was even lower.

A major mass transit handicap is sprawling growth. Mass transit works most effectively in a hub and spoke manner, and the decentralization of urban areas makes it harder and harder to design effective mass transit. This mismatched segregation of home, work, and leisure—particularly in cities such as Houston, Atlanta, and Los Angeles—would require a long time to reverse.

Even where mass transit offers better speed and comparable comfort and convenience, the love of the automobile results in uneconomic decisions affecting energy consumption. For someone driving 30 miles to New York City, tolls, fuel, and parking can easily exceed $20 a day. In comparison, a monthly train pass for the same 30 mile trip runs less than $10 a day. This premium for the privilege of driving is even greater if the yearly $6,000 to $10,000 cost of leasing, insurance, and maintenance is included. Because of congestion, few can claim to drive for speed reasons. The roadway trip often takes more time than via mass transit. Getting more people out of single-passenger cars and into more-energy-efficient mass transit or car pools is going to take more than improving mass transit and the price of gasoline tripling. Americans, like most others in the industrial world, have an emotional attachment to the automobile.

Another reason for choosing the automobile is its development as a mobile office. Cellular phones, laptop computers, and satellite linkups to hand-held communication devices—all of which are getting smaller and smaller—have made it possible to be more productive from the roadways.

Aside from work, the automobile is central for shopping and pleasure. Most people who lived in the suburbs in the 1990s had grown up in the suburbs rather than in central cities. The automobile, and the mobility it affords its owner, is a central aspect of suburban life. Shopping centers and businesses, lured by cheaper land in the exurban areas, do not fear locating away from city centers because Americans show an eagerness to drive greater distances. The auto also caters to the needs of family life. Few suburban parents drive directly to and from work anymore. The growth in extracurricular sports and activities, for parents as well as children, requires the convenience of the automobile.

Higher incomes, higher automobile ownership, and a decline in the population and workplaces that can be served by mass transit has lead to the declining mass transit demand. Criticism of this shift toward the private automobile comes mainly because the individual driver receives the short-term benefits (privacy, comfort, speed, and convenience), while the negative social consequences (air pollution, traffic jams, and resource depletion) are shared by all. Moreover, if people drove less, and drove more-fuel-efficient vehicles, the positive national goal of less dependence on imported oil would be achieved.

Flight

The explosive growth in air travel from 1980 to 2000 occurred because deregulation reduced air fares, disposable incomes rose, and travelers desired to get places faster. People are willing to pay a large premium for speed by flying instead of driving. The premium is largely a reflection of the much greater energy costs of flying, yet it is not always a greater cost. On a passenger-miles-per-gallon basis, usually more energy can be conserved by flying a full aircraft rather than having each passenger drive solo to a given destination.

People make decisions to fly based on speed, the price of the flight, and an airline's on-time arrival. If these factors are the same, then secondary concerns such as comfort become a factor—fewer seats with more spacious seating, and more room to walk about. Since greater comfort means fewer paying passengers, the airlines' decision to cater to the desire for comfort will adversely affect fuel economy per passenger.

The widespread disfavor toward prop planes is another preference adversely affecting fuel economy. Prior to the late 1990s, the 30- to 50-seat plane market was dominated by the more-energy-efficient turboprop planes, yet regional airlines ordered jet engine replacements, citing strong customer preference. Airlines are turning to jets because people want to get places faster. Turboprop planes, which attain near-jet speeds and near-jet performance with less fuel consumption, are noisier and thought to be less safe, even though safety records do not warrant this belief.

The future demand for jet travel is very uncertain. Businesses have traditionally felt a need to travel for face-to-face meetings, but the new communications technology revolution might in the future make business travel less necessary. However, any dropoff in business travel is likely to be replaced by growth in leisure travel as more affluent Americans decide to take more but shorter vacations, and fly for more weekend getaways.

Freight

Individuals and companies make shipping decisions based on price, speed, and reliability. If reliability and speed are equivalent, the decision usually comes down to price. Because energy costs of shipping by air, railway, waterway, and truck vary tremendously, so does the price. Whereas commodities and basic materials are moved by rail, just-in-time components and finished product are mainly moved by truck, accounting for over 80 percent of all freight revenue. Rail is cheaper and a much more energy-efficient means to move freight, and has been approaching trucking for speed; yet businesses continue to pay a premium for trucking because rail historically has been less reliable.

THE RESIDENTIAL SECTOR

The American home is widely perceived as a good investment that appreciates. Any additional improvement to the home is considered wise for two reasons: the enjoyment of the improvement, and greater profit when the time comes to sell. This belief, in part, explains the preference for bigger new homes with higher ceilings over the large stock of older, smaller homes built from 1950 to 1970. The average new home grew from 1,500 square feet in 1970 to over 2,200 by 1997, and the inclusion of central air conditioning grew from 34 to 82 percent. These new homes are usually farther from city centers, and indicates a general willingness to endure the inconvenience and higher energy cost of longer commutes. The continuing trend of bigger suburban homes farther from city centers is attributable to the affordable automobile, the expansive highway system, and the pride of owning one's own home. Whereas a generation ago a family of eight felt comfortable sharing 2,000 square feet of living space, the generation of the 1990s located twice as far from work so that they could afford twice the space for a family of four. But home size alone can be a poor indicator of energy use. Energy consumption can often vary by a factor of two or three for similar families living in identical homes (Socolow, 1978).

Many people approach home ownership investments with only two major concerns: how much down and how much a month. They may be tuned into the present energy costs of a new home, but usually give little thought to how much energy will cost 10 or 20 years into the future. If energy-efficient features push up how much down and how much a month beyond what the buyer can afford, buyers must forgo energy-efficient features. However, selecting energy efficient features for a new home does not necessarily have to increase the down payment and monthly mortgage payment.

Aesthetics is often more important than energy efficiency to home buyers. Energy efficiency that is economically efficient is welcomed if it does not come at the expense of aesthetics. Tightening up a home with better insulation and caulking is fine, but solar collectors on the roof are thought by many to be an unsightly addition to a home. Prior to the 1960s, buildings in the sunbelt were usually built with white roofs. But as air conditioning became widespread, the "cooler" white roof grew in disfavor. Darker roofing shingles were perceived to be more attractive, and did a better job of concealing dirt and mold. Aesthetics won out over energy conservation. In a city like Los Angeles, replacing a dark roof with a white roof can save more than $40 in air-conditioning bills for the hot summer months. Eventually, through education, the aesthetic benefits of a dark roof might be deemed less important than its energy wastefulness.

The energy consequences of bigger homes filled with more power-hungry technology is a need for more energy. Because of federal standards for appliances and heating and air conditioning equipment, the new, larger homes have only incrementally increased consumption. However, the desire for ever-bigger refrigerators and freezers, and the continued introduction of more plug-in appliances, many of which run on standby mode, promises to keep increasing the residential demand for electricity.

By the 1990s, after decades of extremely reliable service, most customers have come to expect electricity to be available when they want it and how they want it, and feel the utility has the obligation to supply the power. If the customer's preference is to keep the air conditioning on all day, it is the obligation of the utility to supply the power for this preference. Customers have become so accustomed and reliant on electricity that the 1990s consumer felt service without interruption was a right. This was far different than a few decades earlier. New York City residents tolerated the major blackout of 1965 and the inconveniences it caused, yet the minor blackout of 1999 was widely believed to be inexcusable and that Con Edison (the local utility) should compensate its customers for damages, and it did.

During prolonged heat waves which can cause blackouts and brownouts, electric utilities always request that customers limit energy consumption by raising thermostats and turning off the air conditioning when not home. It is uncertain if these requests are heeded, and if they are, it is uncertain whether conformity is out of altruism (help the utilities) or self-interest (save money).

FOOD

Just as gasoline is the energy source for the automobile engine, food is the energy source for the human engine, yet it is not common for people to think of food in this way. It is more common to look at food as a means to satisfy a hunger craving. Even among those claiming to want to lose weight, 35 percent of men and 40 percent of women are not counting calories according to the American Medical Association.

The minimum average energy requirement for a sedentary adult to survive is around 2,000 kcal a day, which rises to 4,000 or more if the person is engaged in strenuous labor much of the day. Only a very small percentage of the U.S. population is involved in strenuous labor, yet much of the population consumes over 4,000 calories each day, resulting in an America that is 50 percent overweight and 20 percent obese.

Few people do manual labor anymore. Technology has modified behavior so that most people burn 2,500 to 3,000 calories a day, not 4,000 or more calories as was common up until the 1960s. The washboard was replaced by the washing machine; the manual push mower was replaced by the power mower; the snow shovel was replaced by the snow blower; the stairs were replaced by the elevator. Technology has made it easy to be inactive, yet few individuals will blame technology for their obesity.

The tremendous growth in health and fitness clubs since the 1960s caters to those Americans who would rather burn more calories than reduce calorie intake. By working out regularly, they can overindulge. However, only a small minority of the U.S. population exercises regularly; the majority overindulges and remains overweight.

If a sedentary person consumes more than 4,000 calories a day, obesity is likely to result; yet, among this group, a failure to match food energy input with activity energy output is seldom mentioned as the reason for obesity. Because much of the population cannot control the short-term pleasure of overindulging, and despite awareness of the long-term consequences to health and appearance, some politicians have proposed that health insurance, Medicare, and Medicaid should pay for surgical procedures (stomach staple) and diet pills (Redux, Fen-Phen) to combat what they call an "overeating addiction."

Discarded aluminum cans are disposed of in a red collection bin to be recycled and then reclaimed at Portsmouth Recycling Center. (Corbis Corporation)

MATERIALS

When given a choice among materials, determining the most energy-wise material seldom easy. Consider the energy needed to build an aluminum bicycle frame versus a steel tubular frame. The aluminum bicycle frame takes considerably more energy to build, and needs replacement more often. However, if the bicycle is meant to be a substitute for the automobile, the lighter aluminum frame material can dramatically lower the human power output needed to climb hills. The energy saved during the use of the aluminum frame will more than compensate for the greater energy needed to manufacture it.

The energy-wise choice of material for packaging—paper, plastic, tin, aluminum, or glass—can be equally tricky. Although it takes less energy to make glass and plastic bottles than aluminum cans, few buyers think of purchasing a favorite soda because it comes in glass or plastic rather than aluminum. Energy rarely enters into the buyer's decision. Price and quality concerns are foremost. Plastic wins out on price and glass on quality; rightly or wrongly, there is a perception that glass is better at not altering the taste of the product. Few consumers would be receptive to fine wines coming in plastic bottles or aluminum cans. Glass could someday compete with plastic at the lower end of the beverage market as well if the price of crude oil increased significantly. But glass must also overcome a higher freight cost (glass is much heavier than plastic).

The recycling of bottles and containers is almost universally viewed as an energy-wise and environmentally sound moral good—one of the best way to conserve energy, resources and landfill space. Usually this is true. However, few people realize that for some materials there are great energy savings, and for others very little. For instance, whereas the net energy savings to recycle aluminum cans is substantial, it can take nearly as much energy, and generate as much pollution and waste, to recycle tin cans as to produce new ones from raw materials.

A modest amount of recycling of metals occurs today. Much more can be done, and much more can be done to encourage energy-wise buying decisions as well (choosing products packaged with materials requiring less energy over those requiring more). Whether the American public is willing to invest more time and effort into recycling, and willing to alter product packaging choices solely for energy and environmental reasons, is highly questionable.

BEHAVIOR CHANGE

Energy-related behavior modification goes on continually. Advertisers market the latest household electrical technology to make life easier; tourism agencies promote exotic cruises and resorts; and governments mandate energy conservation and energy efficiency measures to lower carbon emissions and reduce energy imports. These efforts, either direct or indirect, subtle or overt, are all designed to modify behavior that affects energy consumption.

Government

Governments have taken an active role to alter behavior, to conserve energy, and to use it more efficiently. The rationale is that lower energy consumption reduces the need for additional fossil fuel power plants and the need for less crude oil imports. Since the Energy Crisis of 1973, the U.S. government's behavior modification efforts have been many and can be grouped into five categories: information campaigns, feedback, reinforcement, punishment, and reward.

Information campaigns refer to the broad range of brochures, flyers, billboards/signs, workshops, and television and radio advertisements designed to encourage energy-conserving behavior. The success of an information campaign depends on the audience paying attention and taking the message seriously; moreover, the intended audience must trust the government source, and receive confirmation from friends and associates (Stern, 1992).

A lack of trust and confirmation is one of the reasons that the energy-related informational campaigns of governments and environmental groups have faired so poorly. Despite millions being spent in the 1990s to promote conservation, energy efficiency, and renewable energy to combat global warming, the sales of energy-guzzling vehicles skyrocketed as economy car sales declined, the number of vehicles and average miles per vehicle increased, and the average home size and the number of electric appliances in each home kept rising. In a 1997 *New York Times* poll, when asked to rank environmental issues, only 7 percent ranked global warming first (47 percent said air and water pollution), and a CNN poll in that same year found 24 percent of Americans concerned about global warming, down from 35 percent in 1989. Since earlier informational campaigns to change behavior (for example, the supposed energy supply crisis of the 1970s was projected to only get worse in the 1980s) turned out to be erroneous, the American public viewed the government's global warming campaign much more skeptically.

Feedback modification efforts are targeted information programs that address the lack of awareness of people about the consequences, the belief being that if people are aware or educated about the negative consequences of such behavior, it is more likely that they can be convinced to engage in behavior more beneficial to the environment. For example, if the public better understood that all forms of energy consumption come with unfavorable environmental side effects, people would be more likely to conserve and use energy more efficiently. Once made aware of the deleterious environmental consequences, people would be more likely, in theory, to buy a more fuel-efficient minivan than a gas-guzzling sports utility vehicle that marketing research shows customers rarely or ever take off-road.

Reinforcement efforts include rewarding people for engaging in energy- and environmentally-beneficial behavior. High-occupancy vehicle lanes reinforce carpooling because the driver and passengers benefit with a faster trip for the ride-sharing sacrifice, and the EPA-DOE Energy Star label affixed to energy-efficient products is a sign to others that the purchaser uses energy in an environmentally friendly way. Punishments that have been tried include a fine or premium for energy use that causes pollution: the taxation of gas-guzzling automobiles. Rewards tried include tax credits for more environmentally friendly renewable energy and conservation measures, the construction of carpool park-and-ride lots, and government-mandated employer bonuses for employees who do not drive and therefore do not take advantage of subsidized parking (probably the most effective behavior modification program.

It is very difficult to determine the effectiveness of efforts to change behavior. There are those who feel a need to adopt an energy-conserving or efficient lifestyle for appearance sake, yet will assume high-energy behavior when no one is watching (Bell et al., 1990). Moreover, no universally effective method has been found for getting people to reduce energy consumption, and there is great debate as to whether it is a policy worth pursuing. First, many policy changes entail encroachments on freedoms. Since the freedom to choose is a major right that most are very reluctant to give up, passing policy that curtails freedom is controversial. Second, there is disagreement about whether human beings are rational about energy use, and actually modify behavior for any reason except self-interest (say, altruism). Third, sometimes the cost of the program exceeds the value of the energy saved, or the cost-to-benefit ratio is unfavorable. Finally, nonenergy policy has a greater impact on energy consumption than any energy policy itself. Government subsidies for home ownership, the public funding of

highways, and policies encouraging suburban sprawl were far more responsible for per capita energy consumption being much greater in the United States than Japan or Europe than was any energy policy.

Corporate

From the corporate end, almost all of the behavior modification efforts directed at consumers are to get people to use more technology. It usually follows that greater technology use results in greater energy use.

To combat a corporate image of a single-mindedness toward greater consumer consumption, companies such as Dupont and 3M make great efforts at what they refer to as eco-efficiency—improving processes to reduce energy use, waste, and air pollution. The message to consumers is yes, we produce the technology you demand, but we do so in an energy-efficient way that is friendly to the environment. Dupont has been aggressively seeking to hold energy use flat, relying more on renewable sources such as wind and biomass, and is trying to reduce its 1990 levels of greenhouse gas emissions by 45 percent by 2001, and by 65 percent by 2010. The U.S. government encourages this behavior, and would like more corporations to no longer think of energy as just another cost, equal in importance with all other costs. Except for the EPA-administered punishments for pollution, most of the Federal effort is toward reinforcing and rewarding good corporate citizenship actions, not mandating them.

When questioned by electric utilities, a majority of residential customers show a willingness to consider paying a modest amount more per month for electricity powered from nonpolluting renewable energy sources, despite not knowing much about them. How many would actually choose to pay a premium for renewable energy is very uncertain. For those who are skeptical or ambivalent about the possibility of fossil fuel resources exhaustion, air pollution and global warming, the primary interest is the lowest price and best service. Early green-energy marketing efforts have shown promise in reaching those who are concerned about environmental issues. In 1999, Mountain Energy of Vermont signed up over 100,000 Pennsylvanians and Californians who will pay a 5 to 35 percent premium for electricity generation not involving nuclear power or coal. It is a surprisingly good start for Mountain Energy, who cannot actually get green power to the home, but instead must sell the concept—the green power the customer buys displaces the traditionally generated electricity in the area. Because many consumers buying the green energy are in a different region from where it is produced and therefore will not "receive" the cleaner air for which they paid a premium, it is uncertain whether green marketing programs will provide measurable environmental improvements. The result may be significant new renewable energy development, or it may just support the renewable base already in operation.

Food

Because of the huge food surplus in the United States, attributed to the high productivity of mechanized agriculture, there is probably no area of the energy consumption picture where more behavior modification takes place than food. The huge agricultural surplus keeps food products cheap, and would increase tenfold if Americans cut meat consumption in half and ate grains instead. Humans get from grain-fattened cattle only about 5 percent of the food energy they could get by eating the grain the cattle are fed.

There are thousands of special-interest groups trying to modify consumer behavior to eat more, or eat higher up the food chain. This would not be possible without a huge food surplus. It has never been easier to overeat. Even for those who lack the time or motivation to cook, the fast-food industry and microwaveable meals have made food more convenient and widely available than ever before. At the other extreme is the $33 billion a year weight loss industry promoting products to lose the weight gained "eating down" the food surplus. In the middle is the U.S. Department of Agriculture (USDA) with an inherent conflict of interest: One arm of the organization promotes the consumption of food while another arm publishes the highly political and ubiquitous food pyramid of good eating.

Since Americans enjoy the cheap food supply made possible from a huge agricultural surplus, and there seems to be no desire to reject technology and go back to a reliance on manual labor, the food surplus problem of an overweight and obese America is likely to remain for years to come.

OUTLOOK

Easy access to inexpensive energy has come to be viewed as a basic right. The American public goes about its daily life largely optimistic, feeling that any fossil-fuel shortage will be alleviated by new breakthroughs in developing supplemental sources, and that new end-use technology will be developed that can be powered by these new energy sources. Energy conservation and energy efficiency are lifestyle options. However, if the day ever comes when energy conservation or an accelerated adoption of energy-efficient products will need to be mandated, the American public would be more likely to choose the energy efficiency route because it does not necessarily entail sacrifice.

John Zumerchik

See also: Air Travel; Bicycling; Capital Investment Decisions; Communications and Energy; Economically Efficient Energy Choices; Economic Growth and Energy Consumption; Freight Movement; Green Energy; Government Agencies; Industry and Business, Productivity and Energy Efficiency in; Materials; Propulsion; Traffic Flow Management.

BIBLIOGRAPHY

Bell, P. A.; Baum, A.; Fisher, J. D.; and Greene, T. E. (1990). *Environmental Psychology,* 3rd ed. Fort Worth, TX: Holt, Rinehart and Winston Inc.

DeSimone, L. D., and Popoff, F. (1997). *Eco-Efficiency: The Business Link to Sustainable Development.* Cambridge, MA: MIT Press.

Energy Information Administration. (1997). *Household Vehicles Energy Consumption, 1994.* Washington, DC: U.S. Department of Energy.

Engel, J. F.; Blackwell, R. D.; and Miniard, P. W. (1998). *Consumer Behavior,* 8th ed. Hinsdale, IL: Dryden Press.

Farhar, B. C. (1994). "Trends: Public Opinion about Energy." *Public Opinion Quarterly* 58(4):603-632.

Farhar, B. C., and Houston, A. (1996, October). "Willingness to Pay for Electricity from Renewable Energy." National Renewable Energy Laboratory, NREL/TP-461-20813.

Gardner, G. T., and Stern, P. C. (1996). *Environmental Problems and Human Behavior.* Englewood Cliffs, NJ: Allyn & Bacon.

Kempton, W. (1995). *Environmental Values in American Culture.* Cambridge, MA: MIT Press.

Kempton, W., and Neiman, M., eds. (1987). *Energy Efficiency: Perspectives on Individual Behavior.* Washington, DC: American Council for an Energy Efficient Economy.

Nye, D. E. (1998). *Consuming Power: A Social History of American Energies.* Cambridge, MA: MIT Press.

Smil, V. (1999). *Energies: An Illustrated Guide to the Biosphere and Civilization.* Cambridge, MA: MIT Press.

Socolow, R. H. (1978). "The Twin Rivers Program on Energy Conservation in Housing: Highlights and Conclusions." In *Saving Energy in the Home*. Cambridge, MA: Ballinger Publishing Company.

Stern, P. C. (1992). "What Psychology Knows About Energy Conservation." *American Psychologist* 47:1224-1232.

U.S. Census Bureau. (1992). Census data, 1980 and 1990, Journey-to-Work and Migration Statistics Branch, Population Division, Washington, DC.

Veitch, R.; Arkkelin, D.; and Arkkelin, R. (1998). *Environmental Psychology: An Interdisciplinary Perspective*. Englewood Cliffs, NJ: Prentice-Hall.

BERNOULLI, DANIEL (1700–1782)

Daniel Bernoulli, the son of Johann Bernoulli, was born in Groningen while his father held a chair of mathematics at the university. He was born into a dynasty of mathematicians who were prone to bitter rivalry. His father tried to map out Daniel's life by selecting a wife and a career for him. By the time Daniel was thirteen, his father was reconciled to the fact that his son would never be a merchant, but absolutely refused to allow him to take up mathematics, decreeing that Daniel would become a doctor. Daniel gained his baccalaureate in 1715 and master's degree in 1716 at Basle University, but, while studying philosophy at Basle, he began learning about the calculus from his father and his older brother Nikolas. He studied medicine at Heidelberg in 1718, Strasbourg in 1719, and then returned to Basle in 1720 to complete his doctorate. About this time, he was attracted to the work of William Harvey, *On the Movement of Heat and Blood in Animals,* which combined his interests in mathematics and fluids. By 1720 his father had introduced him to what would later be called "conservation of energy," which he applied in his medical studies, writing his doctoral dissertation on the mechanics of breathing.

After completing his medical studies in 1721, he applied for a chair at Basle, but like his father before him, he lost out in a lottery. Disappointed with his lack of success, he accepted an invitation from Catherine I, Empress of Russia, to become Professor of Mathematics at the Imperial Academy in St.

Daniel Bernoulli. (Corbis-Bettmann)

Petersburg in 1725. Catherine was so desperate to secure Daniel that she agreed to offer a second chair to his brother, Nikolas. Unfortunately, Nikolas died of tuberculosis shortly after arriving in Russia. Despondent over his death, Daniel thought of returning home, but stayed when his father suggested that one of his own students, Leonard Euler, would make an able assistant.

Bernoulli and Euler dominated the mechanics of flexible and elastic bodies for many years. They also investigated the flow of fluids. In particular, they wanted to know about the relationship between the speed at which blood flows and its pressure. Bernoulli experimented by puncturing the wall of a pipe with a small, open-ended straw, and noted that as the fluid passed through the tube the height to which the fluid rose up the straw was related to fluid's pressure. Soon physicians all over Europe were measuring patients' blood pressure by sticking pointed-ended glass tubes directly into their arteries. (It was not until 1896 that an Italian doctor discovered a less painful method that is still in widespread use.) However, Bernoulli's method of measuring air pressure is still used today to measure the airspeed of airplanes. Around the same time, he made yet another fundamental discovery when he showed that the movements of strings of musical instruments are composed of an infinite number of harmonic vibrations, all superimposed on the string.

Another major contribution that Bernoulli made while in Russia was the discovery that whereas a moving body traded its kinetic energy for potential energy when it gained height, a moving fluid traded its kinetic energy for pressure. In terms of mathematical symbols, the law of conservation of energy becomes:

$$P + \rho v^2 = \text{constant},$$

where P is pressure, ρ is the density of the fluid and v is its velocity. A consequence of this law is that if the pressure falls, then the velocity or the density must increase, and conversely. This explains how an airplane wing can generate lift: the air above a wing travels faster than that below it, creating a pressure difference.

By 1730 Bernoulli longed to return to Basle, but despite numerous attempts, he lost out in ballots for academic positions until 1732. However, in 1734 the French Academy of Sciences awarded a joint prize to Daniel and his father in recognition of their work. Johann found it difficult to admit that his son was at least his equal, and once again the house of Bernoulli was divided.

Of all the work that Bernoulli carried out in Russia, perhaps the most important was in hydrodynamics, a draft account of which was completed in 1734. The final version appeared in 1738 with the frontispiece "*Hydrodynamica*, by Daniel Bernoulli, Son of Johann." It is thought that Daniel identified himself in this humble fashion in an attempt to mend the conflict between himself and his father. *Hydrodynamica* contains much discussion on the principle of conservation of energy, which he had studied with his father since 1720. In addition, it gives the basic laws for the theory of gases and gave, although not in full detail, the equation of state discovered by Johannes Van der Waals a century later. A year later, his father published his own work, *Hydraulics*, which appeared to have a lot in common with that of his son, and the talk was of blatant plagiarism.

Hydrodynamica marked the beginning of fluid dynamics—the study of the way fluids and gases behave. Each particle in a gas obeys Isaac Newton's laws of motion, but instead of simple planetary motion, a much richer variety of behavior can be observed. In the third century B.C.E., Archimedes of Syracuse studied fluids at rest, hydrostatics, but it was nearly 2,000 years before Daniel Bernoulli took the next step. Using calculus, he combined Archimedes' idea of pressure with Newton's laws of motion. Fluid dynamics is a vast area of study that can be used to describe many phenomena, from the study of simple fluids such as water, to the behavior of the plasma in the interior of stars, and even interstellar gases.

After the dispute with his father in 1734, Daniel Bernoulli lost much of his drive to study mathematics and turned his attention to medicine and physiology. Finally, in 1750, Daniel was appointed chair of physics at Basle, where he taught until his death on March 17, 1782.

Douglas Quinney

BIBLIOGRAPHY

Bell, E. T. (1965). *Men of Mathematics*. London: Penguin.

Cannon, J. T., and Dostrovsky, S. (1981). *The Evolution of Dynamics: Vibration Theory from 1687 to 1742*. New York: Springes.

Fauvel, J., and Gray, J. (1987). *The History of Mathematics*. Houndmills, United Kingdom: Macmillan.

Hollingdale, S. (1983). *Makers of Mathematics*. London: Pelican.

BETHE, HANS ALBRECHT (1906–   )

Hans Bethe, an only child, was born on July 2, 1906, in Strasbourg, when Alsace was part of the Wilhelminian empire. His father was a widely respected physiologist who accepted a professorship in Frankfurt when Hans was nine years old; his mother was a gifted musician who was raised in Strasbourg where her father had been a professor of medicine. The high school Bethe attended in Frankfurt was a traditional Humanistisches

Hans Albrecht Bethe. (Library of Congress)

Gymnasium with a heavy emphasis on Greek and Latin. While there, he learned Latin and Greek, read Kant, Goethe, and Schiller, and also learned French and English and a good deal of science. Classes were from 8 A.M. to 1 P.M., six days a week, with much homework assigned daily.

Bethe's talents, particularly his numerical and mathematical abilities, manifested themselves early. By the time he had finished Gymnasium he knew he wanted to be a scientist and his poor manual dexterity steered him first into mathematics, and then into theoretical physics. In the fall of 1926, after completing two years of studies at the University in Frankfurt, Bethe went to Arnold Sommerfeld's seminar in Munich. Sommerfeld was a forceful and charismatic figure, and among his students were many of the outstanding theorists of their generation: Peter Debye, Paul Epstein, Paul Ewald, Max von Laue, Wolfgang Pauli, Werner Heisenberg, Gregor Wentzel, and Fritz London. In Munich, Bethe discovered his exceptional talents and his extraordinary proficiency in physics and Sommerfeld gave him

indications that he was among the very best students who had studied with him.

Bethe obtained his doctorate in 1928 summa cum laude and became Paul Ewald's assistant in Stuttgart. Ewald—whose wife was the niece of a famous and influential reform rabbi—opened his home to the young Bethe and he became a frequent visitor. Ten years later he married Rose, one of Ella and Paul Ewald's daughters. After his brief stay in Stuttgart, Bethe returned to Munich to do his Habilitation with Sommerfeld.

During the academic year 1930-1931 Bethe was a Rockefeller fellow at the Cavendish in Cambridge and in Rome in Enrico Fermi's Institute. In 1932 he again spent six months in Rome working with Fermi. Fermi and Sommerfeld were the great formative influences on Bethe. Bethe's craftsmanship is an amalgam of what he learned from these two great physicists and teachers, combining the best of both: the thoroughness and rigor of Sommerfeld with the clarity and simplicity of Fermi. This craftsmanship is displayed in full force in the many "reviews" that Bethe has written.

By 1933 Bethe was recognized as one of the outstanding theorists of his generation. His book length *Handbuch der Physik* articles on the quantum theory of one- and two-electron systems and on the quantum theory of solids became classics as soon as they were published. In April 1933, after Adolf Hitler's accession to power, he was removed from his position in Tübingen because he had two Jewish grandparents. He went to England, and in the fall of 1934 he accepted a position at Cornell University and remained there for the rest of his career. At Cornell Bethe built a school of physics where he trained and influenced some of the outstanding theoretical physicists of their generation including Emil Konopinski, Morris Rose, Robert Marshak, Richard Feynman, Freeman Dyson, Richard Dalitz, Edwin Salpeter, Geoffrey Goldstone, Robert Brout, David Thouless, Peter Carruthers, Roman Jackiw, and John Negele.

In 1938 Bethe formulated the mechanism for energy generation in stars. This research grew out of his participation at the third Washington conference on theoretical physics in April 1938. The reaction

$$^1H + {}^1H \rightarrow {}^2H + \beta^+ + neutrino$$

had earlier been suggested by Carl von Weizsäcker as a possibility for energy generation and the production of deuterium in stars. The rate of this reaction in stars was calculated by Bethe and Charles L. Critchfield before the conference. Their conclusion was that the rate of such a reaction under the conditions in stellar interiors would be enough to account for the radiation of the sun, though for stars much brighter than the sun, other more effective sources of energy would be required. Until Bethe tackled the problem nucleosynthesis was conflated with the problem of energy generation. Bethe, on the other hand, *separated* the two problems. He advanced two sets of reactions—the proton-proton and the carbon cycle—that were to account for energy production in stars like the sun. The second depended on the presence of carbon in the star. At that time there was no way to account for the abundance of carbon in stars, that is, it was not at all clear what nuclear reactions in stars between elements lighter than carbon could produce this element. However, the presence of carbon in stars had been corroborated by their spectral lines in stellar atmospheres. Bethe accepted this fact and proceeded to compute the characteristics of stars nourished by the two cycles, and found that the carbon-nitrogen cycle gives about the correct energy production in the sun.

During World War II, Bethe worked on armor penetration, radar, and helped design atomic weaponry. He was a member of the Radiation Laboratory at the Massachusetts Institute of Technology from 1942 till the spring of 1943 when he joined Oppenheimer at Los Alamos and became the head of the theoretical division. Bethe is the supreme example why theoretical physicists proved to be so valuable in the war effort. It was his ability to translate his understanding of the microscopic world—that is, the world of nuclei, atoms, molecules—into an understanding of the macroscopic properties and behavior of materials, and into the design of macroscopic devices that rendered his services so valuable at Los Alamos and later on to industry. Bethe's mastery of quantum mechanics and statistical mechanics allowed him to infer the properties of materials at the extreme temperatures and pressures that would exist in an atomic bomb. Bethe, Fermi, and the other physicists on the Manhattan Project converted their knowledge of the interaction of neutrons with nuclei into diffusion equations, and the solutions of the latter into reactors and bombs.

After the war, Bethe became deeply involved in the peaceful applications of nuclear power, in investigating the feasibility of developing fusion bombs and bal-

listic missiles, and in helping to design them. He served on numerous advisory committees to the government including the President's Science Advisory Committee (PSAC) and was influential in getting the United States and the Soviet Union to sign a Nuclear Test Ban Treaty in 1963. In 1967 he won the Nobel Prize for his 1938 theoretical investigations explaining the mechanism of energy production in stars. Beginning in the mid-1970s Bethe collaborated with G. E. Brown, and this association resulted in exceptional productivity. He contributed importantly to the elucidation of supernovae explosions and to the solar neutrino problem. His most recent researches were concerned with the life cycle of supernovas and the properties of the neutrinos involved in the fusion processes in the sun. He has been and continued into his mid-nineties (at the time this article was written) to be an enormously productive scientist.

Silvan S. Schweber

See also: Nuclear Energy; Nuclear Energy, Historical Evolution of the Use of; Nuclear Fission; Nuclear Fusion.

BIBLIOGRAPHY

Bernstein, J. (1979). *Hans Bethe, Prophet of Energy.* New York: Basic Books.

Rhodes, R. (1987). *The Making of the Atomic Bomb.* New York: Simon and Schuster.

Schweber, S. S. (2000). *In the Shadow of the Bomb.* Princeton, NJ: Princeton University Press.

BICYCLING

Bicycling is a simple, affordable, and energy-efficient means of transportation. Of all human-powered locomotion, it is the fastest and least energy-demanding.

The bicycle has stunning efficiency advantages over other vehicles for several reasons:

- It weighs roughly one-fifth of its payload weight. (By comparison, even a small motorcycle weighs more than the rider.)
- With the exception of avid sport and competitive use, it is typically operated at low speeds that do not cause high aerodynamic drag.
- Its internal mechanical efficiency can be nearly perfect—a 1999 test at Johns Hopkins University

Mode	Vehicle Miles Per Gallon of Gasoline or Food Equivalent	Energy Use (BTU's) Per Pasenger Mile
Bicycle	1560	80
Auto-high economy	50	600 (4 pass.)
Motorcycle	60	2100
Bus-Intercity	5	600 (45 passengers)
Subway Train		900 (1000 passengers)
747 Jet plane	0.1	3,440 (360 passengers)

Table 1.
Energy Use of Various Forms of Transportation

showed chain efficiencies as high as 98.6 percent—and energy losses due to rolling resistance are far less than for other vehicles.

The light weight and mechanical efficiency not only allow the bicycle to be powered by a nonathletic human, but it can be walked over extreme terrain, or laden with heavy cargo, or picked up and carried. These options make the bicycle more versatile than any other vehicle, and allow a bicycle user door-to-door, on-demand transport.

For any given speed, the energy demands are close to half that of running, and 15 to 20 percent that of ice skating or roller skating. Moreover, better bikes, better techniques, and better athletes continually help set new speed and endurance records. The record speeds accomplished with recumbent two-wheeled bikes enclosed in an aerodynamic shield have surpassed 110 kilometers per hour (68 miles per hour). The record distance that cyclists can cover in an hour on an unfaired (no separate process to make it more aerodynamic), upright bike has risen from 35 kilometers (21.7 miles) in the 1890s to over 51 kilometers (31.7 miles) in the 1990s.

The bicycle's energy efficiency superiority extends beyond human locomotion and beyond all other forms of transportation. By converting food into the energy equivalent of gasoline, the kilocalories of food energy needed by a human to pedal a bicycle is only a fraction of that needed to propel planes, trains, and automobiles (see Table 1).

HISTORY

Bicycles have been around since the early part of the nineteenth century. In 1817 Karl Von Drais invented a walking machine to get around the royal gardens faster. Made entirely of wood, the rider straddled two same-size, in-line wheels, steering with the front wheel and locomoting by pushing against the ground. This steerable hobby horse, which could surpass runners and horse-drawn carriages, never became a viable transportation option because of the need for smooth pathways, which were rare at that time.

A second major effort at self-propelled transportation came when Pierre Michaux invented in 1861 the velocipede ("fast foot") that applied pedals directly to the front wheel. To achieve greater speed with every pedal revolution, designers tried larger and larger front wheels, with some reaching almost two meters in diameter. Despite garnering interest from hobbyists, the velocipede had three major deficiencies as transportation: First, lacking gears, it was difficult to climb even a modest grade; second, because the construction was entirely of wood, with metal tires coming slightly later, the cobblestone roads of the day made for an extremely uncomfortable ride; third, the big front wheel created problems. Riding was extremely dangerous and inaccessible to most women and children.

A fresh start and the true beginning of bicycles becoming a popular means of transportation can be traced to around 1886 and the efforts of John Kemp Starley and William Sutton. With equal-sized wheels, tubular steel diamond-shaped frame geometry, and a chain-and-sprocket chain drive to the rear wheel, the "safety bike" looked much like the modern version.

During the late 1890s, bicycles were the worldwide focus of invention and technical innovation, much as biotech engineering and computers are today. We owe many of today's industrial manufacturing processes, designs for bearings, axles, and gearing mechanisms, and the knowledge of lightweight structures, to the explosion of inventions that bicycles produced.

In the United States, England, and other major nations, patents were awarded at the rate of about 5,000 per year per nation. In one peak year, bicycle-related patents comprised close to one-third of all patent-writing activity. Many of these patents were decades ahead of the technology to manufacture them; for example, suspension systems invented a century ago became viable only with modern-day elastic materials and manufacturing technology.

The two most important single inventions of this long-ago era were Starley's tension-spoked wheel and John Dunlop's pneumatic tire. The tension-spoked wheel was and is a marvel of lightweight structures; it allows four ounces of spokes, on a wheel weighing a total of three or four pounds, to support a 200-pound rider. (Today's carbon fiber wheels have yet to show a clear advantage over the wheel made from humble carbon steel spokes.) The pneumatic tire, which Dunlop invented in 1888, vastly improved the bike's comfort, and it also shielded the lightweight working mechanisms from excess vibration and fatigue. The coaster brake appeared in 1889, and it has been a staple of children's bikes ever since.

Because a bicycle uses a low-power engine (the rider) and because that rider can only apply power only over a small rpm range, gearing is essential to match the rider's output to the riding conditions. The first patent for bicycle gears was granted in France in 1868; the rider pedaled forward for one gear and backward for the other gear.

An 1869 patent by France's Barberon and Meunier foresaw today's derailleurs. It described a mechanism that would shift a belt or chain sideways among three sprockets or discs. That same year, Barberon and Meunier also patented a primitive gear hub.

These technical innovations dramatically improved performance. Bicycles of that era were hand-made in cottage industries, and were highly sought after and expensive. So impassioned were their owners that the League of American Wheelmen, founded in 1880, was for several years around the turn of the century the strongest political lobby in the United States, with a membership in the hundreds of thousands. The League's "Good Roads Movement" was the first political movement to lobby for a network of high-quality, paved roads throughout the nation.

Between the development of the chain-drive bicycle in the 1880s, and before Henry Ford popularized the automobile in the 1910s and 1920s, bicycling was extremely popular as an efficient means to quickly get around. It was over three times as fast as walking, and more convenient and cheaper than having a horse-drawn carriage at one's disposal. Unfortunately, the quickly improving performance, reliability and affordability of the automobile made it a formidable competitor to the bicycle.

The automobile ascended to dominance in North America starting in the 1920s, and in Europe by the 1950s. After World War II, most nonauto ground transport in the United States rapidly disappeared. The decades following World War II saw vast reductions in train travel, bus service, and public transportation systems in major cities.

The shift to personal autos was slower in many European countries, where many of the population could not afford autos, and where the governments placed very high taxes on gasoline and automobiles. Bicycles, motorcycles and public transportation continued to be widely used in these countries. In addition to the economic factor, there was a cultural reason for Europe's slower embrace of the automobile. Europeans have long lived with high population density within finite borders. The United States of the 1940s was a far more rural nation, with sprawling farmland inside the borders of major cities. That autos took up lots of space in a city was an obvious drawback to the European mind, but irrelevant in Texas.

The poorer nations of the planet had no choice, and used bicycles and public transportation exclusively. For example, when China was under Mao Tse Tung's rule, the number of private automobiles was only in the hundreds.

Beginning in the 1970s, the bicycle saw a resurgence of interest in North America. Reasons included environmental concerns with the internal combustion engine, the popularity of bicycling as a multifaceted sport (racing, touring, mountain-bike riding, family cycling), the desire for fitness, and the need for alternative ways of commuting in crowded cities.

Bicycling to work is viewed as the most environmentally friendly means of travel, the best way to avoid congested roadways (in some cases, it turns out to be quicker than mass transit or driving), and a means of turning commuting time to exercise time. The bicycle is far more efficient than the automobile in making good use of city streets. Traffic counts taken during the 1980 New York City transit strike showed that a single street lane devoted to bicycles

This Malaysian street scene shows two ways in which bicycles are used worldwide. (Cory Langley)

could carry six to eight times the number of people per hour than as a lane for auto traffic.

During these past few decades, the bicycle itself has vastly improved. The bicycles of the early 1970s usually had mechanical idiosyncrasies. Competition among manufacturers, led by the Japanese companies that entered the U.S. market, resulted in vastly improved quality control. The mountain bike, first made available on a widespread basis in 1983, offered a delightful alternative that mushroomed in popularity, and a decade later, traditional "road" bicycles had all but disappeared from stores. The mountain bike has become most people's vehicle of choice for city riding as well as recreational trail riding.

It has also helped that the bicycle buyer is continually offered more bicycle for less money. No longer are most bikes steel. While steel continues to be a fine material, the dominant material reported by bike enthusiast is now aluminum, which, in the hands of most designers, yields a lighter but more rigid frame.

Well-heeled bicyclists opt for titanium or carbon-fiber composites.

There has also been a huge improvement in the value of today's mountain bikes. A mountain bike in the $300 to $500 price range typically has a suspension fork and aluminum frame that would have made it a $1,500 bike a decade earlier. The $800 dual-suspension bike of 2000 far outclasses the early 1990s $2,500 offering.

THE HUMAN ENGINE

The bicycle's advantages as the world's most mechanically efficient means of transportation are clouded by the limitations of the human engine. To put it in power output terms, the human body can produce sustained power only at modest levels. For most people, 100 watts would be too much, and for an elite athlete, 400 watts is the approximate ceiling. (The athlete may manage a brief burst of 1.1 kilowatts.)

The lower power output is inevitable because a body cannot long produce more power than it can simultaneously convert from the chemical mixing of blood glucose and oxygen (aerobic exercise). The higher brief bursts of power do not rely on real-time glucose/oxygen consumption, which is why the athlete is out of breath for minutes afterwards. The athlete's muscles have "borrowed" the results of future oxygen and glucose consumption in anaerobic exercise. For long-distance travel, yet another limitation appears: the body stores about a two-hour supply of glucose. After that glucose is exhausted, the body has to revert to the far less desirable mechanism of burning fat. The muscles produce less power with fat than they do with glucose, and only a trained endurance athlete can comfortably exercise beyond the glucose barrier into the fat-burning zone.

When the automobile became preeminent in the early twentieth century, it did so with good reason. Whether the energy to power a bicycle is anaerobic or aerobic in nature, it is still minuscule in comparison to what an automobile's internal combustion engine can deliver. In the United States, almost all subcompact cars are equipped with engines that can generate 100 or more horsepower (74,600 watts), and can sustain this output all day long.

The human engine cannot match this power output, yet the mechanical efficiency of the bicycle helps tremendously because a very small amount of horsepower can generate great speed. For example, 0.4 horsepower (298 watts) of output can result in 25 mph (40 kph) speeds or better. One set of calculations shows that if a cyclist rode on level ground, with no rolling resistance, and aided by a 25 mph tailwind, it would require only around 0.2 horsepower (150 watts) to sustain a 25 mph pace.

Gravitational Resistance

If not for the need to climb steep hills, bicycling at a 15 mph (24 kph) clip would never be a strenuous exercise. It takes approximately 82 watts, or 0.11 horsepower, on an efficient bicycle, to ride 15 mph on flat ground. But ground is seldom flat. That same 82 watts achieves only 8 mph climbing a barely discernible two percent grade. A five percent grade slows one down to just over 4 mph. Most riders don't want to slow down that much, so they work harder to maintain some speed. On descents, they work less hard, while going still faster.

Any weight reduction helps. Gravitational forces do not discriminate between bike and rider mass, but the human body does. If one spends an extra thousand dollars to shave ten pounds off a bike, there will be a 10-pound advantage, but if one sheds the 10 pounds from the belly, the body will not have to nurture those ten pounds of living tissue, and one will be a more efficient engine.

Air Resistance

Air resistance is a greater factor than most people realize. Even at 10 mph on flat ground, almost half the rider's energy goes to overcoming wind resistance. Rolling resistance is almost nil at the speed. At 15 mph, two thirds of the energy is need to overcome wind resistance. At 25 mph, about 85 percent of the rider's energy is devoted to overcoming wind resistance, with the remainder overcoming rolling resistance and the tiny frictional losses within the bicycle itself. So sensitive is the wind resistance to the rider's aerodynamic profile that riders who race time trials at these speeds feel the bike slow down dramatically if they sit up to take a drink from their water bottles.

For these reasons, most riders cannot, or will not, increase their speed much above their personal comfort levels, even with lots of training. The additional speed just costs the rider too much energy. A rider going 15 mph must double his power output to ride 20 mph. Why these dramatic numbers? Wind resistance varies with the square of the rider's airspeed, but the energy to overcome wind resistance increases with the cube of the rider's airspeed.

Of course, the air is rarely still, just as the ground is rarely flat. Tailwinds do speed the rider, but headwinds have a direct effect on the rider's speed, too: A rider traveling at 20 mph into a 15 mph (24 kph) headwind encounters as much air resistance as a rider traveling at 35 mph (56 kph) under windless conditions. A headwind will slow one down by half its own speed. If one normally rides 15 mph, and then steers into a 10 mph headwind, the resulting speed will be 10 mph.

Because of the significant increase in drag a rider encounters as speed is increased, small adjustments of bike and body contours can significantly alter energy expenditures. These alterations can be to clothing, frame design, handlebars, wheels (spokes), rider profile, and the race strategy of drafting. Of the two resistance factors, the rider and the bike, the rider accounts for approximately 70 percent of the wind resistance encountered, while the bicycle accounts for only 30 percent. Unlike a sleek automobile, the high upright bike and rider is a very inefficient aerodynamic profile, so encircling a rider and bike fully or partially in a streamlined fairing can drop the drag coefficient by 0.25, resulting in a top speed increase from 30 to 36 mph (48 to 58 kph). However, this gain could be somewhat negated on hot days because of the body overheating, or from instability caused by gusty crosswinds.

Fully faired bicycles and faired adult tricycles are today the stuff of cutting edge inventors and hobbyists. Virtually all of them are recumbent bikes, because it makes sense to start out with a smaller frontal area to begin with. Some are reasonably practical for daily use, with lights, radios and ventilation systems; others are pure race machines.

Rolling Resistance

Rolling resistance is almost directly proportional to the total weight on the tires. It is the sum of the deformation of the wheel, tire, and road surface at the point of contact. Energy loss occurs when the three do not return all of the energy to the cycle.

Rolling resistance varies tremendously by tire. Greater air pressure and less contact area is the reason the rolling resistance that a tops of-the-line racing tire encounters on smooth pavement is half or one-third that of a heavily-knobbed mountain bike tire.

But unlike air resistance, rolling resistance varies directly with speed, which means that as speed increases, the rolling resistance factor becomes less important relative to the air resistance. A bike going 20 mph has twice the rolling resistance of a bike going 10 mph. If the bike has good-quality tires with proper inflation pressure, neither rolling resistance number is high enough to be particularly significant.

Rolling resistance declines with smoother and harder road surfaces, larger diameter wheels, higher tire pressures, smoother and thinner tread, and narrower tires. In the case of rough, pot-holed roads, energy is lost in bounce. For soft surfaces such as gravel or sand, energy is robbed and absorbed by the surface. Anyone who has ridden with severely underinflated tires or through mud can attest to the extremely wasteful loss of human energy.

Over rough surfaces, an opposite effect, not easily measured in the laboratory, becomes apparent to the rider. A tire inflated to very high pressures (for example, 120 pounds) bounces off the peaks of the road surface, making the bike harder to control, and negating any theoretical decrease in rolling resistance. For

that reason, top racers often use moderate inflation pressures (85 to 90 pounds). Studies have found that superinflatable tires (120 to 140 pounds) offer no noticeable advantage over high-inflation tires because they do not appreciably decrease rolling resistance.

Wheel size can have as dramatic an effect on rolling resistance as tire inflation. On paper, a smaller wheel size has more rolling resistance, a rougher ride, and poorer handling over bumps than a larger wheel size. Rolling resistance is inversely proportional to the radius of the cylinder, that is, given the same conditions, smaller-wheel bikes experience more resistance to motion than larger-wheel bikes.

Fortunately for the makers of small-wheel folding bikes, several factors can mitigate these shortcomings, such as by using wider tires to compensate for the smaller diameter. The use of improved modern tire technology and the use of suspension in combination with small wheels also help. In modern times, the father of small-wheel suspension bikes, Alex Moulton, began designing these bikes in the 1950s, and his most recent designs have taken this bicycle type to a new level. Hot on the heels of the Moulton bicycle are bicycles such as Germany's sophisticated dual-suspension Birde. Many riders believe these bikes completely negate the alleged disadvantages of small-diameter wheels.

Pedal Cadence

Numerous physiological studies have addressed the optimum pedal cadence, but these studies usually miss the point because they focus on the seemingly important factor of efficiency.

Efficiency is measured in power output per oxygen consumed. However, the rider's supply of oxygen is, for all practical purposes, unlimited, and most riders do not pedal at an effort level that leaves them constantly breathless. Even if they did, they would not want to follow the results of efficiency studies. These studies consistently show that recreational cyclists produce the best sustained performances (lowest metabolic rate and highest efficiency) when the seat is raised 4 or 5 centimeters above the normal height, the pedal cranks are slightly longer to make it possible to use higher gears, and the pedal cadence is in the 40 to 70 rpm range. However, experienced riders, whether fast or slow, virtually always choose to ride differently than the studies recommend.

Why? Efficiency does not matter if one has unlimited oxygen. What does matter is long-term comfort.

Pedal power has not only been found to be the least energy-demanding and fastest form of ground-based human locomotion, but pedal power excels in the water and air as well. In 1987, a water craft pedaled like a bicycle, called the Flying Fish II, reached a speed of 6.5 meters per second (14.5 miles per hour, which is slightly faster than the top speed of a single rower), and a pedal-powered plane, the Massachusetts Institute of Technology's Monarch B, completed a 1,500-meter triangular course with average speed of almost 10 meters per second (22 miles per hour).

Pedaling at the rate shown to be most efficient uses up blood glucose the fastest. This leaves the rider more susceptible to that sudden loss of energy known as "the bonk," and it also tends to leave more lactic acid and other waste products in the muscles, increasing discomfort and extending recovery time.

By contrast, pedaling faster (90 to 100 rpm) in a lower gear at a lower effort level allows the body to burn some fat along with the glucose, thereby extending the glucose reserves. The more rapid leg motion promotes blood circulation, the better to remove waste products from the muscles. This faster cadence is undeniably less efficient, because the body uses energy just to spin the legs around, but it results in increased long-term comfort.

Novice cyclists will often prefer the "more efficient" slower cadence and higher saddle because this configuration uses the leg muscles in a manner more similar to the way walking uses those muscles. In addition, the novice cyclist is often unable to benefit from a higher pedal cadence because of an inability to apply force perpendicular to the crank, resulting in excessive body motion. A bike rider needs to gain some experience with the "less efficient" faster cadence so his/her muscles develop the coordination to function smoothly in this new discipline. Having done that, the rider is unlikely to go back to the slower, less efficient way.

Cleated shoes and toeclips are also advocated on efficiency grounds. Every world-class cyclist uses toeclips today because studies have shown significant aerobic and anaerobic benefits. Toe clips often give elite riders a false sense of power production during the stroke recovery phase. Elite riders feel that toeclips double their deliverable pedal power—a coordinated push-pull effort, exerting an upward force, with the trailing leg alongside the downward force of the forward leg. In reality, however, the cadence is too fast to create a pulling-up force. The importance of cleated shoes and toe clips is in their ability to stabilize the foot and more effectively generate a pushing force rather than in generating a pulling force during the recovery stage. Recreational cyclists benefit from this foot stabilization as much as elite cyclists, do but many of them are uncomfortable with toe clips. Moreover, toe clips are undeniably clumsy in the frequent start/stop environment of crowded city traffic.

Alternative Propulsion Systems

Throughout the history of the bicycle, inventors have questioned whether the century-old circular sprocket design is the most efficient. Many inventors have built elliptical chainwheels, usually to have a "higher gear" during the power stroke and a "lower gear" during the dead spots at the very top and bottom of the pedal stroke. Numerous studies show that even elite riders are unable to apply propulsive force to the pedals during these portions of the power stroke, much as they are unable to lift on the backside of the power stroke.

These elliptical chainwheels have never been widely popular. Sophisticated cyclists tend to shun them because cyclist develop a riding rhythm and a comfort pattern from years of experience. Novices do not even know they exist. Most bike designers fear that elliptical chainwheels would tend to make novice cyclists less inclined to develop a smooth pedaling style.

During the 1980s, Japan's Shimano Corporation invented a radically different alternative to round chainwheels. The development of Shimano Biopace chainwheels began with a very sophisticated study of the biomechanical performance of bicyclists, and Shimano discovered two flaws that it wanted to correct. The first flaw was that the leg was speeding downward as the foot approached the 6 o'clock (bottom) position, and this downward momentum of the leg mass was not being harnessed and converted into forward motion. The second flaw occured during the upstroke phase of the pedal path. Shimano discovered that at this point in the pedal path, the knee joint was switching from flexion to extension, and the switch was so fast that both sets of muscles were being energized at once—so that the body was fighting itself.

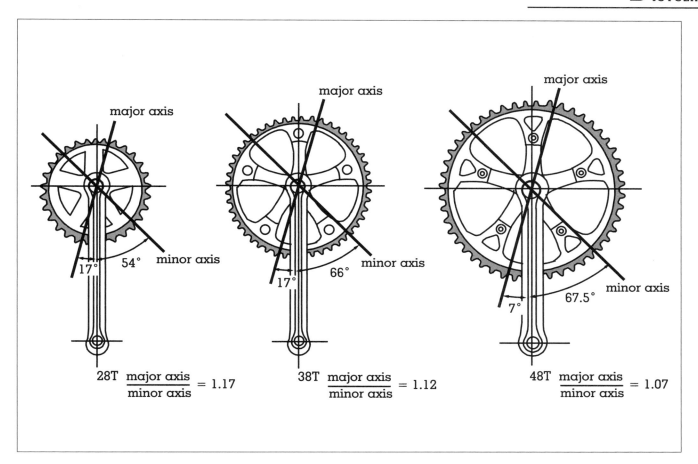

Figure 1.
Biopace chainwheels.

Shimano addressed the first problem with a design that seemed counterintuitive. The chainwheel was shaped so that the rider experienced the feeling of a lower gear during the power stroke. Then, at the bottom of the pedal stroke, the gear got higher, to absorb the energy from the leg mass's downward momentum (see Figure 1). The second problem was addressed by a change in the chainwheel shape, which slowed down that portion of the pedal stroke, giving the leg a few additional hundredths of a second to switch between flexion and extension.

As good as Shimano's research was, the product bombed in the notoriously conservative bicycle marketplace, and Biopace is history today. It's doubtful that bicycles will ever come with nonround chainwheels. The round ones do too good a job, and are too easy to make and market.

Lever propulsion—requiring an up-and-down stair-climbing motion—is another propulsion system long proposed, on efficiency grounds, as a replacement for the standard pedal-and-crank system. Though promising in theory, studies have shown that the muscle efficiency for pedaling a chain wheel is not inferior to that associated with stepping and steep-grade walking. Lever systems maximize the problem of harnessing the leg mass's downward momentum, and work against the smoothness that experienced riders have come to enjoy so much.

There is no basis for the theory that only when pushing the whole stroke vertically do the muscles work efficiently, and that the backward-and-forward foot movement over the top and bottom wastes energy. Certainly, there is some efficiency loss, but it is minimal. Toe clips and better variable gear systems have further minimized "top-dead-center problems" associated with the standard circular sprocket design.

THE BICYCLE AS TRANSPORTATION

Bicycles are the number-one mode of transportation in the world. More than 100 million new bicycles enter the market each year and, in Southeast Asia alone, around 700 million bicycles are used daily as a means of transportation.

China is the world's biggest producer and user of bicycles. Since 1985, China has been manufacturing more than 30 million bicycles each year. In Shanghai, more than 4 million people bicycle to work and school every day. Bicycles far outnumber automobiles. Whereas there is one automobile for every 1.7 people in the United States, in China there is only about one automobile for every 680 people. However, the developing nations seem determined to become overdependent on the automobile as well. In the quest for greater status and mobility, motorized vehicles are quickly becoming the most sought-after possessions in Asia and other developing nations. This pro-auto movement quickens with economic growth that creates more capital to purchase the vehicles. Paying for the energy to run the vehicle is a much smaller concern. As a larger segment of the Asian population rapidly becomes affluent and develops an addiction to the automobile—similar to that is the United States and other developed nations—world petroleum consumption will soar. The U.S. Department of Energy projects Asian transportation petroleum demand will nearly triple from 1995 to 2020, going from 9.6 to 26 quadrillion Btu's.

Although the bicycle is primarily used for recreation in North America, some urban areas (e.g., San Francisco, New York, and Toronto) are experiencing an upswing in the number of workers commuting by bicycle instead of car. Widespread use of bicycles for commuting would have a profound effect on energy supplies and society. If half of the U.S. labor force biked to work instead of driving, the United States could significantly curtail petroleum imports; moreover, the exercise attained by pedaling to work could shrink the obesity and overweight population, and thereby dramatically improve the overall health of many. Despite these benefits, a massive switch to bicycling is highly unlikely for one major reason: It is impractical. There are too many inconveniences involved that are simply unacceptable to the majority of an increasingly affluent population, particularly when the alternative is a climate-controlled, high-speed drive to work in a very comfortable automobile.

The primary inconvenience is the relative slowness and lack of comfort, especially during days of inclement weather. Many millions of Americans live great distances from where they work. Some might consider bicycling five miles to work each day, but few live that near to work. The typical 20-or-30-mile commute makes bicycling an unrealistic option for most people. And for those who do live close to work, few are willing to brave the elements to bicycle year around. Rain, snow, ice, high winds, extreme cold, and extreme heat that are minor inconveniences in a vehicle become major inconveniences on a bicycle. Thus, for backup, bicycle riders usually must own a vehicle for bad weather days, or have access to convenient mass transit.

By contrast, the bicycle survives as basic transportation in the Netherlands and Germany because those nations have a social infrastructure built to make it possible. Those countries have a neighborhood-centered way of life, and the trip distances on a bicycle are often two or three kilometers or less. People cycle slowly out of politeness to others on the crowded streets. Public transit is also far better than in the United States. It is amusing to see a German commuter train station with virtually no auto parking available, but hundreds of bike parking spots.

Intermodal commuting with a bicycle may someday be a way around congested roadways. One of the biggest problems facing mass transit developers is trying to provide service to a sprawling, residential community that needs to commute to equally decentralized work sites. The majority of the population resides a driving distance from mass transit stops, and mass transit service ends too far from a work site to make walking practical. If buses and trains were equipped to carry bicycles, or folding bicycles gained wider acceptance, the bicycle might someday serve as the intermodal link. The bike could be pedaled to the train, loaded for mass transit, and then pedaled the final leg of the commute. It is uncertain whether this alternative means of commuting will ever become attractive enought to overcome the drawbacks of weather, bicycle storage, and safety concerns.

Some people believe that bicyclists require dedicated bike lanes and trails. This is a half-truth. Certainly, these facilities are widespread in Germany and the Netherlands, but they are far less common in other cycling countries such as the United Kingdom, France, and Cuba. In those countries, bicyclists have long known that they can share the road with auto-

Bicycle use on Fifth Avenue in New York City helps to decongest the already crowded streets. (Corbis Corporation)

mobiles as long as all road users respect the same set of rules.

Dedicated facilities are often cited for safety purposes, but this, too, is not a simple truth. Most accidents occur at intersections, and dedicated facilities make intersections far more complex. When the Netherlands allowed their moped riders to travel in the auto lanes instead of the bike lanes, the moped accident rate fell by an astounding 70 percent. The bicycle accident rate would be higher if bicyclists tried to ride at brisk speed in these separated bike lanes. The sub-10-mph speeds that are considered polite in these countries largely allow bicyclists to compensate for the facilities' shortcomings, at the expense of travel time.

In the United States, advocates for bicycling are divided on the question of special facilities. It is doubtful that Dutch-style facilities would create much greater ridership in most U.S. locales, because the trip distances are too long. The best reading of accident statistics shows that adult riders well-schooled in sharing the road with automobiles and respecting the same rules ("vehicular cycling") have about one-fifth the accident rate of a control group of otherwise-well-educated adults.

Aside from the comfort and convenience disadvantages of bicycling, perhaps the biggest obstacle is the price of energy. Energy is far too cheap to make pedal power worthwhile, except as a hobby. For most people, the cost of the one or two gallons of gasoline used commuting each day is paid for in less than 10 minutes of work time. Of course, the pedal power equivalent of electrical energy is even less. To travel at 15 mph (24 kph) on flat ground requires only around 82 watts of power. If one ran on electrical energy instead of food energy, the cost of the electricity to generate that 82 watts to travel 14 miles to work is about 0.7 cent.

Sweating is another problem. Few people want to arrive at work with a sweat-soaked body. There are usually no showers at work, and few want the additional burden of lugging work clothes that would need pressing on arrival. For most riders, sweating is inevitable, particularly when climbing hills. The human body is about one-third efficient in converting the energy of glucose into muscle movement. The other two-thirds is waste heat, and sweat is how the body gets rid of excess heat.

Because of the desire to avoid perspiring, and the limited energy available to propel a bicycle, there is growing interest in electric motor-assisted bicycles that can make that 15-mile commute to work a near-effortless experience. The electric motor allows the rider to maintain a reasonable speed without exerting enough energy to perspire. The U.S. market saw at least a dozen brands of electric-assist bicycle during the 1990s. The more sophisticated offer regenerative braking 15 to 20 miles of range, and recharging cost of 2 to 3 cents for a 15-mile distance (at 10 cents kWh).

These bikes could be a wonderful transportation option, combining the efficiency, quiet and nonpollution of a bicycle with the ease of riding a small motorcycle. For these reasons, electric-assist bikes sell at the rate of over 200,000 per year in Japan. But in the United States, these bikes run up against a cultural barrier. Bicycle enthusiasts, who are less likely to object to their four-figure price tags, do not want a motor and do not want a bike that weighs 60 pounds. Others do not want to be seen riding to work. So U.S. sales have been poor.

As the number of motor scooters and motorcycles in Asia has grown at an alarming rate, policies to encourage the use of electricity-assisted bicycles are seriously being considered as a way to curtail vehicle emissions. In Bangkok, Thailand alone, over one million people needed treatment for smog-related respiratory problems in 1990. Greater affluence and the demand for faster transportation in China have resulted in yearly sales of motorcycles increasing from less than half a million sold in 1991 to over 10 million by 1996, further contributing to air pollution that is already some of the worst in the world. Unless zero-emission transportation, such as electric bicycles, gains in popularity, the number and severity of smog-related respiratory problems are certain to worsen in the most densely populated urban areas of Asia.

John Schubert
John Zunerchik

See also: Aerodynamics; Biological Energy Use, Cellular Processes of; Culture and Energy Usage; Flywheels.

BIBLIOGRAPHY

Allen, J. S. (1981). *The Complete Book of Bicycle Commuting*. Emmaus, PA: Rodale Press.

Cross, K. D. (1977). *A Study of Bicycle/Motor-Vehicle Accidents: Identification of Problem Types and Countermeasure Approaches*. Santa Barbara, CA: Anacapa Sciences.

Faria, I., and Cavanaugh, P. (1978). *The Physiology and Biomechanics of Cycling*. New York: Wiley.

Forester, J. (1994). *Bicycle Transportation*. Cambridge, MA: MIT Press.

Hunt, R. (1989). "Bicycles in the Physics Lab." *Physics Teacher* 27:160–165.

Romer, R. (1976). *The Energy Fact Book*. Amherst, MA: Department of Physics, Amherst College.

Schubert, J. (1988). *Cycling for Fitness*. New York: Random House.

Whitt, F., and Wilson, D. (1974). *Bicycle Science*. Cambridge, MA: MIT Press.

BIG BANG THEORY

The Big Bang Theory is the prevailing theory of the origin of the universe, and it is based on astronomical observations. According to this theory, about 15 billion years ago all the matter and energy in the visible universe was concentrated in a small, hot, dense region, which flew apart in a gigantic explosion.

Before the twentieth century, most scientists believed the universe was static in the sense that it was neither growing nor shrinking as a whole, although individual stars and planets were moving. In 1915 Albert Einstein proposed the general theory of relativity, which is a theory of gravity that has superseded Isaac Newton's theory of gravity for very massive objects. Since general relativity was invented, its equations have been used to describe the possible ways in which the universe might change as time goes on. Einstein, like others before him, thought the universe was static, but the equations of general relativity do not allow for such a thing; according to the equations, the universe has to grow or shrink. In 1917, in order to allow for a static universe, Einstein changed the equations of general relativity by adding a term called "the cosmological constant."

AN EXPANDING UNIVERSE

In the 1920s, cosmologists examined Einstein's original equations without the cosmological constant and found solutions corresponding to an expanding universe. Among those cosmologists was the Belgian Georges Lemaitre, who proposed that the universe began in a hot, dense state, and has been expanding ever since. This proposal came before there was any substantial evidence of an expanding universe.

Nearly all stars in the visible universe are in large clusters called galaxies. The Milky Way galaxy, the galaxy containing the sun and about 100 billion other stars, is one of about 50 billion galaxies that exist in the visible universe. In 1929 the astronomer Edwin Hubble, after making observations with a powerful telescope, discovered that distant galaxies are moving away from the earth and the Milky Way (and from one another). The farther these galaxies are from the earth, the faster they are moving, their speed being approximately proportional to their distance. Galaxies at the same distance from the earth appear to be moving away from us at the same speed, no matter in what direction in the sky the astronomers look. These observations do not mean that the earth is at the center of the universe; astronomers believe that if they made observations from any part of the visible universe that they would find the same general result.

If the galaxies are moving away from each other, then in the past they were closer to one another than they are now. Furthermore, it can be calculated from

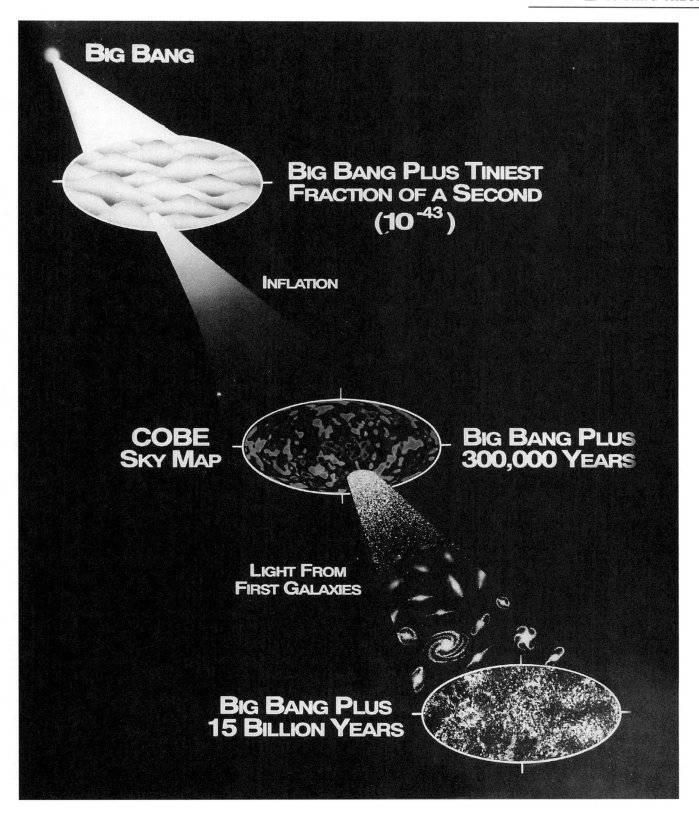

Big Bang Theory, as conceptualized by NASA, 1992. (AP/Wide World Photos)

An artist's impression of galaxies being formed in the aftermath of the Big Bang. The spiral clouds of gas have already started condensing into the shapes of future galaxies. (Photo Researchers Inc.)

the present speeds and distances of the galaxies, that about 15 billion years in the past, all the matter and energy in the visible universe must have been in the same place. That is when the Big Bang happened. Scientists do not know what the universe was like "before" the Big Bang or even whether the concept of earlier time makes sense. The galaxies were formed out of the original matter and energy perhaps a billion years or more after the Big Bang.

THE THEORY GAINS ACCEPTANCE

Fred Hoyle, an astronomer and cosmologist who had a rival "steady state" theory of the universe, coined the name "Big Bang" in order to make fun of the theory in which the universe began in an explosion. The name stuck. Today, nearly all scientists prefer the Big Bang Theory because it can account for more observed properties of the universe than the steady state theory can. In particular, the observed microwave background radiation that appears everywhere in the sky is a remnant of the Big Bang. This radiation cannot be accounted for in a natural way by the steady state theory.

According to present theory, the galaxies are not flying apart into empty space, but space itself is growing larger. Another way of putting this is to say that the universe itself is expanding. Although the universe is expanding, one should not think that everything in the universe is expanding with it. Individual galaxies are not expanding, because their stars are prevented from flying apart by their mutual gravitational attractive forces. Likewise, other forces of nature keep the stars, the sun, Earth, and objects on Earth—down to atoms and nuclei—from expanding along with the universe.

THE FATE OF THE UNIVERSE

What will be the ultimate fate of the universe? Will the expansion go on forever or will gravity slow and then reverse the expansion into a collapse? According to general relativity, whether or not the universe will continue to expand or eventually collapse depends on the amount of matter and energy in the universe. If this matter and energy together are greater than a certain critical amount, their mutual gravitational attraction will reverse the expansion, and the universe will end with what astronomers call the "Big Crunch." If the sum of the matter and energy is below the critical amount, then, although gravity will slow the expansion, the universe will continue to expand forever. At the present time, most observations seem to favor a universe that will expand forever, but the uncertainties are large.

Astronomical observations made in the late 1990s, which are still preliminary, indicate that the expansion of the universe is not slowing down, as required by the attractive gravitational force of general relativity, but is speeding up. One way to account for this speeding up is to put back the cosmological constant into the equations of general relativity. If the cosmological constant has a certain value, general relativity allows for the speeding up that astronomers think they are

seeing. When Einstein first learned that the universe was expanding, he abandoned the cosmological constant, calling it his greatest mistake. If he were alive today, what would he think about the possibility that his constant might be needed after all, but for an entirely different reason? In any case, astronomers continue to make better and better observations with their telescopes and are hoping to obtain more definite answers about the universe during the first decades of the twenty-first century. However, based on the recent history of discoveries in astronomy, it is probable that more surprises are in store.

Don Lichtenberg

See also: Matter and Energy; Particle Accelerators.

BIBLIOGRAPHY

Guth, A. (1997). *The Inflationary Universe*. Reading, MA: Addison-Wesley.

Rees, M. (1997) *Before the Beginning*. Reading, MA: Addison-Wesley.

Weinberg, S. (1977). *The First Three Minutes*. New York: Basic Books.

BIOFUELS

Biofuels are biomass (organic matter) or biomass products used for energy production. Energy created from the use of biofuels is often termed bioenergy. Biomass crops grown for the primary purpose of use in biofuels are called energy crops. Biofuels include wood and wood wastes, domestic wastes, agricultural crops and wastes, animal wastes, peat, and aquatic plants. Almost any type of combustible organic matter can potentially be used as an energy source.

Plants store solar energy by photosynthesis. During photosynthesis, carbon dioxide (CO_2) and water (H_2O) in the presence of light are converted into glucose ($C_6H_{12}O_6$) by the following chemical equation:

$$6CO_2 + 6H_2O + light \rightarrow C_6H_{12}O_6 + 6O_2.$$

Further processes in the plant make more complex molecules from the glucose. The exact makeup of biomass varies with type, but in general it has the chemical formula of $(CH_2O)_n$ and on average is about

75 percent carbohydrates or sugars and 25 percent lignin, a polymer that holds plant fibers together.

Biofuels are used to create a wide variety of energy sources. Ever since the harnessing of fire, biomass has been used for heating and cooking. Residential burning of biomass continues to be a primary source of fuel in less industrialized nations, but also has been used as fuel for electricity generation, and converted to liquid transportation fuels.

CURRENT USE OF BIOFUELS

Despite the fact that the world's biomass reserves are declining due to competing land use and deforestation, worldwide there remains more energy stored in biomass than there is in the known reserves of fossil fuels. Trees account for the largest amount of biomass. Currently biomass is the source of about percent of the energy used worldwide, primarily wood and animal dung used for residential heating and cooking. In developing countries, where electricity and motor vehicles are more scarce, use of biofuels is significantly higher (approximately 35 percent on average). At the higher end are countries such as India, where about 55 percent of the energy supply comes from biomass. Geography also is a determining factor; in some industrialized countries that have large sources of natural biomass forests near urban cities, such as Finland, Sweden, and Austria, there is a relatively high utilization of bioenergy (18, 16, and 13 percent, respectively). Municipal waste, which can be incinerated for energy production, also can be a large source of biomass for developed regions. France, Denmark and Switzerland recover 40, 60, and 80 percent of their municipal waste respectively.

At the low end is the United States, where biomass energy accounted for only about 3 percent (2.7 quadrillion Btus) of the total energy consumption in 1997. However, biomass use had been rising over the previous five years at an average rate of about 1 to 2 percent per year, but fell in 1997 due to a warmer-than-average heating season. Bioenergy produced in the United States is primarily from wood and wood waste and municipal solid waste.

These divergent energy production patterns between the developing world and the United States are understandable. Heating and cooking are the major uses of biomass in the developing world because of affordability, availability, and convenience.

In the United States, where clean and convenient natural gas, propane, and electricity are widely available and affordable, biomass use has limited potential. Nevertheless, U.S. biomass energy production has been increasing because of technological advances for new and improved biomass applications for electricity generation, gasification, and liquid fuels.

The sources of biofuels and the methods for bioenergy production are too numerous for an exhaustive list to be described in detail here. Instead, electricity production using direct combustion, gasification, pyrolysis, and digester gas, and two transportation biofuels, ethanol and biodiesel, are discussed below.

ELECTRICITY GENERATION

In the United States about 3 percent of all electricity produced comes from renewable sources; of this a little more than half comes from biomass. Most biomass energy generation comes from the lumber and paper industries from their conversion of mill residues to in-house energy. Municipal solid waste also is an important fuel for electricity production; approximately 16 percent of all municipal solid waste is disposed of by combustion. Converting industrial and municipal waste into bioenergy also decreases the necessity for landfill space.

These applications avoid the major obstacles for using biomass for electricity generation: fluctuation in the supply, and the type of biomass available. Seasonal variations and differing quality of feedstock are the biggest barriers to more widespread use. This is especially true for biomass wastes.

COMBUSTION

Combustion is the burning of fuels to produce heat. To produce energy, the heat from the combustion process is used to create steam, which in turn drives turbines to produce electricity.

Most electricity from biofuels is generated by direct combustion. Wood fuels are burned in stoker boilers, and mill waste lignin is combusted in special burners. Plants are generally small, being less than 50 MW in capacity. There is considerable interest in combustion of biomass in a process called cofiring, when biomass is added to traditional fuels for electricity production. Cofiring is usually done by adding biomass to coal, but biomass also can be cofired with

Wood chips, coal, and water—almost any type of combustible organic matter—can potentially be used as an energy source. (U.S. Department of Energy)

oil. There are several biomass cofiring plants in commercial operation in the eastern United States. The U.S. Department of Energy estimates that by 2020 the capacity for biomass cofiring could reach 20 to 30 GW. Cofiring has the advantage of requiring very little capital cost since most boilers can accommodate approximately 5 to 10 percent of biomass without modifications.

Estimates for delivery fuel costs for woody bio-mass range between $1.25 and $3.90 per million Btus compared to $0.90 to $1.35 per million Btus for coal. The cost associated with biomass electricity depends largely on the proximity of the plant to the biomass source and whether the feed is a waste material. At 10,000 Btu/kWh generation heat rate, each $1 per million Btus translates to 1 cent per kWh electrical cost. Thus biomass electricity costs can range from competitive with coal to several cents per kWh more expensive.

Cofiring biomass has environmental benefits in addition to lowering greenhouse gases. Since biomass has little or no sulfur, sulfur dioxide (SO_2) emissions are less when biomass fuels are used. In the United States, power plants have allowable sulfur dioxide levels for each gigawatt of power produced. If they produce less than the allowable amount of sulfur dioxide, they receive credits with which they can trade on the open market. The price for these sulfur dioxide credits is about $70 to $200 per ton.

Biomass also has lower levels of nitrogen than fossil fuels, leading to lower nitrogen oxide formation. The high water content in biomass also lowers the combustion temperature, decreasing the formation of thermal nitrogen oxides. In some cases this can lead to nonlinear reductions; for example, in one study when 7 percent wood was cofired with coal, nitrogen oxides emissions decreased by 15 percent. However, such reductions are not seen in all cases. Reburning is possible when using most biomass feedstocks and also can lower emissions.

Use of some biomass feedstocks can increase potential environmental risks. Municipal solid waste can contain toxic materials that can produce dioxins and other poisons in the flue gas, and these should not be burned without special emission controls. Demolition wood can contain lead from paint, other heavy metals, creosote, and halides used in preservative treatments. Sewage sludge has a high amount of sulfur, and sulfur dioxide emission can increase if sewage sludge is used as a feedstock.

GASIFICATION

Gasification of biofuels, which is in the early developmental stage, has been the focus of much recent research, since it has the potential of providing high conversion. During gasification, biomass is converted to a combustible gas by heating with a substoichiometric amount of oxygen. The biomass can be heated either directly or with an inert material such as sand. In some cases steam is added. The product gas consists of carbon monoxide, methane and other hydrocarbons, hydrogen, and noncombustible species such as carbon dioxide, nitrogen, and water; the relative amount of each depends on the type of biomass and the operating conditions. Generally the product gas has an energy content about one-half to one-quarter that of natural gas. The gas is cleaned by removing tars, volatile alkali, ash, and other unwanted materials. The gas is then sent to a steam boiler or combustion turbine for electricity production by a Rankine cycle or a combined cycle (IGCC). Use of gasification technology with an IGCC can double the efficiency of average biomass electricity production using advanced turbine technology.

The capital cost of an IGCC plant for biomass or coal is in the range of $1,500 to $2,000 per installed kW. A comparable natural gas fire facility costs about $750 to $1,000. The economics of biomass electricity based on IGCC technology depend on the relative cost of natural gas and biomass fuels. Biomass must be lower in cost than gas to pay back the additional capital cost of gas production and cleaning. A 1999 estimate suggests that the biomass would have to be $3 per million Btus cheaper than natural gas for biomass to be economical.

PRYOLYSIS

Another emerging area in biofuels is pyrolysis, which is the decomposition of biomass into other more usable fuels using a high-temperature anaerobic process. Pyrolysis converts biomass into charcoal and a liquid called biocrude. This liquid has a high energy density and is cheaper to transport and store than the unconverted biomass. Biocrude can be burned in boilers or used in a gas turbine. Biocrude also can be chemical by altered into other fuels or chemicals. Use of pyrolysis may make bioenergy more feasible in regions not near biomass sources. Biocrude is about two to four times more expensive than petroleum crude.

BIOGAS PRODUCTION

Biogas is composed primarily of methane (CH_4) and carbon dioxide. Biogas is a by-product from anaerobic bacteria breaking down organic material. Large amounts of biogas can be released from areas such as livestock waste lagoons, sewage treatment plants, and landfills. Since biogas is primarily methane, it is similar to natural gas and can be used for energy generation, especially electricity using stationary engine-generators. The goals of capturing biogas are often to prevent these greenhouse gases from being released into the atmosphere, to control odor, and to produce fertilizer; energy production is secondary. Methane is a potent greenhouse gas, with twenty-one times the global warming potential of carbon dioxide. However, when methane is burned, it produces less carbon dioxide per Btu than any other hydrocarbon fuel.

Economics for generating electricity from biogas can be favorable. Landfill gas from municipal solid waste can of supply about 4 percent of the energy consumed in the United States. In 1997, a total of 90 trillion Btus were generated by landfill gas, about 3 percent of total biomass energy consumption.

TRANSPORTATION FUELS

Although biomass used directly for heating and cooking is the thermodynamically most efficient use, followed by use for electricity generation, the economics are much more favorable to convert to a liquid fuel. Economic considerations outweigh thermodynamics; as an electricity generator, biomass must compete with relatively low-priced coal, but as a liquid fuel the competition is higher-priced oil.

Transportation fuels are the largest consumers of crude oil. Petroleum-based transportation fuels are responsible for 35 percent of greenhouse gas emissions in the United States. Only percent of transportation fuels comes from renewable nonpetroleum-based sources, primarily from the use of corn-based ethanol blended with gasoline to make gasohol. Increased use of biofuels could lower some of the pollution caused by the use of transportation fuels.

ETHANOL

The chemical formula for ethanol is CH_3CH_2OH. Ethanol is less toxic and more biodegradable than gasoline. For its octane boosting capability ethanol can be use as a fuel additive when blended with gasoline.

Demand for gasoline is 125 billion gals (473 billion l) per year according to 1998 estimates. The Clean Air Act Amendment of 1990 mandates the use of oxygenated fuels such as ethanol blends with up to 3.5 percent oxygen by weight in gasoline (E-10 or

Corn can be used to produce ethanol. (U.S. Department of Energy)

gasohol). Reformulated gasoline (RFG) is required year-round in areas that are not in compliance for ozone, and oxyfuels are required in the winter in areas that are not in compliance for carbon monoxide. These "program gasolines" total about 40 billion gals (151 billion l) per year.

In 1997 a total 1.3 billion gals of ethanol fuel was produced in the United States. Proposed new low sulfur conventional gasoline standards could greatly increase the demand for ethanol since desulfurization may lower gasoline octane. Almost all fuel ethanol is used as gasohol, but some is used to make E-85 (85% ethanol and 15% gasoline). E-85 can be used in flexible-fuel vehicles (FFVs) which can operate on gasoline or ethanol blends of to 85 percent ethanol.

Eighty-seven percent of the ethanol produced in the United States comes from corn. The remainder comes from milo, wheat, food wastes, and a small amount from wood waste. In Brazil, the largest producer of transportation biofuels, sugar cane is converted into ethanol at the rate of 16 billion l per year. There are 3.6 million cars in Brazil that run on 100 percent ethanol.

Ethanol is more costly to produce than gasoline. The cost of production of ethanol from corn ranges from about $0.80 per gal ($0.21 per l) for large depreciated wet mills to $1.20 per gal ($0.32 per l) for new dry mills. Better engineering designs, the development of new coproducts, and better uses for existing coproducts will help to lower the production cost. For example, recovering the corn germ in dry mills, which is currently in the development stage could lower ethanol production costs by $0.07 to $0.20 per gal ($0.02 to $0.05 per l). However, ethanol currently used for fuel is not competitive with gasoline without a federal excise tax exemption.

While the corn-to-ethanol industry is mature, conversion of energy crops to ethanol is in the commercial development stage. Engineering studies in 2000 estimate the cost of production per gallon for biomass ethanol at $1.22 per gal ($ 0.32 per l). The U.S. Department of Energy projects that technical advances can lower the cost to $0.60 per gallon. This would make ethanol competitive (without a tax exemption) on an energy basis with gasoline when petroleum is $25 per barrel.

To use biomass material, ethanol needs to be produced from the cellulose portion of the biomass, not just from the starch or sugars. Cellulose is more resistant to breakdown than starch or sugars, so different production methods are required. Acid-catalyzed reactions can be used for the breakdown of cellulose into products that can be converted into alcohol. This process, however, is expensive, and there are problems with the environmental disposal of dilute acid streams. Research for the development of an enzyme to break down cellulose began after World War II. It was discovered that a specific microbe, *Trichoderma reesei,* was responsible for the decomposition of canvas (cellulose) tents in tropical areas. Research on this microbe and others is being conducted. Using genetic engineering, new enzymes are being produced with the primary goal to increase efficiency of alcohol production from cellulose.

BIODIESEL

Biodiesel is diesel fuel produced from vegetable oils and other renewable resources. Many different types of oils can be used, including animal fats, used cooking oils, and soybean oil. Biodiesel is miscible with petroleum diesels and can be used in biodiesel-diesel blends. Most often blends are 20 percent biodiesel and 80 percent traditional diesel. Soy diesel can be used neat (100%), but many other types of biodiesel are too viscous, especially in winter, and must be used in blends to remain fluid. The properties of the fuel will vary depending on the raw material used. Typical values for biodiesel are shown in Table 1.

Biodiesel does not present any special safety concerns. Pure biodiesel or biodiesel and petroleum diesel blends have a higher flash point than conventional diesel, making them safer to store and handle. Problems can occur with biodiesels in cold weather due to their high viscosity. Biodiesel has a higher degree of unsaturation in the fuel, which can make it vulnerable to oxidation during storage.

To produce biodiesel, the oil is transformed using

Density (@298 K), kg/m³	860-900
Net heating value, MJ/kg	38-40
Viscosity @ 40 °C mm²/s (cSt)	3.5-5.0
Cold Filter Plugging Point, K	269-293
Flash Point, K	390-440
Cetane Number	46-62

Table 1.
Typical Values for Biodiesel

a process of transesterification; agricultural oil reacts with methanol in the presence of a catalyst to form esters and glycerol. These monoalkyl esters, otherwise known as biodiesel, can operate in traditional diesel combustion-ignition engines. Glycerol from the transesterification process can be sold as a coproduct. Low petroleum prices continue to make petroleum-based diesel a more economical choice for use in diesel engines.

Current consumption of transportation diesel fuel in the United States is 25 billion gal (94.6 billion l) per year. The total production of all agricultural oils in the United States is about 2 billion gal (7.6 billion l) per year of which 75 percent is from soybeans. Total commodity waste oils total about 1 billion gal (3.8 billion l) per year. The amount of other truly waste greases cannot be quantified. Sewage trap greases consist of primarily free fatty acids and are disposed of for a fee. Trap greases might amount to 300 million gal (1.1 billion l) per year of biodiesel feedstock. The production of biodiesel esters in the United States in 1998 was about 30 million gal (114 million l). The most common oil used is soybean oil, accounting for 75 percent of oil production used for most biodiesel work. Rapeseed oil is the most common starting oil for biodiesel in Europe.

Production costs for biodiesel from soybean oil exceeds $2.00 per gal ($0.53 per l), compared to $0.55 to $0.65 per gal ($0.15 to $0.17 per l) for conventional diesel. The main cost in biodiesel is in the raw material. It takes about 7.7 lb (3.5 kg) of soybean oil valued at about $0.25 per lb ($0.36 per kg) to make 1 gal (3.8l) of biodiesel. Waste oils, valued at $1 per gal ($3.79 per l) or less, have the potential to provide low feedstock cost. However, much "waste oil" is currently collected, reprocessed as yellow and white greases, and used for industrial purposes and as an animal feed supplement. Production of biodiesel from less expensive feedstocks such as commodity waste oil still costs more than petroleum diesel. Research has been done to develop fast-growing high-lipid microalgae plants for use in biodiesel production. These microalgae plants require high amounts of solar radiation and could be grown in the southwestern United States.

In addition to greenhouse benefits, biodiesels offer environmental advantages over conventional diesel. Biodiesels produce similar NO_x emissions to conventional diesel, fuel but less particulate matter. Biodiesel is more biodegradable that conventional diesel making any spills less damaging in sensitive areas. In general biodiesel provides more lubrication to the fuel system than low-sulfur diesel.

ENERGY INPUT-ENERGY OUTPUT OF BIOFUELS

Since the Sun, through photosynthesis, provides most of the energy in biomass production, energy recovered from biofuels can be substantially larger than the nonsolar energy used for the harvest and production. Estimates on conversion efficiency (energy out to non-solar energy in) of ethanol can be controversial and vary widely depending on the assumptions for type of crop grown and farming and production methods used. Net energy gain estimates for converting corn to ethanol vary between 21 and 38 percent. Conversion efficiencies can be increased if corn stover (leaves and stocks) is also used and converted to ethanol. Research is being conducted on converting other crops into ethanol. Switchgrass, a perennial, is one of the most promising alternatives. It has a net energy gain as high as 330 percent since it only has to be replanted about every ten years and because there are low chemical and fertilizer requirements. Net energy gains for the production of biodiesel are also high, with estimates ranging between 320 and 370 percent.

FUTURE USE OF BIOFUELS

One of the main benefits from future use of biofuels would be the reduction of greenhouse gases compared to the use of fossil fuels. Carbon dioxide, a greenhouse gas that contributes to global warming, is released into the air from combustion. Twenty-four percent of worldwide energy-related carbon emissions in 1997 were from the United States. Carbon

and due to rising energy consumption, are expected to increase 1.3 percent per year through 2015.

When plants grow, they adsorb carbon dioxide from the atmosphere. If these plants are used for biofuels, the carbon dioxide released into the atmosphere during combustion is that which was adsorbed from the atmosphere while they were growing. Therefore the net balance of carbon dioxide from the use of biofuels is near zero. Since some fossil fuel use is required in both the planting and the production of bioenergy, there are some net carbon dioxide and other greenhouse gases released into the atmosphere. In determining the net carbon dioxide balance, important variables include growth rates, type of biomass, efficiency of biomass conversion, and the type of fossil fuel used for production. The amount of carbon accumulated in the soil and the amount of fertilizers used also have a large effect on the carbon balance. In particular, nitrous oxide (N_2O), a powerful greenhouse gas, can be released as a result of fertilizer application. Estimates for the amount of greenhouse emissions recycled using biomass for energy production range from a low of 20 to a high of 95 percent. Wood and perennial crops have higher greenhouse gas reduction potential than annual crops. Using biomass to replace energy intensive materials also can increase the carbon balance in favor of energy crops. It is estimated that of the nation's annual carbon dioxide emissions could be reduced by 6 percent if 34.6 million acres were used to grow energy crops.

There is some greenhouse gas benefit from planting forests or other biomass and leaving the carbon stored in the plants by not harvesting. However, over the long term, increased carbon dioxide benefits are realized by using land that is not currently forested for growing some energy crops such as fast-growing poplar. The added benefits come from the displacing fossil fuels by the use of biofuels, since energy crops can be repeatedly harvested over the same land.

In the calculation of greenhouse gas benefits of planting energy crops, many assumptions are made. Among them is that the land will be well managed, appropriate crops for the region will be used, there will be careful use of fertilizers and other resources, and efficient production methods will be employed to get the maximum amount of energy from the biomass. Most importantly, it is assumed that biomass is grown in a sustainable manner. Harvested biomass that is not replanted increases greenhouse gas emissions in two ways: Carbon dioxide that had been previously stored in trees is released in the atmosphere, and future carbon fixation is stopped.

To comply with carbon reduction goals, some countries impose taxes on carbon dioxide emissions. Since biofuels have lower full-cycle carbon dioxide emissions than fossil fuels, biofuels are more cost-competitive with fossil fuels in regions where these taxes are imposed.

Another advantage to using biomass as an energy source is a possible increase in energy security for countries that import fossil fuels. More than two-thirds of the oil reserves are in the Middle East. More than half of the oil consumed in the United States is imported and oil accounts for approximately 40 percent of the trade deficit of the United Sates. A substantial biofuels program could help to the increase energy independence of importing nations and lessen the impact of an energy crisis.

There are some disadvantages with the use of bio-fuels as well. Some of the high-yield energy crops also have significant removal rates of nutrients from the soil. Each year the cultivation of row crops causes a loss of 2.7 million metric tons of soil organic matter in the United States. However, there are exceptions: Through the use of good farming practices, Brazilian sugarcane fields have had minimal deterioration from the repeated planting of sugarcane. Moreover, using switchgrass and other grasses increases soil organic matter and thus can help in reducing the soil erosion caused by the cultivation of rowcrops. Research is being conducted into improving sustainable crop yield with a minimal of fertilizer application. Possible solutions include coplanting energy crops with nitrogen-fixing crops to maintain nitrogen levels in the soil.

It is estimated that biomass is cultivated at a rate of 220 billion dry tons per year worldwide. This is about ten times worldwide energy consumption. Advocates suggest that by 2050, better use of cultivated biomass could lead to biomass providing 38 percent of the world's direct fuel and 17 percent of electricity generation. However, a large increase in bioenergy seems unlikely. When the U.S. Energy Information Administration (EIA) does not include any new greenhouse gas legislation into its energy utilization projections, only limited growth for renewable energy is predicted. The EIA estimates an average increase of 0.8 percent per year for fuels through 2020 and an average increase of 0.5 percent for renewable electrical generation without new legislation. Most of the

increase comes from wind, municipal solid waste, and other biomass. The reason for low expected growth in biofuels is that natural gas and petroleum prices are expected to remain relatively low over the next few decades; in 2020 the average crude oil price is projected to be $22.73 a barrel (in 1997 dollars). The average wellhead price for natural gas is projected to increase from $2.23 per thousand cu ft ($2.17 per million Btus to $2.68 per thousand cu ft ($2.61 per million Btus in 2020 (prices in 1997 dollars). Low fossil fuel prices make it difficult for alternative fuels to compete. Projections for the amount of biomass energy use do rise, however, if it is assumed that the Kyoto protocols limiting greenhouse gases will be adopted, since biofuels contribute fewer greenhouse emissions than do fossil fuels. In the case where greenhouse gas emissions are kept to 1990 levels, renewable energy could account for as much as 22 percent of electricity generation in 2020. Even under this scenario, the biggest change in greenhouse gas emissions comes from a decrease in coal use and an increase in natural gas use.

While considerable amounts of biomass exist as wastes, the costs of collection, storage, transportation, and preparation are high. The largest obstacle for the wider us of biofuels is economics, but niche opportunities exist. Strategies to improve economics include extracting high-valued coproducts from the cellulosic matrix, offsetting disposal costs and mitigating environmental problems by using the waste.

Agricultural wastes such as corn stover (stalks, leaves, etc.) have been proposed as bioenergy sources. The annual planted corn acreage is near 80 million acres, and up to 1.5 tons of stover per acre could be collected. In many farm locations stover has a competitive use as animal feed, but in areas where higher-valued uses do not exist, it may be collected and used as an industrial feedstock. In California, rice straw presents a disposal problem, since burning has been disallowed, and the rice straw could be used for ethanol production. Alfalfa growers in Minnesota are developing a technology to separate stems from the protein-containing leaves. Since protein sources are economically valued on a ton-of-protein basis, the stems are available at essentially no cost for electricity generation. Diversion of demolition wood collected in urban areas from landfills also could yield low-cost fuels. However, if biomass is to become a large component of U.S. energy use, it will have to be grown commercially as an energy crop.

Because the energy density of biomass is much

Ethanol-powered snowplow in Hennepin County, Minnesota. (U.S. Department of Energy)

lower than that of fossil fuels, most cost analyses suggest that in order for conversion of biomass to fuels to be economical, the biomass source needs to be close to the processing facility, usually within fifty miles. Lower energy density also means that storage costs can be higher than with fossil fuels, and unlike fossil fuels, it is wholly important that storage time is minimized because weather and bacteria can lower the energy quality of the biomass.

The U.S. Department of Agriculture reports that in 1997 there were 432 million acres of cropland in the United States, of which 353 million acres were planted. Idled cropland accounted for 79 million acres, of which 33 million acres were in the Conservation Reserve Program (CRP). Some planted cropland as well as some or all the idled cropland may be available for energy crops depending on the ability of energy crops to compete economically with traditional crops and on public policy related to the use of CRP land. A 1999 study from University of

Tennessee's Agricultural Policy Analysis Center and Oak Ridge National Laboratory used the POLYSYS (Policy Analysis System) model to estimate the amount of land that might be used for energy crops in 2008 based on two different scenarios. Under both scenarios it is assumed that producers are allowed to keep 75 percent of the rental rate paid by the U.S. government for CRP acreage. In both cases, switchgrass was the energy crop with the most economic potential. In the first scenario, it is assumed that the price for energy crops is $30 per dry ton ($2 per million Btus) and there are strict management practices in the CRP; in this case it is estimated that switchgrass would be competitive on 7.4 million acres. In the second scenario, it is assumed that the price for energy crops is $40 per dry ton ($2.70/per million Btus) and that there are lenient management practices in the CRP; under this scenario it is estimated that switchgrass would be competitive on 41.9 million acres. This would result in an increased annual ethanol production on the order of 4 billion to 21 billion gal (15 billion to 79 billion l) compared to the current corn ethanol production of about 1.5 billion gal (5.7 billion l) per year, or sufficient fuel for 6,000 to 36,000 MW of electrical generating capacity. Such a program could provide additional benefit to farmers by reducing the supply of commodity crops and in turn raising crop prices.

With dedicated feedstock supply systems, energy crops are grown with the primary purpose of energy generation. This means that fuel processors and growers will need to enter into long-term fuel supply contracts that provide early incentives to growers to tie up land. Woody species require four to seven years from planting to harvest. Switchgrass crops require approximately two years from planting to first harvest. High-growth species of poplar, sycamore, eucalyptus, silver maple, and willow are all being tested as energy crops. Hybrid species are being developed for pest and disease resistance. Willows have the advantage that common farm equipment can be modified for harvesting. Selection of biomass depends on many factors including climate, soil, and water availability.

Research is being done in the United States and worldwide to lower some of the barriers to biofuels. Researchers hope to develop high-yield, fast-growing feedstocks for reliable biomass fuel supplies. Research is also being done to improve the efficiency of energy conversion technologies so that more of the biomass is utilized.

Deborah L. Mowery

See also: Agriculture; Biological Energy Use, Cellular Processes of; Biological Energy Use, Ecosystem Functioning of; Diesel Fuel; Environmental Economics; Environmental Problems and Energy Use; Fossil Fuels; Gasoline and Additives; Geography and Energy Use; Green Energy; Hydrogen; Methane; Nitrogen Cycle; Renewable Energy; Reserves and Resources; Residual Fuels; Waste-to-Energy Technology.

BIBLIOGRAPHY

Bhattacharya, S. C. (1998). "State of the Art of Biomass Combustion." *Energy Sources* 20:113-135.

Bridgwater, A. V., and. Double, J. M. (1994). "Production Costs of Liquid Fuels from Biomass." *International Journal of Energy Research* 18:79-95.

Hinman, N. D. (1997). "The Benefits of Biofuels." *Solar Today* 11:28-30.

Hohenstein, W. G., and Wright, L. L. (1994). "Biomass Energy Production in the United States: An Overview." *Biomass and Bioenergy* 6:161-173.

Johansson, T. B.; Kelly, H.; Reddy, A. K. N.; Williams, R. H.; and Burnham, L. (1993). *Renewable Energy Sources for Fuels and Electricity.* Washington, DC: Island Press.

Kendall, A.; McDonald, A.; and Williams, A. (1997). "The Power of Biomass." *Chemistry and Industry* 5:342-345.

Oritz-Canavate, J. V. (1994). "Characteristics of Different Types of Gaseous and Liquid Biofuels and Their Energy Balance." *Journal of Agricultural Engineering Research* 59:231-238.

Sampson, R. N.; Wright, L. L.; Winjum, J. K.; Kinsman, J. D.; Benneman, J.; Kürsten, E.; Scurlock, J. M. O. (1993). "Biomass Management and Energy." *Water, Air, and Soil Pollution* 70:139-159.

Schlamdinger, B., and Marland, G. (1996). "The Role of Forest and Bioenegy Strategies in the Global Carbon Cycle." *Biomass and Bioenergy* 10(5/6):275-300.

Scholz, V.; Berg, W.; and Kaulfuβ, P. (1998) "Energy Balance of Solid Biofuels." *Journal of Agricultural Engineering Research* 71:263-272.

Wright, L. L., and Hughes, E. E. (1993). "U.S. Carbon Offset Potential Using Biomass Energy Systems." *Water, Air, and Soil Pollution* 70:483-497.

BIOLOGICAL ENERGY USE, CELLULAR PROCESSES OF

Just as an internal combustion engine requires fuel to do work, animals need fuel to power their body processes. Animals take in complex molecules as food and break them down to release the energy they contain. This process is called "catabolism." Animals use the energy of catabolism to do work and to assemble complex molecules of their own from simple building blocks, a process called "anabolism." The sum of anabolism and catabolism is "metabolism," a broad term that includes all chemical reactions in the body.

LIVING SYSTEMS FOLLOW THE RULES OF THERMODYNAMICS

Living organisms are extremely complex. Perhaps this is the reason we often forget that all animals, including people, are made up entirely of chemicals and that these chemicals react with each other according to the same rules that govern chemical reactions in test tubes. Indeed, as recently as the 1800s some scientists believed that living organisms contained a "vital force" not found in inanimate objects that was necessary for life and controlled life processes. This idea, known as vitalism, is now rejected by science because this vital force has never been found and we can explain the chemical reactions in the body without resorting to the mystical thinking inherent in vitalism.

Two energy laws that apply to both living and non-living systems are the first and second laws of thermodynamics. The first law states that energy can be neither created nor destroyed, but can only be changed from one form to another. The second law states that the "entropy" (randomness, given the symbol S) of a closed system will increase spontaneously over time. At first glance, this second law would seem to make life itself impossible because living organisms increase in order and complexity (negative S) as they develop, and then maintain this order throughout adulthood. However, living organisms are not closed systems. They are able to maintain and even decrease their entropy through the input of energy from food (animals) or sunlight (plants).

The amount of energy contained in the bonds of a chemical is called the "free energy" of that chemical (given the symbol G). To understand how free energy and entropy are related, consider the following chemical reaction:

$$AB \rightarrow A + B. \qquad (1)$$

The complex substrate molecule (AB) is broken down to simpler product molecules (A and B). The substrate and the products each have both a free energy and an entropy. For this, and for all chemical reactions the following relationship applies:

$$\Delta G = \Delta H - T\Delta S \qquad (2)$$

where H stands for heat given off or taken up during the reaction and T stands for the absolute temperature of the reaction. G and S stand for free energy and entropy, and the Δ symbol means "change in" the variable during the reaction. Thus, Equation (2) can be stated as follows: for any chemical reaction, the change in free energy between the substrates and the products (ΔG = free energy of products - free energy of substrates) is equal to the amount of heat given

off or taken up during the reaction (ΔH) minus the product of the reaction temperature times the change in entropy between the substrates and the products (ΔS = entropy of products - entropy of substrates). Reactions that give off heat are called "exothermic" reactions and have negative ΔH values. Reactions that take up heat are called "endothermic" reactions and have positive ΔH values.

According to the second law of thermodynamics, for a reaction to proceed spontaneously it must produce an increase in entropy ($\Delta S > 0$). Because most spontaneous chemical reactions in the body are exothermic ($\Delta H < 0$), most spontaneous chemical reactions will have ΔG values less than zero as well. This means that if, in the reaction shown in Equation (1) above, we begin with equal amounts of substrates and products ([AB] = [A] × [B]), the reaction will proceed spontaneously (AB will be converted spontaneously to A and B) because the free energy contained in the bonds of AB is greater than the free energy contained in the bonds of A and B ($\Delta G < 0$). The more negative the value of ΔG, the greater the fraction of the available AB that will be converted to A and B.

In a practical sense, we can make use of Equation (2) above to understand this process in the following way. When a large complex molecule is broken down to smaller simpler molecules, energy is released because the smaller molecules contain less energy in their chemical bonds than the complex molecule ($\Delta G < 0$). Assuming the reaction is exothermic ($\Delta H < 0$), this energy will be released partially as heat and partially as an increased randomness in the chemical system ($\Delta S > 0$) and the reaction will occur spontaneously.

COUPLED REACTIONS CAPTURE ENERGY AND DRIVE "UNFAVORABLE" PROCESSES

While heat is vital to the human body, the reader may (quite correctly) suspect that the main reason we are interested in the energy released in chemical reactions such as Equation (1) is that this energy can also be captured, stored and used later to do useful work in the body. The energy of a chemical reaction is captured when an energy-releasing reaction ($\Delta G < 0$) is "coupled" to a reaction that requires energy. Consider the coupled set of reactions below:

$$AB \rightarrow A + B + energy$$
$$ADP + P_i + energy \rightarrow ATP. \qquad (3)$$

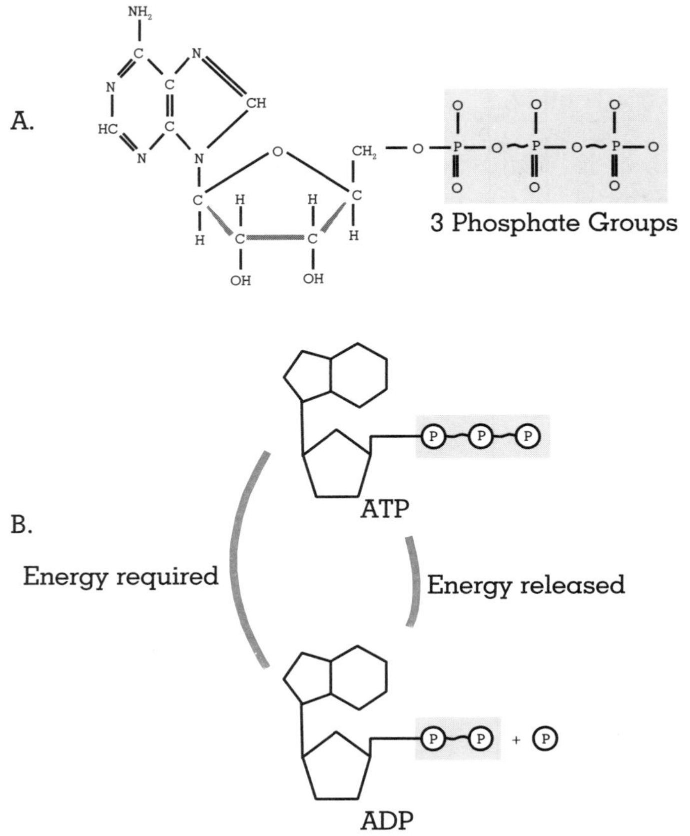

Figure 1.
The structure and formation of ATP. (A) The chemical structure of adenosine triphosphate (ATP). "C" indicates carbon, "N" nitrogen, "O" oxygen, "H" hydrogen and "P" phosphorus. Note the negative charges on the phosphate groups (PO_3^-). (B) ATP can be formed from adenosine diphosphate (ADP).

In these simultaneous reactions, the energy released when the complex molecule AB is broken down is immediately used to build a molecule of adenosine triphosphate (ATP) from a molecule of adenosine diphosphate (ADP) and an inorganic phosphate (P_i). ATP is a high energy compound. It is called the "energy currency" of the body because once it is formed, it provides energy that the body can "spend" later to drive vital reactions in cells (Figure 1).

ADP consists of an adenosine group bound to two phosphates, while ATP is the same molecule with a third phosphate bound. The phosphate groups have negative charges and repel each other as two magnets would if their negative poles were placed close together. Thus there is considerable energy in the bond that holds the second and third phosphates of ATP together. In the coupled reactions shown in Equation (3), the energy released from the breakdown of the chemical bonds in AB is transferred to the high-energy bond between the second and third phosphate groups of ATP.

Once a molecule of ATP is formed, it can be used by a cell to do work or to build complex molecules. Let us say that the cells require a complex molecule (XY). This molecule can be formed from its simpler parts (X and Y) in the reaction below:

$$X + Y \rightarrow XY. \tag{4}$$

However the formation of XY will not proceed spontaneously because the free energy of the product (XY) exceeds the free energy of the substrates (X and Y). We refer to the formation of XY as being an "unfavorable" process because, for Equation (4), $\Delta G > 0$. Cells can form the XY they need only by coupling its formation to a reaction, such as the breakdown of ATP, that provides the energy required to build the chemical bonds that hold X and Y together. This process is shown in the coupled reaction below:

$$\text{ATP} \rightarrow \text{ADP} + \text{Pi} + \text{energy}$$
$$\text{X} + \text{Y} + \text{energy} \rightarrow \text{XY} \qquad (5)$$

The energy released from the breakdown of ATP has been used to drive an unfavorable process. A reaction (the formation of XY) that would not have occurred spontaneously has taken place. Of course, the amount of energy required for the formation of one molecule of XY must be less than the amount released when one ATP is broken down, otherwise the system would have gained total energy during the coupled reaction, and violated the first law of thermodynamics.

ENZYMES INCREASE THE RATE OF CHEMICAL REACTIONS

Thus far we have discussed whether a chemical reaction will occur spontaneously or only with the addition of energy. We have said nothing about the rate of chemical reactions—how fast they occur. If we need to release the energy stored in our food to power the pumping of our heart and allow us to move, we need to release that energy rapidly. We cannot afford to wait hours or days for the energy-releasing reactions to occur.

Enzymes are complex molecules, usually proteins, that speed up chemical reactions. Figure 2 illustrates in graphic form how enzymes function. To fully understand Figure 2, imagine a chemical reaction in which a part of one compound is transferred to another compound:

$$\text{C} + \text{DE} \rightarrow \text{CD} + \text{E} \qquad (6)$$

This reaction occurs spontaneously ($\Delta G < 0$), however it will only occur when a molecule of DE collides with a molecule of C with sufficient energy. The amount of energy that must be present in this collision is called the activation energy (E_a) of the reaction. If it is unusual for a molecule of DE to collide with a molecule of C with sufficient energy, the reaction will proceed very slowly. However, if an enzyme is present that binds both C and DE, the substrates will be brought closely together, reducing the activation energy. The rate at which products are formed will increase and the reaction will proceed more quickly in the presence of the enzyme. Note that enzymes have active sites with very specific shapes that bind substrate molecules and are not used up or altered during the reaction.

To understand activation energy, consider a boulder at the top of a hill. Imagine that the boulder has both the potential energy imparted by its position at the top of the hill and additional kinetic energy that causes it to "jiggle" randomly back and forth around its location. The potential energy of the boulder is analogous to the free energy of a chemical while the random motion is analogous to the random thermal motion that all molecules have. Just as chemical reactions with $\Delta G < 0$ proceed spontaneously, the boulder will have a natural tendency to roll down the hill because its potential energy state is lower at the bottom of the hill than at the top. However, if there is a small rise between the boulder and the slope of the hill, the boulder must have enough "activation" energy from its random motion to get over that rise. We could increase the likelihood that the boulder will roll down the hill either by adding more kinetic energy (giving it a push) or by lowering the rise. Enzymes work by lowering the rise (activation energy).

In thermodynamic terms, a spontaneous reaction ($\Delta G < 0$) may proceed only slowly without enzymes because of a large activation energy (E_a). Adding enzymes to the system does not change the free energy of either the substrates or products (and thus does not alter the ΔG of the reaction) but it does lower the activation energy and increase the rate of the reaction.

ANIMAL CELLS EFFICIENTLY CAPTURE THE ENERGY RELEASED DURING CATABOLISM

Animal cells obtain much of their energy from the breakdown (catabolism) of the six-carbon sugar glucose ($C_6H_{12}O_6$). The overall reaction for the catabolism of glucose is:

$$C_6H_{12}O_6 + 6\,O_2 \rightarrow 6\,CO_2 + 6\,H_2O + \text{energy} \qquad (7)$$

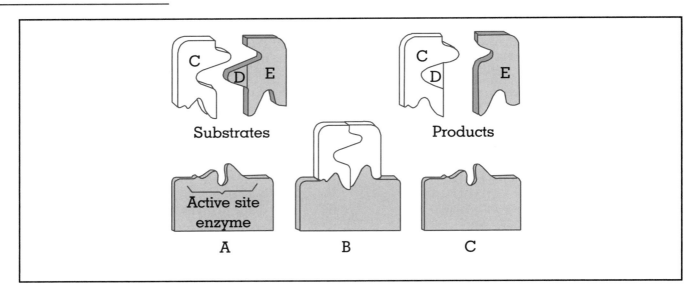

Figure 2.
The action of an enzyme. (A) An enzyme with substrate molecules C and DE. Note that the specific shape of the enzyme's active site matches the shape of the substrate. (B) The enzyme with the substrate molecules bound. (C) The enzyme, unchanged from its original form with the product molecules (CD and E).

In the presence of oxygen (O_2), glucose is broken down to carbon dioxide (CO_2) and water (H_2O). Energy is released because the free energy in the chemical bonds of the products is less than the free energy in the bonds of the glucose. It might seem simplest to couple the energy-liberating breakdown of glucose directly to each energy-requiring process in the body, much as the two chemical reactions in Equation (3) are coupled. However this is not practical. When glucose is broken down in a single step (such as by burning) a large amount of energy is released from every glucose molecule. If the catabolism of a glucose molecule were coupled directly to a process that required only a small amount of energy, the extra energy released from the glucose would be lost as heat. Thus, for efficiency, animal cells break glucose down by a multistep process. Cells release the energy in the bonds of the glucose molecule in a controlled way and capture this energy by using it to produce ATP. The breakdown of ATP, which releases energy in smaller amounts, is then coupled to energy-requiring reactions as in Equation (4).

The first segment of glucose catabolism is called "glycolysis." This process begins when glucose is transported into a cell. In a series of reactions within the cell, each of which requires a specific enzyme, a single six-carbon glucose molecule is converted to two molecules of pyruvic acid (three carbons each). For each molecule of glucose that undergoes glycolysis, two molecules of ADP are converted to ATP, and two molecules of nicotinamide adenine dinucleotide (NAD) accept a hydrogen atom and become NADH. The overall reaction of glycolysis is:

$$\text{Glucose} + 2\,\text{NAD} + 2\,\text{ADP} \rightarrow$$
$$2\,\text{Pyruvic Acid} + 2\,\text{NADH} + 2\,\text{ATP.} \quad (8)$$

The discerning reader will recognize that, while Equation (8) is written correctly, it does not explain one very interesting aspect of glycolysis. In the first two steps of glycolysis, phosphate groups are donated by ATP to glucose. This may seem odd because the goal of glucose catabolism is to liberate energy in the form of ATP but these first steps actually consume ATP! These steps have an important function, however. By adding a charged (polar) phosphate group to the glucose, they make this energy-rich molecule very insoluble in the lipid (nonpolar) cell membrane trapping the glucose inside the cell.

The next steps of glucose catabolism are called the "citric acid cycle." The pyruvic acid formed in glycolysis is transported into the mitochondria, which are subcellular organelles with double (inner and outer) membranes. They are referred to as the "powerhouses" of the cell because they produce most of the ATP. Inside the mitochondria, each three-carbon pyruvic acid molecule is converted to a two-carbon molecule of acetyl-coenzyme-A (acetyl CoA). A molecule of CO_2 is released and a molecule of NADH is generated. The acetyl CoA combines with a four-carbon molecule of oxaloacetic acid, forming the six-carbon molecule citric acid. Then, via a complex set of reactions, each of which requires its own enzyme, the citric acid is reconverted to oxaloacetic acid. Additional molecules of CO_2, NADH, $FADH_2$ (another hydrogen atom acceptor) and ATP are formed in the process. The overall reaction of the citric acid cycle is:

$$\text{Pyruvic Acid} + 4\,\text{NAD} + \text{FAD} + \text{ADP} \rightarrow$$
$$3\,CO_2 + 4\,\text{NADH} + FADH_2 + \text{ATP} \tag{9}$$

The CO_2 generated when pyruvic acid is consumed in this cycle is the CO_2 product seen in Equation (7).

Thus far glucose catabolism has generated only a modest amount of ATP. It has, however, added a substantial number of hydrogen atoms to the hydrogen acceptor molecules NAD and FAD. The NADH and $FADH_2$ that result now pass their hydrogen atoms to a series of proteins in the mitochondrial membrane called the "electron transport system." This system splits the hydrogen atoms into a hydrogen ion (H^+) and an electron. The electron is passed from one protein to the next down the electron transport system. With each transfer, the electron gives up some energy and the protein of the transport system uses this energy to pump hydrogen ions from inside the mitochondrion to the space between the inner and outer membranes. These hydrogen ions then reenter the inner mitochondria through special hydrogen ion channels that capture the energy released in this hydrogen ion movement and use it to convert ADP to ATP. In the final step of glucose catabolism, the hydrogen ions and electrons are combined with oxygen to form water. These are the oxygen and water molecules seen in Equation (7).

Each time NADH gives up an electron to the electron transport system enough H^+ is pumped from the mitochondria to generate three molecules of ATP. However the energy of one ATP must be used to transport the NADH produced during glycolysis into the mitochondria, so this NADH generates a net gain of only two ATP for the cell. For each $FADH_2$ produced, an additional two ATP are generated.

Thus, we find that for each molecule of glucose broken down the cell obtains:

2 ATP produced directly in glycolysis,
4 ATP from the 2 NADH produced in glycolysis (1 NADH per pyruvic acid),
24 ATP from the 8 NADH produced in the citric acid cycle (4 NADH per pyruvic acid),
2 ATP produced directly in the citric acid cycle, and
4 ATP from the 2 $FADH_2$ produced in the citric acid cycle (1 per pyruvic acid).

This yields a total of 36 ATP molecules produced per molecule of glucose consumed. The reader can now appreciate why it is vital that cells release the energy of glucose slowly in a multistep process rather than all at once in a single step. If glucose were broken down in a single reaction, the cell could never couple this breakdown to so many ATP-producing reactions and much energy would be lost. We can also see now why oxygen is vital to cells. In the absence of oxygen, glycolysis can proceed because the pyruvic acid generated is converted to lactic acid. (It is this lactic acid that makes muscles "burn" during heavy exercise.) This generates a small amount of ATP. However, in the absences of oxygen the electron transport system of the mitochondria backs up because it has no oxygen to accept the electrons and form water. Thus, lack of oxygen greatly reduces the ATP available to the cell and can result in cell death.

CARBOHYDRATE, FAT AND PROTEIN METABOLISM ARE CONNECTED

Both fats and proteins can also be catabolized for energy (i.e., ATP production). Dietary fats enter the bloodstream primarily as triglycerides (a three-carbon glycerol backbone and three fatty acids with sixteen or eighteen carbons in each). In cells, the fatty acids are split from the glycerol. The glycerol is converted to pyruvic acid and enters the mitochondria where it takes part in the citric acid cycle. The fatty acids undergo a process known as "beta-oxidation." During beta-oxidation, which occurs in the mitochondria, a molecule called coenzyme A is attached to the end carbon of the fatty acid chain. Then the last two carbons of the chain are cleaved with the CoA attached, producing a molecule of acetyl CoA,

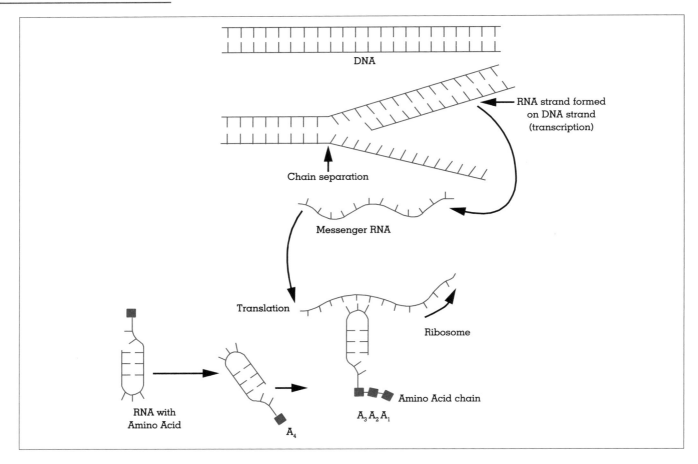

Figure 3.
A chemical reaction that will occur spontaneously because the energy level of the products (P) is less than
  the energy level of the substrate (S). (a) In the absence of an enzyme, activation energy is high. Few molecules have sufficient energy to overcome this barrier and the reaction proceeds slowly if at all. (b) In the
  presence of an enzyme, activation energy is lower and the reaction proceeds more quickly.

and a shortened fatty acid chain. This process is repeated until the entire fatty acid chain has been converted to acetyl CoA. The acetyl CoA enters the citric acid cycle. The catabolism of a single triglyceride molecule with three eighteen-carbon fatty acids yields over 450 molecules of ATP.

Dietary proteins are absorbed into the blood as amino acids, small molecules made up of a carbon backbone and a nitrogen-containing amino group (NH₂). Because protein is relatively rare and difficult to obtain, it is reasonable that the body should metabolize amino acids for energy primarily when other sources (such as sugars and fats) are unavailable. In times of great need, body cells can remove the amino group from amino acids, converting them to a form

that can enter the citric acid cycle. Because muscle is made up primarily of protein, "crash" dieting causes the body to digest muscle tissue for energy.

ANIMAL CELLS USE ATP TO BUILD COMPLEX MOLECULES

Proteins are complex molecules that give cells structure and act as both enzymes and "motors" within cells. Proteins are long strings of amino acids folded in specific three-dimensional formations. There are twenty different amino acids in our bodies. DNA, the genetic material located in the cell nucleus, carries information for the order of the amino acids in each protein. Indeed, in the simplest sense, a "gene" is the section of DNA that carries the information for the construction of a single protein.

We have twenty-three pairs of chromosomes in our cells. Each chromosome is made up of a single huge molecule of DNA and contains many thousands of genes. The process by which the information in a gene instructs the cell in the formation of a protein is illustrated in Figure 3. DNA has the shape of a ladder. The ladder rungs are made up of four different molecules called "nucleotides." The information that the DNA carries is coded in the order of the ladder rungs. When a cell needs a particular protein it begins by "unzipping" the DNA ladder at the gene for that protein, exposing the information on the rungs. Then the cell makes a "messenger RNA" molecule (mRNA) that carries the same information as the gene. This process, called "transcription," requires that the cell build the messenger RNA from nucleotides. The mRNA then leaves the cell nucleus for the cell cytoplasm.

In the cytoplasm, the mRNA attaches to a ribosome and acts as a template for the construction of a protein with the proper amino acid sequence (a process known as "translation"). Single amino acids are brought to the ribosome by "transfer RNA" molecules (tRNA) and added to the growing amino acid chain in the order instructed by the mRNA. Each time a nucleotide is added to the growing RNA strand, one molecule of ATP is broken down to ADP. Each time a tRNA binds an amino acid and each time the amino acid is added to the protein, additional ATP is broken down to ADP. Because proteins can contain many hundreds of amino acids, the cell must expend the energy in 1,000 or more ATP molecules to build each protein molecule.

ANIMAL CELLS USE ATP TO DO WORK

Muscles can exert a force over a distance (i.e., do work). Thus, muscle contraction must use energy. The contractile machinery of muscle is made up of thin filaments that contain the protein "actin" and thick filaments that contain the protein "myosin" (Figure 4a). The myosin molecules have extensions known as "crossbridges" that protrude from the thick filaments. When muscle contracts, these crossbridges attach to the thin filaments at a 90-degree angle, and undergo a shape change to a 45-degree angle (power stroke) that draws the thin filaments across the thick filament. The crossbridge heads then detach, recock to

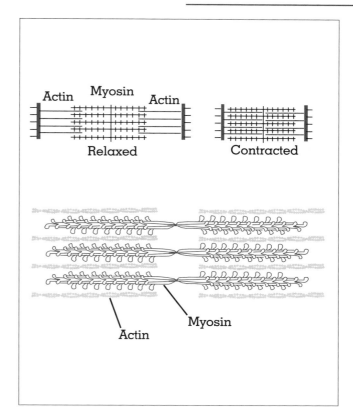

Figure 4a.

The structure and arrangement of the actin and myosin filaments in muscle. During muscle contraction the cyclic interaction of myosin crossbridges with actin filaments draws the actin filaments across the myosin filaments.

90 degrees, reattach to the thin filament and repeat the process. This entire process of myosin interaction with actin is known as the crossbridge cycle (Figure 4b).

Each time a myosin crossbridge goes through its cycle it breaks down one molecule of ATP to ADP and uses the energy released to do work. It would be easier to understand this process if the energy release of ATP breakdown occurred simultaneously with the work performing step—the power stroke; however, a careful examination of Figure 4b reveals that this is not the case. The binding of ATP to myosin allows the myosin crossbridge to detach from the actin-containing thin filament. The breakdown of ATP to ADP with its energy release occurs when the crossbridge is detached and recocks the crossbridge, readying it for another power stroke.

"Efficiency" is the ratio of work done divided by

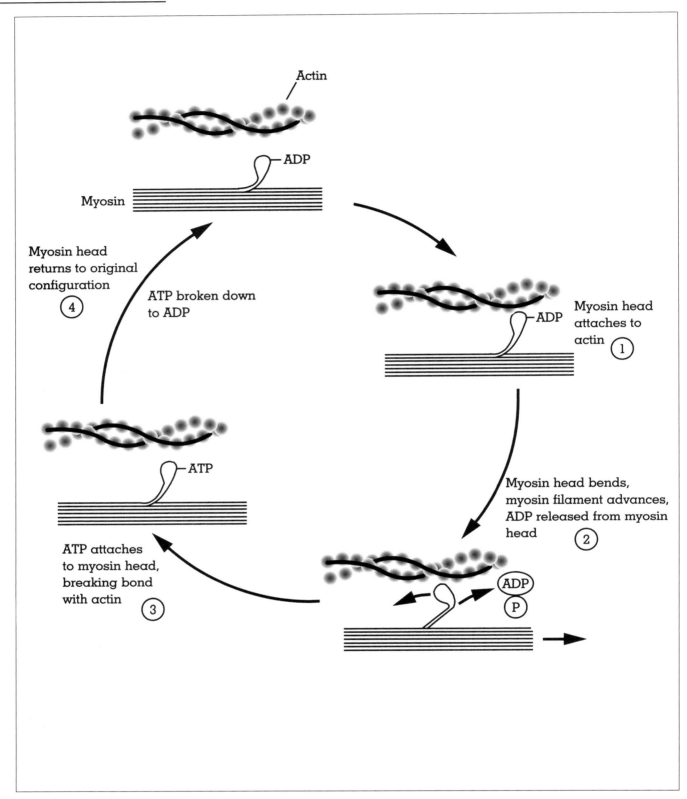

Figure 4b.
The crossbridge cycle in muscle. Myosin crossbridges interact cyclically with binding sites on actin filaments. Note that the energy release step—when ATP is broken down to ADP—recocks the crossbridge head.

energy expended. The efficiency of muscle's conversion of the chemical energy from ATP into mechanical work depends upon the rate of muscle contraction. Imagine an isolated muscle in a laboratory contracting against a weight. If the weight is too heavy for the muscle to lift, the muscle uses energy to develop force but does no work because it cannot move the weight. (Recall that work is equal to force times distance.) Thus, for contractions in which the muscle develops force but does not move a weight (isometric contractions), the muscle has an efficiency of zero. When a muscle applies a constant force to lift a weight through a distance (isotonic contractions), energy use efficiency is greatest (about 50 percent) when the muscle is contracting at one-third its maximum velocity and falls to lower levels when the muscle contracts either faster or more slowly than this. This may seem like a great waste of energy. However, much of the energy that does not do work ultimately appears as heat. This heat may not add to the strict definition of efficiency, but it is not wasted in a biological sense because it serves to maintain our body temperature.

THE METABOLIC RATE IS THE ENERGY OUTPUT OF THE ENTIRE BODY

A "calorie" is the amount of heat energy needed to raise the temperature of one gram of water by 1 degree celsius. Because this is a very small unit compared to the energy needs of the body, we use the kilocalorie, or dietary calorie (1 kilocalorie = 1,000 calories), when discussing total body metabolism. The term "Calorie" (capitalized) refers to kilocalories. The energy output of the entire body is called the "metabolic rate." This rate, expressed as Calories expended per unit time, has several components: 1) the energy needed to maintain life at rest—the basal metabolic rate or BMR, 2) the additional energy needed to digest food, and 3) any additional energy expended to perform exercise and work.

The basal metabolic rate for adults is 1 to 1.2 Calories/minute or 60 to 72 Calories/hour. This energy powers the movement of the chest during respiration and the beating of the heart—processes that are obviously necessary for life. However, a surprisingly large fraction of the BMR is used by cells to maintain ionic gradients between their interior and the fluid that surrounds them (the interstitial fluid or tissue fluid).

The interior of all body cells has a high concen-

tration of potassium ions (K^+) and a low concentration of sodium ions (Na^+). The interstitial fluid and the blood plasma have a high Na^+ concentration and a low K^+ concentration. When electrical signals, known as "action potentials," pass along nerves, protein channels or gates in the nerve cell membrane open and allow sodium to enter the nerve cell and potassium to leave. It is the current carried by these ionic movements that is responsible for the action potential. Once the action potential has passed, the sodium that entered the nerve cell must be pumped back out and the potassium that left must be pumped back in, both against a concentration gradient. Another protein, known as the "sodium-potassium pump" does this pumping, at substantial energy cost. The work of the sodium-potassium pump comprises a significant part of the two-fifths of the BMR resulting from activity of the brain and spinal cord.

In addition to the BMR, the body uses energy to digest and store food. Digestion requires muscular contraction for the motion of the stomach and intestines, as well as the production of digestive enzymes, many of them proteins. The storage of food energy in the form of large molecules also requires energy. For example, glucose subunits are combined and stored as the large molecule glycogen in the liver and muscle. The production of glycogen from glucose requires energy input in the form of ATP.

The energy expenditure needed to produce glycogen is worthwhile for the body because glycogen serves as a ready source of glucose during periods of low food intake and high energy output. Glycogen can be broken down to glucose 6-phosphate (glucose with a phosphate group attached, see Figure 5). In muscle, glucose 6-phosphate is broken down to pyruvic acid through glycolysis and then enters the citric acid cycle. This process provides ATP for muscle contraction (Figure 4b). In the liver, the glucose 6-phosphate is converted to glucose. Without its charged phosphate group, glucose can leave the liver cells and provide for the energy requirements of other tissues, including the brain.

Body activity also adds to the metabolic rate. In general, the more strenuous the activity, the more work is done and the greater the increase in metabolic rate. For an adult male of average size, the BMR (measured lying down) accounts for 1,500-1,600 Calories per day. If this subject sat still but upright in a chair, he would use over 2,000 Calories per day, and if he engaged in prolonged strenuous activity he might expend as much as 10,000 Calories per day. Young people generally have higher metabolic rates than do elderly individuals, partially because younger people have, on average, more muscle mass than the elderly.

The metabolic rate is increased by several hormones including thyroid hormone, adrenalin and male sex hormones. The increase in metabolic rate caused by male sex hormones explains why males have slightly higher average metabolic rates than females of the same size and age. Living in a cold climate increases the metabolic rate because the cold stimulates thyroid hormone production and this hormone increases heat output of the body, while living in a warm climate causes the metabolic rate to decrease.

Training increases the body's ability to perform physical activity. The basic structure of muscle and crossbridge cycle are not altered by training. However, performing strength exercises makes muscle stronger by adding more thick and thin filaments (and thus more crossbridges) in parallel with those that already exist. Cardiovascular training increases the number and size of the blood vessels that supply oxygen to muscles, strengthens the heart and lungs, and even increases the ability of muscle cells to produce ATP by increasing the number of mitochondria they contain. As a result, a trained athlete can achieve a much higher metabolic rate and perform far more work when they exercise than can an untrained individual.

Most physical activity includes moving the body through a distance. In general, larger animals expend less energy to move each gram of body tissue at fixed velocity than do small animals. This difference probably results from the fact that small animals need a faster rate of muscle shortening (and therefore a faster crossbridge cycle) to achieve a given velocity of motion, and means that small animals are inherently less efficient in their locomotion than are large animals. On the other hand, because small animals have less inertia and experience less drag, they can accelerate to maximum speed more quickly and with less energy expenditure than larger animals.

Different animals employ different forms of locomotion, and these forms also differ in efficiency. Most swimming animals are at or near neutral buoyancy in water, and thus do not need to expend energy working against gravity. For this reason, swimming is inherently more efficient than flying even though the swimming animal must move through a medium (water) that is much denser than

air. Running is the least efficient form of locomotion because running animals (including people) move their body mass up and down against gravity with every stride.

The metabolic rate can be measured in several ways. When no external work is being performed, the metabolic rate equals the heat output of the body. This heat output can be measured by a process called direct calorimetry. In this process, the subject is placed in an insulated chamber that is surrounded by a water jacket. Water flows through the jacket at constant input temperature. The heat from the subject's body warms the air of the chamber and is then removed by the water flowing through the jacketing. By measuring the difference between the inflow and outflow water temperatures and the volume of the water heated, it is possible to calculate the subject's heat output, and thus the metabolic rate, in calories.

Another method of measuring the metabolic rate, and one that allows measurements while the subject is performing external work, is indirect calorimetry. In this process, the subject breathes in and out of a collapsible chamber containing oxygen, while the carbon dioxide in the subject's exhaled air is absorbed by a chemical reaction. The volume decrease of the chamber, equivalent to the amount of oxygen used, is recorded. Because we know the total amount of energy released in the catabolism of glucose, and the amount of oxygen required for this process (Equation 7) it is possible to calculate the metabolic rate once the total oxygen consumption for a period of time is known. Of course, our bodies are not just breaking down glucose; and other nutrients (fats and proteins) require different amounts of oxygen per Calorie liberated than does glucose. For this reason, indirect calometric measurements are adjusted for the diet of the subject.

THE BODY CAN STORE ENERGY-RICH SUBSTANCES

Our bodies must have energy available as ATP to power chemical reactions. We must also store energy for use during periods of prolonged energy consumption. When we exercise, ATP powers the myosin crossbridge cycle of muscle contraction (Figure 4b). However, our muscle cells have only enough ATP for about one second of strenuous activity. Muscle also contains a second high-energy

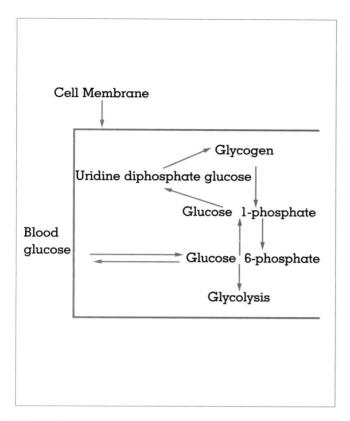

Figure 5.
The breakdown of glycogen to glucose. The glucose then enters glycolysis and the citric acid cycle, providing energy in the form of ATP.

compound called "creatine phosphate" that can give up energy to reconvert ADP to ATP by means of the coupled reactions below:

$$\text{creatine phosphate} \rightarrow \text{creatine} + \text{inorganic phosphate}$$
$$\text{ADP} + \text{inorganic phosphate} \rightarrow \text{ATP}$$

(10)

Muscles contain enough creatine phosphate to power contraction for about ten seconds.

For muscle contraction to continue beyond this brief period, we must rely on the stored energy reserves of glycogen and fats. Glycogen, the storage form of carbohydrate (Figure 5), is present in muscle and liver. Even in the absence of sufficient oxygen, muscle glycogen can be broken down through glycolysis to provide enough energy for an additional five minutes of muscle contraction. When oxygen is present, the glycogen in muscle and liver provide enough energy (via the citric acid cycle) to power

muscle contraction for three hours or more. The "carbo loading" athletes engage in before a marathon race (often including a large pasta dinner) is an attempt to "top off" their glycogen stores, and a depletion of glycogen may contribute to the phenomenon of "hitting the wall" in which athletic performance declines substantially after several hours of intense exercise. When glycogen stores are depleted, we must rely on fats to power muscle contraction.

In a prolonged fasting state, liver glycogen can supply glucose to the blood, and ultimately to tissues such as the brain that preferentially use glucose, for only about twelve hours, even at rest. Thereafter fat and protein stores are broken down for energy. Even people of normal weight have an average of 15 percent (for men) to 21 percent (for women) body fat. So, as long as their fluid intake is sufficient, a healthy person may survive as much as two months of fasting. However fasting has significant negative effects. Fats are broken down to glycerol and fatty acids. Glycerol is converted to glucose by the liver in a process called "gluconeogenesis," and the glucose is released into the blood to provide energy for the brain. The fatty acids are metabolized by a variety of tissues, leaving keto acids that cause the acid level in the blood to rise (fall in pH). Some protein is broken down to amino acids and these, like glycerol, takes part in gluconeogenesis. The metabolic rate falls as the body attempts to conserve energy. This reduction in metabolic rate with fasting is one reason crash dieting is so ineffective. When fat stores are used up, protein catabolism accelerates. Body muscle is digested and the person develops the stick-thin extremities characteristicsc of starving children and those with anorexia nervosa. If they do not receive nourishment, these people will soon die.

Those with a normal diet take in food in the forms of carbohydrates, fats and proteins. Because it has a low water content and produces so many ATP molecules, fat yields 9.3 Calories per gram while carbohydrates and proteins yield less than half as much (4.1 and 4.3 calories per gram respectively). Thus, we get a huge number of calories from a small quantity of fat eaten. The average person in the United States has a diet with 50 percent of the calories in the form of carbohydrates, 35 percent in the form of fat and 15 percent in the form of protein. We need about 1 gram of protein per kilogram of body weight per day to replace body proteins that are broken down. A 70 kg person on an average 5,000-Calorie per day diet receives over twice this amount (5,000 Calories per day × 0.15 of Calories as protein/ 4.3 Calories per gram of protein = 174 grams of protein per day). Thus, most of us are in little danger of protein deficiency. Most of us would probably be healthier if we ate less fat as well. A high fat diet is a known risk factor for diseases of the heart and blood vessels, as well as for colon cancer. Because autopsies on young otherwise healthy soldiers killed in combat indicate that fatty deposits in arteries can be well established by twenty years of age, it is never too early to begin reducing fat intake. A modest goal might to have no more than 30 percent of dietary calories from fat. However, studies from cultures where people consume little red meat indicate that it is possible, and almost certainly healthy, to reduce our fat intake far more than this.

We have seen that energy flow in the body's chemical reactions follows the same basic rules as does energy change in nonliving systems. Energy is taken in as food, then either stored as fat or glycogen, or released in an orderly manner through a multistep enzyme-controlled process and converted to ATP. The ATP is then used to synthesize large molecules needed by the body and to power body processes that do work. The body's overall metabolic rate can be measured and is affected by a variety of internal and external factors. Diet affects the body's energy stores, and insufficient or excess food intake influences metabolic processes. We can use an understanding of metabolism to match our food intake to our body needs and in so doing to maximize our health.

David E. Harris

See also: Biological Energy Use, Ecosystem Functioning of.

BIBLIOGRAPHY

Berne, R. M., and Levy, M. N. (1998). *Physiology*, 4th ed. Boston, MA: Mosby Year Book.

Eckert, R.; Randall, D.; and Augustine, G. (1988). *Animal Physiology*, 3rd ed. New York, NY: W. H. Freeman and Company.

Guyton, A. C., and Hall, J. E. (1996). *Textbook of Medical Physiology*, 9th ed. Philadelphia, PA: W. B. Saunders Company.

Marieb, E. N. (1999). *Human Anatomy & Physiology*, 5th ed. Reading, MA: Benjamin/Cummings Publishing Company.

Martini, F. M. (1998). *Fundamentals of Anatomy & Physiology*, 4th ed. Upper Saddle River, NJ: Prentice-Hall.

Schmidt-Nielsen, K. (1997). *Animal Physiology*, 5th ed. Cambridge, England: Cambridge University Press.

Segel, I. H. (1976). *Biochemical Calculations*, 2nd ed. New York, NY: John Wiley & Sons.

Woledge, R. C.; Curtin, N. A.; and Homsher, E. (1985). *Energetic Aspects of Muscle Contraction*. Monographs of the Physiology Society, No. 41. Boston, MA: Academic Press.

BIOLOGICAL ENERGY USE, ECOSYSTEM FUNCTIONING OF

THE NEED FOR ENERGY

The bald eagle has been the national symbol of the United States since 1782, representing freedom, power, and majesty. The eagle's impressively large beak and talons, combined with its ability to detect details at great distances, make this bird an imposing predator. Interestingly, the most important factor determining what this bird looks like, how it behaves, the prey it seeks, how it interacts with the environment, and the number of bald eagles that are supported by the environment, is energy. In fact, energy is probably the most important concept in all of biology.

Energy is the ability to do work, and work is done when energy has been transferred from one body or system to another, resulting in a change in those systems. Heat, motion, light, chemical processes, and electricity are all different forms of energy. Energy can either be be transferred or converted among these forms. For example, an engine can change energy from fuel into heat energy. It then converts the heat energy into mechanical energy that can be used to do work. Likewise, the chemicals in food help the bald eagle to do the work of flying; heat allows a stove to do the work of cooking; and light does the work of illuminating a room. Biological work includes processes such as growing, moving, thinking, reproducing, digesting food, and repairing damaged tissues. These are all actions that require energy.

The first law of thermodynamics states that ener-

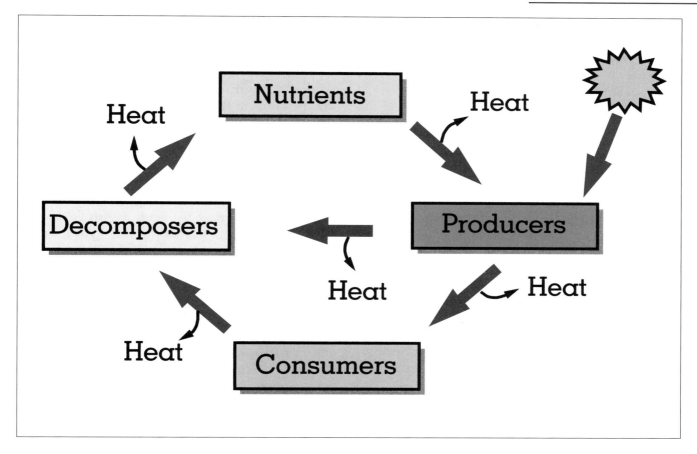

Figure 1.
Energy from the Sun enters the food chain through plants (the producers). At each stage of the food chain, energy is lost as heat.

gy can neither be created nor destroyed. This seems to imply that there is an abundance of energy: If energy cannot be destroyed, then there must be plenty available to do biological work. But biological work requires high-quality, organized energy.

Energy can be converted from one form to another. According to the second law of thermodynamics, each time energy is converted, some useful energy is degraded into a lower-quality form—usually heat that disperses into the surroundings. This second law is very interesting in terms of biology because it states that every time energy is used, energy quality is lost: The more energy we use, the less there is to do useful work. As we shall see, this principle influences every biological event, from interactions between predators and their prey to how many species can live in a habitat.

Plants and Animals Get Energy from the Sun

Living systems are masters of energy transformation. During the course of everyday life, organisms are constantly transforming energy into the energy of motion, chemical energy, electrical energy, or even to light energy. Consequently, as dictated by the laws of thermodynamics, each living system is steadily losing energy to its surroundings, and so must regularly replenish its supply. This single fact explains why animals must eat and plants must harvest light energy through photosynthesis.

Where does this supply of energy come from? Most life on Earth depends on radiant energy from the Sun, which delivers about 13×10^{23} calories to Earth each year. Living organisms take up less than 1 percent of this energy; Earth absorbs or reflects most of the rest. Absorbed energy is converted to heat, while energy is

The population size of a particular species that can be supported in any given ecosystem depends on the resource needs—ultimately, the energy needs—of that species.

Before the 1800s, the North American prairie was able to provide sufficient energy and other resources to support an incredible 30 million to 60 million bison. During the late 1800s all but a few hundred of these extraordinary animals were killed for their skins, meat, tongues, for sport, or to impact the Plains' Native American populations. An intensive breeding program was undertaken—this was among the first times that zoos took an active role to save a species from becoming extinct in the wild—and captive-bred bison were released back into the prairie. But most of the prairie has since been converted to farmland or extensively grazed by livestock. Therefore it would be impossible for bison herds to ever again reach the large numbers seen in the past, simply because there is not enough energy available in the system.

Similarly, widespread control of prairie dogs has been a major factor in the near-extinction of the black-footed ferret.

reflected as light (Figure 1). Solar energy helps create the different habitats where organisms live, is responsible for global weather patterns, and helps drive the biogeochemical cycles that recycle carbon, oxygen, other chemicals, and water. Clearly, solar energy profoundly influences all aspects of life.

Solar energy also is the source of the energy used by organisms; but to do biological work, this energy must first be converted. Most biological work is accomplished by chemical reactions, which are energy transformations that rely on making or breaking chemical bonds. Chemical energy is organized, high-quality energy that can do a great deal of work, and is the form of energy most useful to plants and animals. Solar energy can be used to create these chemical bonds because light energy that is absorbed by a molecule can boost that molecule's electrons to a higher energy level, making that molecule extremely reactive. The most important reactions involve the trans-ferring of electrons, or oxidation-reduction (REDOX) reactions because, in general, removing electrons from a molecule (oxidation) corresponds to a release of energy. In other words, organisms can do work by oxidizing carbohydrates, converting the energy stored in the chemical bonds to other forms. This chemical energy is converted from solar energy during the process of photosynthesis, which occurs only in plants, some algae, and some bacteria.

How Plants Get Energy: Photosynthesis

A pigment is a material that absorbs light. Biologically important pigments absorb light in the violet, blue, and red wavelengths. Higher-energy wavelengths disrupt the structure and function of molecules, whereas longer (lower-energy) wavelengths do not contain enough energy to change electron energy levels. Photosynthesis is restricted to those organisms that contain the appropriate pigment combined with the appropriate structures such that the light energy can trigger useful chemical reactions. If an organism cannot perform photosynthesis, it cannot use solar energy. The Sun's energy enters the food chain through photosynthesis. Plants are the most important photosynthesizers in terrestrial systems, while photosynthesis in aquatic systems generally occurs in algae.

In plants, chlorophyll is the pigment that absorbs radiant energy from the Sun. This allows the transfer of electrons from water to carbon dioxide, creating the products glucose and oxygen. The equation for photosynthesis is:

$$6CO_2 + 6H_2O + light \rightarrow C_6H_{12}O_6 + 6O_2$$

Photosynthesis takes atmospheric carbon dioxide and incorporates it into organic molecules—the carbon dioxide is "fixed" into the carbohydrate. These molecules are then either converted into chemical energy or used as structural molecules. The first powers living systems; the second is what living systems are composed of.

To release energy, the electrons can be removed from glucose and used to create ATP, a molecule that supplies a cell's short-term energy needs. This latter occurs in a series of reactions known as respiration. (Body heat is a by-product of these reactions.) The most efficient respiration reactions are those that use oxygen to accept the electrons removed from glucose. Thus respiration is the reverse of photosynthesis: Organic compounds are oxidized in the presence of oxygen to produce water and carbon dioxide. Photosynthesis captures energy, and both products of photosynthesis are required in the energy-releasing reactions of respiration.

By converting radiant energy from the Sun into stored chemical energy, plants essentially make their own food: All that a plant needs to survive is water, carbon dioxide, sunlight, and nutrients. Plants need not rely on any other organism for their energy needs. In contrast, the only forms of life in terrestrial systems that do not depend on plants are a few kinds of bacteria and protists. Every other living thing must eat other organisms to capture the chemical energy produced by plants and other photosynthesizers.

How Animals Get Energy: Energy Flow Through an Ecosystem

A heterotroph is an organism that relies on outside sources of organic molecules for both energy and structural building blocks. This includes all animals and fungi and most single-celled organisms. An autotroph can synthesize organic molecules from inorganic elements. Most autotrophs are photosynthetic; a few, occurring in very restricted areas such as deep ocean trenches, are chemosynthetic.

Consequently, each organism depends in some way upon other organisms; organisms must interact with each other to survive. For example, a heterotroph must eat other organisms to obtain energy and structural compounds, while many important plant nutrients originate from animal wastes or the decay of dead animals. The study of these interactions is the subject of the science of ecology. Interactions that involve energy transfers between organisms create food chains. A food chain portrays the flow of energy from one organism to another.

Three categories can describe organisms in a community based on their position in the food chain: producers, consumers, and decomposers. These categories are also known as trophic levels. Plants "produce" energy through the process of photosynthesis. Consumers get this energy either directly or indirectly. Primary consumers eat plants directly, whereas secondary consumers get their energy from plants indirectly by eating the primary consumers. Energy enters the animal kingdom through the actions of herbivores, animals that eat plants. Animals that eat other animals are carnivores, and omnivores are animals that eat both plants and animals. Decomposers

get their energy from consuming nonliving things; in the process releasing inorganic nutrients that are then available for reuse by plants.

A North American prairie food chain begins with grass as the producer. The grass is eaten by a prairie dog, a primary consumer. The prairie dog falls prey to a secondary consumer, such as a black-footed ferret. Some food chains contain a fourth level, the tertiary consumer. In this example, the ferret could be eaten by a golden eagle, a top predator. Because organisms tend to eat a variety of things, food chains are generally linked to form overlapping networks called a food web. In addition to the ferret, golden eagles eat ground squirrels and rabbits, each of which are primary consumers in separate food chains. A prairie food web would depict all of the interactions among each of these food chains.

Not all of the energy produced by photosynthesis in the grass is available to the prairie dog. The grass requires energy to grow and produce structural compounds. Similarly, the prairie dog eats food for the energy it needs for the work of everyday life: It needs to find and digest the grass; to detect and avoid predators; to find a mate and reproduce. Each time one of these processes uses energy, some energy is lost to the environment as heat. By the time the predator eats the prairie dog, very little of the original energy from the grass is passed on (Figure 1).

The fact that energy is lost at each stage of the food chain has tremendous influence on the numbers of producers and consumers that can be supported by any given habitat. This can best be illustrated graphically by ecological pyramids. Figure 2 is an example of an energy pyramid. Two other types of ecological pyramids are a pyramid of numbers and a pyramid of biomass. Each section of these pyramids represents a trophic level for the represented community—the producers form the base of the pyramid, the primary consumers the second level, and the secondary and tertiary consumers the third and fourth levels, respectively. Decomposers are also often included. The size of each section represents either the number of organisms, the amount of energy, or the amount of biomass.

Biomass refers to the total weight of organisms in the ecosystem. The small amount of solar radiation incorporated into living systems translates into the production of huge amounts of biomass: On a worldwide basis, 120 billion metric tons of organic matter are produced by photosynthesis each year. However,

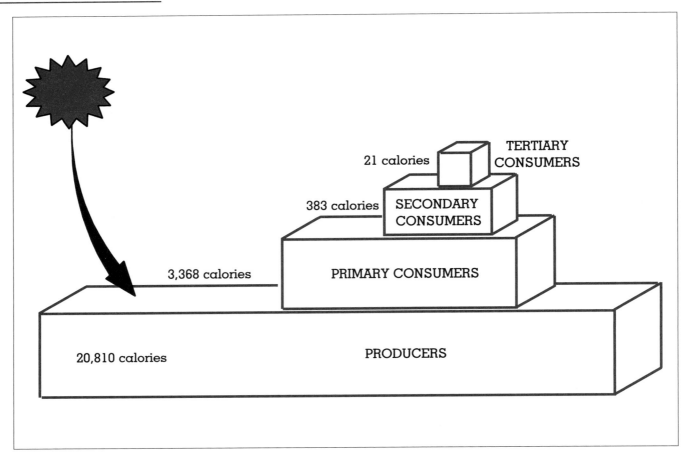

Figure 2.
Pyramids are useful tools for wildlife conservationists in determining the sustainable populations of particular species in any given ecosystem.

a plant uses only a fraction of the energy from photosynthesis to create this biomass; similarly, only a fraction of the energy consumed by an animal is converted to biomass. The rest is used up for metabolic processes and daily activities. The amount of biomass, the amount of energy, and the number of species present at each level of the pyramid are therefore less than in the trophic level before it. On the Serengeti Plain in Africa, there are many more blades of grass than there are zebras, and many more zebras than there are lions.

As shown in Figure 2, when a herbivore eats a plant, only about 10 percent of the energy stored in that plant is converted to animal biomass; the rest is used up in everyday activities. The same is true for each succeeding trophic level. Note that the commonly cited "10 percent energy transfer" figure is only a rough average based on many studies of different ecosystems. Scientists have found that actual transfer rates vary from 1 to 20 percent.

In a temperate forest ecosystem on Isle Royale, Michigan, ecologists found that it takes 762 pounds (346 kg) of plant food to support every 59 pounds (27 kg) of moose, and that 59 pounds of moose are required to support every one pound (0.45 kg) of wolf. The basic point is that massive amounts of energy do not flow from one trophic level to the next: energy is lost at each stage of the food chain, so there are more plants than herbivores and more herbivores than carnivores.

As links are added to the food chain, the amount of energy becomes more and more limited; this ultimately limits the total number of links in the food chain. Most habitats can support food chains with three to four trophic levels, with five being the usual limit. One study of 102 top predators demonstrated that there are usually only three links (four levels) in a food chain.

These factors lie behind the idea that adopting a vegetarian diet is a strategy in line with a sustainable lifestyle. Take the example of the plant → moose → wolf food chain given above. More energy is available to the wolf if it eats the plant rather than the moose. If a person is stuck on a desert island with a bag of grain and ten chickens, it makes little energetic sense to first feed the grain to the chickens, then eat the chickens. More energy is conserved (because fewer energy conversions are required) if the person first eats the chickens, then the grain. Similarly, if everybody were to adopt a vegetarian diet, much more food would be available for human consumption.

Thirty to 40 percent of the calories in a typical American diet come from animal products. If every person in the world consumed this much meat, all the agricultural systems in the world would be able to support only 2.5 billion people. In 1999 these same systems supported 6 billion people, primarily because the majority of the people living in less developed countries consume fewer animal products.

The energy flow through an ecosystem is the most important factor determining the numbers, the types, and the interactions of the plants and animals in that ecosystem.

Where an Animal's Energy Goes

During an animal's lifetime, about 50 percent of the energy it consumes will go to general maintenance (everyday metabolic processes), 25 percent to growth; and 25 percent to reproduction.

Animal nutritionists have developed formulas to guide them in recommending the amount of food to feed animals in captive situations such as in zoos. First, the number of calories needed to maintain the animal while at rest is determined—this is called the basal metabolic rate (BMR). In general, a reptile's BMR is only 15 percent that of a placental mammal, while a bird's is quite a bit higher than both a reptile's and a mammal's. For all animals, the number of calories they should receive on a maintenance diet is twice that used at the basal metabolic rate. A growing animal should receive three times the number of calories at the BMR, while an animal in the reproductive phase should receive four to six times the BMR.

Recall that during respiration, animals gain energy from glucose by oxidizing it—that is, by transferring the electrons from the glucose to oxygen. Because molecules of fat contain more hydrogen atoms (and therefore electrons) than either glucose or proteins, the oxidation of fat yields almost twice the calories as that of carbohydrates or proteins. Each gram of carbohydrate yields four calories, a gram of protein yields five calories, while a gram of fat yields nine calories.

THE EFFECTS OF ENERGY

Adaptations

Adaptations are features an organism has or actions it does that help it survive and reproduce in its habitat. Each organism has basic needs, including energy, water, and shelter; adaptations allow an organism to obtain its basic needs from its habitat. Body structures, behavioral characteristics, reproductive strategies, and physiological features are all examples of adaptations. Different habitats pose different challenges to organisms, and adaptations can be thought of as solutions to these challenges. For example, important adaptations in a desert habitat would include those that allow an animal to conserve water and store energy.

Energy affects adaptations in two fundamental ways. First, plants and animals need energy to survive, and adaptations may allow them to obtain, use, and, in many cases, conserve energy. Not only do plants and animals need a great deal of energy to fuel the chemical reactions necessary for survival, but also this energy can be in limited supply, particularly to consumers. Accordingly, many adaptations seen in different species revolve around the conservation of energy.

Second, solar radiation can result in striking differences among habitats. A rain forest is fundamentally different from a tundra first and foremost because there is a much greater amount of energy available to a rain forest. The average daylight hours in Barrow, Alaska, vary from about one hour in January to nearly twenty-four hours in June. The reduced growing season limits the amount of energy that plants can produce. Contrast this with Uaupés, Brazil, which receives twelve hours of sunlight each month of the year. The year-long growing season contributes to the fact that rain forests are the most productive habitats on Earth.

Regulating Body Temperature. Metabolism refers to all the chemical reactions that take place within an organism. To occur at rates that can sustain life, metabolic reactions have strict temperature requirements, which vary from species to species. There are two ways by which an organism can achieve the appropriate temperature for metabolism to occur. The first is to capture and utilize the heat generated by various energy conversions; the second is to rely on external energy sources such as direct sunlight. In other words, body temperature can be regulated either internally or externally.

An endothermic animal generates its own body temperature, while an ectothermic animal does not. In general, endothermic animals have constant body temperatures that are typically greater than that of the surrounding environment, while ectothermic animals have variable temperatures. Ectotherms rely on behavioral temperature regulation—a snake will move from sun to shade until it finds a suitable microclimate that is close to its optimal body temperature. When exposed to direct sunlight, an ectotherm can increase its body temperature as much as 1°C (32.8°F) per minute.

Endothermic animals can achieve and sustain levels of activity even when temperatures plummet or vary widely. This can be a huge advantage over ectothermy, especially in northern latitudes, at night, or during the winter. In colder climates, an ectothermic predator such as a snake will tend to be more sluggish and less successful than an endothermic predator. There are no reptiles or insects in the polar regions.

However, endothermy is a costly adaptation. An actively foraging mouse uses up to thirty times more energy than a foraging lizard of the same weight; at rest, an endothermic animal's metabolism is five to ten times greater than that of a comparably sized ectotherm. In certain habitats, this translates to a substantial advantage to ectothermy. Because of their greater energy economy and lower food requirements, tropical ectotherms outnumber endotherms in both number of species and number of individuals.

Birds and mammals are endothermic vertebrates. Not coincidentally, they are the only vertebrates with unique external body coverings—feathers and hair, respectively. For both groups, these body coverings evolved as an adaptation to reduce heat loss. A bird's feathers were originally adaptive because they helped keep the animal warm, not because they helped it to fly.

The Energetics of Body Size. Larger animals have lower energy requirements than smaller ones. Gram for gram, a harvest mouse has twenty times the energy requirements of an elephant. Part of the advantage of size probably stems from the fact that a larger animal has proportionately less surface area than a smaller one. When heat leaves the body, it does so through the body surface. It is more difficult for a smaller object to maintain a constant body temperature because it has a greater amount of surface area relative to its volume. One consequence of this relationship is that a whale, because of its lower metabolic rate, can hold its breath and thus remain underwater for longer periods of time than a water shrew.

The Costs of Locomotion. Because oxygen is required for energy-producing metabolic reactions (respiration), there is a direct correlation between the amount of oxygen consumed and the metabolic rate. Not surprisingly, metabolic rates increase with activity. During exercise, a person will consume fifteen to twenty times more oxygen than when at rest.

A comparison of different species reveals that a larger animal uses less energy than a smaller one traveling at the same velocity. This seems to be related to the amount of drag encountered while moving. Small animals have relatively large surface-area-to-volume ratios and therefore encounter relatively greater amounts of drag. Adaptations that reduce drag include streamlined body shapes. However, because a larger body must first overcome a greater amount of inertia, there is a greater cost of acceleration. Therefore, small animals tend to be able to start and stop abruptly, whereas larger animals have longer start-up and slow-down periods.

Interestingly, two different species of similar body size have similar energetic costs when performing the same type of locomotion. Differences in energy expenditure are seen when comparing the type of locomotion being performed rather than the species of animal. For a given body size moving at a given velocity, swimming is the most energetically efficient mode of locomotion, flying is of intermediate cost, and running is the most energetically expensive. An animal swimming at neutral buoyancy expends less effort than an animal trying to stay aloft while flying; running costs the most because of the way the muscles are used.

Nonetheless, birds have higher metabolic rates than mammals of similar size. Most small mammals reduce energy costs by seeking protected environments; birds spend much of their time exposed. Also, because fat is heavy, the need to fly restricts a bird's ability to store energy. Even with a high-protein diet, a bird must eat as much as 30 percent of its body weight each day. This factor alone probably accounts for some birds' summer migratory journey from the tropics to northern latitudes, which, because of their longer days, allow a bird more daylight hours in which to feed itself and its young.

The Energetics of Mating. Sexual reproduction requires a considerable amount of energy, including the energy invested while competing for mates; mating itself (including the production of gametes); and caring for the offspring. Three main mating systems are found in the animal kingdom: polygyny (one male, more than one female), monogamy (one male, one female), and polyandry (one female, more than one male. Each system can be defined in terms of the relative energetic investments of each sex. In a monogamous pair bond, both sexes invest approximately equal amounts of energy; consequently, courtship behaviors tend to be rather involved, and competition is equal between the sexes. Males in a polygynous system spend a great deal of energy competing for mates, while females invest more heavily in parental care, and therefore tend to be very particular with whom they mate. A polyandrous system is the opposite: Females invest in competition and mating, while males invest in parental care.

Adaptations to Habitats. Because of Earth's geometry and the position of its axis, the equator receives more solar energy per unit area than the polar regions. Because Earth's axis is tilted relative to the plane of Earth's orbit around the Sun, this angle of incident radiation varies seasonally. These factors, combined with Earth's rotation, establish the major patterns of temperature, air circulation, and precipitation.

A habitat, or biome, is made up of interacting living and nonliving elements. The major terrestrial habitats include deserts, temperate forests, grasslands, rain forests, tundra, and various types of wetlands. The boundaries of these different habitats are determined mainly by climatological factors such as temperature, precipitation, and the length of the growing season. These conditions are created by the influence of solar radiation, often in conjunction with local factors such as topography.

The amount of biomass produced in a habitat—the productivity of the habitat—is determined by the types of plants (some species are more efficient photosynthesizers than others), the intensity and duration of solar radiation, the amount of nutrients available, and climatic factors such as temperature

and precipitation. Aquatic habitats tend to be less productive than terrestrial ones, largely because there is less sunlight available at depth and there is a scarcity of mineral nutrients. Tropical rain forests have conditions that favor high productivity; one result is that they also have the highest biodiversity of any habitat.

Cold Habitats. Because of considerations of surface area relative to body mass, animals that live in cold habitats tend to have larger body sizes and smaller extremities (especially ears and legs) compared to their counterparts in warmer habitats. Animals that live in cold habitats also have a greater amount of insulation, such as fat, fur, or feathers. Behavioral adaptations include gathering in groups, which effectively decreases the exposed surface area of each individual.

If energy resources are seasonally low, some animals adopt the strategy of migrating to areas with greater resources. A bird that is insectivorous is more likely to be migratory than one that is a seed-eater. Hibernation is another adaptation in response to seasonal energy shortages. The body temperature of a true hibernator closely matches that of its surroundings. The heart slows (a ground squirrel's heartbeat drops from 200 to 400 beats per minute to 7 to10 beats per minute), and metabolism is reduced to 20 to 100 times below normal. Hibernators tend to have much longer lifespans than non-hibernators.

Differences also can be seen in human populations living in cold habitats. Among the Inuit, the body maintains a high temperature by burning large amounts of fat and protein. Increased blood flow to the arms, legs, fingers, and toes helps prevent frostbite. An Australian Aborigine may sleep in below-freezing temperatures with little clothing or shelter, yet conserves energy by allowing the temperature in the legs and feet to drop. Heat is maintained in the trunk, where it is needed most.

Adaptations to Warm Habitats. When water evaporates into the surroundings, the vaporized molecules carry a great deal of heat away with them. One of the best ways to cool an animal's body is to evaporate water from its surface. Adaptations that take advantage of this property include sweating, panting, and licking the body. But water often is a limited resource in warm habitats such as deserts, so many desert animals have adaptations that reduce the amount of water that evaporates from the body. Most

A depiction of how food chains become interconnected to form food webs. (Gale Group, Inc.)

small desert animals avoid the heat and reduce water loss by being nocturnal and living in burrows. Large extremities, particularly ears, help to bring heat away from the body and dissipate it to the surroundings.

Other adaptations are perhaps best exemplified by examining the camel, which is able to conserve water by excreting a very concentrated urine. Also, the upper lip is split, so moisture dripping from the nose reenters the body through the mouth. More importantly, the camel can tolerate dehydration: It can lose an impressive 25 percent of its body weight in water with no ill effects. Its internal body temperature can fluctuate by as much as 6°C (10.8°F). By increasing its temperature during the day and dropping it at night, it can more closely track changes in external temperatures. This also helps to reduce water loss by as much as five liters of water per day. The fat stored in the camel's hump represents an important energy supply in the sparsely vegetated habitat in which it lives.

ENERGY AND HUMANS

The Costs of Technology

Paradoxically, organisms must use energy to get energy: A lion must hunt to eat, while a zebra must sometimes move long distances to find food. Most organisms can gain 2 to 20 calories in food energy for each calorie they use to obtain that energy. This holds for both a hummingbird, whose metabolic rate is 330 calories per minute, and a damselfly, which uses less than 1 calorie per day.

This is also true for human hunter-gatherer societies. Without technology, people use about 1 calorie to gain 5 to 10 calories. The energy return increases to 20 calories through the use of shifting agricultural practices.

Ironically, the cost of getting our food has increased with technological advances. In 1900 we gained a calorie for each calorie we used, while in 1995, for each calorie we invested we got only 0.1

calorie in return. Some of the energy costs associated with food include human labor, the cost of fertilizer, the cost of fuel for the farm machinery, and the cost of transportation of the food.

More developed countries rely heavily on burning fossil fuels to meet energy needs. Fossil fuels are the remains of plants that, over millions of years, have been transformed into coal, petroleum, and natural gas. Just like the natural systems examined in this article, our energy ultimately comes from the Sun and photosynthesis. Although more developed countries have less than 20 percent of the world's population, they use more than 80 percent of the world's energy.

People in less developed countries burn wood, plant residues, or animal dung to fuel stoves and lanterns. About 2 billion people rely on wood to cook their daily meals. Typically, four to five hours per day are spent gathering wood fuels from the surrounding habitat. Because these countries tend to have high population growth rates, there has been an ever-increasing demand for more wood, resulting in a significant amount of habitat degradation in many areas.

The Effects of Human Energy Use

The two major ways by which humans get energy is to either burn fossil fuels or to burn wood for fuel. Both contribute substantially to air pollution, and both can have serious effects on (1) the health of plants and animals and (2) the workings of Earth's atmosphere.

Burning fossil fuels can release air pollutants such as carbon dioxide, sulfur oxides, nitrogen oxides, ozone, and particulate matter. Sulfur and nitrogen oxides contribute to acid rain; ozone is a component of urban smog, and particulate matter affects respiratory health. In fact, several studies have documented a disturbing correlation between suspended particulate levels and human mortality. It is estimated that air pollution may help cause 500,000 premature deaths and millions of new respiratory illnesses each year.

Physiological effects of air pollution are dependent on dosage, the ability of the exposed organism to metabolize and excrete the pollution, and the type of pollutant. Many pollutants affect the functioning of the respiratory tract; some change the structure and function of molecules; others can enter the nucleus and turn genes on or off; and some cause chromosomal aberrations or mutations that result in cancer.

For example, exposure to the air toxin benzene can increase the risk of getting myelogenous leukemia or aplastic anemia, while exposure to ground-level ozone can cause a 15 to 20 percent decrease in lung capacity in some healthy adults.

Air pollution affects plant health as well. Acid rain and ozone can directly damage a plant's leaves and bark, interfering with photosynthesis and plant growth. More serious effects can occur if soil nutrients are leached away and heavy metals are mobilized in the soils upon which plants depend. Without proper nutrients, plants become susceptible to a variety of diseases. The overall result is a decrease in the amount of energy produced by plants. Acid rain affects over 345,960 square miles (900,000 sq. km) of Eastern Europe, where it has taken a significant toll on cities, forest, lakes, and streams (Kaufman and Franz, 1993). Moreover, air pollution is reducing U.S. food production by 5 to 10 percent, costing an estimated $2 billion to 5 billion per year (Smith, 1992).

Burning fossil fuel releases carbon into the atmosphere—more than 6.3 billion tons in 1998 alone. Significant amounts of carbon also come from burning of live wood and deadwood. Such fires are often deliberately set to clear land for crops and pastures. In 1988 the smoke from fires set in the Amazon Basin covered 1,044,000 square miles. By far the most serious implication of this is the significant threat to Earth's ecosystems by global climate change.

Like all matter, carbon can neither be created nor destroyed; it can just be moved from one place to another. The carbon cycle depicts the various places where carbon can be found. Carbon occurs in the atmosphere, in the ocean, in plants and animals, and in fossil fuels. Carbon can be moved from the atmosphere into either producers (through the process of photosynthesis) or the ocean (through the process of diffusion). Some producers will become fossil fuels, and some will be eaten by either consumers or decomposers. The carbon is returned to the atmosphere when consumers respire, when fossil fuels are burned, and when plants are burned in a fire. The amount of carbon in the atmosphere can be changed by increasing or decreasing rates of photosynthesis, use of fossil fuels, and number of fires.

Scientists have been able to compare the seasonal changes in atmospheric carbon dioxide to the seasonal changes in photosynthesis in the Northern Hemisphere. Plants take up more carbon dioxide in the summer, and animals continue to respire carbon dioxide in the winter, when many plants are dormant. Correspondingly, atmospheric carbon dioxide increases in the winter and decreases in the summer.

Atmospheric carbon dioxide, water vapor, methane, and ozone are all "greenhouse gases." When solar energy is reflected from Earth's surface, the longer wavelengths are trapped in the troposphere by these greenhouse gases. This trapped radiation warms Earth. In fact, without greenhouse gases, which have been present for several billion years, Earth would be too cold to support life.

Habitat destruction is another important contributing factor to increased atmospheric carbon dioxide levels. The world's forests are being cut, burned, or degraded at an astounding rate: More than half of the world's tropical rain forests have been lost in the past one hundred years. The forests supply fuel wood for energy; land for crops or pastures; and the wood demands of the global economy. Burning the forests releases carbon dioxide into the atmosphere; cutting or degrading the forests results in fewer plants available to take carbon dioxide out of the atmosphere.

Although scientists have been able to measure increasing levels of carbon dioxide, it is difficult to predict what the effects will be. For example, some models predict that warmer temperatures and the greater availability of atmospheric carbon dioxide will stimulate productivity, which in turn will remove carbon dioxide from the atmosphere, thus neutralizing the problem. However, the availability of carbon dioxide is not what generally limits plant growth. Rather, plant productivity tends to be restricted by the availability of resources such as nitrogen, water, or sunlight. Therefore, increasing the amount of carbon dioxide available to plants probably will have little effect on productivity, especially because it likely will result in greater evaporation rates and changed weather patterns. Although increased levels of evaporation may actually increase rain in some parts of the world, it may not be in those places currently containing rain forests.

A 1999 study by the Institute of Terrestrial Ecology predicts that tropical rain forests will be able to continue to absorb carbon dioxide at the current rate of 2 billion tons per year until global temperatures rise by 8°F (4.5°C). At this point, evaporation rates will be high enough to decrease rainfall for the forests, leading to the collapse of tropical ecosystems. This collapse will decrease the amount of carbon dioxide leaving the atmosphere and have dire consequences for all life.

Allison Brody

See also: Acid Rain; Agriculture; Air Pollution; Atmosphere; Biological Energy Use, Cellular Processes of; Climatic Effects; Environmental Problems and Energy Use; Green Energy; Thermodynamics.

BIBLIOGRAPHY

Brown, L. R. (1999). *Vital Signs*. New York: W. W. Norton.

Curtis, H., and Barnes, N. S. (1989). *Biology*, 5th ed. New York: Worth.

Eckert, R., and Randall, D. (1983). *Animal Physiology*, 2nd ed. San Francisco: W. H. Freeman.

Kaufman, D., and Franz, C. (1993). *Biosphere 2000: Protecting Our Global Environment*. New York: Harper Collins.

Miller, G. T., Jr. (1996). *Living in the Environment*, 9th ed. Belmont, CA: Wadsworth.

Raven, P. H.; Berg, L. R.; and Johnson, G. B. (1995). *Environment*. Fort Worth, TX: Saunders.

Smith, Z. (1992). *The Environmental Policy Paradox*. Englewood Cliffs, NJ: Prentice-Hall.

Wildlife Conservation Society. (1995). *Voyage from the Sun*. New York: Author.

BLACK, JOSEPH (1728-1799)

Joseph Black was born in Bordeaux, France, the fourth child of parents of Scottish extraction. His father was a native of Belfast engaged in the Bordeaux wine trade; his mother was a daughter of an Aberdeen man who had settled in Bordeaux. In all, Black's parents had twelve children. At the age of twelve Black was sent to school in Belfast, and around 1744 proceeded to the University of Glasgow. Black followed the standard curriculum until pressed by his father to choose a profession. He opted for medicine. Black began to study anatomy and chemistry. William Cullen had recently inaugurated lectures in chemistry that were to have a decisive influence on Black's career. Recognizing Black's aptitude, Cullen employed Black as his laboratory assistant.

In 1752 Black transferred to Edinburgh University

to continue his medical training, receiving his M.D. in 1754. In 1756 Cullen arrived at Edinburgh as Professor of Chemistry, and Black at the age of twenty-eight was appointed to the Chair of Anatomy and Chemistry at Glasgow. He became dissatisfied with his qualifications for that position and exchanged duties with the Professor of Medicine. During the ensuing ten years, Black lectured on the subject of medicine and also carried out an active and demanding medical practise. In 1766 Cullen vacated the Chair of Chemistry to become Professor of the Institutes of Medicine, and Black took over Cullen's Chair. Black remained at Edinburgh, but now limited his medical practice to a few close friends.

Joseph Black suffered all his life with ill health and had something of a reputation for parsimony, being as methodical in his financial affairs as in his science. Black never married but was no recluse. A prominent member of Edinburgh intellectual society, he regularly frequented the Royal Society of Edinburgh and the dining clubs for which Edinburgh was then famous. Due to worsening health Black gave his last lectures in 1795-1796, and died in December 1799 in his seventy-second year.

Black was the founder of the measurement of quantities of heat or calorimetry. During his period of tenure at Glasgow, Black carried out experimental research into the nature of heat and discovered what he termed "latent heat." In 1754, Cullen communicated to Black his observations concerning the intense cold produced by the evaporation of volatile substances such as ether. Black was also aware of Fahrenheit's observations concerning the freezing of water. Black reflected that solidification or evaporation required the transfer of quantities of heat not detectable by the thermometer—heat that Black termed "latent." Black concluded that the heat in water at 32°F is equal to the heat in ice at 32°F plus the latent heat of liquefaction. Not satisfied with demonstrating that there was such a thing as latent heat, Black proposed to measure it. He had to first prove the reliability of the thermometer as a measuring tool. After carrying out a number of ingenious but simple experiments involving the mixing of hot and cold water, Black concluded that the expansion of mercury was a reliable indicator of the temperatures of heat.

Black recognized a distinction between quantity of heat and temperature. Although not the first to make this distinction, he was the first to sense its funda-

Joseph Black. (Library of Congress)

mental importance and make use of it: a thermometer can be used to measure temperature but how can the quantity of heat be measured? In conducting his experiments Black was probably guided by Newton's experiments and law of cooling. In the winter of 1761 Black carried out experiments to determine the latent heat of fusion. A mass of water was cooled to 33°F and the time taken to raise the temperature 1°F was noted. As a comparison, Black also determined the time required to melt an identical mass of ice. Conversely, Black determined the time required to lower the temperature of a mass of water and compared it with the time it took to freeze it. From this Black obtained a value for the latent heat.

Black soon realized that latent heat must also play a part in vaporizing water. In 1762 Black carried out a series of investigative experiments. The time to heat water from 50°F to the boiling point was compared with the time it took the water to boil away. From these experiments Black calculated that the amount of heat required to evaporate water was equal to that required to raise the water to 810°F, were this to be possible. Black went on to make a second but closely related discovery: Different substances have different heat capacities. In doing so Black was building up the work of Fahrenheit and George Martine, who had made observations concerning the different rates of temperature rise when heat was applied to equal quantities of water and mercury. Black concluded, since mercury increased in temperature almost twice as fast as water, it must have a smaller store of heat than water. He realized the significance of this effect, but did not initially pursue the matter.

James Watt, then a mathematical instrument maker, was called upon by Black to make items he needed for his experiments. Watt, in turn, was engaged in his own experiments concerning the Newcomen steam engine. Large quantities of water were required to condense the steam in the engine cylinder, and he turned to Black for an explanation. Black in turn explained to Watt his ideas about latent heat. Watt repeated Black's experiments with a smaller improved still. James Watt went on to develop his separate condenser, and later insisted his discovery had not been suggested by the doctrine of latent heat. Watt did, however, readily credit Black with clarifying the problems he had encountered and with teaching him to "reason and experiment with natural philosophy."

The problems raised by Watt revised Black's interest in heat. With his assistant, William Irvine, he set out to determine a more accurate value for the latent heat of steam using a common laboratory still as a water calorimeter. In 1764 it occurred to Black that his knowledge of the latent heat of fusion of ice could be used to measure the latent heat of steam. Plans to put this idea to the test were abandoned, however, when James Watt began to obtain values that Black thought sufficiently precise. Watt was the first to investigate specific heat experimentally and was responsible for drawing Black's attention to the significance and practical importance of such research. Black and his assistant investigated the specific heats of various solids by determining the heats communicated to water, using the method of mixtures. Black ceased his investigations into heat when he transferred to Edinburgh University in 1766. For reasons, that remain unexplained, Black was reluctant to publish his ideas and details of his experiments, his lecture notes were only printed after his death, by John Robinson.

Robert Sier

See also: Water Heating; Watt, James.

BIBLIOGRAPHY

Guerlac, H. (1970). "Joseph Black." In *Dictionary of Scientific Biography*, ed. C. C. Gillispie. New York: Scribners.

Ramsey, W. (1918). *Life and Letters of Joseph Black, M.D.* London: Constable.

Robinson, J., ed. (1803). *Lectures on the Elements of Chemistry, delivered in the University of Edinburgh*, 2 vols. Edinburgh: W. Creech.

BOILER

See: Heat and Heating; Steam Engines

BOMBS

See: Explosives and Propellants

BRITISH THERMAL UNITS

See: Units of Energy

BROOKHAVEN NATIONAL LABORATORY

See: National Energy Laboratories

BUILDING DESIGN, COMMERCIAL

More than 3.3 billion square feet of new commercial buildings were constructed from 1988 to 1998, with 170 percent more expected by the year 2030. Because this new stock is expected to have a lifetime of fifty to one hundred years, this has a dramatic impact on energy consumption not only today, but also for many years to come. This article will discuss the history of commercial building design, how technology has impacted commercial building design, and what impact this has had on building energy use.

HISTORY

The first commercial buildings, built in about 2000 B.C.E., were simple structures that represented the beginnings of architecture—a series of columns, walls, and roofs. Columns represented the upright human stance, walls represented human territoriality, and roofs both kept the rain out and created a crown, or head, for the structure. Walls also represented a separation between the plant and animal world and the human world. The walls of a courtyard formed a human space that became the city. Although the form of buildings has evolved over time, buildings today fundamentally provide these same basic human functions: artistic expression, separation, definition, and shelter.

Modern buildings are fundamentally defined by the mechanical principles that drive their utility. Technology has defined both the form of the buildings and energy use. Electrical lighting, mechanical ventilation, curtain-wall systems, air conditioning, and office equipment all contribute to a modern building's energy consumption. Two of these technologies that have had the most influence on the energy consumed by commercial buildings have been the lightbulb and the air conditioner.

Energy use in buildings has risen dramatically since 1900 because of technologies that enabled the creation of man-made indoor environments. It began with the invention and proliferation of the electric lightbulb in the 1880s. Fluorescent lighting became popular in the late 1930s, and by 1950 had largely replaced incandescent lamps in commercial buildings. However, incan-descent lamps still are used in approximately 17 percent of the pre-1970 building stock.

Tasks such as health care, office work, manufacturing, studying, and other tasks requiring visual acuity all benefited from electric lighting. It also meant that workers didn't need access to a window to be productive. Designers could use artificial illumination to light tasks away from windows. Interior spaces that didn't require skylights or cleristories were now possible in commercial buildings.

The next major advance was elevators in the early 1900s followed by air conditioning in 1920s. Air conditioning changed where we lived and the way we lived, worked, and spent our leisure time. Prior to air conditioning, commercial buildings required natural ventilation. This defined the shape of the building, as each office or room required an operable window T-, H-, and L-shaped floor plans, which allowed the maximum number of windows to provide natural light and ventilation, are still visible in New York, Chicago, Boston, and Denver. During the pre-air-conditioning era, large cities developed in northern latitudes because workers could remain productive performing office work for most of the year.

The New York Stock Exchange was the first "laboratory" for air conditioning, in 1901. The American public first experienced air conditioning in the Olympia Theater in Miami, Florida, in 1920. Department stores and movie theaters realized the potential for air conditioning to increase their business. And the first high-rise office building to be completely air-conditioned was the Milam Building, built in 1928 in San Antonio, Texas this building demonstrated that office workers were also more productive in a temperature-controlled environment. For manufacturers, air conditioning benefited the manufacturing of textiles, printing, and tobacco by allowing precise temperature, humidity, and filtration controls.

This new technology also allowed greater architectural freedom. Post-World War II construction incorporated sealed aluminum and glass facades, or curtain walls. The United Nations Building was the first major post-World War II project to be designed with a curtain wall system and air conditioning. Larger floor cross sections were possible, and interior offices could be created with man-made environments.

The structure and energy use of post-World War II commercial buildings was re defined as a direct result of air conditioning.

Modern buildings are fundamentally defined by the mechanical principles that drive their utility. Technology has defined both the form of the buildings and energy use. (Archive Photos, Inc.)

Electrical energy use in buildings rose from the late 1800s through the 1970s while natural gas and fuel oil use declined. Pre-1970 buildings represent 30 billion square feet of floor area, and have an average size of roughly 12,000 square feet. Approximately 70 percent of these buildings are air conditioned and are illuminated with fluorescent lighting. Overall energy intensity declined from about 1910 to 1940, but increased after 1950 as air conditioning became more common. More than 70 percent of the energy used by pre-1970 buildings came from electricity or natural gas.

THE ENERGY CRISIS

In the late 1960s a host of new technologies allowed architects to expand the horizons of their designs. The boundless promise of low-cost energy and a real-estate boom created a certain freedom of expression. This expression also created some of the most energy-intensive buildings ever built, as technology could overcome the shortcomings of a building design that failed to properly shelter occupants from heat, cold, sun, and glare.

The Arab oil embargo and ensuing "energy crisis" in late 1973 began to change the way commercial buildings were designed and operated. The first reaction to the energy crisis was to conserve energy. Conservation of energy means using less energy regardless of the impact that has on the levels of amenities the building provides. This impacted three areas of building comfort; thermal comfort, visual comfort, and ventilation air

The combination of lower thermostat settings, reduced ventilation, and lower lighting levels created environmental quality problems in many buildings. Even today, some building owners associate energy conservation with poor indoor environmental conditions.

Starting in the early 1980s, Energy efficiency gradually replaced energy conservation as the mainstream approach to saving energy. Energy efficiency relies on three key principles:

1. Life-cycle cost analysis. Energy-efficient buildings are typically designed to be cheaper, on a life-cycle basis, than wasteful buildings.

2. Comprehensive design. Often, energy-efficient buildings are cheaper on a first-cost basis as well as on a life-cycle cost basis. This frequently results from approaching the design of all of the energy-consuming systems of a building comprehensively and finding synergies among the various energy efficiency measures. For example, introducing daylight and energy-efficient artificial lighting can reduce the internal heat loads in a building, which reduces the size of the heating, cooling and ventilating ducts. This in turn reduces the floor-to-floor height and the cost of the elevators, building skin, etc., and enables the developer to add more rentable floor space in a given building volume (which is often constrained by zoning restrictions).

3. Environmental quality. In contrast to being dark and cold in the winter and hot in the summer, well-designed, energy-efficient buildings usually are more enjoyable to inhabit. This is particularly true of well-lit buildings, with low glare, balanced luminance, and visual cues from the lighting as to how to most comfortably inhabit

the space. Well-designed efficient buildings also allow easy and balanced temperature control, avoiding problems such as some zones being too hot while others are too cold.

The energy crisis also created a rush to develop new technologies and practices to reduce the energy consumption of buildings. One popular technology was solar energy. Commercial building solar technologies focused on harnessing the Sun's energy to cool the building through absorption cooling, provide hot water, or illuminate building interiors.

The Frenchman's Reef Hotel in St. Thomas, U.S. Virgin Islands, was the first large-scale commercial demonstration of solar absorption cooling. Other popular solar applications in the 1970s and early 1980s included swimming pool heating, commercial laundry hot water heating, and the use of natural daylight to offset electrical lighting in buildings.

It took several years before real improvements in commercial building energy efficiency occurred. This was because first-cost decisions drive a large segment of the new construction market. If a construction project begins to run over budget, energy-efficient features may be the first items cut, since they are less visible to the building tenant. In the 1980s, marketplace dynamics helped to change the market for energy efficiency because of the following factors:

- Utilities offered financial incentives to customers to reduce energy consumption because it was less expensive than constructing new power plants to meet the growing demand for electricity.
- Manufacturers incorporated energy-efficient design features into their product lines, making new products and services available to commercial building owners.
- Widespread adoption of building energy codes requiring minimum levels of energy efficiency to be included in new buildings.

The commercial building market began to respond to these dynamics by significantly reducing energy use.

CURRENT COMMERCIAL BUILDING DESIGN

A recent study conducted by California utilities identified the efficiency of buildings constructed in the 1990s with respect to what was required by the state building code. The study showed that most newer buildings use 10 percent to 30 percent less energy than buildings barely meeting the code. This finding is noteworthy in part because California has one of the most stringnet building codes in the country. The technologies and practices that created this result will be discussed in more detail below.

The building envelope in a commercial building plays a very different role than in a residential building. Figure 1 shows the complex interactions of heat flow in a modern commercial building. Energy (Q) enters the building from direct solar gain and heat from people, equipment, and lights. Energy leaves the building through the walls, roof, and windows and by heating ventilation air. An efficient building minimizes energy entering the building and balances that with the energy leaving the building.

The Commercial Building Envelope

The building envelope is one of the keys to both building energy use and thermal comfort. A high-performance building envelope will require a smaller mechanical system, provide natural lighting, and shelter occupants from heat and glare. Building envelopes in a modern building have four key elements that impact energy use. These elements are

- thermal performance
- building orientation
- permeability (air and moisture)
- daylighting.

Envelope construction characteristics for a modern building are shown in Table 1. The predominant stock of buildings is small masonry or metal buildings. The building envelope is an integral part of the way a building is illuminated. Windows and skylights can deliver a significant portion of the lighting needed for building occupants to be productive. Proper daylighting requires that glare and direct sun be minimized while maximizing the use of diffuse light. Architecture and glass property selection accomplish this. Architectural features include aperture areas, overhangs, fins, and light shelves or horizontal surfaces that redirect light entering windows deep into a space. Glass properties include a low solar heat gain coefficient and a high visible light transmittance. When properly designed, daylighting can deliver improved learning rates in schools and higher sales in retail stores.

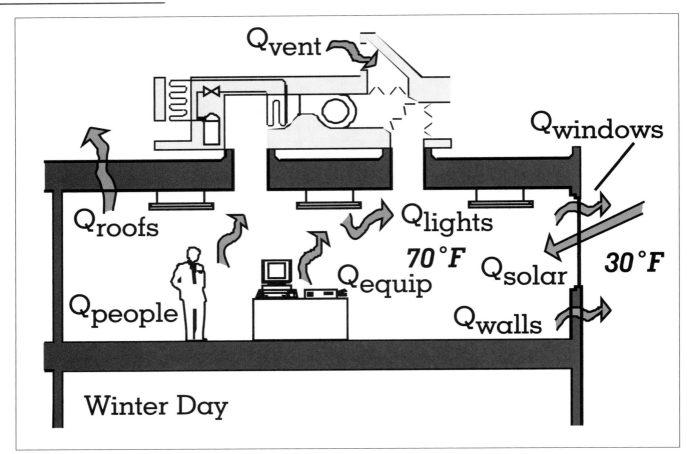

Figure 1.
Heat flow in a modern building.

Mechanical Systems

Building comfort is one of the key elements of a successful commercial building. The thermal comfort of a building is often as compelling as the aesthetics of the design. Modern building mechanical systems have two primary functions: maintain spaces within a predefined comfort range and deliver outdoor air to each space to assure proper ventilation. They have to do this in a quiet and efficient manner.

Heating, ventilating, and air-conditioning (HVAC) systems in modern buildings fall into two general categories: single-zone systems and multiple-zone systems. Both systems use air as the primary transfer mechanism to deliver heating and cooling to a space.

Single-zone systems deliver conditioned air to a single thermal zone. These systems are popular in small buildings (fewer than 10,000 square feet) and in single-story larger buildings. Usually they are vapor compression systems that cool air before it is delivered to a space. Single-zone systems serve 57 percent of post-1980 buildings.

Multiple-zone systems deliver conditioned air to more than one thermal zone. These systems typically have a direct expansion compressor or cold-water chiller to deliver cooling to the system. A central boiler or warm-air furnace at the system level, or electric heating elements at each zone, provide heating capabilities. This collection of components is controlled to maximize comfort and minimize energy use by computer-based controls. Roughly half of these buildings have an energy management control system that is operated by a trained energy manager.

A multiple-zone system must be able to deliver heating to a perimeter thermal zone in the winter while cooling an interior zone to offset internal

Building Characteristic	Predominant Construction	Percent of Post-1980 Buildings
Size	10,000 sq. ft. or less	74%
Exterior Walls	Concrete Masonry	54%
	Metal Siding	27%
Roofing	Metal Roof	40%
	Non-wood Shingles	30%
Insulation	Roof	79%
	Walls	71%
Windows	More than One Pane	51%
	Tinted Glass	36%

Table 1.
Post-1980 Envelope Construction Characteristics

gains. To accomplish this, modern multiple-zone systems use a technique called variable air volume (VAV).

A VAV system controls the temperature in a thermal zone by the volume of air delivered to that zone. If a zone thermostat demands heat, the system would first reduce the volume of air to the zone to the minimum required to meet outdoor air ventilation requirements and then begin to heat the air. This helps reduce energy use in two ways: first, the volume of air moving through the system is reduced to a minimum, lowering fan energy use, and second, reheating previously cooled air is minimized. Variable-speed drives are the most efficient method of controlling VAV fans and are used in 8 percent of post-1980 buildings.

The advantages of VAV are that temperature can be controlled while minimizing energy consumption. Also, the system can be smaller because the maximum demand for cooling never occurs simultaneously in all spaces. The disadvantages are the additional space required for the air-handling plants and ductwork.

Another energy-saving feature of modern HVAC systems is the ability to cool the building using outdoor air. This is accomplished through a control and damper arrangement called an outdoor air economizer. An economizer varies the outdoor air supply from the minimum setting to up to 100 percent outdoor air, provided the outside air temperature is less than the supply air temperature required to maintain comfort conditions. Outdoor air economizers are present in 85 percent of post-1980 buildings.

Lighting Systems

Lighting has a significant impact on building occupants, for better or for worse. Lighting also is a significant energy user and is rich with potential energy savings. For some time there has been a good deal of attention and effort invested in mining the energy savings from lighting systems. Preliminary studies into the ancillary benefits of energy-efficient lighting show that quality lighting can have positive effects such as improved productivity, reduced health complaints, and reduced absenteeism.

Figure 2 shows how buildings built after 1980 use energy. Lighting is the most significant energy expenditure in a modern building. The key qualities of an effective lighting system are

- energy efficiency
- room surface brightness
- reduction of glare
- adequate task illumination
- uniform light distribution
- good color lamps
- visual interest
- lighting controls.

Key technologies that are used in modern lighting include electronic ballasts, more efficient tubular fluorescent lamps, compact fluorescent lamps, and lighting controls. Fluorescent lighting is the predominant lighting system installed in post-1980 buildings and is used in 71 percent of floor space. Specialty retail stores use a combination of fluorescent and

195

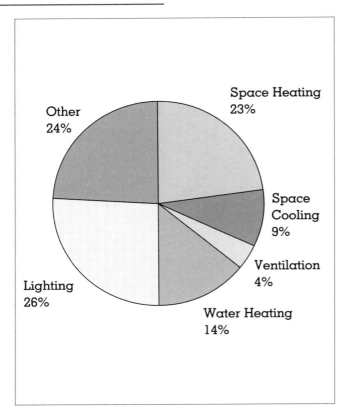

Figure 2.
Post-1980 energy end uses.

incandescent, with halogen lights accounting for 2 percent of floor space.

Lighting controls used in post-1980 buildings include automatic shutoff controls through an energy management system (6%) and an occupancy sensor (2%). Daylighting automatically dims artificial lighting in areas where light entering windows and skylights provides adequate illumination. Daylighting, while a promising technology, has not received widespread application.

Office Equipment

All of the advances in modern building design have not dramatically reduced the energy intensity of modern buildings. In fact, they have become more energy-intensive during the 1990s. Buildings constructed from 1990 to 1995 typically used 15 percent more energy per square foot than buildings constructed during the 1980s.

In 1995 an estimated 43 million PCs and computer terminals were used in commercial buildings. More than half of the 4.6 million buildings in the

United States had at least one PC or computer terminal. The more PCs and computer terminals used in a given building, the greater the impact on the building's energy consumption. The proliferation of personal computers, printers, copiers, and other types of "plug loads" is the main cause of the rise in energy intensity in recent years.

THE FUTURE OF COMMERCIAL BUILDINGS

The workplace of the future has captured the imagination of researchers and designers internationally and across a wide range of disciplines, from computer science and furniture design to organizational systems. Noticeably missing from these discussions and visions is the energy research community. With more and more businesses operating under tighter and tighter margins, building owners and occupants are going to demand increased attention to the delivery and management of energy in buildings as other changes take place.

Trends in the Workplace

We are moving in a direction in which the information technologies associated with work processes are almost totally decoupled from energy and architectural technologies and systems. Not only is the workplace of the future going to demand more flexibility and rapid reconfiguration of space, it also is increasingly moving toward a radically different use of facilities. In addition, in the interest of meeting shareholder and organizational concerns for increasing profit margins, the workstation footprint is shrinking rapidly. All of these trends have implications for energy.

For example, current trends show telecommuting increasing 20 percent annually. According to one estimate, New York City will have an additional 180 million square feet of empty office space as a direct result of telecommuting. This could shift patterns in energy demand as well as require the development and deployment of new technologies that are suited to renovating facilities rather than for use in new buildings. Furthermore, although many telecommuting advocates see working at home as a way to reduce transportation impacts, there is some indication that the demand for transportation will actually increase. This is because many of the workers who have home offices are independent consultants who spend a good deal of their time in their cars visiting clients and developing work. Further, the introduction of new information technologies is opening up new markets in many areas, which, in turn, are associated with rapid delivery of products, which increase the demand for vehicles.

Many organizations are moving to open-planned, high-communication, team-centered layouts with little understanding of the organizational and performance implications of this fad, and with even less understanding of the implications for the design and delivery of energy where it is needed, when it is needed. Furthermore, the workstation "footprint" is becoming ever smaller as furniture manufacturers reduce the size of the cubicle setting in response to demand from businesses to put more people into smaller areas.

Energy Technology Development Implications

These trends have an impact on both energy use and the technologies that are developed to deliver thermal and visual comfort to the workplace. For example, how does an existing building with a fixed comfort delivery system meet the needs of these new and varying layouts? Current practice suggests that increased airflow is the answer, resulting in higher levels of reheating and recooling of air in a typical office system. Higher occupant density also translates into greater demand for outdoor air ventilation. Higher outdoor air rates can increase total building energy consumption from 20 percent to 28 percent in buildings built prior to 1989. These trends could have the following impacts leading to greater use in the workplace of the future:

- increased air movement to accommodate high occupant densities resulting in increased fan energy consumption;
- individuals bringing in desk lamps, resulting in the use of low-efficacy light sources (i.e., incandescent lamps) in addition to the high, efficacy light sources installed in the buildings;
- electric resistance heaters to warm areas of high air flow (too much cold air) and low air flow (not enough warm air) resulting from high-density loads with controls designed without a high enough degree of zone resolution;
- personal fans to create air movement that is impacted by typical partitions;
- increased ventilation requirements resulting in greater heating and cooling energy demand.

To avoid significantly higher energy consumption, building designers of the future must integrate the knowledge and expertise of the energy-efficiency community. There are several areas of integration, including:

- enhanced personal and team control over ambient conditions, including lighting, temperatures, ventilation, and acoustics;
- greater integration of energy technologies into the design of furnishings;
- glazing materials and window technologies that promote views and natural ventilation while reducing heat gain and glare;
- greater attention to understanding the energy demands that will be required to support the high-technology office of the future;
- development of ways to humanize windowless and underground spaces through features such as sensory variability, borrowed daylight through light tubes, and simulated windows;
- intelligent building systems that can identify and correct potential problems before they become large and more difficult to manage.

Jeffrey A. Johnson
William Steven Taber, Jr.

See also: Air Quality, Indoor; Building Design, Energy Codes and; Building Design, Residential; Cool Communities; Economically Efficient Energy Choices; Efficiency of Energy Use; Lighting; Office Equipment; Solar Energy; Windows.

BUILDING DESIGN, ENERGY CODES AND

Energy efficiency codes and standards for new buildings are the most cost-effective ways to improve the efficiency of a new building. Improvements can be made when a building is being designed and constructed at a fraction of the cost that it would take to make similar improvements once it is built. Energy codes require all new buildings to be built to a minimum level of energy efficiency that is cost-effective and technically feasible. Considering the average lifespan of a building, investments in energy saving technology ultimately pay for themselves many times over.

Building codes play a role in supporting the welfare of the community. They protect life, health and safety. They also protect the investment that has been made by bankers, insurance companies, businesses, and individuals. Finally, building codes promote economic development by protecting the value of the built environment.

This section will discuss the history of building construction regulations, the role of energy in building construction regulations, how these regulations have evolved into our current system, and the future of building construction regulations and energy.

BUILDING CONSTRUCTION REGULATIONS

The need to develop standards, codes, and other rules for the design, construction, operation, and maintenance of buildings has been driven over the years by a number of factors. These factors include comfort, fire safety catastrophic events, egress, sanitation, and health, among others. The first recorded building code was that of the Amorite king, Hammurabi, in 1763 B.C.E. His Code of Laws, one of the first codes of law in world history, contained 282 rules including the principles of "an eye for an eye" and "let the buyer beware." The Code of Laws stated: "If a builder build a house for some one, and does not construct it properly, and the house which he built fall in and kill its owner, then that builder shall be put to death."

Many of the building codes used initially in North America were imported from Europe by the early settlers. In 1630, the City of Boston building code stated: "No man shall build his chimney with wood nor cover his roof with thatch." In 1865, the City of New Orleans adopted an ordinance requiring the inspection of buildings for public use. In 1905 the National Board of Fire Underwriters published the first national building code in the United States, the Recommended National Building Code.

Some notable evolutionary processes occurred over the past during the twentieth century. While local government has generally retained the authority to enforce building construction regulations, the development of the codes and standards they adopt has shifted away from "home grown" criteria of various state and local agencies towards voluntary sector activities. In 1927 the Pacific Coast Building Officials (a precursor to the International Conference of Building Officials) conference published the

---

MILLER SQA

The 290,000-square-foot Miller SQA building was designed by William McDonough, FAIA to be a state-of-the art "green" building. Miller SQA, a wholly owned subsidiary of Herman Miller, Inc., is a remanufacturer, manufacturer, and vendor of office furniture that provides "just in time" furniture products for small businesses and nonprofit institutions. The building is a manufacturing plant, warehouse, and headquarters housing approximately 600 workers in a manufacturing plant and 100 workers in the office portion. The SQA building also has a lunchroom; rest areas at each end of the manufacturing area; and a fitness center, including a full-size basketball court.

Energy-efficient aspects of the building include large-scale use of energy-efficient lighting, daylight controls, and state-of-the art digital HVAC controls, including sensors, controllers, and data loggers. Green components include environmentally sensitive materials throughout the building, minimally invasive site utilization (including a wetlands and use of natural field vegetation rather than planted and mowed grasses), enhanced indoor air quality, and extensive recycling. In addition, building materials were obtained locally whenever possible to reduce transportation costs and energy.

Studies showed that the new SQA building was associated overall with a higher quality of work life than the old building. For example, 16 percent of the office workers said they had headaches often or always in the old building, while only 7 percent did in the new building. In addition, there were small (typically less than 2%) increases in worker performance, can be significant to an organization in a competitive market. Finally, the new building uses 18 percent less energy than the old building.

Uniform Building Code. Then, in 1945, the Southern Building Code Congress International published the Standard Building Code.

Federal Legislation	Scope	Date Enacted
Energy Conservation and Production Act	Develop performance standards for all new buildings	8/14/76
New Buildings Act	HUD shall promulgate, implement and enforce energy performance standards	5/4/77
Department of Energy Organization Act	Transfer authority from HUD to DOE	8/4/77
Housing and Community Development Act of 1980	DOE shall promulgate interim standards that apply only to Federal buildings	10/8/80
Omnibus Budget Reconciliation Act	Standards to be developed through private sector.	8/13/81
Cranston-Gonzales National Affordable Housing Act	Energy efficiency standards for public housing	11/28/90

Table 1.
Codes and Standards Legislative History

Groups such as the American Society of Mechanical Engineers (ASME), National Fire Protection Association (NFPA), American Society of Heating, Refrigeration, and Air Conditioning Engineers (ASHRAE) and Underwriters Laboratories (UL) develop codes and standards for use by industry members. Model codes organizations including Building Officials and Code Administrators International (BOCA), International Conference of Building Officials (ICBO), and Southern Building Code Congress International (SBCCI) are code official organizations that develop model codes specific to their region of the country.

FEDERAL LEGISLATIVE HISTORY

Energy codes and standards stem from policy directives in the early 1970s that were hastened by the Arab oil embargo. The Secretary of the Department of Housing and Urban Development was directed in 1971 to reduce maximum permissible energy loss by about one-third for a typical home and to revise the insulation standards for apartments. The resulting thermal envelope criteria in HUD's Minimum Property Standards became the first energy code.

The Arab oil embargo and ensuing "energy crisis" in late 1973 began to change the scope of building construction regulations. The public became quickly aware of the true cost of energy as gasoline prices sky-rocketed. The oil embargo created a ripple effect that demanded a larger public policy role to reduce wasteful energy use.

Table 1 describes the ensuing legislation that was promulgated to promote energy efficient construction as a result of the Arab oil embargo.

Development of standards, codes, and other regulations to address energy in buildings began in earnest in the early 1970s. The Omnibus Budget Reconciliation Act prohibited the government from promulgating regulations that would apply to non-public construction. This left the task of energy codes and standards development up to two key processes. They are the model code processes set up by the building regulation community and the consensus processes set up by the American Society of Heating, Refrigeration and Air-conditioning Engineers (ASHRAE) and their co-sponsor, the Illuminating Engineering Society of North America (IESNA).

ASHRAE Consensus Process

Some states recognized the need and the role that they could play if energy criteria were added to their building codes. Through the National Conference of States on Building Codes and Standards (NCSBCS) a request was made by the states for the National Bureau of Standards (NBS) to develop some criteria that could be used to address energy issues through building codes. The NBS released suggested energy

Title	Scope	Description
ASHRAE/IES Standard 90A-1980	All Buildings	Updated Sections 1 through 9 of ASHRAE/IES Standard 90-1975 with more stringent provisions.
ASHRAE/IES Standard 90.1-1989	Commercial and High Rise Residential	Updated entire commercial and high-rise residential standard with more stringent provisions.
ASHRAE Standard 90.2-1993	Low Rise Residential	Updated low-rise residential standard with more stringent provisions.
ASHRAE Energy Code for Commercial and High Rise Residential Buildings	Commercial and High Rise Residential	Standard 90.1-1989 in mandatory language for use by model code organizations.
ASHRAE/IESNA Standard 90.1-1999	Commercial and High Rise Residential	Updated entire commercial and high-rise residential standard with more stringent provisions

Table 2.
History of ASHRAE/IES Standards

requirements for new building designs in early 1974. Since the NBS criteria had not received widespread input from and agreement by the building community the states looked to a voluntary sector standards development organization to develop a standard for building energy design. Using the NBS criteria as a starting point the ASHRAE established a Standard 90 project and committee and began the process of writing a standard in cooperation with the IESNA. Released in August of 1975, ASHRAE/IES Standard 90-75 addressed energy conservation in new building design.

The DOE sponsored two major projects with ASHRAE as a result of the Omnibus Reconciliation Act; Special Project 41 and Special Project 53. These projects led to a series of changes and has continued to be amended from time to time. Table 2 outlines the evolution of the ASHRAE/IES Standard through 1999.

Model Codes Development Process

Because the ASHRAE/IES standard contained both mandatory requirements and recommendations, it was difficult to adopt and implement as a uniform building code. Such codes must clearly state minimum requirements that can be uniformly interpreted and applied. For this reason it was determined that a model code was needed that contained only the

mandatory requirements of Standard 90-75 and placed them in enforceable code language that would fit within a building code. Working with NCSBCS the three U.S. model code organizations (Building Officials and Code Administrators, International Conference of Building Officials, and Southern Building Code Congress) developed and released a model energy code for new buildings in late 1997: Model Conservation Energy Code (MCEC) 1977. Table 3 shows the evolution of the model codes from the MCEC to the International Energy Conservation Code (IECC).

RECENT TRENDS IN CODE ADOPTION AND ENFORCEMENT

In 1975, the Energy Conservation and Production Act tied the availability of federal funds to states for their energy conservation efforts to the adoption by states of an energy code for new building construction. Beginning in the mid-1970s states began to adopt energy codes in earnest. This was accomplished in three ways:

1. Through individual state legislation, wherein the legislature adopted by reference a particular energy code or wrote the energy code requirements directly in the legislation.

Title	Scope	Description
Model Energy Code (MEC) 1983	All Buildings	Technically equivalent to ASHRAE/IES Standard 90A-1980
MEC 1986	All Buildings	Low-rise residential thermal envelope provisions updated to improve stringency.
MEC 1988	All Buildings	Minor upgrades.
MEC 1992	All Buildings	Low-rise residential thermal envelope provisions updated to improve stringency.
MEC 1993	All Buildings	Reference to ASHRAE Energy Code incorporated into Chapter 6.
MEC 1995	All Buildings	Window provisions added to reference National Fenestration Rating Council.
International Energy Conservation Code 1998	All Buildings	Simplified commercial provisions added as Chapter 7 and low-rise residential Solar Heat Gain Coefficient requirement.

Table 3.
Evolution of the Model Energy Code

2. To convey the authority to adopt an energy code to a state regulatory agency such as the state energy office or state office with authority for a state building code.

3. Take no action; leaving the decision up to local government.

Ultimately, the responsibility for properly implementing these legislative actions fell upon the local governments of counties and incorporated cities, towns, boroughs, and so on. In total there are over 2,000 counties in the United States and over 40,000 independent units of local government that have some ability to adopt, implement, and enforce energy codes.

Since 1975 numerous states have adopted energy codes through legislative or regulatory mechanisms. They range from those that apply to all new buildings to those that only apply to state-owned buildings, non-residential buildings, or only buildings in localities that have adopted a building code. Where the state has taken no action or only partial action on a selected building type, such as state-owned buildings, then local government is free to take action if it so chooses. In some cases state and local government are prohibited or preempted from taking any action.

Other notable adopters of energy codes are the U.S. Department of Housing and Urban Development, Department of Defense, and public utilities.

Once adopted there are many ways to implement and enforce energy codes. Where a state or local building code exists, there is already an adopted code and enforcement mechanism in place to ensure compliance with the energy code. Where no such building code infrastructure exists, then implementation in the absence of such support must rely on the following mechanisms:

- Builder certification as a condition for utility connection
- Warranty with penalties if non-compliance is verified
- Contractor licensing tied to code compliance
- Energy rating schemes that pull the market
- Professional architect and engineer certification
- Third party certification and inspection

Over time more states have secured authority to adopt and implement building codes and over time those codes have tended to be more uniform. Current activities are likely to keep the focus on national uniformity in the absence of Federal preemptive authority.

MEC Version or State Code that's Equivalent	States Adopted
98 IECC, For state-owned and stated-funded buildings	1 State (NE)
Exceeds 95 MEC, Statewide adoption/equivalence	4 States (CA, FL, OR, MN)
Exceeds 95 MEC, Partial adoption or equivalence (i.e. only state funded bldgs, dependent on local jurisdiction, etc.)	2 States (AK, WA[1])
95 MEC, Mandatory statewide adoption/equivalence	13 States (CT, GA, MD, MA, NC, NH, RI, SC, OH, VA, VT, UT, WI)
95 MEC, Partial adoption/equivalence	3 States (OK, LA[2], HI[2])
93 MEC, Mandatory statewide adoption/equivalence	1 State (DE)
93 MEC, Partial adoption/equivalence	5 States (TX[3], ND, MT, AL, KS)
92 MEC, Mandatory statewide adoption/equivalence	7 States (AR, IN, IA, KY, NM, TN, NY)

[1] Code exceeds 95 MEC for electrically heated buildings, but is less stringent for non-electrically heated buildings.
[2] LA and HI have 95 MEC adopted for multi-family low rise only.
[3] TX is listed twice because 93 MEC is mandatory only for state funded low-rise bldgs. Local jurisdictions are adopting 92, 93, 95 MEC on their own.

Table 4.
Status of State Adoption as of August 1999

CURRENT CODE POLICY AND STATE ADOPTION

The Energy Policy Act of 1992 (the Act) was a major policy action to promote the improved use of the nation's energy resources. The Act includes both improving the supply of energy resources and promoting the efficient use of those resources. One aspect of the Act focuses on improving the efficiency of new buildings through upgrading and adoption of energy efficiency codes and standards. The Act, combined with technical and financial support, provides states with an unprecedented level of federal support to improve the efficiency of new buildings.

State and locally adopted energy codes are supported by the Federal government in three ways:

- Federal law requires that States act to review and upgrade their codes.

- Technical support is provided by DOE in the form of tools, training programs, code user support through a toll-free number, and analysis directed by the states.
- DOE provides financial support in the form of over $4 million in special projects funding annually.

Since the adoption of the Act in 1992, state-of-the-art energy codes have been extended to cover an additional 39 percent of residential construction and 26 percent of commercial construction. Two-thirds of new U.S. residential construction (1 million homes annually) fall under federal, state, and local energy codes that meet or exceed the 1995 version of the Model Energy Code (MEC). An additional 975 million square feet of commercial construction falls under codes that meet or exceed ASHRAE/IES Standard 90.1-1989. Table 4 shows the status of state residential code adoption as of August 1999.

On December 9, 1994, the International Code

Council (ICC) was established as a nonprofit organization dedicated to developing a single set of comprehensive and coordinated national codes. The ICC founders—the Building Officials and Code Administrators (BOCA), the International Conference of Building (ICBO), and the Southern Building Code Congress International (SBCCI)—created the ICC in response to technical disparities among the three sets of model codes now in use in the United States.

Since 1972, the Council of American Building Officials (CABO) has served as the umbrella organization for BOCA, ICBO, and SBCCI. In November 1997, it was agreed to incorporate. CABO into the ICC. Responsibility for developing and maintaining the Model Energy Code (MEC) was transferred from CABO to the ICC in order to provide proper interface with the International Codes.

The first version of the International Energy Conservation Code (IECC) was published in 1998.

THE FUTURE OF ENERGY CODES

As with any market, the new construction market has a wide range of efficiencies within its participants. In theory, energy codes can shift the average efficiency of the market by eliminating the option of building to an efficiency level lower than that mandated by the code. This effect can produce significant savings even when the code minimum is set at the market "average" efficiency level.

Evidence from numerous code evaluations suggests that energy codes have transformed markets in three ways:

1. In areas where codes are well enforced, the stock of poor performing new buildings has been reduced to a minimum, and

2. In areas where utility incentive programs were successful, the overall efficiency of a typical building exceeds code.

3. Codes have brought more efficiency technologies into widespread use in the market (e.g., vinyl window frames, T-8 fluorescent lamps).

The focus of codes and standards will shift from specifying the installation of prescriptive measures to the actual performance of the final building. This shift is consistent with an overall emphasis on objective or performance-based codes within the building code community. The next generation of energy codes will

- Incorporate new technologies and practices into the standard that replace less efficient technologies and practices,
- Assure proper performance of measures once specified and installed thereby assuring that energy and environmental benefits to energy codes are realized by building owners and occupants, and
- Improve enforcement through partnerships and support of innovative enforcement practices by local governments.

Jeffrey A. Johnson
David Conover

See also: Building Design, Commercial; Building Design, Residential.

BIBLIOGRAPHY

American Society of Heating, Refrigerating, and Air Conditioning Engineers (ASHRAE). (1999). *ASHRAE/IESNA Standard 90.1-1999 Energy Standard for Building Except Low Rise Residential Buildings.* Washington, DC: Author.

Building Codes Assistance Project.*BCAP—Building Codes Assistance Project.* Washington, DC: Author. <http://www.bcap-energy.org/>.

International Energy Conservation Code 2000. (2000). Falls Church, VA: International Code Council.

U.S. Department of Energy. (July 20, 2000). *Building Standards & Guidelines Program (BSGP).* Washington, DC: Author. <http://www.energycodes.org/>.

BUILDING DESIGN, RESIDENTIAL

From early cave dwellings to the well-appointed suburban homes of today, the most fundamental reason for shelter remains to provide protection from weather and other possible dangers. Beyond this most basic purpose, modern culture has forged a number of other expectations for homes, including affordability, comfort, health, durability, and peace of mind. This article explains the concepts behind energy-efficient homes—how they meet our basic needs and expectations; gives an overview of the history of energy-efficient innovations in American housing; and provides projections for the future.

HISTORICAL DEVELOPMENT

The history of energy technology in homes from the turn of the twentieth century until the 1970s is dominated by developments that contributed to energy consumption, including modern heating and cooling systems and the proliferation of appliances and lighting systems. The trend to increased energy efficiency began after World War II with the development of individual technologies that, only in recent years, have been integrated into systems solutions.

The evolution of energy-efficient homes began in the 1830s with the advent of wood-framed construction, which is still the dominant building technique. At about the turn of the twentieth century, advances in glass manufacturing allowed builders to add windows in increasing size and quantity to both residential and commercial buildings. During the fuel shortages in World War II, there was a demand for increased insulation and some consideration given to passive solar design. But low oil prices following that war and the development of central heating and cooling systems curtailed the development of energy-efficient designs and led to increased reliance on mechanical systems to create home comfort.

During the OPEC oil embargo of the 1970s, energy efficiency became a national priority. Levels of insulation increased, and double-glazed windows became standard in colder climates. As a result of the oil crisis, there was a push to develop new energy-efficient technologies. Once wall insulation had improved and the building envelope air-sealed, losses due to windows, equipment inefficiency, and system design became more critical. New window technology featured low-E coatings and insulating gases between the two panes. These windows have became increasingly popular since the mid-1980s. Concurrent improvements in heating and cooling equipment included condensing furnaces and water heaters, heat pumps, and dramatic increases in air-conditioner efficiency.

Energy efficiency experienced some important setbacks in the mid-to late 1980s due to a number of prominent technology failures. For example, early pulse combustion gas boilers, compact fluorescent lighting, and some triple-glazed windows experienced a variety of performance problems because products were introduced before all technical issues were resolved. In some cases, energy efficiency just suffered from bad press. For instance, air-sealed homes were wrongly associated with bad indoor air quality.

By the early 1990s, the results of research and technology improvements began to be more effectively passed down to builders. In addition, other technology improvements were introduced or increased market penetration, including advanced wall system alternatives to stick framing (e.g., structural insulated panels and insulated concrete forms), geothermal heat pumps, combined space and water heating systems, mechanical ventilation with heat and recovery, and a new generation of compact fluorescent lights. Another critical development was advanced computer technology, which made it easy to model the complex dynamics of residential energy use. Energy audits and modeling have become widely available to assess the most cost-effective improvement measures. Finally, diagnostic procedures such as duct and infiltration testing and infrared imaging have enabled building scientists to refine their concept of the house as a system.

RESIDENTIAL ENERGY EFFICIENCY

Each aspect of energy use in a home influences the comfort, health, and safety of the occupants as well as their utility bills. An energy-efficient home properly utilizes systems solutions to effectively reduce energy use while improving the quality of life for its occupants. A systems solutions approach examines the interactive effects of all of the components within the house. While there is no definitive set of features, the building science community is converging on eight common elements to reduce home energy use when properly incorporated into the systems solution: air sealing, insulation, windows, duct sealing, heating and cooling, lighting and appliances, mechanical ventilation, and diagnostic testing.

Air Sealing

If outdoor air can easily leak into and through homes, both comfort and energy-efficient performance will be difficult to maintain. Today, off-the-shelf technologies that contribute to airtight construction include a variety of house wraps, sealants, foams, and tapes. In energy-efficient homes, builders use these tools to seal the myriad of cracks

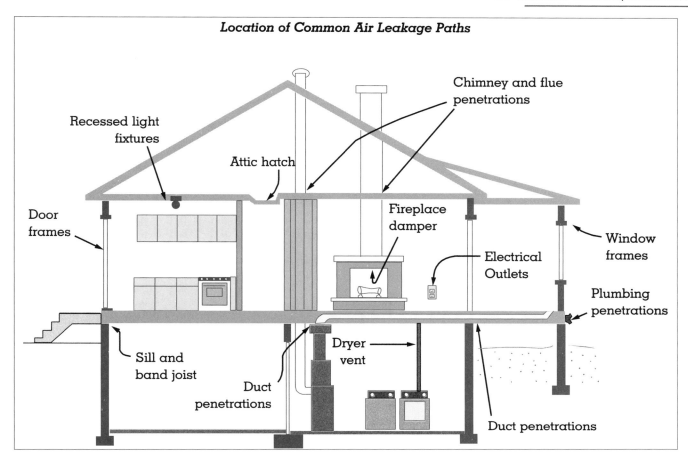

Location of Common Air Leakage Paths

Figure 1.
SOURCE: U.S. Environmental Protection Agency

and gaps in framing along with hundreds of holes for plumbing, mechanical equipment, and electrical wiring (Figure 1). Builders have adopted new job-site management techniques to assure that air sealing is properly completed. One approach is to have a single individual responsible for a comprehensive regimen of airtight construction details. Another approach requires subcontractors to be contractually responsible for sealing all holes they make in the construction process. Diagnostic procedures include blower door testing to measure how well homes are sealed.

Insulation

A home must be correctly insulated to be energy-efficient. Table 1 shows U.S. Department of Energy (DOE) recommended minimum insulation values for different climates throughout the country. The "R-value" is a measure of insulation effectiveness—the higher the R-value, the more insulation is pro-

vided. Although there are advantages and disadvantages to each of the wide variety of insulation systems available, the most important criterion for performance is the care with which it is installed. Gaps, holes, voids, and compressions can substantially reduce insulation effectiveness. Again, diagnostic procedures can help builders and contractors maintain quality in the insulation process.

Windows

Window technology has changed dramatically since the mid-1980s. Today, advanced windows are available that approximate the thermal performance of an insulated four-inch wall and use low-emissivity (low-E) coatings, microscopic layers of metallic material applied to interior glass surfaces of double-glazed windows. These coatings prevent undesirable solar heat gain in the summer and excessive heat loss in the winter.

Cost Effective Insulation R-Values

If you live in a climate that is...	and your heating system is a...	then insulate to these levels in the...			
		ceiling	wood frame walls	floor	basement/ crawl space walls
Warm with cooling and minimal heating requirements (i.e.,FL & HI; coastal CA; southeast TX; southern LA, AR, MS, AL & GA).	gas/oil or heat pump	R-22 to R-38	R-11 to R-15	R-11 to R-13	R-11 to R-19
	electric resistance	R-38 to R-49	R-11 to R-22	R-13 to R-25	R-11 to R-19
Mixed with moderate heating and cooling requirements (i.e., VA, WV, KY, MO, NE, OK, OR, WA, & ID; southern IN, KS, NM, & AZ; northern LA, AR, MS, AL, & GA; inland CA & western NV).	gas/oil or heat pump	R-38	R-11 to R-22	R-13 to R-25	R-11 to R-19
	electric resistance	R-49	R-11 to R-28	R-25	R-11 to R-19
Cold (i.e., PA, NY, New England, northern Midwest, Great Lakes area, mountainous areas (e.g., CO, WY, UT, etc.)).	gas/oil	R-38 to R-49	R-11 to R-22	R-25	R-11 to R-19
	heat pump or electric	R-49	R-11 to R-28	R-25	R-11 to R-19

Table 1.

SOURCE: U.S. Environmental Protection Agency

Note: (a) Adopted from the U.S. Department of Energy 1997 Insulation Fact sheet. (b) Insulation is also effective at reducing cooling bills. These levels assume that you have electric air-conditioning. (c) R-Values are for insulation only (not whole wall) and may be achieved through a combination of cavity (batt, loose fill, or spray) and rigid board materials. (d) Do not insulate crawl space walls if crawl space is wet or ventilated with outdoor air.

Duct Sealing

American homes often utilize duct systems to distribute heated and cooled air to each conditioned room. These ducts can leak as much as 25 to 35 percent of the air they are supposed to deliver to the living space. Figure 2 shows some of the common areas of duct leakage. In addition to wasting energy and money, leaky ducts can lead to durability and indoor air quality problems.

Duct mastic has replaced duct tape as the most effective sealing material. Mastic is a fibrous, elastomeric compound that permanently seals duct connections and seams. In addition, the "boot" connections between ducts and floors and between walls and ceilings should be fully caulked and/or foamed airtight. Lastly, air handlers that house the heating and cooling coils and circulation fans are often extremely leaky and should be fully sealed while still allowing access for service and filter replacement. Use of new airtight airhandlers provides even better performance.

Heating and Cooling

Bigger is not better when it comes to heating and cooling equipment. Oversized equipment is not only more expensive to purchase but it also can reduce comfort. The most prevalent examples are the rapid temperature fluctuations and humidity control problems caused by frequent on/off cycling of oversized equipment. Poor dehumidification occurs because the cooling coils do not sustain cold temperatures long enough to effectively condense moisture out of the air.

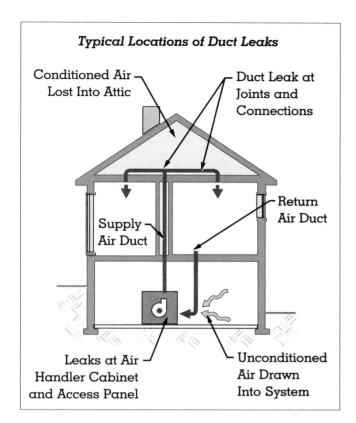

Typical Locations of Duct Leaks

Conditioned Air Lost Into Attic

Duct Leak at Joints and Connections

Return Air Duct

Supply Air Duct

Leaks at Air Handler Cabinet and Access Panel

Unconditioned Air Drawn Into System

Figure 2.

SOURCE: U.S. Environmental Protection Agency

The impact of incorrect equipment sizing practices is magnified in an energy-efficient home that has dramatically reduced heating and cooling loads.

Once right sizing is addressed, new high-efficiency equipment, including furnaces, boilers, heat pumps, and air conditioners provide further energy savings and other quality improvement benefits. Geothermal heat pump equipment offers further efficiency gains by taking advantage of much more stable year-round ground temperatures (warmer than ambient air in the winter and cooler than ambient air in the summer) with buried heat exchanger loops.

Lighting and Appliances

High-efficiency lighting and appliances also can provide significant energy savings. Whole-house lighting designs typically include hard-wired compact fluorescent lighting (e.g., recessed, sconce, surface-mounted and exterior-wall-mounted fixtures), Thinner diameter (1 inch is 1/4 inch) high-efficiency fluorescent lighting with electronic ballasts, and motion sensors in rooms where lights might normal-

ly be left on during long periods of non use (e.g., laundry rooms and children's bathrooms and bedrooms). These efficient fixtures and controls provide high-quality lighting at lower cost than incandescent bulb fixtures with conventional controls. High-efficiency refrigerators and horizontal-axis clothes washers, clothes dryers, and dishwashers also offer significant energy savings. The U.S. Environmental Protection Agency (EPA) and Department of Energy work with lighting and appliance manufacturers to offer the ENERGY STAR label on high-efficiency models so that consumers can easily identify them.

Mechanical Ventilation

Ventilation is required to maintain indoor air quality and comfortable moisture levels. American Society of Heating, Refrigerating, and Air-Conditioning Engineers (ASHRAE) guidelines require that a properly ventilated home replace its full volume of air approximately every three hours. Without proper ventilation, moisture generated indoors from human respiration, bathing, cooking, indoor plants, and cleaning can condense on wall and window surfaces, eventually leading to mold and mildew as well as structural damage to the house. Poorly balanced air flows due to duct leakage and inadequate air sealing can exacerbate the problems. In addition, there are internal sources of air pollution, such as off-gassing of construction materials and furnishings, solvents, combustion appliances, candles, dust mites, and pets. In most homes, particularly energy-inefficient ones, ventilation is provided accidentally by leakage of air through cracks and holes in the building envelope. Depending on wind and weather conditions, this may or may not provide adequate ventilation. Moreover, air drawn in from attics, crawl spaces, garages, or even outdoors can bring a wide range of undesired pollutants indoors (e.g., dust, pollen, radon, mold, mildew, moisture, carbon monoxide, and solvent fumes).

To provide a constant, controlled supply of fresh air to maintain proper indoor air quality and prevent moisture damage, there are three basic ventilation strategies: constant supply, constant exhaust, and balanced systems. Constant supply systems provide a continuous flow of fresh air to living areas and force stale air out through air leaks and kitchen and bathroom fan vents. Constant exhaust systems continuously remove stale air while allowing fresh air to enter either through air leaks or special vents in walls or windows. Balanced systems provide the greatest control of indoor air quality by combining both supply and exhaust functions. Some systems also incorporate a heat exchanger to pre heat or pre cool the incoming air, and dehumidifiers to remove moisture from humid outdoor air. There are new developments in lower-cost, balanced systems that utilize heating and cooling system ducts to deliver fresh air while stale air is removed by exhaust fans.

Diagnostic Testing

To build a home with maximum speed and quality, builders often hire subcontractors to install plumbing, insulation, heating and cooling systems, gypsum board, and other building components. Each trade brings specialized experience to a particular job but often fails to realize the full impact of its work on other parts of the home building process. Since the builder has the ultimate responsibility for the performance of the home, whole house diagnostics such as blower door and duct blaster tests or infrared imaging can catch problems before the buyer moves in. A blower door depressurizes the house to measure whole house infiltration and locate leaks in the building envelope. Duct blaster tests measure duct leakage and identify potential sources of comfort complaints due to reduced air flow. Infrared cameras take a thermal picture of a house to identify insulation voids and gaps, thermal bridging, and hot or cold spots. By implementing diagnostic procedures as part of a quality assurance program, the builder can avoid costly customer callbacks. Diagnostic testing services as well as advice on efficient design are available from a wide variety of experts, including local utility staff, energy consultants, home energy raters, weatherization specialists, and some mechanical, insulation, or air sealing subcontractors.

SYSTEMS SOLUTIONS

A house is a complex system of interacting parts that contribute to overall performance including comfort, energy use, health, maintenance, and longevity. For example, a common air distribution system utilizes supply ducts running through the attic and return ducts tied directly to the air handler inside the home. If the ductwork is not properly sealed and there are combustion appliances in the home, this configuration can lead to health and fire hazards because the

return ducts in this case draw more air than the leaky supply ducts are able to provide to the home. The resulting negative pressure can draw combustion appliance exhaust back into the home and even create flame rollout at the equipment combustion chambers. A systems approach properly integrates tight construction, sealed ductwork, and properly functioning ventilation systems to save energy and eliminate indoor air pollution dangers.

Systems thinking also takes advantage of cost reduction opportunities. For instance, energy-efficient homes can be heated and cooled with smaller, lower-cost equipment. In some cases, energy-efficient homes have such small heating requirements that a single water heater can be used for space and water heating. In addition, a home with air-sealed construction, a well-insulated envelope, and high-performance windows no longer needs duct systems extended to outside walls to maintain comfort. A much more compact duct system will cost less, be quieter, and increase efficiency while delivering superior performance. All of these savings can be used to offset the extra cost for other energy features. As a result, an energy-efficient home based on systems solutions delivers improved weather protection, affordability, comfort, indoor air quality, and durability.

WEATHER PROTECTION

To varying degrees, code-compliant homes offer reasonable protection from the rigors of harsh weather conditions. However, features that improve the energy efficiency of homes also offer better weather protection. For example, increased protection from severe weather such as tornadoes and hurricanes can require building systems with greater structural integrity, such as six-inch framed walls, insulated concrete form (ICF) walls, and structural insulated panels (SIP). These systems are also much more energy-efficient because they allow for a better insulated building envelope and tighter construction. Protection from another, more pervasive weather condition, moisture, requires energy-efficient building practices. Building an air-sealed and properly insulated home provides obvious advantages by blocking water vapor from entering the home. Mechanical ventilation systems prevent excessive accumulation of moisture inside the home. In addition, right-sized air conditioners operate efficiently, and more effectively remove moisture from the air.

	Standard Code Home	Energy Efficient Home
Initial Cost	$130,000	$132,500
Mo. Mortgage:	$1,075	$1,095
Mo. Utility Cost:	$110	$80
Total Mo. Cost:	**$1,185**	**$1,175**
Mo. Cost Advantage:	**$0**	**$17**

Table 2.
Cost Advantage of Energy Efficient Homes
Cash flow example for an "average" home.

Affordability

Homes are typically purchased with 15- or 30-year mortgages. This long-term financing allows most new homes to be built much more efficiently than required by code while costing less. The reason is that monthly utility bill savings can easily exceed small increases in the monthly mortgage for the additional energy features. Table 2 shows an example where $2,500 of energy efficiency improvements cost only $20 more on the monthly mortgage but reduce monthly utility bills by $30 for a positive cash flow every month of $10, beginning the day the buyer moves in. This cost advantage can increase substantially as builders apply systems solutions to construct homes that are 30 percent to 50 percent more efficient than code at little or no additional cost.

Comfort

Drafts, condensation on windows, ice damming, excessive noise from outdoors or equipment operation, and rooms that are cold in winter and hot in summer will diminish comfort in a home. Air-sealed construction, improved insulation, high-performance windows, right-sized, efficient heating/cooling distribution systems, and mechanical ventilation commonly found in energy-efficient homes all work together to effectively eliminate these problems.

Health

Good indoor air quality, critical for a healthy home, requires effective control of pollutants and moisture. To some extent this can be accomplished by thoughtfully selecting materials that contain low levels of volatile organic compounds and formaldehyde. In addition, hard surfaces that can be easily maintained can help prevent dust mite activity and accumulation of organic matter. Beyond these basic material choices, the path to improving indoor air quality is through energy-efficient building practices. Air-sealed construction and duct systems block a wide array of pollutants from attics, garages, crawl spaces, and basements from penetrating indoors. In addition, because energy-efficient homes avoid moisture problems, they also avoid molds and mildew that can cause serious health problems associated with a number of allergies and asthma. Lastly, mechanical ventilation replaces stale indoor air with fresh outdoor air.

Durability

The features that contribute to the energy efficiency of the home also yield maintenance and durability benefits. High-performance windows block out ultraviolet (UV) radiation that can accelerate wear on carpets, interiors, and furnishings. Correctly sized heating and cooling equipment will operate near its design conditions and will likely last longer. More importantly, proper systems solutions protect against moisture damage arising from poor duct sealing, ventilation, or air sealing. A house designed as a system will likely last longer and be easier to maintain.

PROJECTIONS FOR THE FUTURE

Systems Applications

Future builders will continue the current trend of integrating systems solutions into the construction process to reduce costs, conserve resources, increase energy efficiency, provide greater control of the indoor environment, and reduce the cost of home ownership. Mechanical systems such as heating, ventilation, and hot-water heaters will be combined to improve comfort and efficiency. Homes will become healthier and "greener" as awareness of and concerns about indoor and outdoor environments grow. Builders will adopt diagnostic procedures to provide quality assurance, reduce liability, and increase customer satisfaction.

There are a surprising number of computer-controlled systems in homes today, all acting independently to control heating, cooling, security, lighting, appliances, entertainment equipment and even coffee makers. Houses of the future will integrate all of these computerized functions into centralized home automation systems that will help optimize energy efficiency, comfort, and function at minimal cost to homeowners.

Construction Applications

New building systems include structural insulated panels, insulated concrete forms, and autoclaved concrete walls. These building systems are inherently better insulated, have better air sealing, and are stronger, less tool-intensive, and less wasteful while speeding and simplifying the construction process. Their superior performance results in a home with increased energy efficiency and durability and greater comfort for the residents. All of their advantages, combined with the rising cost of quality lumber, will position building systems at the forefront of the future housing market.

Factory-Made Housing

The two major types of factory-made housing are manufactured (HUD code) and modular. The major distinction between the two is that a manufactured home has a permanent chassis so it can be moved and complies with the national HUD code, while a modular home is permanently installed on a traditional foundation and adheres to the relevant local codes. As of 2000, manufactured and modular homes represent nearly half of all new homes sold, and this fraction is growing. The advantages of a factory-made housing include greater control of the production process, full protection from adverse weather, less uncertainty about materials and labor availability, reduced construction time, and a less wasteful production process. In addition, factory-made housing can be built to higher efficiency standards because the factory setting allows greater consistency and quality control of key measures such as air infiltration reduction, duct design and sealing, and insulation installation.

Indoor Air Quality

Home owners are increasingly aware of the importance of indoor air quality. This is especially true as connections are being made between rapidly increasing cases of allergies and asthma and indoor environments. High-quality filters on air distribution systems and mechanical ventilation solutions that provide both high-quality fresh air and dehumidification will emerge as standard features. Building materials will be selected with better consideration for air-quality impacts. For example, carpets are being developed that protect against dust mite and mold problems.

Environmental Factors

The average American home is responsible for more annual pollution than the average car. This often comes as a surprise because the pollution attributed to homes is produced miles away at a power plant or out-of-sight from roof exhaust flues. However, every time someone flips a switch, activates the air conditioning, or takes a shower, pollution is being produced. There will be growing appreciation for energy efficient homes that help prevent pollution.

In both the materials used and the construction process, sustainability and efficiency will become standard. Improvements in job site management and building design will reduce waste and cost of construction. Materials such as engineered wood, recycled carpeting, and cellulose insulation, which have lower environmental impact, will become cheaper and more widely available. Not only are buildings constructed with such material friendlier to the environment, they also provide higher-quality, lower-cost solutions.

Information Technology

The home sale process will drastically change in response to the information technology revolution. Internet sites are already being established as alternative methods for selecting neighborhoods and homes, completing purchases, and arranging for financing. Thus consumers will have much greater access to information about the comfort, quiet, durability, indoor air quality and resale benefits of energy-efficient housing. As home buyers learn to distinguish between the asking price and the actual cost of homes, builders will incorporate more energy-saving features to drive down total ownership costs.

Sam Rashkin
Japhet Koteen

See also: Air Conditioning; Air Quality, Indoor; Building Design, Commercial; Building Design, Energy Codes and; Domestic Energy Use; Insulation; Windows.

BIBLIOGRAPHY

Carmody, J.; Selkowitz, S.; and Heschong, L. (1996). *Residential Windows*. New York: W. W. Norton.

Gardstein, C., and Zuckman, W. (1999). "Don't Just Raise the Bar, Redefine It." *Spotlight on SIPA* 9(1).

Heede, R. (1995). *Homemade Money: How to Save Energy and Dollars in Your Home.* Snowmass, CO: Rocky Mountain Institute.

Jump, D. A.; Walker, I. S.; Modera, M. P. (1996). "Field Measurements of Efficiency and Duct Retrofit Effectiveness in Residential Forced Air Distribution Systems." In *Proceedings of the ACEEE 1996 Summer Study on Energy Efficiency in Buildings,* Vol. 1, pp. 147–155. Washington, DC: American Council for an Energy Efficient Economy.

Nevin, R. and Watson, G. (1998). "Evidence of Rational Market Valuations for Home Energy Efficiency." *The Appraisal Journal* (October): 401–409.

U.S. Environmental Protection Agency. (1993). *U.S. EPA National Vehicle and Fuel Emissions Laboratory Fact Sheet: Annual Emissions and Fuel Consumption for "Average" Passenger Car.* Washington, DC: Author.

U.S. Environmental Protection Agency. (1998). *U.S. EPA Energy Star Homes Fact Sheets: Duct Insulation.* Washington, DC: Author.

BUTANE

See: Liquefied Petroleum Gas

CAPACITORS AND ULTRACAPACITORS

Capacitors store electrical energy in the form of an electric field between two electrically conducting plates. The simplest capacitor is two electrically conducting plates separated spatially. By inserting a dielectric material (a poor conductor of electricity) between the two plates the capacity can be greatly increased (Figure 1). The dielectric material used determines the major characteristics of the capacitor: capacitance, maximum voltage or breakdown voltage, and response time or frequency. The first capacitor, the Leyden jar accidentally discovered in 1745, is a glass jar coated with copper on the inside and outside. The inside and outside copper coatings are electrically connected to a battery. The two spatially separated copper plates are the electrodes, and the glass is the dielectric of the Leyden jar capacitor. The capacity to store electrical energy at certain frequencies and to provide high-power discharges makes a capacitor an essential component in most electrical circuits used in electronics, communication, computers, manufacturing, and electric vehicles.

Capacitance is related to the area of the plates (*A*), the distance between the plates (d), and the dielectric constant (ε) of the material between the plates (Figure 2, equation I). The dielectric constant or permittivity of a material is the increased capacitance observed compared to the condition if a vacuum was present between the plates. Common dielectric materials are polystyrene ($\varepsilon = 2.5$), mylar ($\varepsilon = 3$), mica ($\varepsilon = 6$), aluminum oxide ($\varepsilon = 7$), tantalum oxide ($\varepsilon = 25$), and titania ($\varepsilon = 100$). In the Leyden jar the dielectric is silica.

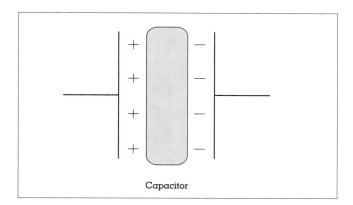

Figure 1.

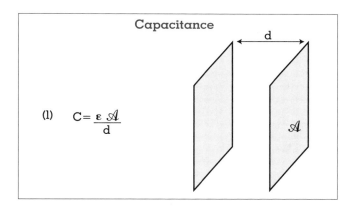

(1) $\quad C = \dfrac{\varepsilon \, \mathscr{A}}{d}$

Figure 2.

A capacitor, previously called a condenser, stores electrical energy based on the relationship between voltage (V) and stored charge (Q) in coulombs as shown in the equation C=QU. One farad of capacitance is a coulomb per volt of stored charge. The voltage limit of a capacitor is determined by the breakdown potential of the dielectric material.

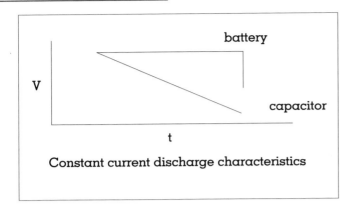

Figure 3.

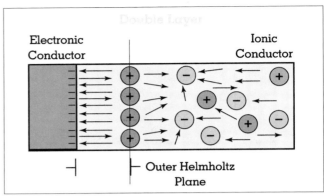

Figure 5.

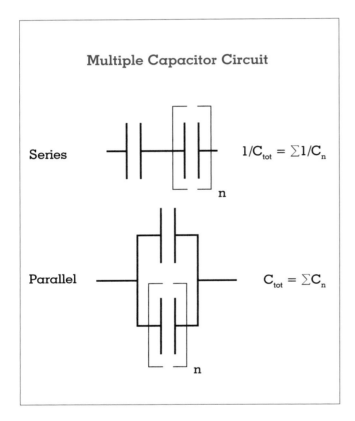

Figure 4.

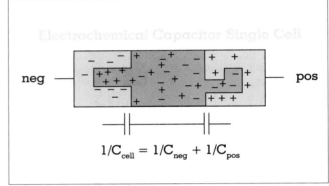

Figure 6.

Like a battery, a capacitor is an electrical energy storage device. There are, however, significant differences in how a battery and a capacitor store and release electrical energy. A battery stores electrical energy as chemical energy and can be viewed as a primary source. Capacitors need to be charged from a primary electrical source. During a constant current discharge, a battery will maintain a relatively constant voltage. In contrast, a capacitor's voltage is dependent on the "state of charge," decreasing linearly during a constant current discharge (Figure 3). The energy of a capacitor in joules is defined in the equation

$$E = \tfrac{1}{2}CV^2.$$

Capacitors are often combined in series or parallel, with the resulting circuit capacitance calculated as depicted in Figure 4. An important relationship is the time constant of a capacitor. The time constant is based on the product of the resistance and capacitance and is known as the RC time constant. A capacitor in a dc circuit will charge or discharge 63.2 percent in one RC time constant. The time dependence of a capacitor is shown in the equations.

$$V(t) = V_i e^{-t/RC}$$

and

$$I = CdV/dt$$

Electrochemical capacitors are also known as double layer capacitors, ultracapacitors, or supercapacitors. These devices are based on either double-layer charge storage or pseudocapacitance. Electrochemical double-layer capacitors, originally developed by Standard Oil Company during the 1960s, store charge at the interface between an electrically conducting electrode such as carbon and an ionically conducting electrolyte such as sulfuric acid. The double layer, first described by Hermann von Helmholtz in 1853, can be considered the equivalent of a parallel plate capacitor wherein the distance of charge separation is given by the ionic radius of the electrolyte, while the solvent continuum is the dielectric (Figure 5). The large charge storage offered by electrochemical capacitors is due to amplifying the double-layer capacitance ($\approx 15 \times 10^{-6}$ F/cm^2) by a large surface area electrode ($\approx 2 \times 10^7$ cm^2/g). Electrochemical capacitors typically have capacitance values of millifarads to tens of farads, in contrast to electrolytic capacitors, which typically have values in the range of picofarads to microfarads.

The single cell of an electrochemical capacitor consists of two electrodes separated by an electrolyte (Figure 6). The cell voltage is limited to the oxidation and reduction limit of the electrolyte, about 1.2V for aqueous and 3-4V for organic electrolytes. To obtain high-voltage electrochemical capacitor devices, single cells are connected in series to achieve the desired voltage. In contrast, electrolytic capacitors can have single-cell voltages of several hundred volts, depending on the dielectric.

Pseudocapacitance is used to describe electrical storage devices that have capacitor-like characteristics but that are based on redox (reduction and oxidation) reactions. Examples of pseudocapacitance are the overlapping redox reactions observed with metal oxides (e.g., RuO_2) and the p- and n-dopings of polymer electrodes that occur at different voltages (e.g. polythiophene). Devices based on these charge storage mechanisms are included in electrochemical capacitors because of their energy and power profiles.

A Ragone plot (Figure 7) compares the power and energy density of electrical energy storage devices. Electrolytic capacitors, based on an oxide dielectric, for example, are associated with high-power densities

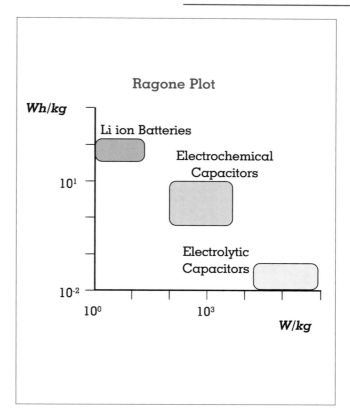

Figure 7.

and low energy densities. Batteries, on the other hand, have high-energy density but limited power. Electrochemical capacitors have good energy and power densities.

Capacitors are used in many applications. Every type of capacitor has an optimum performance, depending on the voltage, capacitance, weight and volume, and frequency criteria. Optimization of circuit design requires knowledge of the performance attributes and limitations of each type of capacitor. Typically electrolytic capacitors are high-voltage, low-capacitance devices used as filters in circuits or for fast-time constant (<10 mS) circuits. Ultracapacitors are lower-voltage high-capacitance devices used as standby power for random access memory devices, power sources for actuators, circuit elements in telephone equipment, and for long-time constant (>10 mS) circuits. Ultracapacitors are used in electric vehicles and cellular phones. The rapid charging characteristics and high energy make ultracapacitors useful in smart-card applications.

Alan B. McEwen

See also: Batteries; Electricity; Electric Motor Systems; Electric Powers, Generation of; Helmholtz, Hermann von.

BIBLIOGRAPHY

Horowitz, P., and Hill, W. (1981). *The Art of Electronics.* Cambridge, Eng.: Cambridge University Press.

Conway, B. (1999). *Electrochemical Supercapacitors: Scientific Fundamentals and Technological Applications.* New York: Kluwer Academic/Plenum.

CAPITAL INVESTMENT DECISIONS

Capital investment decisions in equipment, buildings, structures and materials have important economic implications over the life of the investment. Examples of capital investment decisions include selecting the insulation for a building, choosing a heating and air conditioning system, purchasing a fuel-efficient vehicle that satisfies an individual's needs, and selecting an energy-generation facility for an industrial plant.

Capital investment decisions are best made within the context of a life-cycle cost analysis. Life-cycle cost analysis focuses on the costs incurred over the life of the investment, assuming only candidate investments are considered that meet minimally acceptable performance standards in terms of the non-monetary impacts of the investment. Using life-cycle analysis, the capital investment decision takes into account not just the initial acquisition or purchase cost, but maintenance, energy use, the expected life of the investment, and the opportunity cost of capital. When revenue considerations are prominent, an alternative method of analysis such as net benefit or net present value may be preferred.

The economic problem is to identify the best capital investment from a set of possible alternatives. Selection is made on the basis of a systematic analysis of expected costs, and revenues if they differ, over time for each project alternative.

APPROACH

A systematic approach for economic evaluation of projects includes the following major steps:

1. Generate a set of project or purchase alternatives for consideration. Each alternative represents a distinct component or combination of components constituting investment. We shall denote project alternatives by the subscript x, where x = 1, 2, ... refers to alternatives 1, 2, and so on.

2. Establish a planning horizon for economic analysis. The planning horizon is the set of future periods used in the economic analysis. The planning horizon may be set by organizational policy (e.g., 5 years for new computers or 50 years for new buildings), by the expected economic life of the alternatives (i.e., the period over which an alternative remains the cost-effective choice), by the period over which reasonable forecasts of operating conditions may be made, or by the period of interest of the investor. The planning horizon is divided into discrete periods, usually years, but sometimes shorter units. We shall denote the planning horizon as a set of t = 0, 1, 2, 3, ..., n, where t indicates different points in time, with t = 0 being the present, t = 1 the end of the first period, and t = n the end of the planning horizon. When comparing mutually exclusive alternatives using either life-cycle costing or net-benefits methods, the same planning horizon must be used.

3. Estimate the cash-flow profile for each alternative. The cash-flow profile should include the costs and revenues if they differ, for the alternative being considered during each period in the planning horizon. For public projects, revenues may be replaced by estimates of benefits for the public as a whole. If revenues can be assumed to be constant for all alternatives, only costs in each period are estimated. Cash-flow profiles should be specific to each alternative. We shall denote revenues for an alternative x in period t as $B(t,x)$, and costs as $C(t,x)$. By convention, cash flows are usually assumed to occur at the end of the time period, and initial expenditures to occur at the beginning of the planning horizon, that is, in year 0.

4. Specify the discount rate, or minimum attractive rate of return (MARR) used to discount cash flows to a common time basis. Discounting recognizes that revenues and costs incurred at different times in the future are generally not valued equally to revenues and costs occurring in the

present. Money received in the present can be invested to obtain interest income over time. The MARR represents the trade-off between monetary amounts in different time periods. The MARR is usually expressed as an annual percentage rate of interest. The value of the MARR may be set for an entire organization based upon the opportunity cost of investing funds internally rather than externally in the financial markets, or it may be set for different classes of investment depending on their riskiness. For public projects, the value of the MARR, often called the social rate of discount, is specified in accordance with public policy. The future equivalent value of a dollar one period out is calculated as $(1 + MARR)$, and the equivalent value two periods in the future is $(1 + MARR) \times (1 + MARR) = (1 + MARR)^2$. In general, if you have Y dollars in the present (denoted Y(0)), then the future value in time t (denoted Y(t)) is:

$$Y(t) = Y(0) \times (1+MARR)^t$$

and the present value, Y(0) of a future dollar amount Y(t) is:

$$Y(0) = Y(t)/(1+MARR)^t$$

5. Establish the criterion for accepting or rejecting an alternative and for selecting the best among a group of mutually exclusive alternatives. When both revenues and costs are to be considered, the mutually exclusive alternative with the greatest net heights is selected. For example, the alternative heating system for a building might be selected on the basis of lowest life-cycle cost, and the alternative airport configuration might be selected on the basis of highest net benefits. When all alternatives are assumed to be feasible and have equal benefits, the alternative with the smallest discounted total cost is selected.

6. Perform sensitivity and uncertainty analysis. Calculation of life-cycle costs and net benefits assumes that cash-flow profiles and the value of MARR are reasonably accurate. In most cases, uncertain assumptions and estimates are made in developing cash flow profile forecasts. Sensitivity analysis can be performed by testing how the outcome changes as the assumptions and input values change.

CALCULATIONS

Calculations of life-cycle costs, net benefits, or other measures of economic performance are commonly performed on electronic calculators, commercial spreadsheet software, or by hand. The calculation approach for life-cycle costs is to first compute the net cost amount in each period for each alternative, C(t,x). The life-cycle cost (LCC) of each alternative is then calculated as the sum of the discounted values of C(t,x) over the entire planning horizon:

$$LCC(x) = \Sigma^n_{t-0}C(t,x) / (1 + MARR)^t$$

For a set of mutually exclusive alternatives, the project with the smallest LCC is the most cost-effective.

Other discounting conventions can be used for selecting capital investments. Time-adjusted cash flows can be expressed not only in terms of present value, but also as future, and equivalent annual values.

Capital investments can also be selected on the basis of other measures of performance such as return on investment, internal rate of return, and benefit-cost ratio (or savings-to-investment ratio). However, care must be taken in the application of these methods, as an incremental analysis is required to ensure consistent comparison of mutually exclusive alternatives. Also, rather than requiring a separate value to be calculated for each alternative, as in the case of the life-cycle cost method, these other methods incorporate the difference between two mutually exclusive alternatives within a single measure. For example, the net benefits measure directly pressures the degree to which one alternative is more economically desirable than another.

Special cases of capital investment decisions include lease or buy decisions, when-to-replace decisions, which design to choose, and comparison of alternatives with unequal service lives. These special cases are covered in Park (1997) and in Ruegg and Marshall (1990), as are the other methods for capital investment decisions.

Example 1.

Consider two alternative building designs (Ruegg and Marshall, 1990). One is a conventional building design; the other is an energy conserving design. Table 1 summarizes cost for both designs.

Assume the expected life of the building is 20 years and the discount rate or MARR is 8 percent. The net

	Conventional Design	Energy Conserving Design
Construction cost	$9,130,000	$9,880,000
Annual maintenance and operation (nonfuel)	$90,000	$60,000
Major repair (every 10 years)	$100,000	$50,000
Annual energy consumption	$55,000	$12,000

Table 1.
Conventional Design and Energy Conserving Design

present value of the costs over the life of building—that is, the life-cycle cost—for the conventional design is $9,834,635 and for the energy-conserving design, $9,834,068. Assuming both designs have the same functionality, the energy-conserving design is slightly preferable from a cost standpoint. Sensitivity analysis, however, may reveal little or no significant cost difference in the two choices.

Example 2.

Assume that a consumer is interested in purchasing a compact car. The consumer plans to keep the car for 10 years, at which point he or she assumes it will be worth nothing. A manufacturer offers a conventional model averaging 36 mpg, and a gasoline-electric hybrid car offering 70 mpg. The conventional car costs $13,000 and the hybrid car costs $19,000. The typical consumer drives 20,000 miles per year and gasoline costs $1.50 per gallon. The cars provide the same performance, comfort and reliability, and

their costs of repair are approximately the same. Assume the consumer has a discount rate of 5 percent. Capital investment analysis can be used to compare the costs of the two cars over their lives.

For the conventional car, $833—(20,000 miles per year / 36 miles per gallon) × ($1.50 per gallon)—will be spent each year on gasoline. Life-cycle costs of the car for the purchase of the car and gas are

$$\text{LCC (conventional)} = \$13,000 + \Sigma^{10}_{t-1}\$833 / (1.05)^t$$
$$= \$13,000 + \$6,435$$
$$= \$19,435$$

For the hybrid car, $429—(20,000 miles per year / 70 miles per gallon) × ($1.50 per gallon)—will be spent each year on gasoline. A similar calculation to that for the conventional car reveals that LCC (hybrid) = $22,309. The life-cycle cost analysis indicates that from an economic point of view, the conventional car is the better purchase.

We can also explore the sensitivity of this conclu-

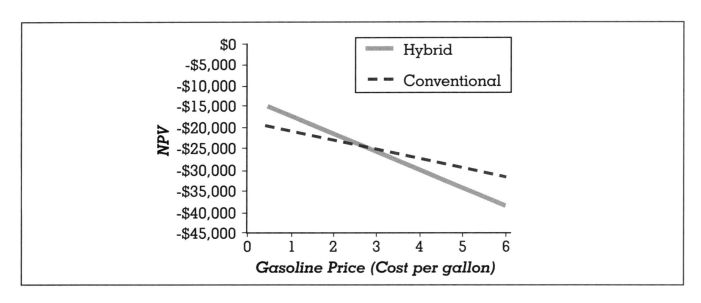

Figure 1.
Sensitivity of life-cycle cost analysis to gasoline prices.

sion to our assumptions about the initial purchase price and the cost of gasoline. Figure 1 shows the LCC of the hybrid and the conventional car over the ten-year period as a function of the cost of gasoline. When gas prices are approximately $3 per gallon, the two cars cost about the same. This value is referred to as the break-even point. If gas prices reach $3.75 per gallon, the approximate cost in Japan, the hybrid car is more economical. Sensitivity analysis can also be conducted for other input variables, such as initial purchase price, miles driven per year and actual fuel economy.

Not all decisions are made on the basis of economics alone. Consumers choose capital investments on the basis of efficiency, aesthetics and perceived benefits. An important environmental benefit of hybrid cars is the reduction in emissions. This is of particular relevance to an environmentally conscious consumer, and may shift the choice even though the benefit to the consumer is difficult to measure in dollars.

Sue McNeil

BIBIOGRAPHY

Au, T., and Au, T. P. (1992). *Engineering Economics for Capital Investment Analysis*, 2nd ed. Englewood Cliffs, NJ: Prentice-Hall Inc.

Hendrickson, C., and Au, T. (1989). *Project Management for Construction*. Englewood Cliffs, NJ: Prentice-Hall Inc.

Park, C. S. (1997). *Contemporary Engineering Economics*. Menlo Park, CA: Addison Wesley.

Ruegg, R., and Marshall, H. E. (1990). *Building Economics*. New York: Van Nostrand Reinhold.

CARBON DIOXIDE

See: Air Pollution; Climatic Effects

CARNOT, NICOLAS LÉONARD SADI (1796–1832)

Sadi Carnot, French physicist and engineer, was born in Paris during the French Revolution. He was the eldest son of Lazare Carnot, a military engineer known for writings on mathematics, mechanics, military strategy and politics. Lazare was a major government official under the First Republic that had begun in 1789, and often was described as the "Organizer of Victory" for his training and equipping of the raw recruits called to arms to defend France. Lazare had a great interest in the theory and building of machines of all kinds, an interest he passed on to his son.

Sadi Carnot's early education in mathematics, physics, languages and music was provided at home by his father. In 1812 he entered the prestigious Ecole Polytechnique, a college intended mainly for future military officers, and received a rigorous training in mathematics and science. In 1813 Carnot petitioned Napoleon to allow the Polytechnique students to help defend France against their European attackers. Napoleon granted this request and in March 1814 Carnot and his comrades from the Polytechnique fought in vain to keep the attacking armies out of Paris. Later that year Carnot graduated tenth in his class of sixty-five from the Polytechnique. He then studied military engineering at the Artillery College in Metz, and from 1816 to 1818 served as a second lieutenant in charge of planning fortifications.

Carnot soon realized that he did not have the temperament of a soldier and in 1818 left the army. After leaving the army Carnot took up residence in his father's former Paris apartment, and was presumably supported by his family whiile he attended classes at Sorbonne, the College de France, and the Conservatoire des Arts et Metiers. He also frequently visited factories and workshops, both to see steam engines actually in use, and to learn more about the economics of such industrial use of energy. There were rumors that he did at least on a few occasions receive some consultant's fees for his advise, but there was no clear documentary evidence of this. In 1827 he returned to active military service with the rank of captain, but this lasted only a little more than a year. He resigned in 1828 and died of cholera four years later in Paris.

Sadi Carnot has been called "the founder of the science of thermodynamics." In 1824, when he was twenty-eight, he first became interested in steam engines. All that time, Great Britain led the world in the design and improvement of such engines for industrial purposes. Always the French patriot, Carnot wanted his country to surpass the British, who had spawned the Industrial Revolution. He thought

Nicolas Léonard Sadi Carnot. (Library of Congress)

that a more scientific discussion of steam engines based on sound physical principles might reveal more about such engines than had the highly practical, engineering-type approach of the British pioneers.

Carnot developed the concept of an ideal "Carnot engine" (i.e., one in which all physical processes were completely reversible) to study engine efficiency and to apply his results to practical steam engines. Carnot demonstrated that this idealized engine would be more efficient than any practical steam engine ever built or that could be built. Carnot's engine consisted of a gas in a cylinder fitted with a frictionless piston. He imagined a "cycle," in which the gas absorbed heat from a hot source at temperature T_H, expanded and did work by pushing the piston outward, gave up heat to a colder condenser at temperature T_C, and then contracted, returning to exactly the same state as at the beginning of the cycle. This cycle later came to be called a "Carnot" cycle by researchers in the fields of heat and thermodynamics.

From his study of this cycle, Carnot concluded that the engine efficiency was independent of the working substance (e.g., steam or air). He also found that the maximum possible efficiency of an ideal engine (i.e., the ratio of the work done to the heat delivered from the hot source) was in every case

$$e_{max} = 1 - T_C/T_H .$$

There is a problem with Carnot's analysis, however, since at that time almost all physicists (including Carnot) thought heat consisted of a substance called "caloric," which could not be created or destroyed. As a result, the amount of heat taken from the hot source at temperature T_H would have to be the same as that delivered to the cold reservoir at temperature T_C. Because no heat was converted into work, the efficiency of such an engine would be zero.

It is noteworthy, however, that some of Carnot's notes that were published together with his classic treatise *Réflexions sur la puissance motrice de feu* (Reflections on the Motive Power of Fire), written in 1824 but only formally published in 1878, contain the following sentences: "When a hypothesis no longer suffices to explain phenomena, it should be abandoned. This was the situation with caloric, which physicists regarded as matter, as a subtle fluid." He goes on to say that Count Rumford's experiments (1798) had shown that heat is produced by motion, for example, by the rubbing of two objects together, which increases the molecular motion in the objects. When this concept of heat was introduced into Carnot's 1824 manuscript, there emerged a lucid statement of the conservation of energy principle—what became the first law of thermodynamics.

Carnot's research also made a major contribution to the second law of thermodynamics. Since the maximum efficiency of a Carnot engine is given by 1 - $T_{C/T}H$, if the engine is to be 100 percent efficient (i.e., e_{max} = 1), T_C must equal zero. This led William Thomson (Lord Kelvin) to propose in 1848 that T_C must be the absolute zero of the temperature scale later known as the "absolute scale" or "Kelvin scale."

Because Carnot's 1824 manuscript remained unpublished at the time of his death in 1832, it was left to Kelvin and Rudolf Clausius to show how the second law of thermodynamics was implicit in Carnot's work. For this reason Kelvin once referred to Carnot as "the profoundest thinker in thermodynamic philosophy in the first thirty years of the nineteenth century."

Joseph F. Mulligan

BIBLIOGRAPHY

Challey, J. F. (1971). "Carnot, Nicolas Léonard Sadi". In *Dictionary of Scientific Biography, Vol. 3*, ed. Charles Coulston Gillispie. New York: Scribner.

Gillispie, C. C. (1971). *Lazare Carnot Savant*. Princeton: Princeton University Press.

Harismendy, P. (1995). *Sadi Carnot: L'ingénieur de la Républic*. Paris: Perrin.

Kuhn, T. S. (1955). "Carnot's Version of Carnot's Cycle." *American Journal of Physics* 23:91–95.

Kuhn, T. S. (1977). "Energy Conservation as an Example of Simultaneous Discovery." In *Kuhn: The Essential Tension*, pp. 66–104. Chicago: University of Chicago Press.

Mendoza, E., ed. (1960). *Reflections on the Motive Power of Fire by Sadi Carnot and Other Papers on the Second Law of Thermodynamics by E. Clapeyron and R. Clausius*. New York: Dover Publications, Inc.

CARS

See: Automobile Performance

CARSON, RACHEL (1907–1964)

Nature writer and marine biologist Rachel Carson set off a storm of controversy in 1962 with the publication of her book *Silent Spring*. In her exposé on the dangerous consequences of the indiscriminate use of pesticides, Carson questioned the benefits of the synthetic chemical DDT, condemned scientific conceit, chastised the chemical industry for pursuing dollars at the expense of nature, and chided agriculturists and government officials for polluting croplands and roadsides. Calling such behavior irresponsible, Carson suggested that if people were not careful, they would eventually destroy the natural world so completely that one day they would awaken in springtime to find no birds chirping, no chicks hatching, and a strange shadow of death everywhere.

Agriculturists and chemical officials scorned Carson's jeremiad and argued that she misrepresented the evidence, while conservationists such as Supreme Court Justice William O. Douglas praised her for writing what he called "the most important chronicle of this century for the human race." From the Supreme Court to Ladies' Home Gardening Clubs, Americans discussed Carson's polemic, and in 1963, President John F. Kennedy entered the fray by commissioning a study on pesticides. In sum, *Silent Spring* contributed to a broader discussion of the environment and served as one significant catalyst for the emergence of the modern environmental movement.

Carson joined a growing number of voices that expressed concern about wilderness preservation, clean air and water, and nuclear energy. And, in concert with others, her writings and action—lobbying Congress, giving speeches, writing letters-to-the-editor, and working with conservationist organizations—spawned a host of environmental regulations that helped shape the contours of government policy, industrial action, scientific development, and individual lifestyles. In 1969, Congress created the Environmental Protection Agency. In 1970, the first Earth Day was held. Thus, Carson's work helped revolutionize the ways that many thought about the environment. President Jimmy Carter honored Carson posthumously in 1980 with the Medal of Freedom, saying, "Always concerned, always eloquent, she created a tide of environmental consciousness that has not ebbed."

Pesticide use was not Carson's chosen topic. She preferred to author works that simply fostered a deeper appreciation of nature. A shy and soft-spoken woman, Carson wrote with an Albert Schweitzer-like reverence for life. All was sacred to her. Her style was lyrical, vivid, and romantic, falling mostly within the nature-writing tradition. She gave her creatures anthropomorphic characteristics, set them in dramatic situations, hoping, she said, "to make animals in the woods or waters, where they live, as alive to others as they are to me."

Born in 1907 to Robert and Maria Carson, Rachel developed her admiration for nature in the woods and wetlands of her home in the Allegheny hills of western Pennsylvania. Her mother nurtured this interest with nature-study books. Simultaneously, Rachel cultivated her desire to write, publishing her first piece at eleven in the children's magazine *St. Nicholas*.

One of the greatest influences in Carson's life, next to her mother who was her lifelong companion, was her biology teacher at Pennsylvania College for Women. After a required course from Mary Scott Skinker, Carson switched her major from English to biology and, following her mentor's footsteps, pursued her studies with a master's in marine zoology from Johns Hopkins University. After a short stint of teaching at the University of Maryland, Carson landed a job in 1935 with the U.S. Bureau of Fisheries (later the Fish and Wildlife Department). In this position, Carson pushed for the protection of natural resources in twelve pamphlets she wrote for the Department's "Conservation in Action" series. To supplement her income, Carson wrote nature articles for popular magazines and completed the first two books of her sea trilogy.

The publication of her second book in 1951 brought Carson much acclaim and changed her life significantly. *The Sea Around Us* garnered positive reviews and was on the *New York Times* best-seller list for eighty-six weeks. Among other honors, it won the National Book Award and the John Burroughs Medal for excellence in nature writing. In 1951, Carson gained an esteemed Guggenheim Fellowship, and in 1952 she resigned from her government position. Book royalties from *The Sea Around Us* made it possible for her to live by writing. In 1955, she released her third book on the sea, again to widespread praise and more prizes.

While Carson's favorite topic was the sea and coastal shores, she saw the "contamination of man's total environment" as "the central problem of [her] age." Watching her own Allegheny Hills change in the wake of burgeoning industrial activity fed this concern. The atomic bomb and the dumping of nuclear wastes on the ocean's floor increased her anxiety. But it was the spread of synthetic pesticides that disturbed her the most. Although troubled by research on DDT in the 1940s, Carson did not get embroiled in the issue until 1957 when she followed a trial in Long Island, New York between local citizens and the U.S. Department of Agriculture. At issue was the spraying of pesticides over private land. Disturbed by plaintiff complaints of poisoned bird sanctuaries and gardens, Carson spent the next four years researching and writing about the impact of synthetic chemicals on the ecosystem.

Carson's work on pesticides and her writings on the sea are two parts of the same message. In all, she wanted to communicate the wonder she felt for the natural world, a world she saw as harmonious, balanced, and beautiful. And, in each, she challenged her fellows to reverence nature and act responsibly to preserve and protect natural habitats. Disputing the

Rachel Carson. (Corbis-Bettmann)

notion that humans live separate from and in dominion over the rest of nature, Carson placed people within a "vast web of life" connected to all parts of the ecosystem. Humans were but one small piece. They ought, she emphasized, to respect that place and not squander the world's resources. In the interests of energy use, the message was clear. Harvesting and employing the earth's natural resources was no longer a matter of human need only. Energy officials and policy-makers must consider the requirements of other constituents in the ecosystem, and must heed possible long-term environmental consequences.

Carson succeeded in conveying this view, in part, because she was not the only one fighting for environmental protection. For close to a century, since at least the writings of George Perkins Marsh, the conservationist movement had been building. In the 1940s, conservationists Aldo Leopold and Paul Sears espoused a similar ecological ethic and, for much of the 1950s, David Brower, head of the Sierra Club, fought for public attention in the hopes of saving natural wilderness areas. In 1956, Brower and others successfully resisted the damming of Echo Park in

Colorado, a moment now marked as a turning point in environmental protection, and one that had immediate consequences for developers interested in harnessing the energy of water. In the same year that Carson released *Silent Spring*, Murray Bookchin offered a parallel warning in his book *Our Synthetic Environment*. It is in this context that the success of Carson's work makes sense. Her writings complemented other actions. Still, it was Carson's lay appeal—her refusal to use technical and scientific jargon—that popularized this ecological vision and catapulted concerns about nature into the mainstream of American life, making the environment a crucial part of the agenda in any future technological decisions.

The modern environmentalist's interdependent understanding of the world that tied humans to nature has had multiple implications for technology and energy issues, both nationally and globally. Beginning in the 1950s and 1960s, technological decisions and government, business, and individual energy needs could not easily be divorced from social or ecological concerns. By the 1970s the common assumption was that human actions almost always changed the environment, often in irretrievable ways. Questions of costs and benefits and issues of short term and long-term effects dominated policymaking. This became most apparent in 1973 when the energy crisis intersected with the environmental movement. With the OPEC (Organization of Petroleum Exporting Countries) embargo, costs of oil soared, forcing the nation to focus on the depletion of natural resources and America's high energy-dependent lifestyles. Energy issues became inextricably tied to environmental and economic factors.

Rachel Carson played a crucial role in the ways Americans interpreted these events. Though a quiet person, more a recluse than an activist, her naturalist concerns won out. Propelled away from her refuge by her beloved sea to write a book on pesticides, Carson's activism grew and she spent her last days in the halls of Congress lobbying for environmental legislation. Her fight was short-lived, however. Carson died of breast cancer in 1964, two years after the publication of *Silent Spring*.

Linda Sargent Wood

BIBLIOGRAPHY

Bookchin, M. (1962). *Our Synthetic Environment*. New York: Knopf.

Carson, R. (1941). *Under the Sea-Wind: A Naturalist's Picture of Ocean Life*. New York: Simon and Schuster.

Carson, R. (1951). *The Sea Around Us*. New York: Oxford University Press.

Carson, R. (1955). *The Edge of the Sea*. Boston: Houghton Mifflin Company.

Carson, R. (1962). *Silent Spring*. Boston: Fawcett Crest.

Lear, L. (1997). *Rachel Carson: Witness for Nature*. New York: Henry Holt and Company.

Lear, L., ed. (1998). *Lost Woods: The Discovered Writing of Rachel Carson*. Boston: Beacon Press.

Lutts, R. H. (1985). "Chemical Fallout: Rachel Carson's *Silent Spring*, Radioactive Fallout and the Environmental Movement." *Environmental History Review* 9:210–225.

Melosi, M. V. (1985). *Coping with Abundance: Energy and Environment in Industrial America*. Philadelphia: Temple University Press.

Worster, D. (1977). *Nature's Economy: A History of Ecological Ideas*. New York: Cambridge University Press.

CATALYSTS

A catalyst is a substance whose presence increases the rate of a chemical reaction. The exercise of using catalysts is called catalysis. Today the vast majority of all commercially important chemical reactions involve catalysts, especially in the fields of energy production, petrochemicals manufacture, pharmaceuticals synthesis, and environmental protection.

HISTORY

Catalysis was practiced long before it was recognized as a scientific discipline. The earliest example of catalytic reactions was in the generation of alcoholic beverages through biocatalysis dating from the Neolithic Age. About 2,500 years ago a base-catalyzed (potash lye) process was used to manufacture soap in the Mediterranean area. Although the details are not known, in the 1500s alchemists claimed to have prepared sulfuric acid (sulfuric ether and oil of vitriol they called it) by a mysterious process that would probably be classified as a heterogeneous catalytic reaction today.

During the first third of the nineteenth century, several systematic observations led researchers to conclude that the mere presence of metals induced chemical transformations in fluids that would otherwise not have occurred. Early on, Thenard had observed that ammonia could be decomposed when passed through a red-hot porcelain tube, but only if the tube contained iron, copper, silver, gold, or platinum. Humphrey Davy observed that a warm Pt wire glowed red when placed into a gas-air mixture, and the gas was combusted. His cousin Edmond Davy was able to combust alcohol when exposed to finely-divided Pt particles even at room temperature. Döbereiner combined these discoveries with a hydrogen generator to produce a commercial lighter. Michael Faraday commented on Döbereiner's work in 1823 and, during three months of experiments in 1835, demonstrated catalyst poisoning. Faraday considered catalysts to be just one manifestation of ordinary chemical reactions. Eilhardt Mitscherlich summarized these and other strange results in 1834 and attributed the phenomena to being in "contact" with the substances. Five years before the word was catalysis was coined, in 1831 Peregrine Phillips obtained the first patent in field for an improved method of making sulfuric acid.

In 1836 Jons Jakob Berzelius considered eight seemingly unrelated experimental results and concluded that there was a common thread among them. The commonality he defined as catalysis. In doing this, Berzelius proposed that a "catalytic force" was responsible for catalytic action. The concept of catalysis is today considered by most researchers to be due to Berzelius, probably because of the popularity of his annual *Handbook of Chemistry* where he published his definition of catalytic action. For the next one hundred years many referred to the phenomenon as "contact catalysis" or "contact action," as proposed by Mitscherlich.

Justus von Liebig was another leader in the training of chemists, and many of his students were placed in influential positions throughout the scientific world during the mid-1800s. Liebig was a forceful personality who defended his "turf" with vigor. His concept of catalysis was strongly influenced by purification, a subject poorly understood at that time. Making an analogy with spoilage, Liebig proposed that catalytic action is based on an induced vibration. Just as one rotten apple will eventually cause all apples in a barrel to rot, so Liebig considered that a substance that vibrates at just the right frequency will induce vibrations in certain other molecules through contact and thereby enhance the rate of their reaction. Liebig used this concept as his basis of catalysis and to explain many other phenomena; he even considered catalysis as being analogous

German chemist Wilhelm Ostwald. He was awarded a Nobel Prize for his work on chemical equilibrium. (Library of Congress)

to a perpetual motion machine. Liebig's view was seriously considered for nearly half a century.

Wilhelm Ostwald was also defining physical chemistry during the 1880s. As an editor of a new journal devoted to physical chemistry, he wrote brief critical comments about many papers. Reviewing a paper that used Liebig's vibrational theory to explain results, Ostwald provided a new definition of a catalyst that was widely accepted and led to his being awarded the Nobel Prize in 1909.

Ostwald first came to catalysis through his work on the acceleration of homogeneous reactions by acids. This work was popular at the time although ultimately it would be shown to be incorrect because he believed that the acid, acting as a catalyst, did not enter into the chemical change which it influenced but rather acted by its mere presence (contact catalysis).

Discarding Liebig's theory as worthless because it could not be subjected to experimental verification, Ostwald contended that a catalyst merely sped up a reaction that was already occurring at a very slow rate. He also indicated that a catalyst cannot change the equilibrium composition. By analogy, he considered a catalyst to be like oil to a machine, or a whip to a horse; both caused the rate to increase.

Catalysis was soon divided into two classes: positive ones that accelerate reactions and negative ones that suppress reactions. It is now recognized that what was viewed for many years as negative catalysis was actually a case of catalyst poisoning; that is, some material is so strongly adsorbed that it effectively reduces the number of catalytic sites available to the reactant, thereby decreasing the reaction rate. This led to an understanding that while the catalytic steps themselves regenerate the original catalyst, most catalytic reactions are accompanied by side reactions that irreversibly decrease the catalytic activity. Some of these include sintering of highly dispersed metal particles; chemical poisoning by reactants, products, or impurities in the feed stream; physical blockage of active sites; or mechanical damage to catalytic particles. In spite of these effects, Haensel once calculated that each Pt atom in a petroleum reforming catalyst could convert a staggering 10 million hydrocarbon molecules into higher octane fuels during its lifetime before the catalyst had to be regenerated.

A recent definition of catalysis that is based on thermodynamics was advanced by the Subcommittee on Chemical Kinetics, Physical Chemistry Division, IUPA:

"A catalyst is a substance that increases the rate of a reaction without modifying the overall standard Gibbs energy change in the reaction; the process is called *catalysis*, and a reaction in which a catalyst is involved is known as a *catalyzed reaction*."

While this definition does not address the question of "how" catalysts effect rate increases, it does ensure that a catalyst cannot cause the equilibrium composition to deviate from that of the uncatalyzed reaction.

In 1947 Sir Hugh S. Taylor summarized the state of catalysis in a "Science in Progress" article as follows:

"Catalysis has been employed in science to designate a substance which by its mere presence facilitates or enhances the rate of chemical reactions. As such it was a cloak for ignorance. When the states of an over-all catalytic process can be described in terms of a well-defined succession of chemical and physical processes the details of which are well understood or are quite plausible, then the necessity for employing such a word as catalysis to mask our ignorance no longer exists. . ."

HOMOGENEOUS CATALYTIC REACTION MECHANISMS

Compared with uncatalyzed reactions, catalysts introduce alternative pathways that, in nearly all cases, involve two or more consecutive reaction steps. Each of these steps has a lower activation energy than does the uncatalyzed reaction. We can use as an example the gas phase reaction of ozone and oxygen atoms. In the homogeneous uncatalyzed case, the reaction is represented to occur in a single irreversible step that has a high activation energy:

$$O + O_3 \rightarrow 2\,O_2 \qquad \text{(Reaction 1)}$$

When chlorine acts as a catalyst, the reaction can be considered as two steps with the Cl being depleted in Reaction 2 and regenerated in Reaction 3:

$$O_3 + Cl \rightarrow ClO + O_2 \qquad \text{(Reaction 2)}$$

$$O + ClO \rightarrow Cl + O_2 \qquad \text{(Reaction 3)}$$

The activation energies of Reactions 2 and 3 are each much lower than the activation energy of the uncatalyzed case [1]. Thus, the kinetic definition could be stated along the following lines: A catalyst effects the rate increase by altering the homogeneous reaction pathway to a polystep reaction pathway, wherein each catalyzed step has a lower activation energy than the single homogeneous reaction, thereby increasing the rate of reactant conversion above that of the uncatalyzed reaction.

HETEROGENEOUS CATALYTIC REACTION MECHANISMS

Heterogeneous catalytic reactions always involve more than one phase with an interface separating them. The chemical reactions occur at that interface, as shown in Figure 1. A fluid molecule (e.g., gaseous) to be converted must react with a surface (usually solid) to form a surface adsorbed species. That species then reacts either with another adsorbed molecule (or a molecule from the fluid phase, or it may act unimolecularly as in Figure 1) to be transformed into an adsorbed product molecule, which then desorbs into the fluid phase. Each step (dashed lines) must have an activation energy that is lower than the homogeneous barrier height (solid curve). The depth of the potential energy curve indicates the strength with which

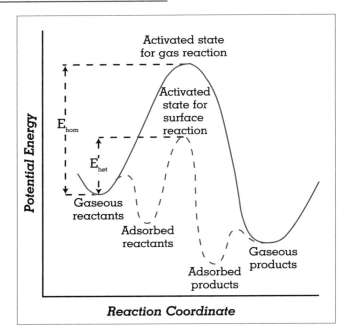

Figure 1.
Potential energy curves for a reaction proceeding
homogenously (full curve) or on a surface (dotted
line).

each of the species is adsorbed. If the energy decrease
is very large, the molecules are strongly adsorbed.
When the strongly adsorbed molecule is the reaction
product, it may be difficult to remove and can cover
the active sites to the point where the reaction rate is
actually inhibited by its presence.

Heterogeneous catalytic systems offer the advantage that separation of the products from the catalyst
is usually not a problem. The reacting fluid passes
through a catalyst-filled reactor in the steady state, and
the reaction products can be separated by standard
methods. A recent innovation called catalytic distillation combines both the catalytic reaction and the separation process in the same vessel. This combination
decreases the number of unit operations involved in a
chemical process and has been used to make gasoline
additives such as MTBE (methyl tertiary butyl ether).

KINETICS

All catalytic reactions involve chemical combination of
reacting species with the catalyst to form some type of
intermediate complex, the nature of which is the subject of abundant research in catalysis. The overall reaction rate is often determined by the rate at which these
complexes are formed and decomposed. The most
widely-used nonlinear kinetic equation that describes

homogeneous reactions involving enzyme catalysts was
developed by Leonor Michaelis and Maude Menten:

$$\text{Rate} = k\,\theta_A = k\,C_{Eo}\,C_A/(C_M + C_A)$$

where A is the reacting species, E the enzyme catalyst,
C_i is the fluid phase concentration of each species i, k
is the temperature dependent reaction rate constant,
and C_M is the Michaelis constant that has the same
dimensions as do the concentration terms. θ_A is the
fraction of enzyme molecules tied up in the intermediate complex at any time. Note that when C_A is much
smaller than C_M (weak binding) the rate depends linearly on both the enzyme concentration C_E and the
reactant concentration C_A. However, if C_A is much
larger than C_M (strong binding), then θ_A approaches
unity and the reaction rate depends only on the concentration of the enzyme and is independent of the
reactant concentration.

A similar nonlinear equation for heterogeneous
catalytic systems was developed empirically by Olaf
Hougen and Kenneth Watson and derived on a more
scientific basis by Irving Langmuir and Cyril
Hinshelwood. When applied to fluid reactants and
solid catalysts, the nonlinear equation in its simplest
form becomes

$$\text{Rate} = k\,\theta_A = k\,K_A\,C_A/(1 + K_A C_A)$$

where again k is the reaction rate constant, θ_A is the
fraction of active sites covered with adsorbed A, and K_A
is the adsorption equilibrium constant (a large value
means A is strongly adsorbed).

If the three-parameter Michaelis-Menten equation
is divided by C_M, it becomes the same as the three-
parameter Langmuir-Hinshelwood equation where
$1/C_M = K_A$. Both these rate equations can become
quite complex when more than one species is competing with the reactant(s) for the enzyme or active
sites on the solid catalyst.

TRANSPORT EFFECTS

It is not unusual for the full chemical potential of a
reaction to be diminished by slower transport processes (i.e., to be transport limited). In fast liquid phase
enzyme reactions, mechanical stirring rates can have a
strong influence on the observed kinetics that may be
limited by the rate of contacting of the reactants and
enzymes. Most heterogeneous catalytic reactions take

place on catalysts with surface areas of 100 to 1,000 m²/g. These high surface areas are usually generated by preparing catalysts in the form porous pellets (a few mm in diameter) containing a network of interconnecting pores that may be in the range of a few nanometers in diameter. Diffusion into these small pores by reacting molecules whose size is the same order of magnitude can be extremely slow. Assuming all pores are uniform cylinders, a typical silica-alumina cracking catalyst with surface area of 300 m²/g and pore volume of 0.25 cm³/g would contain pores the order of 33 Å diameter and more than 17,000,000 miles/g total length if they were all connected end to end.

For reactions taking place in a flow reactor packed with catalyst particles, each reacting molecule must negotiate a series of seven consecutive steps to accomplish its conversion. It must diffuse across the external boundary layer surrounding the pellet; diffuse inside the pores to an active site; adsorb on, react, and desorb from the active sites; then the liberated product molecules must diffuse back out of the pellet and across the boundary layer before being counted in the product stream. Any one of these sequential steps could be the bottleneck that limits the overall performance of the catalyst. Moreover, heat generated (or absorbed) during the reactions must be accounted for in order to avoid damage to the catalyst and/or hazards to personnel and the environment. This is why reaction engineering plays such an important role in optimizing catalytic processes.

MULTIFUNCTIONAL CATALYSTS

Catalytic processes frequently require more than a single chemical function, and these bifunctional or polyfunctional materials must be prepared in a way to assure effective communication among the various constituents. For example, naphtha reforming requires both an acidic function for isomerization and alkylation and a hydrogenation function for aromatization and saturation. The acidic function is often a promoted porous metal oxide (e.g., alumina) with a noble metal (e.g., platinum) deposited on its surface to provide the hydrogenation sites. To avoid separation problems, it is not unusual to attach homogeneous catalysts and even enzymes to solid surfaces for use in flow reactors. Although this technique works well in some environmental catalytic systems, such attachment sometimes modifies the catalytic specifici-

ty of the homogeneous catalyst due to the geometric constraints imposed on the molecules by the solid. With so many factors contributing to the interdisciplinary field of catalysis, it is not surprising that almost all branches of physical science, math, and engineering must be included in the successful development of a catalytic process.

IMPORTANT COMMERCIAL PROCESSES

Industrial catalytic applications comprise four major categories: chemicals manufacturing (25% of money spent on catalytic processes), environmental protection (23%), petroleum processing (26%), and polymers production (26%). In 2003 the total sales of catalysts worldwide is predicted to be $8.9 billion (not including the value of the precious metals and substrates used and includes only manufacturing fees). It has been estimated that about 20 percent of all the world's manufactured products have been touched somewhere along the line by one or more catalytic processes. The field of catalysis has obviously blossomed during the twentieth century, and without any doubt it will be a major factor in the world economy during the foreseeable future.

Burtron H. Davis
Joe W. Hightower

See also: Faraday, Michael; Thermodynamics.

BIBLIOGRAPHY

Boudart, M., and Diega-Mariadassou, G. (1984). *Kinetics of Heterogenous Catalytic Reactions*. Princeton, NJ: Princeton University Press.

Hill, C. G. (1977). *An Introduction to Chemical Engineering Kinetics and Reactor Design*. New York: John Wiley & Sons.

Prettre, M. (1963). *Catalysis and Catalysts*, tr. D. Antin. New York: Dover Publications.

Roberts, W. W. (2000). "Birth of the Catalytic Concept." *Catalysis Letters* 67:1.

Satterfield, C. G. (1970). *Mass Transfer in Heterogeneous Catalysis*. Cambridge, MA: MIT Press.

CATAPULTS

See: Elastic Energy

CHARCOAL

Charcoal is perhaps the oldest known fuel, having been found in archeological sites dating as far back as the Pleistocene era. Charcoal is a relatively smokeless and odorless fuel, and thus ideal for cooking and heating.

HISTORY

As humans entered the Bronze Age, charcoal was the only material that could simultaneously heat and reduce metallic ores. Later, the addition of an air blower made it possible to achieve temperatures high enough to soften or melt iron. During the Industrial Revolution, charcoal was largely displaced in most ironworks by coke derived from coal. However in Brazil, which lacks adequate coking coal resources, most of the charcoal produced is still used to reduce iron ore.

Charcoal was produced in pits, and later in kilns, by burning wood with air insufficient for complete combustion. The heat generated drives off the volatile materials in the wood, leaving a char that contains 60 to 90 percent carbon. In the nineteenth and early twentieth centuries, these volatiles were a major source of raw materials, chiefly acetic acid, acetone, methanol and creosote, for the burgeoning organic chemical industry. However charcoal's utility as a source of starting compounds was short-lived because petroleum-derived feedstocks proved to be cleaner and cheaper sources of these chemicals. By World War II, U.S. charcoal production declined by two-thirds, but since then, the popularity of backyard cooking raised charcoal production to an all-time high of about 1.5 million tons in 1996. As of 1999,

Charcoal typically has the following properties:	
density, as formed	0.3 to 0.5 gm/cubic cm
density, compacted	1.4 gm/cubic cm
fixed carbon	70 - 90 percent
volatile matter	5 - 25 percent by weight
ash	5 percent
heating value	12,000 BTU per pound (30 KJ per gm)

Table 1.
Properties of Charcoal

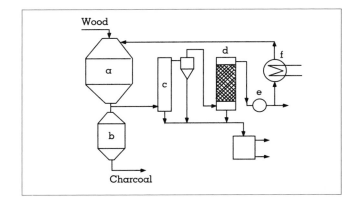

Figure 1.
Typical batch process from a DeGussa patent.

worldwide production is roughly 30 million tons, with about half of this in Asia and India, and one-third in South America. Precise data are not available from developing countries, where production is fragmented and manufacturing information is not systematically monitored.

USE

Charcoal is used in electrically heated furnaces to smelt specialty metals such as ferrosilicon. It is a preferred household fuel in developing countries with adequate forest resources. In the United States 95 percent of charcoal use is for barbecuing, while in Japan and Europe charcoal use is split evenly between cooking and industrial needs.

Much has been said about the backyard barbecue as a major air pollutant. However, most cooking smoke comes from food, not from the fuel used. Charcoal contains almost no sulfur, the major pollutant from burning coal. Of the total U.S. 1996 energy consumption of 93.8 quadrillion Btus, only 3.2 percent came from burning biomass. Most of this was derived from direct combustion of fireplace wood and industrial wastes like sawdust, bark and bagasse. In developing countries charcoal may constitute up to 40 percent of energy use. This energy source may not be sustainable because of conversion of forests to farmland. In Brazil, most of the charcoal destined for metallurgical use is made from fast-growing wood species raised on plantations.

Other markets for charcoal are as a filtration medium, a horticultural soil improver, and an adsorbent. Its large surface area of hundreds of square meters per

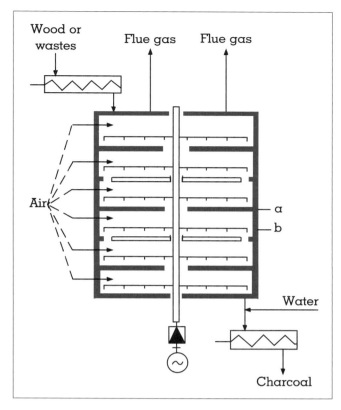

Figure 2.
Typical multiple hearth furnaces of the Hereshoff patent.

gram enables it to adsorb large organic molecules and thus improve color, taste, and/or odor in liquids and gases. Its adsorptivity can be improved by steaming in the presence of certain chemicals. The resulting "activated carbon" is used for a wide variety of applications ranging from water purification to poison-gas masks.

PROCESS

Modern charcoal retorts are charged with wood, biowaste (bark, sawdust, etc.), peat, and sometimes low-rank coals. Yield and properties (hardness, density, surface area, etc.) can vary widely so the desired end use must be considered. Charcoal from coniferous trees is soft and porous, while that from hardwoods is dense and strong. For barbecuing, charcoal is usually compressed into briquettes, with binders and additives chosen to improve handling and ease of ignition.

The manufacturing process usually involves slow heating to about 275°C in a kiln or retort. The reaction is exothermic, and about 10 percent of the heat of combustion of the original woody feed is lost. The gases and volatile liquids generated are usually burned to supply process heat. Then the wood is further carbonized by heating to about 500°C without air. Modern plants generally employ closed retorts where dried wood is heated by external means. This allows for better process control and enhanced pollution abatement. Both batch and continuous methods are used.

There are many process variations. A typical batch process from a DeGussa patent is shown in Figure 1. The reactor is 100 cubic meters and wood pieces are fed by a conveyer belt. Hot, 550°C, wood gas is fed concurrently. A carbonization zone (where the char, mostly carbon, is being formed) travels downward as the wood is pyrolyzed (heated intensely in the absense of air), and the off gases are burned to generate all of the process heat plus some of the energy needed to dry the raw wood. Energy requirements for producing 1 kg. of charcoal, including the drying operation, are 2.5 MJ of heat and 0.25 MJ of electrical energy. The process also requires 0.05 cubic meters of cooling water. A retort of this size and design will produce about 300 tons of charcoal a month.

Continuous production of charcoal is typically performed in multiple hearth furnaces, as illustrated in the Herreshoff patent shown in Figure 2. Raw material is carried by a screw conveyor to the uppermost of a series of hearths. Air is supplied countercurrently and burns some of the wood to supply process heat. As the layers of wood carbonize, they are transported to the lower (hotter) hearths by rakes. The hot charcoal product is discharged onto a conveyor belt and cooled with a water spray.

Fresh charcoal is a strong absorbent for gases, and this is an exothermic process. The heat generated can be enough to cause spontaneous ignition in some cases. Hence it is customary to age charcoal by exposure to air and thus cover the absorption sites with a layer of nitrogen gas. Larger molecules will desorb and replace smaller molecules, so the charcoal will still be effective as a decolorant or deodorizer.

Herman Bieber

See also: Explosives and Propellants.

BIBLIOGRAPHY

Bansal, R. C. (1998). *Active Carbon*. New York: M. Dekker.
DeGussa, German Patent 744,135 (1937).

Johansson, T. B.; Kelly, H.; Reddy, A. K.; and Williams, R. H., eds. (1993). *Renewable Energy: Sources for Fuels and Electricity.* Washington, DC: Island Press.

Nichols, Belgian Patent 309,763 (1968).

Proceedings of the ESCAPE/FAO/UNEP Expert Group Meeting on fuelwood and Charcoal. (1982). Energy Resources Development Series #24. New York: United Nations.

U.S. Bureau of the Census. (1998). *Statistical Abstract of the United States.* Washington, DC: U.S. Government Printing Office.

CHEMICAL ENERGY, HISTORICAL EVOLUTION OF THE USE OF

BEGINNINGS

Electric power, which is produced as a result of chemical reaction, and chemical change that is initiated by the flow of electricity embody the science and technology of electrochemistry. Its history is fairly recent, though electrochemistry might have been used in early historic times for medicinal purposes and electroplating with gold, a suggestion based on vessels containing copper tubes and central iron rods unearthed at various sites in Mesopotamia.

Public awareness of electrochemical processing dates from a meeting of the Royal Society in 1807. English society paid large admission fees to hear Humphry Davy lecture on electrically inducing chemical reactions and to witness him produce sodium and potassium from potash. In a spectacular flourish, Davy dropped the amalgamated product into water, at which moment the alkali metals reacted violently with the water to generate hydrogen, which burst into flame.

The true beginnings of electrochemistry were the experiments of Luigi Galvani and Alessandro Volta in the late eighteenth century. In 1780, Galvani, an Italian anatomist, generated what he believed to be animal electrical energy. Volta, a contemporary Italian physicist, doubted its "animal" nature and, in setting out to disprove it, invented what we now call the "galvanic" or "voltaic" cell (Figure 1), so named to honor these scientific pioneers.

ANIMAL ELECTRICAL ENERGY

Galvani in 1780 noticed that the leg muscles of a freshly killed frog, which was hung from an iron hook, twitched when a nerve was touched with a copper scalpel at the time an electric spark was being produced on a nearby frictional arc. Beginning in 1786, Galvani experimented with this system, inducing muscular contractions by physically contacting copper with iron, with one metal connected to the nerve of a frog, the other to its muscle. Galvani described his frog experiments in 1791 in a publication entitled *De veribus electricitatis in motu musculari commentarius* (Latin: *Commentary on the origins (springs) of electricity in muscular motion*), attributing the phenomenon to "animal electricity." Galvani was influenced by knowledge of electric fish and the effects of electrostatic charges in causing muscles to contract. He likely was also aware of the use of glass rods charged by rubbing with wool to electrify people, first done by Stephan Gray in 1730, with a youth suspended in air to prevent electricity from leaking to the ground.

In 1796, Volta demonstrated that Galvani's electricity was generated by contact between dissimilar metals. Using zinc and silver discs, Volta showed that brine-soaked cardboard or leather could be used in place of the frog. Though Galvani's experiments did not demonstrate "animal electricity," his intuition had validity. Emil du Bois-Reymond, using instruments more sensitive than those available to Galvani, showed in the 1840s that electrical currents do flow in nerves and muscles. Much later, Julius Bernstein in 1902 proposed that nerve cells can become electrically polarized, and David Nachmansohn in 1938 showed that current travels along the nerve to the nerve ending, releasing acetylcholine. The nerve signal, upon reaching a muscle causes a chain of reactions culminating in the breakdown of adenosine triphosphate (ATP) and the release of energy. Today's electrocardiogram (EKG), used in medical examinations, records such electrical impulses generated by the heart.

VOLTA'S PILES

The galvanic cell invented by Volta in 1800 was composed of two dissimilar metals in contact with mois-

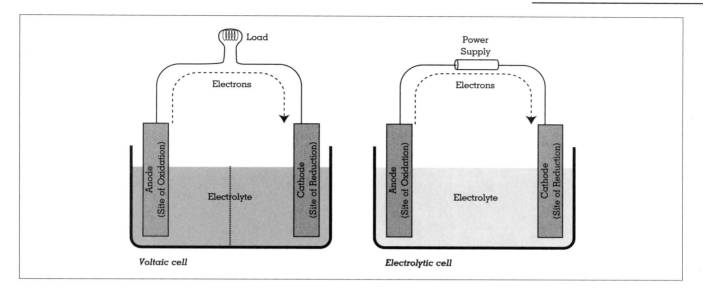

Figure 1.
Comparison of the voltaic/galvanic cell with the electrolytic cell.

tened paper or a salt solution. The first battery, two or more cells in series, consisted of a pile of alternating silver and zinc discs with separators in between soaked in electrolyte, the term now used for a liquid capable of conducting an electric current. Volta soon developed an improved version called "couronne de tasses" (French: crown of cups), which permitted drawing electric current at controlled rates. "Volta's Piles" were the only practical source of continuous electricity until they were displaced near the end of the nineteenth century by dynamos in stationary power applications.

A pile was the first source of energy capable of converting water into its separate elements in significant quantities. Using a seventeen-cell pile, William Nicholson and Anthony Carlisle in 1800 found hydrogen and oxygen evolving in the ratio of two to one. Also in 1800 in other firsts, William Cruickshank generated chlorine from sodium chloride and precipitated metals, notably copper, from solutions of their salts. Davy used the pile to produce an electric arc between two sticks of charcoal, or "poles," spaced three inches apart. Electric furnace poles now can be fifteen feet long and weigh ten tons. Davy deemed the poles positive or negative, depending upon which side of the battery they were connected, and his assistant, Michael Faraday, called the poles electrodes. William Hissinger and Jons Jakob Berzelius in 1803 showed that oxygen and chlorine

evolve at the positive poles, and alkalis, metals, and hydrogen at the negative poles. From this came the later development of the chlor-alkali industry.

ELECTROCHEMICAL THEORY

Although theoretical understanding propels further discovery and application, practical application more often precedes understanding. It took most of the nineteenth century to reach an understanding of the underlying electrochemical processes in which (1) an electric current driven by an imposed electromotive force or "voltage" results in storing the potential chemical energy in reaction products (electrolysis), and (2) a generated voltage induces electrical current (as in batteries).

Electrolysis in electrolytic cells involves the migration of ions through the electrolyte and the subsequent reactions of the ions at the two electrodes. Voltaic (or galvanic) batteries do the reverse, using electrochemical processes to convert the chemical energy stored in the reactants to electrolytic energy. In both cases, electrons flow through the external circuit so that there is no net production or loss of charge in the overall process.

The first electrolytic theory was expounded in 1805 by Christian Grothuss who postulated that an electric field rotates the molecules so that their positive and negative components ("ions") face their

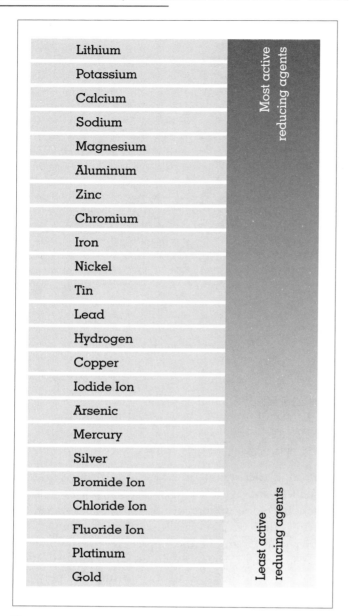

Figure 2.
Volta's Electrochemical Series.

ion, current in an electrolytic cell is forced to flow from the more "electropositive" metal to the less electropositive one. Berzelius in 1811 extended the Electrochemical Series to include non-metals. Volta further differentiated between groups, such as metals, carbon and some sulfides, which conduct electricity without undergoing chemical change, and salt solutions, which are decomposed by the electric current.

The convention that electric current in the external circuit (the wire connecting the electrodes) flows opposite to the direction of electron flow, that is, from negative to positive, proved to be an unfortunate source of confusion. The selection originated in Benjamin Franklin's suggestion in 1747 that rubbing a glass rod with a dry hand causes an electric fluid to pass between hand and rod and results in an excess or "plus" of electric fluid in the rod, and a corresponding deficit or "negative" charge on the hand.

Between 1831 and 1834, using a voltaic pile, Faraday discovered the inductive effects of the electromagnet — the basic principle underlying the operation of electrical machinery — and experimentally quantified the direct proportionality between electrical current and the rates of electrochemical reactions (Faraday's Law). Working with voltaic piles, a great battery of eight Leyden jars charged with static electricity, and using data obtained on other devices by other investigators, Faraday established that electricity was universal and not dependent on its source. During the course of this intense activity, he also invented both the direct and alternating current dynamos.

Seemingly unrelated work by Wilhelm Ostwald in 1884 on a physical phenomenon now called osmotic pressure, provided an important link in understanding electrochemical behavior. Osmotic pressure is generated when a porous but impermeable membrane is interposed between solutions containing differing concentrations of a dissolved substance. The membrane prevents the dissolved substance from diffusing from the more concentrated into the less concentrated zone, thus giving rise to the difference in pressure. Shortly after, as published between 1885 and 1888, Jacobus Henricus van't Hoff noted that osmotic pressure was proportional to concentration, but most important, the pressures were often multiples of predictions. Van't Hoff attributed this observation to the solute dissociating into several ions. Ostwald recognized that this dissociation into ions was the electrolytic process as described by Svante

opposing poles. These "ions" (not ions in the later sense) separate and move toward their opposing pole until they meet and combine with ions of the neighboring molecules. Soon after, Volta deduced that electric current in a battery would naturally flow from electrodes containing more "electropositive" metals to the less electropositive ones when the electrodes are connected to each other. Using this phenomenon, Volta established the hierarchical Electrochemical Series of Elements (Figure 2). In an analogous fash-

Arrhenius in 1884. The three (Ostwald in Germany, van't Hoff in Latvia, and Arrhenius in Sweden) were dubbed "Die Ioner" (German: the ionists) by contemporaries, for their close, friendly, and productive collaboration.

It fell to Hermann Nernst, working alongside Arrhenius in Oswald's laboratory, to develop between 1887 and 1889 the thermodynamic relationships between a cell's open circuit potential (the voltage with no electricity flowing) and the amount of heat released or consumed by the net chemical reaction occurring at the two electrodes in a cell. This was conclusively demonstrated by Fritz Haber in 1905. Haber also developed the glass electrode, a premier measuring tool of analytical chemistry. This invention led to its further use by Walter Wright and Lucius Elder in measuring pH in 1928, and to Arnold Beckman commercializing a sturdy and stable pH meter in 1935. Beckman's pH meter is one of the most useful instruments in all of science.

GOLDEN AGE OF ELECTROCHEMISTRY

The nineteenth century has been termed the golden age of electrochemistry. At first, only limited amounts of energy, sufficient for experimental purposes, were available from the cells patterned after Volta's original pile. In 1836, John Daniell introduced a more powerful battery containing zinc and copper electrodes that was capable of maintaining a reliable steady current for long periods of time. In its wake came electroplating, developed by George Richards Elkington (1836), followed by electrotyping by Moritz Hermann Von Jacobi for reproducing articles such as engravings, medals, and printer type setups (1839). Wright's basis, little changed today except in detail, for selecting solutions to plate anything from eating utensils, art objects, tiny computer electrical contacts to expensive automobile bumpers, came in 1840. Wright also developed cadmium plating (1849), rediscovered after World War I when the metal became available in commercial quantities. Robert Wilhelm von Bunsen produced magnesium in 1852 and lithium in 1855 from their salts, but their exploitation did not begin until well into the twentieth century. Friedrich Wohler made calcium carbide in 1862, but thirty years elapsed before Thomas Willson's accidental rediscovery while attempting to produce aluminum. However, Willson went a step further, reacting the carbide with water to produce acetylene, and launch-

Battery System	Net Electrochemical Reaction
Rechargeable Cells	
Plante or Lead-Acid Cell	$PbO_2 + Pb + H_2SO_4 = 2PbSO_4$
Nickel Cadmium	$2NiOOH + Cd + 2H_2O = Ni(OH)_2 + Cd(OH)_2$
Nickel Hydrogen	$2NiOOH + H_2 = 2Ni(OH)_2$
Non-Rechargeable Cells	
Leclanche or dry cell	$Zn + 2MnO_2 = ZnO + Mn_2O_3$
Alkaline Cell	$Zn + 2MnO_2 = ZnO + Mn_2O_3$
Silver-Zinc	$Ag_2O_2 + 2Zn + 2H_2O = 2Ag + 2Zn(OH)_2$
Reuben Cell	$HgO + Zn + H_2O = Hg + Zn(OH)_2$
Zinc-Air	$Zn + O_2 + 2H_2O = 2Zn(OH)_2$
Fuel Cell	$2H_2 + O_2 = 2H_2O$
Lithium Iodine	$2Li + I_2 = 2LiI$
Lithium-Sulfur Dioxide	$2Li + 2SO_2 = Li_2S_2O_4$
Lithium-Thionyl Chloride	$4Li + 2SOCl_2 = 4LiCl + S + SO_2$
Lithium-Manganese Dioxide	$Li + Mn(IV)O_2 = LiMn(IV)O_2$
Lithium-Carbon Monofluoride	$nLi + (CF)_a = nLiF + nC$

Table 1.
Examples of Commercially Developed Batteries

ing acetylene's use for heat, light and power, and more importantly for metallurgical applications such as producing alloy steels in the 1890s.

A great step forward came in 1859 Gaston Plante learned to store electricity in his lead storage cell batteries. Plante had been investigating the effect of current drain on different metals when he noticed the surprising behavior of the lead plates in dilute sulfuric acid. Plante discovered that he could increase the amount of electricity stored in the cell by allowing the cell to rest after it was discharged. In what we now call "formation," the lead peroxide coating on the positive lead electrode is converted to the sulfate during the rest period. Reversing cell polarities from time to time, and repeating these charging and discharging cycles results in a buildup of electrical storage capacity in the cell. Such a rechargeable battery is called a secondary battery, in contrast to a primary battery, one that cannot be recharged. The lead storage cell battery remains the dominant secondary battery, with 40 percent of the world's lead production used in this application.

The lead storage cell battery was ahead of its time. Battery recharging was tedious and difficult, and electrolytic processing was not economical until electricity became commercially available in the late 1800s. Plante had to use a series of Bunsen mercury-zinc primary cells to recharge his battery. The Bunsen cell, invented in 1850, in which a platinum electrode was replaced with carbon actually was a modification of a cell by William Grove. Grove is better known for his invention of the hydrogen-oxygen fuel cell in 1839,

233

which was proven practical and reliable as part of the Gemini and Apollo space programs in the 1960s.

Another major step was the dry cell, first produced by C. Gassner in 1888. The dry cell was an outgrowth of Georges Leclanche's 1868 cell in which a carbon rod was contained in a porous cup filled with a crushed, then compressed, carbon and manganese dioxide mixture. In its later form, a zinc electrode was suspended in the mixture, then molded together into a cylindrical form using a binder. Billions of such batteries, but improved over the years in reliability, leak-proofing, endurance, and shelf life, are used worldwide to power flashlights, toys, and a host of portable electronic devices.

By the end of the nineteenth century, most battery types had been identified, and several hundred had experienced some commercial application. Batteries became so rugged and reliable that Robinson in 1871 used them in train signaling, and Thomas Alva Edison in powering his quadruplex telegraph (simultaneous messages on one wire) in 1874. Alexander Graham Bell's telephone exchanges, starting in New Haven, have operated on batteries since 1878.

Nickel-cadmium and nickel-iron are prime examples of rechargeable (secondary) batteries, invented in 1901 by Waldemar Jungner and Edison, respectively. In the 1920s and 1930s, such batteries powered radios prior to rural electrification. The sealed nickel-cadmium battery is now the most widely used battery in consumer products.

RISE OF THE ELECTRICAL INDUSTRY

The era of large electrical power generation began in 1875, when Cornell University was lit with electric arc lights using Gramme dynamos, continuous high-voltage electric current generators invented in 1869 by Zenobe-Theophile Gramme. This was followed in 1879 by the introduction of the carbon-filament incandescent lamp by Thomas Edison, and the carbon-arc lamp by Charles M. Brush of Cleveland. Neither application was feasible using batteries alone to supply the power. However, electrochemical technology was instrumental in furthering the growth of the electric power industry. Lead storage cell batteries remained a secondary source of electric power, to accommodate sudden surges and provide incremental power during periods of high demand.

Electrochemical processing proved essential to efficient power transmission. J. B. Elkington, building on family electroplating technology, in 1869 used electrowinning (recovery of metals from solutions by electrolysis) to remove impurities from copper and consequently double the electrical conductivity of commercial copper. Other important metals so refined include silver, gold, nickel, antimony and bismuth. In electrowinning, an imposed electric field causes the targeted metal, in this case copper, to be dissolved from impure slabs, then to be selectively deposited on thin sheets. Sludge, periodically removed from the bath, is stripped of its valuable byproducts of gold, silver and tin.

In 1885, Charles Martin Hall invented his aluminum process and Hamilton Young Castner in 1890 developed the mercury-type alkali-chlorine cell, which produced caustic (sodium hydroxide) in its purest form. Edward G. Acheson in 1891, while attempting to make diamonds in an electric furnace, produced silicon carbide, the first synthetic abrasive, second to diamond in hardness. Four years later, Jacobs melted aluminum oxide to make a superior "emery cloth." Within two decades, these two abrasives had displaced most natural cutting materials, including naturally occurring mixtures of aluminum and iron oxides.

Great achievements came with the introduction of cheap hydroelectric power, starting with a 100,000 hp installation on the Niagara River in 1894. The first contracts were not for lighting, but for the electrolysis of alumina with the Pittsburgh Reduction Company (now Alcoa) and the manufacture of silicon carbide by the Carborundum Co. In 1895, Moorehead produced 50 percent ferrosilicon in an electric furnace, leading to ferroalloys with qualities not possible in blast furnaces. In 1898, the Mathieson Alkali Works and Dow Chemical began the manufacture of bleaching powder from electrolytic chlorine, exemplifying the accelerating opportunity for chemical manufacture in the growing textile, paper, fertilizer, soap, and glass industries.

FUEL CELLS

The concept of the fuel cell, that is, a cell in which inert electrodes immersed in an electrolyte could be intimately contacted with a reacting fuel (e.g., hydrogen) and oxidant (e.g., air) and so generate an electric current, was demonstrated in 1839 by Grove and intensively studied by him during the next decade.

Grove recognized that electrodes above the surface of an electrolyte, (e.g., sulfuric acid) would be wetted by capillary action and so allow the platinum electrodes to catalyze the electrochemical reactions of a fuel and oxidant such as hydrogen and oxygen.

The poor efficiencies of coal-fired power plants in 1896 (2.6 percent on average compared with over forty percent one hundred years later) prompted W. W. Jacques to invent the high temperature (500°C to 600°C [900°F to 1100°F]) fuel cell, and then build a 100-cell battery to produce electricity from coal combustion. The battery operated intermittently for six months, but with diminishing performance, the carbon dioxide generated and present in the air reacted with and consumed its molten potassium hydroxide electrolyte. In 1910, E. Bauer substituted molten salts (e.g., carbonates, silicates, and borates) and used molten silver as the oxygen electrode. Numerous molten salt battery systems have since evolved to handle peak loads in electric power plants, and for electric vehicle propulsion. Of particular note is the sodium and nickel chloride couple in a molten chloroaluminate salt electrolyte for electric vehicle propulsion. One special feature is the use of a semi-permeable aluminum oxide ceramic separator to prevent lithium ions from diffusing to the sodium electrode, but still allow the opposing flow of sodium ions.

It 1932, ninety years after Grove, Francis Bacon developed "gas diffusion" electrodes. These electrodes were unique in having the reacting gases fed to one side of each electrode while the other side was in contact with the alkaline electrolyte. A further improvement was E. Justi's "double skeleton catalyst electrodes" in 1952. Justi used finely divided nickel and silver catalysts incorporated onto nickel skeletons as hydrogen and oxygen electrodes, respectively. In 1953 Karl Kordesch's further improvement in oxygen diffusion electrodes involved inclusion of a water-repellent, such as Teflon, in carbon-based electrodes. Leonard W. Niedrach in 1965 also made improvements using an ion exchange membrane fuel cell. Since the 1960s, a number of fuel cell batteries have been developed for use as remote power sources, fueled with methanol, hydrazine or hydrogen obtained subsequent to partially oxidizing a hydrocarbon mixture (e.g., natural gas) or an alcohol (e.g., methanol). Such indirect hydrogen fuel cell batteries are finding application, by General Electric for one, in providing residential power where transmission lines have limited access and are costly.

VEHICULAR POWER SOURCES

The crowning use of electrochemical energy may come with its broad application to vehicular propulsion. Such engines offer the promise of both superior efficiency and exceptionally clean gaseous emissions. Numerous possibilities have received close scrutiny during the past four decades. Zinc-air, zinc-bromine, and lithium-sulfur batteries were subjects of substantial investigation in the 1960s, then sodium-sulfur, zinc-nickel and nickel-iron in the 1970s and 1980s. Contenders added in the 1990s included sodium-nickel chloride, lithium-ion and nickel-metal hydride batteries.

At least eight electric vehicles using lead-acid, nickel-metal hydride, or sodium nickel chloride batteries were commercially available in 1998, but with limited acceptance because power and capacity were insufficient and at too high a price. Consequently, hybrid engines, that is, internal combustion or diesel engines in combination with batteries or fuel cells, may offer greater prospects near-term. The first mass-produced vehicle with a hybrid engine was the Toyota Prius (Latin: pioneering), a subcompact on sale in Japan beginning in 1997. Larger hybrid-powered vehicles are being tested by Toyota and other manufacturers. Twenty hybrid electric buses and trucks were under study in 1998, typically targeted for 25 percent lower diesel fuel consumption and 30 to 50 percent reduction in emissions.

Prius uses an internal combustion engine operating at constant speed to drive a generator that powers a motor connected to the drive train. Excess energy is used to charge a nickel-metal hydride battery, which is intermittently used to handle the power demands of steep climbs and rapid accelerations. When the vehicle slows, the motor behaves as a generator, converting kinetic energy into electricity.

EMERGING OPPORTUNITY

Research, development and entrepreneurial activities are pervasive, fueled by ever-growing demands for lighter, smaller and more powerful batteries with even more reliability and extended shelf life for industrial, automotive, military, business, medical, and home uses. Mobile communications, computing, data storage and access, instrumentation, intelligent sensors, controls for individual devices and systems, power tools, home appliances, systems backup, and remote standby applications provide countless opportunities for improved battery technology.

Materials science is contributing to better fabrication and use of extremely reactive metals. In the 1960s emerged highly reactive and powerful lithium batteries using carbon, titanium sulfide, and transition metal oxide structures to tie up (intercalcate) the reactive lithium. Notable among these is the lithium-ion battery, widely used in cellular phones, video cameras and laptop computer display panels. In this battery, both electrodes are lithium intercalcating electrodes, with lithium intercalcated into carbon and either nickel or cobalt oxide, respectively. An offshoot of the lithium-ion development is the rechargeable lithium-polymer battery in which complexes of lithium salts with certain polymers (e.g., polyethylene) can serve as electrode separators, but still allow transfer of lithium ions between electrodes.

Proliferation of new materials is also important to improving mature batteries. Lead-coated fiberglass is used to reduce the weight of lead acid batteries, and microporous polyethylene is added as separators to contain electrolyte and so prevent its leakage. The nickel-metal-hydride battery is essentially the nickel-cadmium battery, but with the cadmium replaced with a metal alloy. Hydrogen generated during the charging cycle is stored as a hydride and then released and reacted during the discharge cycle. The metal alloys consist largely of nickel and readily available rare earth, modified with cobalt, manganese and aluminum, to improve corrosion resistance. Commercial introduction of new technology comes at a modest pace in this field, and more chapters of electrochemical history are yet to be written.

Barry L. Tarmy

See also: Faraday, Michael; Nernst, Walther Hermann; Volta, Alessandro.

BIBLIOGRAPHY

Bowden, M. E. (1997). *Chemistry Is Electric!* Philadelphia: Chemical Heritage Foundation.

Design News. (10/5/98). "Out of Juice!" Annual Auto Issue: Battery Technology, pp. 98–108. <http://www.design-news.com>.

Graham, R. W., ed. (1981). *Primary Electrochemical Cell Technology: Advances Since 1977.* Park Ridge, NJ: Noyes Data Corporation.

Liebhafsky, H. A., and Cairns, E. J. (1968). *Fuel Cells and Fuel Batteries.* New York: John Wiley and Sons.

Natishan, P. M., ed. (1997). *What is Electrochemistry?*, 4th ed. Pennington, NJ: The Electrochemical Society.

Mantell, C. L. (1950). *Industrial Electrochemistry.* New York: McGraw-Hill.

Schallenberg, R. H. (1982). *Bottled Energy: Electrical Engineering and the Evolution of Chemical Energy Storage.* Philadelphia, PA: American Philosophical Society.

Stock, J. T., and Orna, M. V., eds. (1989). *Electrochemistry, Past and Present.* Washington, DC: American Chemical Society.

Toyota Press Information. (1997). *Toyota Hybrid System.* Tokyo: Toyota Motor Corporation, Planning Group, International Public Affairs Division.

CHERNOBYL

See: Nuclear Energy; Nuclear Energy, Historical Evolution of the Use of

CIRCUIT BREAKERS AND FUSES

See: Electricity

CLAUSIUS, RUDOLF JULIUS EMANUEL (1822–1888)

Physicist Rudolf Julius Emanuel Clausius was born January 2, 1822, in Koeslin, Pomerania, Prussian province (now Koszalin, Poland). He was the sixth son of eighteen children born to teacher and a Protestant minister. Clausius attended gymnasium (secondary school) in Stettin and from 1840 the University in Berlin, where he studied mathematics and physics mainly as a student of Gustav Magnus. In 1851 he joined The Physics Society, which was formed in 1845 by Magnus's students. In tandem with his university studies, he taught in a Berlin gymnasium to sponsor the education of his younger sisters and brothers. In 1848 he earned a doctorate degree, and in 1850 he became lecturer at Berlin University while simultaneously teaching at the Artillery School of the Prussian army. Clausius left Berlin in 1855 and moved to Zurich, where he became a professor of physics at Polytechnicum, then later at the University of Zurich. Between 1867 and 1869 he resided in Wuerzburg, and finally he settled in Bonn, where he spent the last two decades of his life as a professor at the university. Clausius married in 1859. His wife died tragically in 1875, while giving birth to their sixth child. Again the sense of duty towards his family was a foremost concern in Clausius's life, and he sacrificed his scientific pursuits to supervise the education of his children. A serious knee injury that he had acquired in the Franco-Prussian war further limited his scientific activity, causing him discomfort until he died on August 24, 1888, in Bonn.

From the beginning of his studies Clausius's interest was mathematical physics. He exhibited keen mathematical thinking, although he was not particularly communicative. His textbooks were popular. The works of Julius Mayer, James Joule, and Hermann Helmholtz were the basis of the knowledge of heat nature in the 1840s and the starting point for Clausius's research. In 1850 Clausius gave a lecture in Berlin in which he formulated the basis for heat theory. He stressed that heat is not a substance, but consists of a motion of the least parts of bodies. Furthermore he noted that heat is not conserved. Clausius formulated the equivalence between heat and work and the first law of thermodynamics as follows: mechanical work may be transformed into heat, and conversely heat into work, the magnitude of the one being always proportional to that of the other. During the next fifteen years Clausius concentrated his research on thermodynamics. He gave several formulations of the second law and finally defined and named the concept of entropy. He started his theoretical investigations with the consideration of the idealized process of the so-called Carnot cycle on the basis of the ideal gas. For this purpose, one assumes that the gas is enclosed in an expansible wrap through which no heat exchange is possible. The Carnot cycle consists of four steps:

Rudolf Julius Emanuel Clausius. (Corbis Corporation)

1. The gas of initial temperature T1 in a volume V expands (V rises) isothermally (T1 stays constant). To keep temperature T1, the gas comes in contact with body B1 of constant temperature T1 and receives the quantity of heat Q1.

2. One takes the body B1 away. The gas continues to expand and its temperature decreases to T2.

3. The gas at temperature T2 compresses (V decreases) isothermally (T2 stays constant). The superfluous heat Q2 is transformed to the body B2 of constant temperature T2.

4. One takes the body B2 away. Through compression the gas is brought to the initial state raising the temperature to T1. The cycle is complete. The gas is in the same state as at the beginning of the cycle; it is therefore a reversible process.

During the expansion the gas produces the work, in contrast to the compression where the exterior work is needed. One needs part of the heat to perform work, so Q1 is larger than Q2. Defining the withdrawn heat as negative heat, every reversible cycle system satisfies the relation Q1/T1+Q2/T2=0. In 1854 Clausius further considered processes where the receiving of positive and negative quantities of heat take place at more than two temperatures. Next he decomposed every reversible cyclic process into an infinite number of Carnot processes. In this way, instead of Q/T, d Q/T appears as a differential of a new physical quantity, which depends only on the state parameters of the gas (in general of the body), but it does not depend on the closed path of integration needed to run through the complete cycle. Denoting the new quantity S, one arrives at the relation d Q =T d S.

In 1865, in a talk given in Zurich, Clausius introduced the notion of entropy for the function S, the Greek word for transformation. He meant that the important functions in science ought to be named with the help of the classical languages. After passing through the described reversible cycle process, entropy remains constant. A larger and more common group of thermodynamic processes are irreversible. According to Clausius, the entropy is the measure of the irreversibility of a process and increases constantly to the maximum. In the same lecture he brought the two laws of mechanical heat theory into the neat form: The energy of the universe is constant. The entropy of the universe increases. The definitive physical interpretation of entropy was first made possible in the framework of statistical mechanics by the Austrian physicist Ludwig Boltzmann in 1877.

Clausius was among the first researchers to look for a foundation of thermodynamics in the realm of kinetic theory. In a paper from 1857 he described the qualitative distinction of characteristic features of phases of matter (solid state, liquid, gas) from a microscopic, that is molecular, point of view and derived the formula connecting the pressure of an ideal (i.e. non-interacting) gas to the mean kinetic energy of the molecular constituents of the gas. An 1858 paper was inspired by the critique of Christoph Buys-Ballot, who noted the apparent conflict between the claim of kinetic theory that the molecular constituents of gases move with velocities of the order of some hundred meters per second at room temperature and the everyday experience of the comparatively slow evolution of diffusive processes. (It takes some minutes to smell the effect of overcooked milk outside the kitchen.) Clausius identified the multi-scattering of molecules among each other as

the main reason for the slowing down of diffusive processes. He introduced the notion of the mean free path length, which is the average length a molecule flies between two consecutive collisions with some of its neighbours. This notion has become an integral part of modern transport theory.

In an obituary talk given at the Physical Society of Berlin in 1889, Hermann Helmholtz stressed that Clausius's strict formulation of the mechanical heat theory is one of the most surprising and interesting achievements of the old and new physics, because of the absolute generality independent of the nature of the physical body and since it establishes new, unforeseen relations between different branches of physics.

Barbara Flume-Gorczyca

BIBLIOGRAPHY

Clausius, R. (1850). Poggendorff's Annalen 79:368–397, 500–524.
Clausius, R. (1854). Poggendorff's Annalen 93:481–506.
Clausius, R. (1857). Poggendorff's Annalen 100:353–384.
Clausius, R. (1858) Poggendorff's Annalen 105:239–256.
Clausius, R. (1865). Poggendorff's Annalen 125:353–400.
Helmholtz, H. von. (1889). Verhandlungen der physikalischen Gesellschaft zu Berlin: 1–6.

CLEAN FUELS

See: Alternative Fuels

CLIMATIC EFFECTS

Earth's climate has fluctuated over the course of several billion years. During that span, species have arisen, gone extinct, and been supplanted by others. Humankind and our immediate ancestral species have survived the sometimes drastic fluctuations of the past several million years of our evolving climate quite handily. Although human beings and their ancestors have always changed the environment in which they have lived, human actions since the advent of the Industrial Revolution may be producing impacts that are not only local but also global in

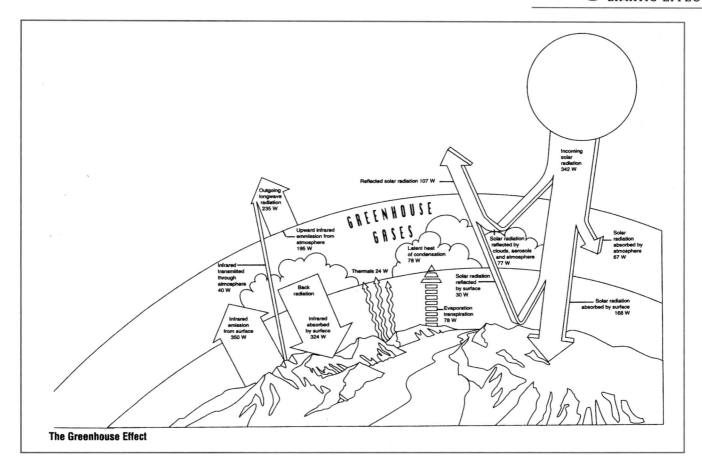

The Greenhouse Effect

Illustration of the greenhouse effect. (Public domain)

scope. Ever-growing fossil fuel use and other activities associated with industrialization have altered the concentrations of several atmospheric gases, including carbon dioxide. Theoretically, these alterations can cause alterations in the overall heat retention of Earth's atmosphere. Scientists around the world are engaged in studies to determine whether these changes in the atmosphere pose a risk of significant, negative environmental and human impacts.

CLIMATE CHANGE THEORY

Climate change means different things to different people. To some, climate change refers to physical changes in climate and little more. To others, climate change is a theory of how certain gases in the atmosphere influence the climate. Still others focus on the human aspect, and consider climate change only in regard to the way human activity influences the climate. Climate change theory is quite complex.

Unlike Albert Einstein's $E = mc^2$, for example, the theory of human-driven climate change is not a singular theory but consists of several interlocking theories, some better defined than others.

At the heart of climate change theory is a more humble theory called the greenhouse effect. This theory, first quantified by mathematician Joseph Fourier in 1824, has been repeatedly validated by laboratory experiments and by millions of greenhouse owners. The greenhouse effect is simple. When sunlight reaches the surface of Earth, some of its energy is absorbed by the surface (or objects on the surface), some is reflected back toward space unchanged, and some is first absorbed by objects or the surface of Earth and then reemitted in the form of heat. Over a bare patch of ground, this dynamic would cause no net increase in the temperature over time because the heat absorbed during the day would be reradiated toward space overnight.

But if there is a greenhouse on that patch of

239

ground, things are different. The sunlight enters as usual, and some of it is reflected back out as usual, but part of the incoming solar energy that was held by the surface and reemitted as heat is prevented from passing back out by the glass, and the greenhouse warms up a bit. If there are water-beating plants or materials inside the greenhouse, water vapor concentration will increase as evaporation increases. Since water vapor also can trap heat, the greenhouse warms still further. Eventually the air reaches maximum humidity, and the system reaches temperature equilibrium.

Scientists have known for a long time that the greenhouse effect applies not only to greenhouses but to Earth as a whole, with certain gases (called greenhouse gases) playing the role of the glass in the example above. The primary greenhouse gases are water vapor (H_2O), carbon dioxide (CO_2), methane (CH_4), nitrous oxide (N_2O), ozone, and chlorofluorocarbons (CFC). When scaled up to the entire planet, the natural greenhouse effect produces a pronounced warming effect. This warming is a natural aspect of Earth's environment, crucial for the maintenance of life on Earth. In fact, without Earth's natural greenhouse effect, and the warming that goes with it, Earth would be a much colder planet, inhospitable to life as we know it. The greenhouse effect has also been seen to maintain warmer planetary atmospheres on Mars and Venus. While the effect is mild on Mars, the high carbon dioxide level on Venus (78,000 times that of Earth) keeps the atmosphere about 500°C (900°F) higher than it would otherwise be.

Against the backdrop of an Earth warmed by its own greenhouse effect, other forces operate that can increase or decrease the retention of heat by the atmosphere. Some of these forces are of human origin, some are produced by nature, and some are produced by mutual feedback reactions.

Which forces have dominated in the recently observed warming of the climate is still an open question. The last landmark report of the United Nations' Intergovernmental Panel on Climate Change (IPPC) concludes:

Although these global mean results suggest that there is some anthropogenic [of human origin] component in the observed temperature record, they cannot be considered as compelling evidence of a clear cause-and-effect link between anthropogenic forcing and changes in the Earth's surface temperature. It is difficult to achieve attribution of all or part of a climate

change to a specific cause or causes using global mean changes only. The difficulties arise due to uncertainties in natural internal variability and in the histories and magnitudes of natural and human-induced climate forcings, so that many possible forcing combinations could yield the same curve of observed global mean temperature change [IPCC, 1995, p. 411].

Nonetheless, the IPCC's chapter on the attribution of climate change concludes:

The body of statistical evidence. . . when examined in the context of our physical understanding of the climate system, now points towards a discernible human influence on global climate. Our ability to quantify the magnitude of this effect is currently limited by uncertainties in key factors, including the magnitude and patterns of longer-term natural variability and the time-evolving patterns of forcing by (and response to) greenhouse gases and aerosols [IPCC, 1995, p. 439].

WARMING AND COOLING FORCES

Human activities (as well as nonhuman biological, chemical, or geological processes) release a variety of chemicals into the atmosphere, some of which, according to climate change theory, could exert a warming or a cooling effect on Earth's climate. In climate change literature, these are referred to as "climate forcings." Some of these forcings are actually secondhand responses to changes in the climate caused by others. In such a case, a forcing might be referred to as a feedback.

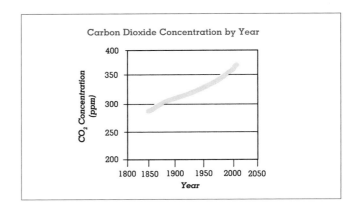

Figure 1.

SOURCE: <http://www.giss.nasa.gov/data/si99/ghgases>.

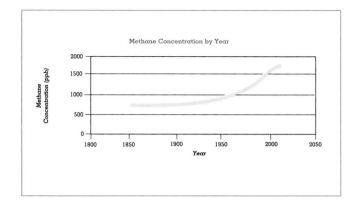

Figure 2.
SOURCE: <http://www.giss.nasa.gov/data/si99/ghgases>.

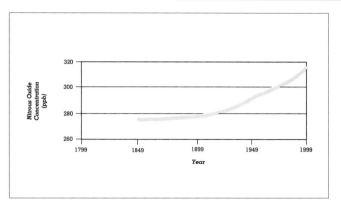

Figure 3.
SOURCE: <http://www.giss.nasa.gov/data/si99/ghgases>.

Carbon Dioxide

Carbon dioxide, considered a warming gas, comprises about 0.036 percent of the atmosphere by volume. As Figure 1 shows, carbon dioxide levels have increased as a component of the atmosphere by nearly 30 percent from the late eighteenth century to the end of the twentieth century, when the level was close to 365 parts per million by volume. Prior to the period of industrialization, carbon dioxide levels were largely stable, at about 280 parts per million, though fluctuations as low as 200 parts per million or as high as 300 parts per million have been observed through analysis of air bubbles trapped in arctic ice cores.

Carbon dioxide is released into the environment by human activities such as fuel burning, cement production, and land use.

Since highly accurate, direct measurement of carbon dioxide levels began only in the late 1950s, most of our understanding of carbon dioxide's historical patterns of fluctuation come from indirect measurements, such as the analysis of gas bubbles trapped in glaciers and polar ice caps. Though indirect measurements carry greater uncertainty than direct measurements of carbon dioxide levels, indirect measurements have contributed to our understanding of Earth's carbon cycle. Still, significant gaps in our understanding remain, specifically involving questions of time lag, the impact of world vegetation on atmospheric carbon dioxide levels, other processes that might lock carbon dioxide away from the atmosphere, and the role of carbon dioxide as a causal agent of climate change.

Methane

Methane is a greenhouse gas up to fifty-six times as powerful a warming agent as carbon dioxide, depending on the time scale one considers. In a twenty year time frame, for example, a given quantity of methane molecules would have fifty-six times the impact of the same quantity of carbon dioxide molecules, but since carbon dioxide has a longer lifespan, this ratio declines over time. As an atmospheric component, methane is considered a trace gas, comprising approximately 0.00017 percent of the atmosphere by volume. As Figure 2 shows, methane levels in the atmosphere increased nearly 120 percent from the middle of the nineteenth century to the end of the twentieth century, when the levels were the highest ever recorded, though the pattern of methane emissions has been highly irregular and actually showed downturns toward the end of the twentieth century for reasons that are not clear. Studies of methane concentrations in the distant past show that methane concentrations have fluctuated significantly, from as few as 400 parts per billion to as many as 700 parts per billion, due to changes in wetlands and other natural sources of methane. Today methane comes from a variety of sources, some of human origin, others of nonhuman origin.

Nitrous Oxide

Nitrous oxide is a long-lived warming gas with a relative warming strength 170 to 310 times that of carbon dioxide, depending on the time scale one considers. Nitrous oxide, like methane, is considered a trace gas in the atmosphere, but at considerably lower

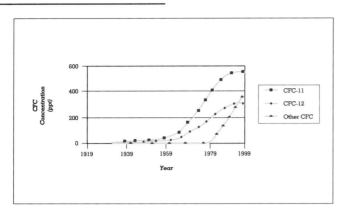

Figure 4.

SOURCE: <http://www.giss.nasa.gov/data/si99/ghgases>.

levels, about 0.00003 percent of the atmosphere by volume. As Figure 3 shows, nitrous oxide concentrations have increased significantly from the middle of the nineteenth century to the end of the twentieth century. Prior to the industrial period, nitrous oxide concentrations fluctuated at an average of 270 parts per billion by volume, though fluctuations as low as 200 parts per billion by volume have been measured from the distant past.

Nitrogen oxides (of which nitrous oxide is the major component) come from a variety of sources, only some of which are of human origin.

Chlorofluorocarbons (CFCs)

Chlorofluorocarbons (CFCs) are man-made compounds used as cooling agents and propellants in a broad range of applications. There are many different species of CFCs, some of which have been banned. CFCs are very powerful warming gases. Some species are more than ten thousand times more capable of trapping heat than is CO_2. Of course, CFCs also are found at much lower concentrations than the other greenhouse gases. Whereas carbon dioxide is measured in parts per million, and methane in parts per billion, CFCs are measured in parts per trillion. Figure 4 shows the concentration of the three major CFCs from 1929 to 1999.

Some CFCs also can break down ozone. Ozone-depleting CFCs can exert either warming or cooling effects, depending on where they are found. In the lower atmosphere, ozone-depleting CFCs exert a warming effect through the absorption of heat reradiated from Earth's surface. In the upper atmosphere, ozone destruction exerts a cooling effect by destroy-

ing some of the high-altitude ozone that can either warm or cool the surface in different circumstances. On a net basis, our current understanding is that the ozone-depleting CFCs (banned by the Montreal Protocol in 1987) exerted a cooling effect. Replacement chemicals for the ozone-depleting CFCs are considered pure warming gases, but with a considerably lower warming potential than the chemicals they replaced. Because of the complexities of ozone chemistry in the atmosphere and uncertainties regarding the warming or cooling potential of remaining ozone-depleting CFCs and replacement compounds, the ultimate impact of CFCs on climate change is highly uncertain.

Aerosols

Aerosols are not gases in the strictest sense of the word, but are actually liquid or solid particles small enough to stay suspended in air. Both human-made and natural processes generate aerosols. Some aerosol particles tend to reflect light or cause clouds to brighten, exerting a cooling effect on the atmosphere. Other aerosol particles tend to absorb light and can exert a warming effect. Most human-made aerosols seem to exert a cooling effect on the climate. On a global basis, some have estimated that this cooling effect offsets about 20 percent of the predicted warming from the combined greenhouse warming gases, but that cooling is not uniform: the offsetting impact varies geographically, depending on local aerosol concentrations.

The omission of aerosol considerations in earlier climate models led to considerable overprediction of projected global warming and predicted regional impacts, though newer models have done much to internalize the cooling effect of aerosols. Aerosols act as cooling agents through several mechanisms, however, some of which are only poorly understood. Besides directly scattering incoming sunlight, most particulates also increase the reflectivity, formation, and lifetime of clouds, affecting the reflection of incoming solar radiation back to space.

Water Vapor

Water vapor is the most abundant of the greenhouse gases and is the dominant contributor to the natural greenhouse effect. About 99 percent of all the moisture in the atmosphere is found in the troposphere, which extends about 10 to 16 kilometers above sea level. Only about one-third of the precipitation that falls on Earth's continents drains to the oceans. The rest goes back into the atmosphere as a result of evaporation and transpiration.

In the lower part of the atmosphere, the water vapor content of the atmosphere varies widely. On a volume basis, the normal range is 1 to 3 percent, though it can vary from as little as 0.1 percent to as much as 5 percent.

Water vapor can be a climate warming force when it traps heat, or can cause either climate warming or cooling when it takes the form of clouds, which reflect incoming solar energy away from Earth.

Most climate models predict that a warming of Earth's atmosphere would be accompanied by an increased level of water vapor in the lower atmosphere, but determining whether this has happened in response to recent climate warming is difficult. Data on water vapor concentrations are limited, and the data suffer from a range of limitations, including changes in instrument type, limited geographic coverage, limited time span, and so on. Data from satellites may offer some relief for these problems, but such data have been gathered only for a few years.

Some researchers have observed what appear to be slight increases in water vapor in various layers of the atmosphere, ranging up to 13 percent. Others have analyzed satellite data, and seen what appears to be a drying of the atmosphere, rather than increased moisture levels.

Solar Activity

Rather than burning with a steady output, the sun burns hotter and cooler over time. Several cycles of increased or decreased solar output have been identified, including cycles at intervals of eleven years, twenty-two years, and eighty-eight years.

Though measurements of solar output have been taken only for the past eighteen years, longer trend patterns can be derived from indirect data sources, such as ice cores and tree rings. Cosmic rays, which fluctuate with the sun's activity, also strike constituents of the atmosphere, creating radioactive versions of certain elements. Beryllium, in particular, is ionized to ^{10}Be by cosmic rays. The ^{10}Be then gets incorporated into trees as they grow, and is trapped in bubbles in ice masses, as is carbon dioxide.

A 1995 reconstruction of historical solar output levels from 1600 to 2000 shows that solar irradiance has risen over time, but with many short-term peaks and troughs in the overall curve of increase, increas-

ing the level of solar output that constitutes the main driver for the climate system's temperature.

Studies suggest that increased solar output may have been responsible for half of the 0.55°C (1°F) increase in temperature from 1900 through 1970, and for one-third of the warming seen since 1970.

Ozone

Ozone is a highly reactive molecule composed of three atoms of oxygen. Ozone concentrations vary by geographical location and by altitude. In addition, ozone exerts a different climate-forcing effect, depending upon altitude.

At lower, tropospheric altitudes, ozone exerts a warming force upon the atmosphere. Tropospheric levels of ozone have been increasing in the Northern Hemisphere since 1970, and may have doubled in that time. Ozone concentrations in the Southern Hemisphere are uncertain, while at the poles, tropospheric ozone concentrations seem to have fallen since the mid-1980s. At higher, or stratospheric altitudes, ozone exerts a cooling force upon the atmosphere. Ozone concentrations in the stratosphere have been declining over most of the globe, though no trend is apparent in the tropics. Much of the decline in stratospheric ozone concentrations has been attributed to the destructive action of the chlorofluorocarbons discussed previously.

Section Summary

It is clear that human action can affect seven of eight of the major greenhouse "forcings": carbon dioxide, methane, nitrous oxide, ozone, CFCs, aerosols, and water vapor. As studies of solar variation have shown, it is also clear that human action is not the only factor involved in determining the impact of these forcings. There is still substantial uncertainty regarding the actual climate impact of the climate forcings.

OBSERVED CLIMATE CHANGES

Part of the concern about global climate change stems from the human tendency to seek meaning in events that may or may not be more than simply a random event. A particularly cold winter, a particularly hot summer, an especially rainy season, or an especially severe drought will all send people off on a search for the greater meaning of the phenomenon. Is it a pattern, or a one-time event? Must we build a dike, or has the danger passed? Since the summer of

1988, virtually all unusual weather events seem to have triggered questions about global climate change.

Our ability to really know what the climate is doing is limited by a short observational record, and by the uncertainties involved in trying to figure out what climate was like in the past, or might be like in the future, for comparison with recent climate changes. While Earth's climate has been evolving and changing for more than four billion years, recordings of the temperature cover only about 150 years, less than 0.000004 percent of the entire pattern of evolving climate. In fact, temperature records are spotty before about 1960 and cover only a tiny portion of the globe, mostly over land. In addition to that 150-year conventional surface temperature record, temperature readings taken from weather balloons cover the years since 1970, and satellite temperature readings cover the years since 1982. Modern, reliable measurements of greenhouse gases are also very recent sources of data, beginning with carbon dioxide measurements at the South Pole in 1957, at Mauna Loa in 1958, and later for methane, nitrogen oxides, and chlorofluorocarbons.

Aside from temperature readings, other climate trends proposed as secondary effects of global warming carry information about the state of the climate. Changes in absolute humidity, rainfall levels, snowfall levels, the extent of snowfall, the depth of snowfall, changes in ice caps, ice sheets, sea ice, and the intensity or variability of storms have all been proposed as secondary effects of global warming. But because the history of recording climate trends is extremely short, most evidence regarding nontemperature-related changes in Earth's climate and atmospheric composition prior to the recent history of direct measurements is gathered from indirect sources such as air bubbles trapped in polar ice, or the study of fossils. This evidence, while interesting as a potential "reality check" for global human-made climate change models, is considered far less reliable than direct observational data.

These limitations in our evidence make it difficult to draw hard-and-fast conclusions regarding what changes have actually occurred recently in comparison to past climate conditions. More importantly, these limitations make it difficult to determine whether those changes are beyond the range of previous climate trends, happening at a faster rate than previous climate trends, or are being sustained for longer than previous climate trends. These are all critical questions when evaluating whether humanity is causing changes to Earth's normal climate patterns.

Nevertheless, scientists have evidence at hand regarding recent changes in both atmospheric composition and global climate trends that suggest that humanity has at least changed Earth's atmospheric composition in regard to greenhouse gases and other pollutants. These changes may or may not be contributing to recently observed changes in global warmth. A quick review of the climate changes suggested by the available evidence follows.

Temperature Trends

Besides readings of Earth's surface temperatures taken with standard glass thermometers, direct readings of atmospheric temperatures have been taken with satellites and weather balloons. In addition to direct measurements of Earth's recent temperatures, proxy measurements of temperatures from farther in the past can be derived from borehole temperature measurements, from historical and physical evidence regarding the extent and mass of land and sea ice, and from the bleaching of coral reefs.

This information is in relatively good agreement regarding what seems to be happening to global temperatures, at least in the recent periods of change spanning the past few hundred years, though there are discrepancies among some of the data sets. Temperatures recorded at ground-based measuring stations reveal a mean warming trend ranging from 0.3°C to 0.6°C (0.5°F to 1.1°F) since about 1850, with 0.2°C to 0.3°C (0.4°F to 0.5°F) of this warming occurring since 1960. The warming is not uniform, either in chronology or in distribution. More of the change occurs over land than over water. More of the warming happens at night, resulting in warmer nighttime temperatures rather than hotter daytime temperatures. More of the warming is noticeable as a moderation of wintertime low temperatures rather than as an increase in summertime high temperatures. Temperatures taken from weather balloons (also called radiosondes) and from satellites span a much shorter period of time (though, arguably, a more rigorously standardized measuring technique), and there is controversy over what they indicate, and how much weight should be given to such a short data set. Some analysts contend that the satellite and balloon recordings show a slight cooling trend in the tropics (about 0.1°C [0.2°F] per decade) since 1982. Others contend that the discrepancy is only an artifact caused by a limited data set, and the recent, unrelated increase in the strength of the El Niño southern oscillation.

And even here, taking the simplest of physical measurements, temperature, uncertainties are present. Temperature readings (satellite or ground station) were not taken specifically for the sake of evaluating the climate patterns of the entire Earth. Consequently the readings were taken from a variety of locations, cover only selected parts of the atmosphere, and are not necessarily well placed to be most informative about the climate as a whole. Further, measurement techniques and stations varied over the course of the temperature record, with data adjustments of a full degree occasionally needed to make the different sets of data compatible with each other. Satellites and balloons measure a different part of the atmosphere than ground stations do, making the comparability of such records questionable. In addition, the shortness of the satellite data record, punctuated as it has been by impacts of volcanic eruptions and the El Niño southern oscillation, further complicate the evaluation of temperature data.

Finally, perspective is important. While the past ten thousand years have been abnormally placid as far as climate fluctuations go, evidence of prior climate changes show an Earth that is anything but placid climatically. Some 11,500 years ago, for example, there is evidence that temperatures rose sharply over short periods of time. In Greenland, temperatures increased by as much as 7°C (12.6°F) over only a few decades, while sea surface temperatures in the Norwegian Sea warmed by as much as 5°C (9°F) in fewer than forty years. There is also evidence of about twenty rapid temperature fluctuations during the last glaciation period in the central Greenland records. Rapid warmings of between 5°C and 7°C (9°F to 12.6°F) were followed by slow returns to glacial conditions over the course of 500 to 2,000 years.

Rainfall Trends

Changes in precipitation trends are, potentially, a form of indirect evidence reflecting whether Earth is currently experiencing man-made climate change. Climate change models suggest that an enhanced greenhouse effect could cause changes in the hydrologic cycle such as increased evaporation, drought, and precipitation. But the IPCC warns that "our ability to determine the current state of the global hydrologic cycle, let alone changes in it, is hampered by inadequate spatial coverage, incomplete records, poor data quality, and short record lengths."

The global trend in rainfall showed a slight increase (about 1%) during the twentieth century, though the distribution of this change was not uniform either geographically or over time. Rainfall has increased over land in high latitudes of the Northern Hemisphere, most notably in the fall. Rainfall has decreased since the 1960s over the subtropics and tropics from Africa to Indonesia. In addition, some evidence suggests increased rainfall over the Pacific Ocean (near the equator and the international dateline) in recent decades, while rainfall farther from the equator has declined slightly.

Sea Level Trends

Changes in sea level and the extent of ice sheets, sea ice, and polar ice caps are still another form of indirect evidence reflecting whether Earth is currently undergoing anthropogenic climate change. Climate change theory would suggest that rising global temperatures would cause sea levels to rise due to a combination of the thermal expansion of water and melting of glaciers, ice sheets, ice caps, and sea ice.

Studies of sea leavels considered to reflect our best understanding of sea-level rise, as summarized in the 1995 reports of the United Nations Intergovernmental Panel on Climate Change, indicate a rise of 18 cm during the twentieth century, with a range of uncertainty of 10 to 25 cm. There is little evidence that the rate of sea-level rise has increased during that time period, though in theory the rate of warming has been accelerating. But thermal expansion of water is only one contributor to sea-level changes. Glaciers, ice sheets, and land water storage all play a role—a highly uncertain role.

Surface waters

Global warming would also be expected to influence surface waters such as lakes and streams, through changes induced in the hydrologic cycle. However, the last published report of the IPCC states no clear evidence of widespread change in annual streamflows and peak discharges of rivers in the world (IPCC, 1995, p. 158). While lake and inland sea levels have fluctuated, the IPCC also points out that local effects make it difficult to use lake levels to monitor climate variations.

Snow and Ice Trends

Global warming would also be expected to influence things such as snowfall, snow depth, and snow coverage (or extent), but studies examining changes in these aspects of the climate are quite mixed. Consistent with the indications of slight warming of the global climate, snow cover has declined in recent years, with a higher percentage of precipitation in cold areas coming down as rain rather than snow. But while the annual mean extent of snow cover over the Northern Hemisphere has declined by about 10 percent since 1979, snowfall levels have actually increased by about 20 percent over northern Canada and by about 11 percent over Alaska. Between 1950 and 1990, snowfall over China decreased during the 1950s but increased during the 1960s and the 1970s. Snowfall over the 45–55-degree-latitude belt has declined slightly. Snow depth levels, which respond both to atmospheric temperature and to the ratio of rainfall to snowfall, show equally mixed changes. Snow-depth measurements of the former Soviet Union over the twentieth century show decreased snow depth of about 14 percent during the Soviet winter, mostly in the European portion, while snow depth in the Asian sectors has increased since the 1960s.

Glaciers, Ice Caps, and Ice Sheets

With regard to glaciers and ice caps, the state of knowledge is even more limited. Glaciers and ice caps may have accounted for 2 to 5 centimeters of the observed sea-level rise discussed above, but the range of uncertainty is high. With regard to ice sheets, data are contradictory: There is not enough evidence to know whether the Greenland and Antarctic ice sheets are shrinking, hence contributing to sea-level rise, or growing, and hence retarding sea-level rise. They may even be doing both—growing on top and shrinking at the margins.

Weather Intensity and Variability Trends

Finally, increases in the intensity or variability of weather are considered another form of indirect evidence reflecting whether Earth is currently undergoing human-driven climate change. Predictions of increased incidence of extreme temperatures, tornadoes, thunderstorms, dust storms and fire-promoting weather have been drawn from basic global climate change theory. However, evidence has not so far borne out these predictions on a global scale. The IPCC concludes:

> [O]verall, there is no evidence that extreme weather events, or climate variability, has increased, in a global sense, through the 20th century, although data and

analyses are poor and not comprehensive. On regional scales, there is clear evidence of changes in some extremes and climate variability indicators. Some of these changes have been toward greater variability; some have been toward lower variability [IPCC, 1995, p. 173].

Section Summary

Evidence regarding changes in Earth's climate in the twentieth century is mixed, and encompasses a range of uncertainties. While the most recently published IPCC report holds that there is a discernible human influence on climate, this conclusion is not dependent on the evidence of actual changes in Earth's climate, as shown in this figure. On that note, the IPCC (1995, p. 411) says, "Despite this consistency [in the pattern of change], it should be clear from the earlier parts of this chapter that current data and systems are inadequate for the complete description of climate change." Rather, this conclusion is based on mathematical modeling exercises and "reality checked" with what hard evidence we have.

UNCERTAINTY AND FUTURE RESEARCH NEEDS

While recent studies of climate have contributed a great deal to our understanding of climate dynamics, there is still much to learn. The process of searching for evidence of man-made climate change, in fact, is both a search for new discoveries about how climate works, and continuing refinement of our understanding of the underlying theories we already have.

While greenhouse effect theory is a relatively uncontroversial issue in the scientific sense, the theory of global, human-driven climate change is at a much younger stage of development. Although there are very few articles in science journals that contradict either the overall theory or details of the core greenhouse effect, the same cannot be said for the theory of human-driven climate change and the consequences of that change. Indeed, nearly every month on the pages of leading science journals, studies jockey back and forth about key elements of human-made climate change.

Current climate change models have acknowledged weaknesses in their handling of changes in the sun's output, volcanic aerosols, oceanic processes, and land processes that can influence climate change. Some of those uncertainties are large enough, by themselves, to become the tail that wags the dog of climate change. Three of the major remaining uncertainties are discussed below.

The Natural Variability of Climate

Despite the extensive discussion of climate modeling and knowledge of past climate cycles, only the past thousand years of climate variation are included in the two state-of-the-art climate models referred to by the IPCC. As discussed earlier, however, the framework in which we view climate variability makes a significant difference in the conclusions we draw regarding either the comparative magnitude or rate of climate changes, or the interpretation of those changes as being either inside or outside of the envelope of normal climate change variations. The IPCC report summarizes the situation succinctly:

Large and rapid climatic changes occurred during the last ice age and during the transition towards the present Holocene period. Some of these changes may have occurred on time-scales of a few decades, at least in the North Atlantic where they are best documented. They affected atmospheric and oceanic circulation and temperature, and the hydrologic cycle. There are suggestions that similar rapid changes may have also occurred during the last interglacial period (the Eemian), but this requires confirmation. The recent (20th century) warming needs to be considered in the light of evidence that rapid climatic changes can occur naturally in the climate. However, temperatures have been far less variable during the last 10,000 years (i.e., during the Holocene) [IPCC, 1995, p. 416].

Until we know which perspective is more reflective of Earth's climate as a whole—the last ten thousand years, or a longer period of time—it will be difficult to put recent warming trends in perspective, or to relate those trends to potential impacts on the climate and on Earth's flora and fauna.

The Role of Solar Activity

At the front end of the climate cycle is the single largest source of energy put into the system: the sun. And while great attention has been paid to most other aspects of climate, little attention has been paid to the sun's role in the heating or cooling of Earth. Several studies in the late 1990s have highlighted this uncertainty, showing that solar variability may play a far larger role in Earth's climate than it was previously given credit for by the IPCC. If the sun has been

heating up in recent times, researchers observe, the increased solar radiation could be responsible for up to half of the observed climate warming of the past century. But as with satellite measurements of Earth's temperature, the short timeline of satellite measurements of solar irradiance introduces significant uncertainty into the picture. Most researchers believe that at least another decade of solar radiation measurement will be needed to clearly define the influence of solar input on the global climate.

Clouds and Water Vapor

Between the emission of greenhouse gases and change in the climate are a range of climate and biological cycles that can influence the end result. Such outcome-modifier effects are called "feedbacks" or "indirect effects" in the climate change literature.

One such feedback is the influence of clouds and water vapor. As the climate warms, more water vapor enters the atmosphere. But how much? And which parts of the atmosphere, high or low? And how does the increased humidity affect cloud formation? While the relationships among clouds, water vapor, and global climate are complicated in and of themselves, the situation is further complicated by the fact that aerosols exert a poorly understood influence on clouds.

Earlier computer models, which omitted the recently validated cooling effect of aerosols, overestimated the global warming that we would have expected to see by now, based only on the levels of greenhouse gases that have been emitted. As discussed earlier, aerosols themselves may offset 20 percent of the expected impact of warming gases. In addition, though direct cooling impacts of aerosols are now being taken into account by climate models, aerosol impact on clouds remains a poorly defined effect with broad implications, given a range of additional cooling potential of up to 61 percent of the expected warming impact from the warming greenhouse gases.

The last published report of the IPCC acknowledges that "the single largest uncertainty in determining the climate sensitivity to either natural or anthropogenic changes are clouds and their effects on radiation and their role in the hydrological cycle ... At the present time, weaknesses in the parameterization of cloud formation and dissipation are probably the main impediment to improvements in the simulation of cloud effects on climate" (IPCC, 1995, p. 346).

THE IMPACTS OF CLIMATE CHANGE

Global warming, and the potential climate changes that might accompany such warming, are estimated using of complex computer models that simulate, with greater or lesser complexity and success, the way Earth's climate would change in response to the level of greenhouse gases in the air. It is widely acknowledged that the potential temperature changes predicted by global warming theory do not pose a direct threat to human life. In fact, since more people die from extremes of cold rather than heat, the actual warming of the atmosphere, on net, could save more lives through warmer winters than it takes through hotter summers.

The major concerns about climate change focus on the second- and thirdhand impacts that would theoretically accompany global warming. Climate change theory suggests that warming of the overall environment could lead to a variety of changes in the patterns of Earth's climate as the natural cycles of air currents, ocean currents, evaporation, plant growth, and so on change in response to the increased energy levels in the total system. The most commonly predicted primary impacts of global warming are increased activity in the hydrologic, or water cycle of Earth, and the possible rise of oceans due to thermal expansion and some melting of sea ice, ice sheets, or polar ice caps. More dynamic activity in the water cycle could lead to increased rainfall in some areas, or, through increased evaporation rates, could cause more severe droughts in other areas. Rising sea levels could inundate some coastal areas (or low-lying islands), and through saltwater intrusion, could cause harm to various freshwater estuaries, deltas, or groundwater supplies.

Some have also predicted a series of thirdhand impacts that might occur if the climate warms and becomes more dynamic. Wildlife populations would be affected (positively and negatively), as would some vegetative growth patterns. The "home range" of various animal and insect populations might shift, exposing people to diseases that were previously uncommon to their area, and so on.

But one need not wade far into the most recently published IPCC report on the potential impacts of climate change before encountering an admission that uncertainty dominates any discussion of such potential impacts:

Impacts are difficult to quantify, and existing studies are limited in scope. While our knowledge has increased significantly during the last decade and qualitative estimates can be developed, quantitative projections of the impacts of climate change on any particular system at any particular location are difficult because regional scale climate change projections are uncertain; our current understanding of many critical processes is limited; and systems are subject to multiple climatic and non-climatic stresses, the interactions of which are not always linear or additive. Most impact studies have assessed how systems would respond to climate changes resulting from an arbitrary doubling of equivalent atmospheric carbon dioxide concentrations. Furthermore, very few studies have considered greenhouse gas concentrations; fewer still have examined the consequences of increases beyond a doubling of equivalent atmospheric carbon dioxide concentrations, or assessed the implications of multiple stress factors [IPCC, 1995, p. 24].

The IPCC report goes on to point out that this extreme uncertainty is likely to persist for some time, since unambiguous detection of human-made climate change hinges on resolving many difficult problems:

Detection will be difficult and unexpected changes cannot be ruled out. Unambiguous detection of climate-induced changes in most ecological and social systems will prove extremely difficult in the coming decades. This is because of the complexity of these systems, their many non-linear feedbacks, and their sensitivity to a large number of climatic and non-climatic factors, all of which are expected to continue to change simultaneously. The development of a base-line projecting future conditions without climate change is crucial, for it is this baseline against which all projected impacts are measured. The more that future climate extends beyond the boundaries of empirical knowledge (i.e., the documented impacts of climate variation in the past), the more likely that actual outcomes will include surprises and unanticipated rapid changes [IPCC, 1995, p. 24].

Uncertainties of this scale do not imply, as some analysts have asserted, that there is no reason to fear negative change, nor does it imply that we must fear drastic impacts. Rather, uncertainties of this scale indicate the need for a sustained research program aimed at clarifying our understanding of Earth's climate and how human activities might or might not translate into negative environmental impacts.

Risk-Reduction Benefits - Near Term

Reduced risk of harm from changing weather patterns:	NONE
Reduced risk of harm from extreme weather events:	NONE
Reduced risk of harm through famine avoidance:	NONE
Reduced risk of harm through disease prevention:	NONE
Reduction in other proposed climate change hazards:	NONE
Reduced risk of harm through avoided economic impacts of climate change:	NONE

Risk-Reduction Benefits - Long Term

Reduced risk of harm from changing weather patterns:	NONE-HIGH
Reduced risk of harm from extreme weather events:	NONE-HIGH
Reduced risk of harm through famine avoidance:	NONE-HIGH
Reduced risk of harm through disease prevention:	NONE-HIGH
Reduction in other proposed climate change hazards:	NONE-HIGH
Reduced risk of harm through avoided economic impacts of climate change:	NONE-HIGH

Risk Reduction Liabilities - Near Term

Induced fatalities from reduced disposable income:	Approx. 15,500/yr.
Lives not saved through other available risk-reduction investments:	Approx. 800,000/yr.

NET RISK-REDUCTION BENEFIT - NEAR TERM	**NONE**
NET RISK-REDUCTION BENEFIT - LONG TERM	**NONE-HIGH**

Table 1.
Risk-Alteration Ledger: The Kyoto Protocol

INTERNATIONAL AGREEMENTS ON RELEASE OF GREENHOUSE GASES

In 1997, a treaty was developed to limit the amount of greenhouse gases released into the atmosphere. Under the auspices of the United Nations Secretariat, the Third Conference of the Parties, held in Kyoto, Japan, produced the Kyoto Protocol (Table 1). This protocol calls for reductions in greenhouse gas emissions by various countries, though developing countries, predicted to become the dominant producers of greenhouse gases in the twenty-first century, were not bound to greenhouse gas reductions. Several countries, such as Australia, Iceland, and Norway, were allowed to increase their levels of greenhouse gas emissions under the treaty. The major reductions in emissions were to come from Europe, where Latvia, Estonia, and Lithuania agreed to an 8 percent reduction in emissions relative to 1990 levels. Japan and Canada agreed to 6 percent reductions from 1990 levels, and the United States agreed to reduce greenhouse gas emissions 7 percent below 1990 levels.

The Kyoto Protocol covers reductions in carbon dioxide, methane, nitrous oxide, and three fluorocarbons: hydrofluorocarbons, perfluorocarbons, and sulfur hexafluoride. The protocol also included mechanisms for considering greenhouse gas reductions stemming from changes in land use, and enshrined the principle of international emissions trading, though not the mechanism or specifics, which were left for later Conferences of the Parties to resolve. Finally, the Kyoto Protocol created a "clean development mechanism" by which developing countries could develop advance credits for taking actions that would limit the release of greenhouse gases in the future.

Several obstacles stand in the way of the Kyoto

Protocol. In 1997, prior to President Clinton's acceptance of the Kyoto Protocol, U.S. Senate Resolution 98, the Byrd-Hegel resolution, which was passed by a vote of ninety-five to zero, imposes specific requirements that must be met before the Kyoto Protocol can be ratified. The resolution calls for a specific timeline and commitments by developing countries to reduce greenhouse gas emissions, and evidence that adoption of the Kyoto Protocol would not result in serious harm to the U.S. economy. In addition, the Fifth Conference of the Parties (1999) failed to resolve numerous outstanding issues held over from the previous conference, and put off critical decision making until the Sixth Conference of the Parties in The Hague, Netherlands, in November 2000.

COST OF REDUCING GREENHOUSE EMISSIONS AND IMPACT ON ENERGY SYSTEMS

Reducing emissions of carbon dioxide and other greenhouse gases is not a trivial problem. Fossil fuels provide the overwhelming majority of energy production globally, and are predicted to do so through 2010. In the United States in 2000, fossil fuels were used to produce 70 percent of all energy generated. Alternative technologies such as nuclear power, solar power, wind power, hydropower, geothermal power, and hydrogen power have promise, but also have significant limitations and are considerably more costly than fossil fuel use. Consider that nearly 25 percent of total U.S. energy consumption in 1996 was for transportation, which is nearly all powered by fossil fuels.

Estimates of the cost of reducing greenhouse gas emissions, and the impact that such reductions would have on energy systems, vary widely. Estimates of greenhouse gas reduction costs are critically dependent on the assumptions used in economic models. Models assuming that greenhouse gas reduction targets will be met using international, multiemission trading systems suggest lower costs than those models that assume less international trading, single-gas approaches, carbon taxes, and so on.

In a comparison of nine economic models, estimated costs to the United States as of the late 1990s ranged from a loss in gross domestic product from $40 billion to $180 billion, with assumptions of no emission trading; from $20 billion to $90 billion with trading only among developed countries; and from

$5 billion to $20 billion with assumptions of global trading systems. Studies showing values at the higher end of the ranges outnumbered those showing costs at the lower end of the spectrum. The largest costs in such models stem from accelerated fuel substitution and the adoption of more expensive nonfossil fuel forms of fuel generation.

Kenneth Green

See also: Acid Rain; Air Pollution; Atmosphere; Environmental Problems and Energy Use; Ocean Energy Systems; Pollution and Energy Efficiency.

BIBLIOGRAPHY

Broad, W. J. (1997). "Another Possible Climate Culprit: The Sun." *New York Times*, September.
Elliot, W. P. (1995). "On Detecting Long-Term Changes in Atmospheric Moisture." *Climatic Change* 31:349–367.
Etheridge, D. M., et al. (1998). "Atmospheric Methane between 1000 A.D. and Present: Evidence of Anthropogenic Emissions and Climate Variability." *Journal of Geophysical Research* 103(DI 3):15979–15993.
Fluckinger, J., et al. (1999). "Variations in Atmospheric N_2O Concentration During Abrupt Climate Changes." *Science* 285:227–230.
Fourier, J. (1824). "Remarques générales sur la temperature du globe terrestre et des espaces planétaires." *Annals de chimie et de physique* 27:136–167.
Green, K. (1997). *A Plain English Guide to the Science of Climate Change.* Los Angeles: Reason Public Policy Institute.
Green, K., et al. (1999). *Climate Change Policy Options and Impacts: Perspectives on Risk Reduction, Emissions Trading, and Carbon Tax.* Los Angeles: Reason Public Policy Institute.
Hansen, J., et al. (1998). "Climate Forcings in the Industrial Era." *Proceedings of the National Academy of Science* 95:12753–12758.
Intergovernmental Panel on Climate Change (IPCC). (1995). *Climate Change 1995, Impacts, Adaptations and Mitigation of Climate Change: Scientific-Technical Analyses.* Cambridge, Eng.: Cambridge University Press.
Intergovernmental Panel on Climate Change (IPCC). (1995). *Climate Change 1995, The Science of Climate Change.* Cambridge, Eng.: Cambridge University Press.
Kerr, R. A. (1995). "Studies Say—Tentatively—That Greenhouse Warming is Here." *Science* 268:1567.
Kerr, R. A. (1996). "A New Dawn for Sun-Climate Links?" *Science* 271:1360.
Kerr, R. A. (1997). "Did Satellites Spot a Brightening Sun?" *Science* 277:1923.
Lean, J., and Rind, D. (1998). "Climate Forcing by Changing Solar Radiation." *Journal of Climate* 11:3069–3094.

Manahan, S. E. (1984). *Environmental Chemistry*, 4th ed. Monterey, CA: Brooks/Cole Publishing Company.

Rind, D. (1998). "Just Add Water Vapor." *Science* 281:1152.

Schroeder, S. R., and McGuirk, J. P. (1998). "Widespread Tropical Atmosphere Drying from 1979 to 1995." *Geophysical Research Letters* 25(9):1301–1304.

Spencer, R. W., and Christy, J. R. (1992). "Precision and Radiosonde Validation of Satellite Gridpoint Temperature Anomalies, Part 11: A Tropospheric Retrieval and Trends during 1979–90." *Journal of Climate* 5:858–866.

Taubes, G. (1997). "Apocalypse Not." *Science* 278:1004–1006.

Watson, R. T.; Meira Filho, L. G.; Sanhueza, E.; and Janetos, A. (1992). "Greenhouse Gases: Sources and Sinks." In *Climate Change 1992*, ed. J. T. Houghton, B. A. Callander, and S. K. Varney. Cambridge, Eng.: Cambridge University Press.

Weyant, J. P., and Hill, J. N. (1999). "Introduction and Overview." In "The Costs of the Kyoto Protocol: A Multi-Model Evaluation." *The Energy Journal* (special issue).

Willson, R. C. (1997). "Total Solar Irradiance Trend During Solar Cycles 21 and 22." *Science* 277:1963–1965.

COAL, CONSUMPTION OF

INTRODUCTION AND BACKGROUND

Coal, essentially fossilized plant material, has been used as an energy source for centuries. As early plants decomposed in the absence of oxygen (for example, at the bottom of deep lakes), oxygen and hydrogen atoms broke off the long organic molecules in the plant material, leaving mostly atoms of carbon, with some other impurities. Formed as long as ago as 300 million years and more, during the Carboniferous Period (named for obvious reasons), coal's main useful component is carbon.

Coal comes in various forms. Older coal is harder, higher in energy content and has a higher proportion of carbon. Anthracite is the hardest, purest version, with a carbon content of about 90 percent. It is also rarer, found in the United States almost exclusively in eastern Pennsylvania. Its energy content is about 25 million Btus per short ton, 67 percent higher than lignite, the softest form of coal, which sometimes contains less than 40 percent of carbon (most of the balance being water and ash—the latter composed mostly of sodium carbonate and potassium carbonate). Lignite is much younger than anthracite. Some

lignite deposits are less than 100 million years old (a third the age of most anthracite), being formed from plants that lived in the Cretaceous and Tertiary Eras. Not all of its impurities, in the form of water, carbonates, or independent atoms of oxygen and hydrogen have had the time to be removed by geological and chemical processes.

Bituminous coal—an intermediate form—is and by far the most widely found and used. Its energy content averages around 21.5 million Btus per short ton. Some forms of coal have a substantial sulfur content and contain other impurities as well, chiefly carbonates. Peat, also used as an energy source through burning, can be viewed as an earlier stage of coal: pressure in the absence of oxygen over a period of tens of millions of years will change peat into coal.

The oxidation of carbon through burning produces a significant amount of heat through an exothermic reaction, a chemical reaction that releases energy. However, since most of the energy gained from burning coal results in the production of carbon dioxide (CO_2), this principal greenhouse emission is higher for coal than for other fossil fuels, namely oil and natural gas, in which other elements are oxidized in addition to carbon. Because the burning of coal produces so much carbon dioxide, its use will be severely affected by compliance with the Kyoto Protocol, discussed in a following section.

The other fuels, when burned, also produce water vapor and—in the case of oil—other hydrocarbons, as well as carbon dioxide. For equal amounts of energy, oil produces about 80 percent of the CO_2 that coal does; natural gas only produces 55 percent of coal's CO_2 level.

EARLY HISTORY OF COAL USE

According to the U.S. Energy Information Administration, although scattered use of coal may have occurred as early as 1100 B.C.E., it was substantially later before coal became widely used. Nevertheless, coal has been utilized for millenia. In China, as early as the fourth century C.E., coal sometimes substituted for charcoal in iron smelting. By the eleventh century, coal had become the most important fuel in China, according to *The Columbia History of the World* (Garrity and Gay, 1972). In the Middle Ages, in various parts of Europe, especially England, coal began to be used as an energy source

Older railroad engines, like the Scottish Railway 4-6-0 engine here, were major consumers of coal. (Corbis-Bettmann)

for smelting, forges, and other limited applications. But it was not until the fifteenth century that it began to be used for residential heating.

England's good fortune in having large deposits of coal was particularly important for that nation, as the English had destroyed most of their forests between the twelfth and sixteenth centuries, in order to produce heat (chiefly residential), as well as charcoal for industrial purposes. By 1840, Britain's coal production was ten times that of Prussia and significantly higher than that of France and other European nations. The availability of coal, along with various inventions, such as the Watt steam engine, which was usually coal-driven, helped the Industrial Revolution to begin in Britain in the last quarter of the eighteenth century. The rest of Europe began catching up a few decades later. But still, around 1870, Britain alone produced over 30 percent of the manufactured goods in the world.

Coal is used to produce coke, manufactured by heating coal in the absence of air. Coke, when heated to high temperatures, yields carbon monoxide, which reduces the iron oxides in ore to iron. Steel contains a small quantity of carbon in an alloy with iron, and coke refining adds carbon to the iron, in addition to refining the ore. As a result of their supply of coal, the British had the capacity to smelt considerably more iron and later, steel, than other countries, resulting in military and economic advantages.

With the development of improvements in the steam engine and other inventions useful for manufacturing—often in the area of textiles—the British were also able to expand new, energy-intensive industries more rapidly than their competitors. At the beginning of the nineteenth century, switching from renewable (animal, wind, hydroelectricity, and mainly wood) energy resources to coal became the hallmark of strong, expanding economies.

HEALTH EFFECTS OF COAL COMBUSTION

Some effects of massive coal burning are quite pernicious, particularly in the absence of strong efforts to reduce the sulfates and other pollutant byproducts. The London "pea soup" fog, conjuring up mystery, intrigue, and respiratory disease, was largely due to the intensive use of coal in conjunction with stagnant and humid meteorological conditions. This local weather effect has virtually vanished in London since the advent of centrally generated electricity as a substitute for coal fires in each individual living area. Much of this electricity is now fueled by nuclear energy among other sources. Another mitigating factor has been the introduction of anti-pollution measures at coal-burning plants, including limestone scrubbers in smokestacks and fluidized bed reactors for coal combustion.

While it was prevalent, the persistent presence of coal-generated smog was found to have serious health effects on the public. This became apparent on days when air pollution was especially bad, and hospital admissions for respiratory ailments increased considerably. The classic example is the "Great London Smog" of December 4, 1952, to which was attributed a net increase of about 4,000 deaths in the following several days. More recently, in Czechoslovakia and East Germany, following the close of the Cold War, researchers discovered that in several severely impacted locations, the rate of upper respiratory disorders—especially among children—was extremely high.

The power plant in the background burned coal as fuel. (Corbis-Bettmann)

REDUCED COAL EMISSIONS THROUGH TECHNOLOGICAL ADVANCES

Over the past decades, advances have been made that reduce environmental impacts of coal burning in large plants. Some are standard and others experimental. Limestone (mainly calcium carbonate) scrubber smokestacks react with the emitted sulfates from the combustion and contain the chemical products, thereby reducing the release of SO_x into the atmosphere by a large factor (of ten or more). Pulverization of coal can also allow for the mechanical separation of some sulfur impurities, notably those in the form of pyrites, prior to combustion. Currently deployed—with more advanced versions in the development stage—are various types of fluidized bed reactors, which use coal fuel in a pulverized form, mixed with pulverized limestone or dolomite in a high temperature furnace. This technique reduces sulfate release considerably. There are pressurized and atmospheric pressure versions of the fluidized bed.

Each technique has its own advantages, but the pressurized version, operating at high temperature, also makes use of jets of the combustion gases to keep the pulverized fuel/limestone mixture in a suspension, later employing both waste heat and a turbine, driven by a jet of combustion gases, to extract more of the energy generated by the reaction. Design engineers hope to increase efficiencies from the standard level of about 33 percent to 40 percent and even, possibly, approaching 50 percent. If successful, this increase in efficiency could have a major positive impact on carbon dioxide emissions, since up to 50 percent more energy could be extracted from the same amount of oxidized coal. Thus, these technologies, while greatly reducing sulfate emissions, can contribute to reducing carbon dioxide emissions by increases in efficiency.

Nevertheless, in the minds of much of the public

and of decision–makers, there is a preference for substituting coal with natural gas (which is primarily methane, CH_4) to the degree that is easily feasible. Natural gas produces far less of most types of pollutants, including SO_x, NO_x, and CO_2. The tendency away from coal is mitigated by the fact that coal is much cheaper as a fuel (although the capital cost of building a coal plant is higher). For electricity plants that are already built and whose capital costs are sunk, coal's advantage is clear: it costs $0.85 per million Btu, compared with $2.18 per million Btu for natural gas and $2.97 per million Btu for crude oil.

U.S. COAL CONSUMPTION: QUANTITATIVE EVOLUTION

Throughout the industrialized world over the past two centuries, coal became relied upon as an energy source for industrial processes and for residential heat. In the United States, all the coal consumed before the year 1800—much of it imported from Britain—amounted to only 108,000 tons, which is one ten-thousandth of current annual U.S. production. Until 1840, wood exceeded coal as an energy source. However, coal then began a slow, steady expansion in usage, and, for over a century, until 1951, it was the chief energy source in the United States, contributing in the area of transportation (railroads) as well as the earlier, familiar sectors of industrial processes and residential heat.

With the discovery of oil in Pennsylvania in 1859, the seeds of a new energy era were sown. After an initial period of growth from a zero base, oil began to contribute significantly to the energy budget around the turn of the century. In the years following, oil slowly substituted for coal in transportation, starting with automobiles (where coal-fired steam engines had never been very successful) and continuing in trains. Later, it also displaced coal in residential heat and many industrial processes as well as, to a degree, in electricity generation. The "cleaner" energy sources that smelled better than coal began to be substituted for it, particularly in the post–World War II period. From 1952 to 1983, coal lost its primary position, and oil and natural gas vied for the title of main energy source in the United States (oil always remained ahead). But since 1984, coal has regained the lead in domestic production, although not consumption, due mainly to its renewed leading role in electricity generation. The development of cleaner combustion meth-

Year	Coal	Gas	Oil	Hydro	Nuclear
1950	12.35	5.97	13.31	1.44	0.00
1955	11.17	9.00	17.25	1.41	0.00
1960	9.84	12.39	19.92	1.61	0.01
1965	11.58	15.77	23.25	2.06	0.04
1970	12.26	21.79	29.52	2.65	0.08
1975	12.66	19.95	32.73	3.22	1.90
1980	15.42	20.39	34.20	3.12	2.74
1985	17.48	17.83	30.92	3.36	4.15
1990	19.11	19.30	33.55	3.09	6.16
1995	20.11	22.16	34.66	3.44	7.18
1996	20.99	22.59	35.72	3.88	7.17

Table 1.
U.S. Energy Consumption by Source, 1950–1996
NOTE: Quads = Btu $\times 10^{12}$
SOURCE: Energy Information Administration, U.S. Department of Energy. (1988). *Annual Energy Review*. DOE/EIA-0384(87); Energy Information Administration, U.S. Department of Energy. (1988). *Monthly Energy Review, July 1988*. DOE/EIA-0035(88/07);. Energy Information Administration, U.S. Department of Energy. (1997). *Annual Review of Energy 1997*.

ods removed some of the political objections to obvious coal pollution, in the form of sulfates and particulates. These methods could be introduced economically, due to the advantages of scale inherent in the centralized nature of electricity production.

Since coal consumption bottomed out in the 1960s, there has been a steady increase in the fraction of energy produced by coal in the United States. About 90 percent of coal use in 1996 was for electricity generation, and most of the rest of its usage was accounted for in industrial coking processes. Nuclear power had a chance to substitute for coal in electricity generation and, to an extent, made serious inroads until encountering significant political and economic problems related to perceptions of public safety and to the increasing cost of nuclear facilities. Following the Three Mile Island nuclear accident in 1979, the use of nuclear power as an energy source was strongly opposed by many public interest groups. No new plants in the United States that were ordered after 1973 have been completed. Thus, it is unlikely that the United States will see widespread replacement of coal–fired electric plants by the nuclear option for the foreseeable future.

The Organization of Petroleum Exporting Countries (OPEC) oil embargo in 1973 resulted in temporary oil shortages in the United States. This crisis focused attention on the need to minimize oil's use except for transportation, for which no other energy source would provide a very effective substitute. As a result, coal use has increased in both fractional and absolute terms since the 1970s and now is the single largest domestically produced energy source in the United States. In 1996, some 21 quads of energy were produced by coal, and in 1997, coal energy consumption rose to 23 quads. About one-third (31%) of all domestic energy production in the United States now comes from coal, which also produced a majority (52%) of the nation's electricity in 1997. Coal accounted for over one–quarter of energy consumption, since a large amount of petroleum is imported from other countries. In addition, nearly 10 percent of U.S. coal production is exported. Table 1 shows the trends in energy consumption by source in the United States.

WORLD COAL CONSUMPTION

The recent history of the world use of coal roughly follows that of the United States for two reasons. First, the United States and the industrial nations have had, in the aggregate, similar energy behavior in terms of energy sources. Second, the United States itself accounts for about one quarter of world energy use. Thus, world energy use patterns reflect, to a considerable degree, those of the United States.

World coal usage, inclusive of the three major types of coal—anthracite, bituminous (by far the most prevalent form) and lignite—reached a plateau in the first decade of the twentieth century and climbed only very slowly in the half century that followed. By 1880, coal use had equaled wood use on a worldwide basis. The usage around the turn of the century was on the order of 2.2 gigatons per year (around 55 quads), of which about 600 million tons were in the United States. World oil production progressively supplemented the use of coal between 1900 and 1950, increasing by more than an order of magnitude in that period of time, from a little over a quad to some 20 quads. Coal's increase over those years was fractionally much less.

After 1930, four other energy sources began to contribute significantly, as wood use continued its slow decline and coal production was relatively flat. These four were oil, natural gas, nuclear power (beginning in the 1950s), and hydroelectricity. The

Energy Source	1970	1995	2010	2020
Oil	97.8	142.5	195.5	237.3
Natural Gas	36.1	78.1	133.3	174.2
Coal	59.7	91.6	123.6	156.4
Nuclear	0.9	23.3	24.9	21.3
Renewables	12.2	30.1	42.4	50.2
Total	206.7	365.6	519.6	639.4

Table 2.
Summary and Projections of World Energy Consumption (1970–2020)
NOTE: Quads = Btu $\times 10^{12}$
SOURCE: Energy Information Administration, U.S. Department of Energy. (1998). *International Energy Outlook 1998*. DOE/EIA-0484(98). Washington, DC: U.S. Government Printing Office. (<http://www.eia.doe.gov/oiaf/ieo98>)

latter two are relatively small players, but oil and natural gas are major sources of energy, with oil energy production actually exceeding coal in the 1960s.

The first oil crisis in 1973 was a politically–driven event. It resulted from production cutbacks by oil–producing nations following the Yom Kippur War between Egypt and Israel and marked a watershed in patterns of energy use in the industrialized world—especially, as previously noted, in the United States. Oil was saved for transportation to the degree possible, when it became evident that there would be an increased reliance on potentially unreliable foreign sources as domestic sources were depleted. Energy conservation achieved a strong boost from this traumatic event, in which oil prices rose sharply and the uncertainty of oil sources became clear to First World nations.

Since the early 1990s the United States has imported more oil than it has produced for its own use. And, as the nuclear option became frozen, coal has become the chief source for generating electricity, which itself accounts for about 35 percent of the energy sector. In 1997, 52 percent of electricity produced in the United States was generated from coal and in other recent years the fraction has approached 56 percent. Since the United States accounts for one–quarter of total world energy usage, the increase in coal use in the United States alone has a significant impact on worldwide statistics.

EFFECTS OF THE KYOTO PROTOCOL

The U.S. Department of Energy (DOE) has analyzed and projected energy use by sector and energy source for a number of years. One recent forecast (see Table 2) analyzed the implications of the international agreement in the Kyoto Protocol to the United Nations Framework Convention on Climate Change. The United States committed, under the Protocol (Annex B), to reduce greenhouse emissions by 7 percent from 1990 levels by the period between 2008 and 2012. This translated to a 31 percent decline in the production of greenhouse gases (chiefly carbon dioxide) relative to the DOE's assessment of the most likely baseline number, for U.S. energy use and related carbon dioxide emissions predicted for that time.

Other industrial states have committed to reductions nearly as large. However, the world's chief user of coal, China, as a developing nation, has not yet made solid commitments to reduction. Neither has India, another major coal burner, although it uses less than one-quarter as much as China. Limitations on emissions by these emerging world powers are still the subject of discussion. It is clear that, without some limitations on CO_2 emissions by major Third World industrializing nations, the goal of rolling back world greenhouse emissions to the 1990 level will be very difficult to achieve.

In 1995, about 92 quads of coal-fired energy were consumed in the world. This constituted close to one-quarter of the 366 quads estimated to be the world's total energy production. The level of coal use will be a major determining factor in whether greenhouse emission goals for 2010 will be met. Assessing the likely state of affairs for coal use at that time requires predicting the state of the world's economies by region as well as estimating probable technological advances by then.

By the year 2010, one DOE model predicts a world energy level of 520 quads per year. Countries already using relatively large quantities of energy will contribute less to the increase in the total energy use than will large, rapidly developing countries, such as China. The projected U.S. excess of carbon dioxide release over that permitted by the Protocol will be on the order of 550 million tons of carbon per year (1803 million metric tons, rather than 1252). Of course, this projected excess would arise from all fossil fuel use, not just that of coal. However, for comparison, in coal equivalent energy, this amounts to some 700 million tons, or about 20 quads, which amounts to 18 percent of the total

projected energy use for the United States in that year. It would come to nearly 90 percent of current coal consumption.

Looking at the allocation of energy production among sources, the Kyoto Protocol is meant to increase preference for those energy sources that do not produce carbon dioxide and, secondarily, for those that produce much less than others. There will be a particular disincentive to use coal, since this source produces the most CO_2 per energy produced.

The most likely substitute for coal is natural gas, which, as noted earlier, releases about 55 percent of the amount of carbon dioxide that coal, on the average, does. In addition, it produces far fewer other pollutants, such as sulfates and polycyclic hydrocarbons, than coal and oil yield on combustion.

Current U.S. coal consumption is just under 1 billion short tons per year—second highest in the world—after China, which produces some 50 percent more. By 2010, the projected U.S. baseline energy case (in the absence of any attempt to meet Kyoto Protocol limits) would raise this level to 1.25 billion tons, in rough numbers. Coal currently accounts for about one-third of all United States carbon dioxide emissions. If the same patterns of energy source use were to hold in 2010—and if one wished to reduce the carbon dioxide emissions by 30 percent, while making no reductions in usage of other fossil fuels—this would mean reducing coal emissions by 90 percent. Such a scenario is clearly highly unlikely, even if one were to take much longer than 2010 to accomplish this goal.

Moreover, substituting 90 percent of coal with natural gas would reduce the level of emissions only by $0.90 \times 0.45 = 0.38$, which is still less than half way to the goal of reducing emissions by an amount equal to that produced by 90 percent of U.S. coal use.

Therefore, merely reducing coal use will not be sufficient to satisfy the Protocol. Any plan to comply with the Protocol needs to assume substitution, first by non-combustion energy sources—that is by renewables or nuclear energy—and second by natural gas. This would have to be accompanied by achievement of far greater efficiencies in energy production (for example by introduction of far more fuel-efficient steam gas turbines, driven by natural gas) and by more efficient use of energy.

The remaining possibility for the United States, under the Kyoto Protocol, would be to compensate for the excess of carbon emissions over the committed goal by either planting more trees, in the United States or elsewhere, or by purchasing carbon "pollution rights," as envisioned by the protocol, from other countries. How either of these schemes will work out is in some question. Both would require bilateral agreements with other countries, probably many other countries. The template of the trade in acid rain "pollution rights" to help all parties meet agreed–upon goals may not be a good analogy for carbon emissions, since acid rain, although international, is generally a regional, not a global problem. Further, carbon sources and sinks are not as well understood as are the sulfate and nitrate sources (chiefly coal) that are responsible for acid rain. This uncertainty will make it more difficult to achieve the international agreements necessary to make the "pollution trade" work as a widely-accepted convention, necessary due to the global nature of the problem.

Anthony Fainberg

See also: Coal, Production of; Coal, Transportation and Storage of; Environmental Problems and Energy Use; Fossil Fuels.

BIBLIOGRAPHY

Bethe, H. A., and Bodansky, D. (1989). "Energy Supply." In *A Physicist's Desk Reference,* ed. Herbert L. Anderson. New York: American Institute of Physics.

Borowitz, S. (1999). *Farewell Fossil Fuels.* New York: Plenum Trade.

Durant, W., and Durant, A. (1976). *The Age of Napoleon.* New York: Simon and Schuster.

Energy Information Administration, U.S. Department of Energy. (1996). *Coal Industry Annual 1996.* Washington, DC: U.S. Government Printing Office.

Energy Information Administration, U.S. Department of Energy. (1997). *Annual Review of Energy 1997.* Washington, DC: U.S. Government Printing Office.

Energy Information Administration, U.S. Department of Energy. (1997). *International Energy Database, December 1997.* Washington, DC: U.S. Government Printing Office.

Energy Information Administration, U.S. Department of Energy. (1998). *International Energy Outlook 1998.* DOE/EIA-0484(98). Washington, DC: U.S. Government Printing Office.

Garrity, J. A., and Gay, P., eds. (1972). *The Columbia History of the World.* New York: Harper & Row.

Howes, R. and Fainberg, A. eds. (1991). *The Energy Sourcebook.* New York: American Institute of Physics.

United Nations Framework Convention on Climate Change. (1997). *Kyoto Protocol.* New York: United Nations.

COAL, PRODUCTION OF

GEOLOGY OF COAL

Coal is a fossil fuel—an energy source whose beginnings can be traced to once-living organic materials. It is a combustible mineral, formed from the remains of trees, ferns, and other plants that existed and died in the tropical forests 400 million to 1 billion years ago. Over vast spans of time, heat and pressure from Earth's geological processes compressed and altered the many layers of trees and plants, slowly transforming these ancient vegetal materials into what we know as coal today. The several kinds of coal now mined are the result of different degrees of alteration of the original material.

It is estimated that approximately 0.9 to 2.1 m of reasonably compacted plant material was required to form 0.3 m of bituminous coal. Different ranks of coal require different amounts of time. It has been estimated that the time required for deposition of peat sufficient to provide 0.3 m of the various ranks of coal was: lignite, 160 years; bituminous coal, 260 years; and anthracite, 490 years. Another estimate indicates that a 2.4 m bed of Pittsburgh Seam (bituminous) coal required about 2,100 years for the deposition of necessary peat, while an anthracite bed with a thickness of 9.1 m required about 15,000 years.

Depending on the environment in which it was originally deposited, coal will have higher sulfur content when it was formed in swamps covered by seawater; generally, low-sulfur coal was formed under freshwater conditions. Although coal is primarily carbon, it's complex chemical structure contains other elements as well—hydrogen, oxygen, nitrogen, and variable trace quantities of aluminum, zirconium, and other minerals.

COAL RANK

Coal is a very complex and diverse energy resource that can vary greatly, even within the same deposit. The word "rank" is used to designate differences in coal that are due to progressive change from lignite to anthracite. Generally, a change is accomplished by increase in carbon, sulfur, and probably in ash. However, when one coal is distinguished from another by quantity of ash or sulfur, the difference is

A longwall shearer cuts swatches of coal 750 feet long and 28 inches thick, 800 feet underground at Consolidation Coal Company's Blacksville #2 mine. (Corbis-Bettmann)

said to be of grade. Thus a higher-grade coal is one that is relatively pure, whereas a higher-rank coal is one that is relatively high on the scales of coals, or one that has undergone devolatization and contains less volatile matter, oxygen, and moisture than it did before the change occurred.

In general, there are four ranks of coal: lignite, subbituminous, bituminous, and anthracite. Lignite, a brownish-black coal with generally high moisture and ash content, as well as lower heating value, is the lowest-rank coal. It is an important form of energy for electricity generation. Some lignite, under still more pressure, will change into subbituminous, the next higher rank of coal. It is a dull black coal with a higher heating value than lignite and is used primarily for generating electricity and space heating. Even greater pressure will result in the creation of bituminous, or "soft" coal, which has higher heating value than subbituminous coal. Bituminous coals are primarily used for generating electricity. Anthracite is formed from bituminous coal when great pressure developed during the geological process, which occurred only in limited geographic areas. Sometimes referred to as "hard" coal, anthracite has the highest energy content of all coals and is used for space heating and generating electricity.

In addition to carbon, all coals contains many non-combustible mineral impurities. The residue from these minerals after coal has been burned is called ash. Average ash content of the entire thickness of a coal seam typically ranges from 2 to 3 percent, even for very pure bituminous coals, and 10 percent or more for many commercial mines. These materials, which vary widely in coal seams with respect to kind, abundance, and distribution, among from shale, kaolin, sulfide, and chloride groups.

WORLD COAL RESERVES AND PRODUCTION

Coal is the most abundant and most economical fossil fuel resource in the world. Proven coal reserves exceed 1 trillion tons, and indicated reserves are estimated at 24 trillion tons. Coal is found in every continent of the world, including Antarctica, although the largest quantities of coal are in the Northern Hemisphere. Coal is mined in some sixty countries in nineteen coal basins around the world, but more than 57 percent of the world's total recoverable reserves are estimated to be in the United States, and China, which together account for more than two-thirds of the world's coal production.

COAL MINING METHODS

Depending on the depth and location of the coalbed and the geology of the surrounding area, coal can be mined using either surface or underground methods. In the United States, coal is usually mined underground if the depth of the deposit exceeds 200 ft.

In surface mining, the covering layers of rock and soil (called "overburden") are first removed using either a power shovel, a dragline (for large surface mines), or bulldozers and front-end loaders (for small mines). Front-end loaders also can be used to load coal. In large mines, coal usually is loaded using power shovels and hydraulic shovels. Depending on the size of the mine, shovels and draglines ranging from 4 cu m to 50 cu m are usually used for loading and excavating. Large-capacity haul trucks, usually in the range of 170 to 240 mt but possibly as big as 320 mt, are then used to transport coal to loading stations for shipping and sold as raw coal, or to a preparation plant for further processing. For post mining reclamation, draglines are used.

Depending on geologic conditions and surrounding terrain, there could be several types of surface mining. If the coal seam is of the same depth in flat or gently rolling land, area mining is developed where the overburden from one cut is used to fill the mined-out area of the preceding cut. Contour mining and mountain-top removal are methods that follow a coalbed along the hillsides. The overburden is cast (spoiled) down-

A dragline at Atlantic Richfield's Black Thunder strip mine in Gillette, Wyoming, loads coal into a dump truck. The coal is transported to cities around the country where it is burned to generate electricity. (Corbis-Bettmann)

hill from this first pit, exposing the coal for loading by trucks. The second pit could then be excavated by placing the overburden from it into the first pit. Digging starts where the coal and surface elevations are the same and precedes toward the center of a hill or mountain until the overburden becomes too thick to remove economically. An open pit combines the techniques of contour and area mining and is used where thick coalbeds are steeply inclined.

After coal is extracted, the pit is backfilled with earth and subsequently reclaimed or restored to its approximately original contour, vegetation, and appearance.

The use of underground mining methods requires integration of transportation, ventilation, ground control, and mining methods to form a system that provides the highest possible degree of safety, the lowest cost per ton of product, the most suitable quality of final product, the maximum possible recovery of coal, and the minimum disturbance of environment. Depending on the location of coal deposits, there can be three different types of underground mines: a drift mine is one in which a horizontal (or nearly horizontal) coal seam crops to the surface in the side of a mountain, and the opening of a mine can be made into the coal seam. Transportation of coal to the outside can be by track haulage, belt conveyor, or rubber-tired equipment.

A slope mine is one in which the coal is of moderate depth and where access is made through an inclined slope (maximum, 16°). This type of mining also may follow the coalbed if the coal seam itself is included and outcrops, or the slope may be driven in rock strata overlying the coal to reach the coal seam. Either a belt conveyor (no more than 30% grade), coal trucks (maximum grade, 18%), or electrical hoist if the slope is steep, can be used to transport coal out of the mine.

When the coal seam is deep, a shaft mine is used because the other two types of access are cost-pro-

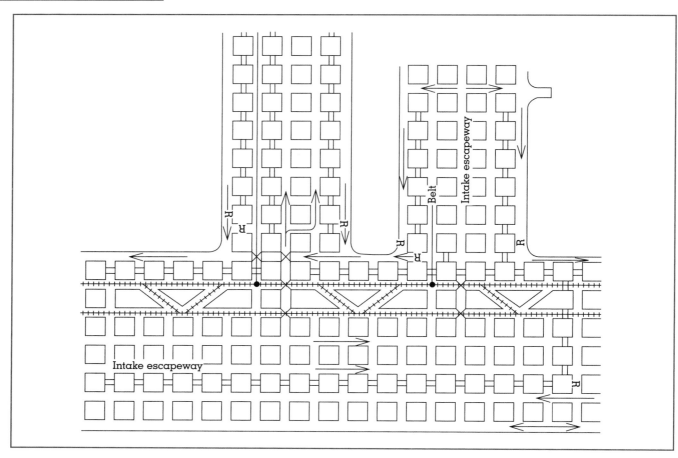

Figure 1.
A room-and-pillar mining system. Pillars are left behind in the rooms to support the roof.

hibitive. Vertical shafts are drilled for both production and ventilation.

Production methods underground are generally classified according to the types of mining equipment used (conventional, continuous mining, or longwall) or by the method in which coal is being extracted (longwall or longwall caving). Both conventional and continuous mining are room-and-pillar systems; even the longwall method uses room-and-pillar during development.

In the conventional mining system, the coal face is first undercut, center cut or top cut using a cutting machine that most nearly resembles a large chain saw on wheels. The outlined coal blocks are drilled in a predetermined drill pattern using a mobile powered drill, with holes charged with explosives, and the coal is dislodged. The broken coal is gathered by a loading machine onto a shuttle car and dumped onto a nearby belt, to be transported out of the mine.

In the continuous mining method, a continuous

mining machine (also referred to as a continuous miner) is employed in the extraction process. This machine combines several extracting functions into one continuous process: cutting, loading, and tramming, thereby tearing the coal from a seam and automatically removing it from the area by a machine-mounted conveyor onto a shuttle car, which is used to transport the mined coal to a dumping station, then transported out of the mine using a conveyor belt. Remote-controlled continuous miners allow an operator to control the machine from a distance, increasing safety. The mine roof is further secured using wooden timbers; steel crossbars on posts; or, most commonly, roof bolts.

Both conventional and continuous mining methods use a room-and-pillar system in which the coal is mined by extracting a series of "rooms" into the coalbed, and leaving "pillars," or columns, of coal to help support the mine roof (Figure 1). Depending on the location, the rooms are generally 20 to 30 ft. wide

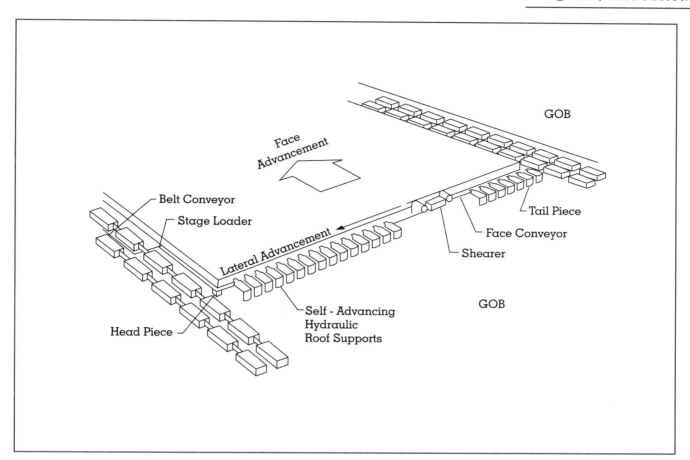

Figure 2.
A typical setup. The panel is 8,000–10,000 feet long. The face is 800–900 feet wide.

and the pillars 20 to 90 ft. wide, with the height determined by the thickness of the coal seam. In the not-too-distant future, robotic versions of these machines, now under development, will allow for enhanced automatic operations and even greater efficiencies than now possible. Although still utilized in stand-alone production operations, continuous miners also are employed for main entry and longwall panel developments.

As a rule of thumb, 50 to 55 percent of coal can be extracted using continuous mining. To improve this extraction ratio, a pillar-recovery process usually is applied when mining reaches the end of the panel and the direction of the mining is reversed. The continuous miner mines into the pillars, recovering as much coal as possible, as the roof is allowed to systematically collapse. Usually this can increase the extraction ratio by up to 5 percent.

Although the development of the continuous mining system in the 1950s consolidated several opera-

tions in one machine and have greatly improved coal production, it is still not fully "continuous," as the face haulage and roof support operations remain as major impediments to truly continuous production.

The introduction of the longwall system has provided not only continuous cutting and loading but also continuous haulage and roof support. In the longwall mining system, large blocks of coal, outlined in the development process, are completely extracted in a single, continuous operation.

The longwall consists of a panel of coal, usually 8,000 to 10,000 ft in length and 800 to 900 ft in width (Figure 2). In the face area, a rotating drum (or a plow) is dragged mechanically back and forth across a wide coal seam. The loosened coal falls onto a conveyor for removal from the mine. The system has its own hydraulic roof supports, which advance with the machine as mining proceeds. The supports provide not only high levels of production but also increased miner safety. Newer versions of the longwall system employ sensors to detect the amount of coal remaining in the seam being mined, as well as robotic controls to enhance efficiency. As the face advances after the coal is mined, the roof is systematically allowed to cave behind to form a gob. In general, longwall systems can provide an extraction ratio of up to 80 percent.

Longwall mining has helped revolutionize underground coal mine operations in the past two decades, with its share of total U.S. underground production increasing from 10 percent to 48 percent, surpassing continuous mining tonnage in 1994, and the trend has held true since then.

Several modified versions of longwall methods also are practiced in areas where the coal seam is either thick or steeply inclined. Since the maximum height a shearer can reach is about 14 ft, thicker coal seams have to be mined using either a multiple pass method, where the top seam is mined followed by a lower pass, or a longwall caving method, where the lowest seam is mined using the traditional longwall method and the upper portion of the seam is allowed to cave under gravity; coal is then collected behind the shield support and shipped out of the mine.

COAL MINING AND THE ENVIRONMENT

It has been said that mining is a temporary use of the land. While mining does disturb the land, modern technologies and increased application of environmentally safe mining methods in the United States and other major mining countries have enabled today's coal mining industry to provide the valuable energy resources modern society requires without destroying the environment in the process. In the United States, stringent environmental regulations mandate specific standards for reclamation, quality of water discharge, and other mining practices that may disturb the land.

While some problems still exist, there is no question that coal mining operations are more efficient and safer for workers and leave less of an environmental footprint than operations several generations ago. As society's demand for energy from coal continues to increase and as coal's price declines (between 1978 and 1996 U.S. mine mouth prices fell from $47.08 to $18.50 per ton in constant 1996 dollars), there is certain to be even greater efforts to limit the environmental impact of mining operations.

Jerry C. Tien

See also: Coal, Consumption of; Coal, Transportation and Storage of.

BIBLIOGRAPHY

Coleman, L. L. (1999). *International Coal*, 1998 ed. Washington, DC: National Mining Association.

Fiscor, S. (1998). "U.S. Longwall Thrive" *Coal Age*, February, pp. 22–27.

Hower, J. C., and Parekh, B. K. (1991). "Chemical/Physical Properties and Marketing." In *Coal Preparation*, 5th ed., ed. J. W. Leonard and B. C. Hardinge. Littleton, CO: SME.

Katen, K. P. (1982). "Modern Mining Methods—Longwall, Shortwall," In *Elements of Practical Coal Mining*, 2nd ed., ed. D. F. Crickmer and D. A. Zegeer. New York: Society of. Mining Engineers/American Institute of Mining and Metallurgy.

National Mining Association. (1998). *Facts about Coal: 1997–1998*. Washington, DC: Author.

Reid, B. (1998). "Longwall Production at Record Pace." *Coal Leader* 32(9):1.

Schroder, J. L., Jr. (1982). "Modern Mining Methods—Underground." In *Elements of Practical Coal Mining*, 2nd ed., ed. D. F. Crickmer and D. A. Zegeer. New York: Society of Mining Engineers/American Institute of Mining and Metallurgy.

Simon, J. A., and Hopkins, M. E. (1982). "Geology of Coal," In *Elements of Practical Coal Mining*, 2nd ed., D. F. Crickmer and D. A. Zegeer. New York: Society of Mining Engineers/Institute of Mining and Metallurgy.

Stefanko, R. (1983) "Geology of Coal." In *Coal Mining Technology: Theory and Practice*, ed. C. Bise. New York: Society of Mining Engineers/American Institute of Mining and Metallurgy.

Thakur, P. (1997). "Methane Drainage from Gassy Mines." In *Proceedings, Sixth International Mine Ventilation Congress*, ed. R. V. Ramani. Littleton, CO: SME.

Tien, J. C. (1998). "Longwall Caving in Thick Seams". *Coal Age*, April, pp. 52–54.

COAL, TRANSPORTATION AND STORAGE OF

Coal competes primarily in the market for low-cost boiler fuels. Coal is also characterized by a relatively low energy content per unit of weight (at best two-thirds that of residual oil). Consequently, low-cost and efficient transportation is essential to the competitiveness of coal.

OCEAN TRANSPORTATION

World trade in coal totaled 576 million tons (524 million tons) in 1998, of which 523 million tons (476 million tonnes) shipped in oceangoing vessels. Coal shipments use the same dry bulk vessels that transport other bulk commodities, such as iron ore and bauxite, so vessel rates for coal shipments are hostage to wider market forces. However, the cyclic pattern observable in vessel rates disguises the long-term trend in which rates have varied little in nominal terms. For example, spot vessel coal rates in the 1998–1999 time period were about the same as in the mid-1980s, varying between $5 and $10 per ton.

Coal is generally shipped either in vessels capable of transversing the Panama Canal (Panamax vessels of 60,000 dwt) or Capesize carriers of 200,000 dwt and greater. Vessels may be designed for self-unloading or be "gearless" carriers that require onshore bulk-handling equipment.

BARGE TRANSPORTATION

Coal-carrying barges move in tows of fifteen to forty barges, pulled by a single towboat of 2,000 to 10,000 hp. A "jumbo"-size barge carries 1,800 tons (1,633 tonnes) of coal, so a large tow can move 72,000 tons (65,304 tonnes) of coal, as much as five unit trains. These large volumes result in significant economies of scale. Barge rates can run (on a cost-per-mile or cost-per-kilometer basis) a quarter or less of rail rates.

The primary cost variable in barge shipments is fuel; a midsize towboat can consume 5000 gal (18.9 kl) of diesel fuel daily. Barge shipments are also dependent on weather conditions; low water or frozen rivers and canals can halt shipments. As with ocean vessels, the barges that move coal also ship other bulk commodities, making the rates and availability of barges for coal shipments dependent on conditions in other markets. Backhauls (i.e., shipment of one commodity to a terminal and return with a different product) can substantially reduce coal rates.

Barges receive coal at a dock to which the coal is initially transported by rail or truck. These transloading facilities can play an important role in the coal supply system as intermediate storage sites and by

A cargo barge travels up the Rhine. Though weather conditions can affect timeliness, the economies of scale usually make water transport cheaper than rail. (Corbis Corporation)

providing facilities where different coals can be blended into a custom product.

RAIL TRANSPORTATION

Rail-transported coal is typically moved in unit trains that operate in dedicated shuttle service between a mine and a destination. Unit trains operating in the western United States and Canada consist of 100 to 120 lightweight aluminum railcars carrying upward of 121 tons (110 tonnes) of coal apiece, or more than 14,000 tons (12,700 tonnes) per train. In the 1990s, distributed power (DP) came into widespread use in the western United States. In this system a remotely controlled engine is put into the middle of a train, allowing greater traction and control of train motion. DP trains can consist of 135 cars and are the most efficient method of rail transportation of coal.

Railroad productivity has increased dramatically since the mid-1970s. In part this reflects reform of outdated labor practices, but the technical sophistication of the rail industry is rarely appreciated. Modern systems use microwave Centralized Traffic Control (CTC) systems to move trains safely with minimal between-train clearance, allowing substantial increases in system capacity. Modern diesel electric locomotives rely on microprocessor and alternating-current motor technology to provide enhanced power (6,000-hp class) and greater traction, allowing two or three engines to do the work of five earlier models. All aspects of the rail system, from operations to invoicing, are heavily dependent on computer processing.

Because rail systems exhibit economies of scale,

there is a tendency toward consolidation manifested either in state ownership or merger of privately owned systems into a handful of competitors. High barriers of entry into the rail business allow the exercise of monopoly power over rates to customers with single-rail access. This has been a persistent issue in nations with deregulated rail industries, such as the United States and Canada. Rate complaints by coal mines and consumers have been common in these countries. On the other hand, railroads have had difficulty earning adequate returns on investment, due to intermodal competition, heavy capital investment requirements, and other factors. The tension between shipper demands for low rates and high-quality service, and railroad efforts to improve profitability, was a political controversy in the nineteenth century and continues to be an unresolved public policy issue in the early twenty-first century.

TRUCK TRANSPORTATION

Truck transportation is used to move coal to a transloader for placement onto a water or rail carrier, or for direct shipment to the customer. Trucks have the advantage of routing flexibility and modest capital requirements, but coal can be economically transported for at most about 100 miles (160 km) one-way or less, due to the high unit cost of moving a low-value product in relatively small batches.

Coal-carrying vehicles are typically end-dump trucks with a carrying capacity of roughly 25 to 50 tons (22.7 to 45.4 tonnes), depending on local road conditions and safety regulations. In the 1990s, strides were made toward increasing the productivity of truck operations, such as higher-capacity vehicles. But while these improvements have enhanced the ability of trucks to compete with railroads for short hauls, they have not significantly increased the maximum radius within which truck shipments are economical.

COAL SLURRY PIPELINES

Coal slurry pipelines have been widely discussed, but few slurry pipelines have been built. In addition to the Black Mesa operation in Arizona, a 38-mile (61-km) pipeline was built by the Soviet Union, and a 108-mile (173-km) pipeline in Ohio was mothballed in 1963 after six years of operation. It is arguable to what extent the limited use of slurry pipelines is due to economics or to political opposition from rail car-

riers and interests concerned with water rights.

The most successful slurry operation is the dedicated pipeline that serves the 1,580-MW Mohave Generating Station in southern Nevada. The plant receives all of its coal via a 273-mile (437-km) pipeline built in 1970 that originates at the Black Mesa mine in Arizona. Coarsely ground coal is mixed with water (the slurry is about 47% solids by weight) and pumped through an 18-inch (46-cm) pipe. At the plant the coal is dewatered using centrifuges. The pipeline has a capacity of about 5 million tons (4.5 million tonnes) annually.

STORAGE AND OXIDATION

Storage is necessary at several points in the coal supply chain. Because coal is transported in batches (e.g., a unit train or a vessel), rather than moved continuously through a network, like natural gas, the supply chain must accommodate surges and lulls in demand at the mine; at the origin and receipt dock or port for water shipment; and at the end user, such as a power plant. The global wave of privatization and deregulation, particularly in the electric sector, has increased pressure on logistics managers to make the coal supply chain as seamless as possible to minimize the amount of coal in storage at any time. Stored coal ties up working capital and, as discussed below, can deteriorate and create safety hazards

About 2.5 million tons (2.3 million tonnes) of coal are burned daily in U.S. power plants. This is equivalent to roughly 21,000 railcars in transit, so it is apparent that coordinating production and consumption is no easy task. Accidents, rail strikes, natural disasters (e.g., floods that take out bridges and rail lines) and severe weather (e.g., deep river freezes that halt barge traffic) can all severely disrupt deliveries for utility customers dependent on a reliable coal supply for base load plants. Nonetheless, to reduce costs U.S. utilities have significantly reduced typical inventory levels over time. Whereas a coal inventory of ninety days of supply was once typical, inventories now frequently run in the range of thirty to forty-five days.

Another reason to keep inventories low is the potential for storage problems. Coal in storage must be carefully handled. Improperly stored coal can oxidize (weather), causing a loss of heat content. And if heat is allowed to build up in a stagnant coal pile (or in a vessel, barge, or railcar), the coal can self-ignite.

Self-ignition is particularly a risk with lower-grade subbituminous coals and lignite. To avoid oxidation, coal piles should be turned frequently so that heat can vent, and piles should be packed and shaped to minimize surface exposure.

Stan M. Kaplan

See also: Coal, Consumption of; Coal, Production of; Locomotive Technology; Transportation, Evolution of Energy Use and.

BIBLIOGRAPHY

Fieldston Company, Inc. (1996). *Fieldston Coal Transportation Manual: 1996–97*. Washington, DC: Author.

Schmidt, R. D. (1979). *Coal in America: An Encyclopedia of Reserves, Production, and Use*. New York: McGraw-Hill.

Singer, J. G., ed. (1981). *Combustion: Fossil Power Systems*. Windsor, Conn.: Combustion Engineering Company.

COGENERATION

Cogeneration is the production of two useful forms of energy in a single energy conversion process. For example, a gas turbine may produce both rotational energy for an electric generator and heat for a building.

During the energy conversion process, an energy converter converts some form of energy to a form having a more suitable use. A light bulb and a gasoline engine are two familiar converters. People invest in electric energy to operate a light bulb because light is useful; likewise, people invest in gasoline for energy to run the automobile internal combustion engine because automobiles are useful. The laws of nature require that there be no loss of energy in the conversion. If 100 joules of energy are converted, then 100 joules remain after the conversion. However, the laws of nature neither require the converted energy to be in the form we desire, nor do they require that the other forms be useful. If the converter of 100 joules were a light bulb, only about 10 joules would emerge as light. The other 90 joules would be heat. Touching an ordinary light bulb when lit attests to the heat that is produced.

Efficiency is a practical measure of the performance of a converter: efficiency is equal to the desired form of energy divided by the total energy converted. If the light converted 100 joules of energy into 10 joules of light energy, we would say its efficiency is $10 \div 100 = 0.1$ or 10 percent.

Heat is always produced to some extent in energy conversion. In fact, when energy has gone through all possible conversions, it ends up as thermal energy in the environment. The efficiency of a steam turbine at a large electric power plant is about 50 percent. This means that 50 percent of the energy converted is rejected as heat to the environment by massive cooling towers that are prominent at power plant sites. Heat is a useful energy commodity, so one must wonder why rejected heat is not put to some use. The idea of cogeneration is to do just that.

Evaluating the practical worth of thermal energy in a substance such as water requires consideration of both temperature and the amount of the substance. To understand this we say the thermal energy of a substance is equal to the number of molecules times the energy per molecule. The thermal energy per molecule (i.e., the second factor) increases with increasing temperature. So, even if the temperature is high, making energy per molecule larger, the total will still be small if there are only a few molecules. Similarly, if the temperature is low and a large number of molecules is involved, the total thermal energy can be large. The temperature of the water removing heat from a steam turbine is relatively low—only 10 to 15°C above the temperature of the environment—but a huge amount of water is needed to remove the heat from the turbine, so the thermal energy transferred to the water must be quite large. The thermal energy, although low-grade (about 80°F or 30°C), is appropriate for heating buildings. In a scheme of relatively small scale called district heating, buildings are heated in some towns and cities. But usually a power plant, especially a nuclear power plant, is well removed from the city and the economics of piping the heat to where it is needed is very unfavorable—requiring not only longer runs of piping, but resulting in greater heat loss from those longer runs. Consequently, for remotely sited plants, the thermal energy is rejected to the environment and goes unused.

Industry needs both electricity and heat. It is possible for an industry to produce its electricity from gas-fired turbogenerators and use the rejected heat for industrial purposes. The rejected heat can be at relatively high temperature, making it more useful if some sacrifice is made in the efficiency of the turbogenerator. It is in areas like this that cogeneration has its greatest potential and one sees commercial cogeneration enterprises evolving to provide a growing share of energy production.

Joseph Priest

BIBLIOGRAPHY

Horlock, J. H. (1997). *Cogeneration—Combined Heat and Power: Thermodynamics and Economics.* Malabar, FL: Krieger Publishing Company.

Spiewak, S. A., and Weiss, L. (1994). *Cogeneration and Small Power Production Manual,* 4th ed. Liburn, GA: Fairmont Press.

COGENERATION TECHNOLOGIES

Cogeneration or combined heat and power is the simultaneous production of heat and power in a single thermodynamic process that has a history going back several centuries. Originally employed to save labor, its inherent fuel economy took it to the forefront of the industrial revolution in the nineteenth century. More recently the environmental benefits derived from reduced fuel consumption have made cogeneration a significant factor in global environmental strategies, while current trends towards utility deregulation and distributed power generation continue to bolster the market for this technology.

Cogeneration encompasses several distinct thermodynamic processes of simultaneous heat and power production. One utilizes air as a medium, another steam, a third employs heat rejected from a separate combustion process, such as an internal-combustion engine, and a fourth utilizes a thermochemical process such as found in a fuel cell. Although each process is distinct, they are often combined together to maximize the energy production in a single thermodynamic system.

The oldest form of combined heat and power is the smokejack, developed in Tibet to turn prayer wheels during religious ceremonies. Captured Tartar slaves introduced this device into Europe by the early fourteenth century and Leonardo da Vinci sketched one around 1480. Commentators as diverse as Montaigne (1580), John Evelyn (1675), and Benjamin Franklin

(1758) mention smokejacks, which were small windmills installed inside a chimney and powered by the hot air rising from fires. The rotary motion of the fan was used to power a spit or lathe. The amount of power produced would be dependent on the velocity and mass flow of the heated air and the efficiency of the blades, but in general use the smokejack delivered approximately one dog-power. Turnspit dogs were specifically bred to turn spits and other apparatus requiring rotary motion, although children, slaves, and servants were also pressed into this labor, which was basically a larger version of a hamster in a wheel. Prior to the widespread electrification of farms in the mid–to–late twentieth century, American farms often had similar devices allowing all members of the farm community to contribute to the domestic workload, reminding us that the current leisurely life of our canine friends is a relatively recent phenomenon.

Franklin also noted that the natural draft of a chimney was also able to turn a smokejack, an idea recently promoted on a generative power using a large natural draft chimney with an air turbine. In 1832, Charles Busby used a smokejack to power a pump to circulate water through pipes for "warming and cooling the interior of buildings." It is uncertain if Busby's "Patent Circulation" achieved wide success, although it would have worked well since the flow of exhaust air through the chimney would be directly related to the amount of circulated needed.

By the end of the nineteenth century, smokejacks had evolved into the hot air turbine, which found application as aircraft turbosuperchargers before evolving into gas turbines. Some engines such as the General Electric LM2500 have a separate hot air power turbine that converts hot exhaust air into mechanical power.

An interesting use of air involved the use of compressed air for power distribution in urban areas. Still widely used as a power source within factories, several cities in the mid- to late-nineteenth century had compressed air public utility systems, with Paris being perhaps the largest example. Simple air motors could be installed on a wide variety of equipment, and at least one clothing factory utilized the air exhausted from sewing machine motors to provide ventilation for the sewing machine operators, which also provided cooling since the air expanded passing through the motor.

Despite the widespread use of smokejacks, the industrial revolution could not be sustained by power measured in dog units. Although water and wind had

Electric power cogeneration site. (Greenpeace)

been useful sources of energy for many centuries, they were geographically limited and offered limited, usually seasonal, availability. The solution was to harness the power of steam, of which the earliest example is Heron's rotating steam engine in ancient Greece. Steam power grew from these small-scale applications to larger uses during the eighteenth century, when Savery and Newcomen introduced larger steam engines to pump water out of mines. Savery's engines operated with low pressure (<2 psig) steam and a thermal efficiency of roughly one percent. Large steam engines required large boilers, which in turn created more steam, which Desaugliers in 1720 adapted to manufacturing applications.

James Watt doubled the efficiency of the steam engine by introducing the separate condenser in the 1760s and created a new unit, the horse power, to measure its output. In 1784, Oxford brewer Sutton Thomas Wood obtained a patent for using waste steam from an industrial process to drive a steam engine and also to use the exhaust steam or hot water from a steam engine for heating or manufacturing, marking the first cogeneration patent.

Despite Watt's contribution to the advancement of the steam engine, his reliance on low pressure steam resulted in safe, but large and inefficient engines. Only after his patent rights expired were other inventors, such as Richard Trevithick in England and Oliver Evans in Philadephia, able to build and market high pressure engines that were more efficient, much smaller, and indeed more dangerous than their low pressure ancestors. Higher pressure steam made it possible to operate these engines economically without a condenser, but the exhaust steam was rapidly put to useful purposes.

The idea of cogeneration, fortunately, slowly found its way into many factories. In the mid 1820s a religious community led by George Rapp built a utopian community, Old Economy, on the Ohio River outside Pittsburgh. The Evans engine that had powered their steamboat was reinstalled in a mill and the engine exhaust steam was distributed through pipes to warm the community's buildings in an early example of district heating. By mid-century exhaust steam was widely used in industrial settings both in Britain and America, and English sanitary reformer Edwin Chadwick proposed using the waste heat from factory engines to heat public baths and worker housing. Private entrepreneurs built several such baths, charging a penny for admission with discounts for frequent bathers.

Interestingly, one of the first and greatest works on thermodynamics, Sadi Carnot's 1824 *Reflexions sur la Puissance Motrice du Feu (Reflections on the Motive Power of Heat)* has been one of the foremost impediments to cogeneration practice. Carnot stated flatly that the efficiency of a heat engine was strictly a function of the temperature difference across the engine, which is entirely correct insofar as the engine itself goes but ignores the common situation where the engine is not acting alone in an isolated process. Take, for instance, two Carnot cycle engines, each with the same high temperature condition. One has a low temperature reservoir such as that produced by a large body of water. The reservoir in the other is at a higher temperature, but the heat in the reservoir is then used for a useful purpose, such as space heating. While the first engine has the higher engine efficiency as Carnot postulated, the second might have an overall system efficiency four or five times higher than the first engine. Many students of thermodynamics even today continue to be misled by Carnot's useful, but limited, theories.

As factories grew in size and complexity, managers began to more closely analyze the various cost elements of their businesses. Many enterprises requiring large amounts of power, such as cotton mills, located their factories based on the availability of water power, but a lively public discussion on the relative costs of steam and water power took place in the early 1840s. Several mill owners reported their surprise at discovering that steam power was often no more expensive than water power and often less so, particularly where the exhaust heat of the engine could be utilized for process or space heating. Steam was also much more dependable, and did not rely on the vagaries of weather and excessive demand on available water supplies.

Post–Civil War America witnessed a tremendous growth in steam power, including the 1882 introduction of the central electric station by Thomas Edison. These stations were initially thought of as curious novelties rather than solid investments, a view undoubtedly shared by many early electric pioneers trying to make a profit. Ever mindful of the direct correlation between reducing waste and increasing profits, many electric light plants sold steam and hot water to residential and commercial customers, in many cases making the difference between a profit and loss. In Pennsylvania, state lawmakers regarded cogeneration as unfair competition to traditional utilities and passed a bill outlawing it, although it was quickly repealed when they discovered how widespread it was. By World War I more than four hundred central stations in the United States utilized cogeneration, most small enough to measure their electric power output in kilowatts.

The introduction of large steam turbines, higher boiler pressures and temperatures, and the effects of a war–time energy crisis in America led most utility managers to prefer large power plants located close to coal mines, obviously reducing opportunities to utilize the plant's heat output. Several utilities, however, built plants that could also supply heat in dozens of cities, including New York, Philadelphia, Boston, and Denver, and many utilities supplied heat in order to keep their customers from installing their own small cogeneration plants.

In Europe, industrial use of cogeneration greatly expanded between the wars, while in the new Soviet Union large-scale combined heat and power plants were incorporated into the formal national planning process as the preferred power plant technology. Beginning with the 1930 World Power Conference in Berlin, a lively debate took place for a decade between Soviet engineers promoting large-scale heat and power systems and German engineers arguing for smaller distributed cogeneration schemes. Steam turbines remain the largest source of thermal electric power today and their fuel flexibility and easy adaptability allows them to be commonly employed in cogeneration schemes of all sizes.

The introduction of the internal combustion engine in the late nineteenth century opened up an entirely new approach to combined heat and power. Rather than using the same fluid for the heat and power process as was the case with hot air and steam processes, the tremendous waste heat generated by the internal combustion process can easily be transformed into useful heat. Cogeneration applications using stationary engines were common in Europe prior to World War I and remain quite popular because the heat is relatively easy to capture and transport. The most widespread application of internal combustion cogeneration is most certainly the automobile heater, which was widely introduced in the mid-1920s and is today virtually ubiquitous. This almost always utilizes hot water as a heat transfer medium, although many owners of the Volkswagen Beetle enjoyed the simplicity of hot air heat from its air-cooled engine. Surprisingly, experts continue to predict the imminent demise of the internal combustion engine due to its thermal inefficiency, failing to recognize that cold weather instantly turns it into a very efficient cogeneration plant and that its touted replacement, the electric car, is essentially useless in cold climates. The hot exhaust from internal combustion engines can also be utilized for useful heat and power purposes using a turbosupercharger or heat recovery apparatus.

During the early twentieth century the hot air turbine was reborn as the gas turbine for stationary power applications but was adapted to aircraft propulsion on the eve of World War II. Heat was being recovered from operating stationary gas turbines before the 1950s and both aeroderivative and industrial gas have undergone substantial development in the last half of the twentieth century. Initially used primarily in industrial applications, small gas turbines (along with their internal combustion cousins) enjoyed a large market for "total energy" applications in the 1960s and 1970s. A total energy plant was designed to operate without a connection to the electricity utility grid, either because it was inaccessible or uneconomical. Many American schools, hospitals, and shopping malls had such plants, most of which were eventually connected into utility grids.

The 1978 Public Utility Regulatory Policy Act (PURPA) introduced the word cogeneration, and required utilities to interconnect with and buy power from cogeneration systems at their avoided incremental production cost. This led to a rapid growth in cogeneration capacity in the United States from about 10,000 Mwe in 1980 to almost 44,000 Mwe by 1993. Most new capacity was installed at large industrial plants such as petroleum refineries, petrochemical plants, and paper mills. But PURPA also had some negative impacts that have given cogeneration a bad reputation. A few states required utilities to buy excess power from cogenerators at very high rates, which later proved to be in excess of incremental market costs. Also, a few systems were optimized for

Steam pours from the cooling tower at a nuclear power plant in Stedman, Missouri. (Corbis-Bettmann)

electric production rather that overall efficiency, with minimal use of the "waste heat" in order to qualify for the high power buyback rates established under PURPA. But there are only a few documented cases of such practices.

Another cogeneration technology with roots in the nineteenth century is the fuel cell, which was first described by Sir William Grove in 1839 but only in the past few years become a viable competitor in the marketplace, largely due to development funded first by the American space program and more recently by the U.S. Department of Energy and Electric Power Research Institute. Although still relatively expensive compared to other generation technologies, fuel cells have been widely demonstrated as reliable and clean energy producers. A few companies plan to introduce and market fuel cell cogeneration systems in the next several years. Their principal competitor appears to be small microturbines and internal combustion engines.

The United States obtained about 9 percent of its electricity from combined heat and power (cogeneration) systems as of 1997. Cogeneration is more prevalent in some European nations than in the

United States, with Germany, the Netherlands, and the Czech Republic obtaining 15 percent or more of their electricity from cogeneration facilities. Denmark leads the world and gets 40 percent of its electricity from combined heat and power systems

Cogeneration, by whatever name it is known, has survived and prospered for centuries because of its adaptability and inherent fuel efficiency. Many combined heat and power plants consistently achieve an overall thermal efficiency of 80 to 90 percent, more than three times more efficient than average utility plants and 50 percent higher than the newest and most efficient combined cycle gas turbine power plant. Cogeneration can be done with any fuel that can be burned, or even without burning a fuel as in Iceland where steam cogeneration is employed in geothermal fields to extract power from the steam before it is transported several miles as hot water to heat cities. Several nuclear cogeneration plants have been built with varying degrees of success, including an air-transportable model designed for remote military radar stations.

The current movement to deregulate electric utilities has created significant new opportunities for all three combined heat and power markets: industrial, district heating, and individual buildings. In the United States nearly 10,000 Mwe of new cogeneration capacity installed during 1994 to 1996, primarily involving large gas turbine plants located at industrial facilities. A number of European countries are actively cogeneration as a means to reduce global warming and to reduce fuel poverty in lower income households. In the United States, many organizations are working to reduce the institutional and regulatory obstacles to cogeneration, so that adoption of cogeneration can continue to expand. The U.S. Department of Energy has set a goal of doubling cogeneration capacity in the United States by 2010.

Morris A. Pierce

See also: Cogeneration; District Heating and Cooling; Fuel Cells; Industry and Business, Energy as a Factor of Production in; Turbines, Gas; Turbines, Steam.

BIBLIOGRAPHY

Hirsh, R. F. (1989). *Technology and Transformation in the American Electric Utility Industry*. Cambridge, Eng.: Cambridge University Press.

Hirsh, R. F. (1999). *Power Loss: The Origins of Deregulation and Restructuring in the American Electric Utility System*. Cambridge, MA: MIT Press.

COKING

See: Coal, Production of

COMBUSTION

"Combustion" is a term often used synonymously with "burning." However, a distinction can be made that explains why combustion is more than just burning. To burn something is to set it on fire. To combust something is to subject the material (or fuel) to the process of rapid oxidation that leads to the consumption of both the material (or fuel) and the oxidizer (usually the oxygen in air) with the release of heat and light. (Usually the oxidizer is oxygen but there can be nonoxygen species, that under certain circumstances fit the definition of an oxidizer being a substance that can accept electrons in a chemical reaction.) Fires and burning involve combustion, but not all combustion involves fire in the form of visible, hot flames. There are flames that are invisible but release heat, and there are flames that emit light but have so little evolution of heat that they are called "cool flames." By making this distinction between burning and combustion, many features of combustion such as ignition, extinction, and flames can each be discussed separately from a scientific perspective.

Combustion is the entire process by which something is oxidized. It is part of the use of gasoline or diesel fuel in automobiles and trucks, as well as part of propulsion in aircraft either in jet engines or propeller engines. This latter association is so often made that the propulsive devices in aircraft are called combustors. Similarly, furnaces and boilers, that often involve flames for the production of heat, are combustion devices involving many of the elements of the complete process. Incinerators, too, are commonly associated with combustion of fuel in the form of waste materials. Other common manifestations of combustion are house, forest, and chemical fires;

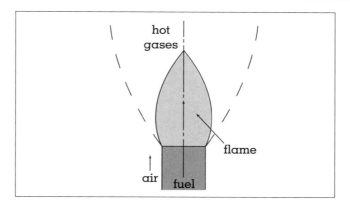

Figure 2.
Laminar diffusion flame jet.

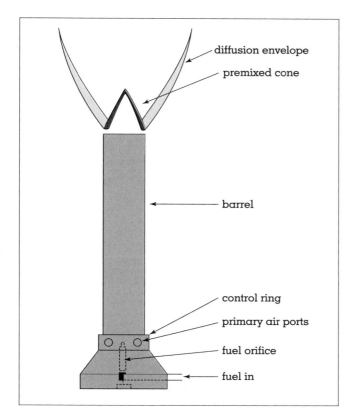

Figure 1.
A Bunsen burner.

explosions of flammables; and air pollution from cars and incinerators.

Because of the beneficial and adverse aspects of combustion, it is necessary to better understand it (i.e., all the components in the process of oxidizing a fuel with attendant heat and light).

COMBUSTION SCIENCE

Flames

A flame is a thin region of rapid, self sustaining oxidation of fuel that is often accompanied by the release of large amounts of heat and light. Flames are what we most commonly associate with combustion. One part of combustion science focuses on the different ways flames can be formed and the scientific and practical consequences of each.

Premixed Flame. For this type of flame, the fuel and oxidizer—both gases—are mixed together before flowing to the flame zone (the thin region of the flame). A typical example is the inner core of a Bunsen burner (Figure 1), or combustion in an auto-

mobile engine cylinder. (In general, a burner is the part of the combustion device that supplies fuel and sometimes air and where the flame is produced and stabilized.)

Diffusion Flame. When the fuel and oxidizer are initially unmixed and then mix in a thin region where the flame is located, the flame is called a diffusion flame (Figure 2). The word diffusion is used to describe the flame because the fuel and oxidizer are mixed on the molecular level by the random thermal motion of the molecules. An example of a diffusion flame is a candle flame or flares at an oil refinery.

Laminar Versus Turbulent Flames. Premixed and diffusion flames can be either laminar or turbulent gaseous flames. Laminar flames are those in which the gas flow is well behaved in the sense that the flow is unchanging in time at a given point (steady) and smooth without sudden disturbances. Laminar flow is often associated with slow flow from small diameter tubular burners. Turbulent flames are associated with highly time dependent flow patterns, often random, and are often associated with high velocity flows from large diameter tubular burners. Either type of flow—laminar or turbulent—can occur with both premixed and diffusion flames.

Droplets/Sprays. Flames can also be established with fuels that are initially liquids. A typical example is the flame around a droplet of hydrocarbon fuel such as diesel fuel. Droplets can burn individually, with a gaseous diffusion flame surrounding the evaporating liquid fuel center (Figure 3). When many droplets are combined into an array, a spray is formed. Burning of the droplets in the spray may

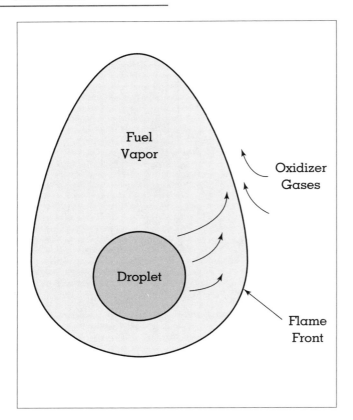

Figure 3.
Shapes of diffusion flames surrounding a burning spherical fuel droplet.

consist of a continuous flame if the droplets completely evaporate or of isolated flames around each droplet if fuel evaporation is slow. An intermediate situation occurs when incomplete evaporation takes place. The types of flames formed depend on the fuel droplet sizes and spacing.

Liquid Pool Flames. Liquid fuel or flammable spills often lead to fires involving a flame at the surface of the liquid. This type of diffusion flame moves across the surface of the liquid driven by evaporation of the fuel through heat transfer ahead of the flame. If the liquid pool or spill is formed at ambient conditions sufficient to vaporize enough fuel to form a flammable air/fuel mixture, then a flame can propagate through the mixture above the spill as a premixed flame.

Solid Fuel Flames. The flames from the combustion of solids such as coal and wood are the result of a combination of processes including the burning of gases that have been released from the heated solid (devolatization) that burn in the gas phase as diffusion flames. The remaining nonvolatile material, the char, then is oxidized on its surface and in its pores as oxygen diffuses into the interior. If the particles are large, devolatization and char burning occur simultaneously.

Very small solid fuel particles such as sawdust, agricultural grains, or coal dust can sustain flames when they are suspended in air. In fact, very serious fires have occurred in grain storage towers and coal mines because of the flammability of suspended dusts. The combustion of the individual particles follows the usual pattern of solid particle burning—devolatization and char burning. The combustion of the whole cloud of particles is similar to spray combustion and its characteristics depend on the nature of the fuel, size of the particles, and the number of particles in a given volume.

Other seemingly solid fuel flames such as those from the burning of plastics are actually more like liquid pool flames because the plastic melts and volatizes ahead of the advancing flame front.

Ignition

Even if a fuel and oxidizer are present in proportions that could sustain combustion, nothing may occur unless the combustible mixture is brought to the right conditions by an ignition source. Typical ignition sources are spark plugs in car engines, pilot flames in gas stoves, and matches for lighting barbecues. In order for an ignition source to be effective, it has to raise the temperature of the combustible mixture enough so that combustion can continue after the ignition source is removed. This means that the amount of heat added during the ignition process must be adequate to overcome any heat loss, to the engine walls for example, and still raise the temperature of the gas region to a value high enough to cause the flame to propagate.

Extinction

Once propagating, flames will continue to propagate unless they are extinguished or quenched. An obvious cause of extinction is the depletion or cessation of fuel flow. Flames can be extinguished by heat loss (e.g., by passage through very small passageways that accentuate heat loss), through smothering by water or chemical fire extinguishers that slow the combustion process, or blowing the flame away with high velocity flows. Flames can also be extinguished by removing one of the reactants, such as air.

Detonations

These are types of combustion waves (actually shock waves—extremely thin regions in which flow properties such as pressure and temperature change enormously—sustained by combustion) that consume fuel at supersonic speeds and create very large pressure and temperature increases. Detonations are formed only under special conditions that convert an ordinary flame propagating through a combustible mixture to a detonation. When these special conditions exist, then the detonation can have devastating consequences (e.g., explosions in buildings containing natural gas from leaks or in mines filled with natural gas as a result of the mining operation).

FUELS FOR COMBUSTION

Fuels for combustion are initially gases, liquids, or solids. A fuel initially in one phase may be transformed into another during the burning process (i.e., liquids vaporized to gases). The factors involved in the selection of the fuel phase or its physical and chemical characteristics for an application such as burning in an automobile or jet aircraft involve many different considerations such as price, availability, and source.

Among the various selection considerations are specific combustion characteristics of different fuels. One of the combustion characteristics of gaseous fuels is their flammability limit. The flammability limit refers to the mixture proportions of fuel and air that will sustain a premixed flame when there is either limited or excess air available. If there is a large amount of fuel mixed with a small amount of air, then there is a limiting ratio of fuel to air at which the mixture will no longer sustain a flame. This limit is called the rich flammability limit. If there is a small amount of fuel mixed with excess air, then there is a limiting ratio of the two at which the flame will not propagate. This limit is called the lean flammability limit. Different fuels have different flammability limits and these must be identified for each fuel.

The combustion characteristics of liquid fuels are similarly determined by measures of their ability to sustain a flame. Two measures of the combustion characteristics of liquid fuels especially related to safety are flash point and autoignition temperature. The flash point is the maximum temperature at which a liquid fuel can be maintained in an open vessel exposed to air before which it will sustain a flame

in the presence of a pilot flame. The autoignition temperature is a similar concept except no pilot flame is present. The autoignition temperature is the maximum temperature at which a liquid fuel can be maintained in an open vessel exposed to air before which the fuel bursts into flame without the presence of an external ignition source.

Solid fuels, unlike gases and liquids, are entirely characterized by their composition. For example, coal can be characterized by its carbon, hydrogen, oxygen, sulfur, and nitrogen content. The water and mineral content of coal are also important means of differentiating coals from various sources.

THE CHEMISTRY OF COMBUSTION AND ITS EFFECT ON THE ENVIRONMENT

The most commonly used fuels for combustion are hydrocarbons, materials that are compounds of only hydrogen and carbon. Occasionally, fuels such as alcohols, that contain oxygen, are burned. When hydrocarbon fuels with or without oxygen are burned in air (combusted) to completion, the products are water, from the hydrogen part of the fuel, and carbon dioxide, from the complete conversion of the carbon part. If oxygen is present in the fuel, it shows up in the final product as part of either the water or carbon dioxide.

Carbon dioxide has been implicated as a contributing factor in global warming. Increased global warming has been associated with increased release of carbon dioxide into the atmosphere attributed in part to an increase in the combustion of hydrocarbon fuels. Carbon dioxide is an inevitable consequence of the complete combustion of hydrocarbons in air. If combustion devices are made more efficient, less fuel is required and less carbon dioxide is released into the atmosphere.

Unlike carbon dioxide and water that are the inevitable by products of complete combustion of hydrocarbons, species such as carbon monoxide, ethene, toluene, and formaldehyde can be emitted because combustion has been interrupted before completion. Many factors lead to emissions from incomplete combustion. Emitted unburned hydrocarbons and carbon monoxide are regulated pollutants that must be eliminated. In automobiles with spark ignited engines, these emissions are almost entirely removed by the catalytic converter.

Soot particles, that are comprised primarily of carbon and hydrogen in an 8 to 1 ratio and are about 20-50 nm in diameter when first formed (coagulation and surface growth ultimately leads to a chain of soot spheres much larger than 50 nm), are the result of incomplete combustion of hydrocarbons. Soot, too, is a regulated combustion pollutant and is a particular problem with diesel engines. The black clouds emitted from the vertical exhaust stacks of trucks are laden with soot particles. The nature of diffusion flames precludes a practical way to reduce soot emissions in the diffusion flame processes that occur in diesel engines.

In contrast to carbon monoxide, small hydrocarbon molecules and soot that result from incomplete conversion of the hydrocarbon fuels, nitric oxide and nitrogen dioxide, are noxious emissions that result from the oxidizer—air. However, fuel components that contain nitrogen may also contribute, in a lesser way, to the formation of the oxides of nitrogen.

The nitrogen component of air, normally inert and unreactive, reacts at the very high temperatures of combustion. It reacts in a series of simple steps with atomic and molecular oxygen to yield NO, nitric oxide, that is subsequently converted to NO_2, nitrogen dioxide, in the atmosphere. Nitrogen dioxide, is a regulated, undesirable emission because it forms a brownish haze, leads to acid rain, and is a component of photochemical smog. Both nitric oxide and nitrogen dioxide can be considered inevitable byproducts of high-temperature combustion in air. The concentration of NO emitted into the atmosphere can be reduced either by lowering the temperature of combustion by various engineering techniques (with negative effects on performance) or through catalytic conversion to molecular nitrogen in postcombustion cleanup as is done in automobiles.

RESEARCH ACTIVITIES

The following synopsis of current research activities in the field of combustion is organized around the list of papers presented at the 27th International Symposium on Combustion (1998) under the auspices of The Combustion Institute.

Elementary Reaction Kinetics, Kinetic Mechanisms, Models, and Experiments

Examining the details involved in the oxidation and pyrolysis (thermal decomposition) of fuel molecules is very important. The results of these research activities will permit predictions about the chemicals emitted during incomplete combustion because reaction rate constants and chemical pathways will be evaluated and determined.

Laminar Premixed Flames

Research in this area focuses on understanding the chemical, thermal, and fluid-mechanical (behavior of fluids) structure of these types of flames. Recent advances in computer based modeled flames requires the knowledge developed in this type of research for calibration, validation, and prediction.

Laminar Diffusion Flames

Here, too, computer based predictions about the nature of these flames require information about the chemicals and science of diffusion flames for the predictions to be accurate. The predictions are made accurate by comparison with measured chemical species concentrations, measured temperatures, and flow characteristics.

Premixed Turbulent Combustion and Nonpremixed Turbulent Combustion

Because many practical flames are turbulent (spark ignited engine flames, oil field flares), an understanding of the interaction between the complex fluid dynamics of turbulence and the combustion processes is necessary to develop predictive computer models. Once these predictive models are developed, they are repeatedly compared with measurements of species, temperatures, and flow in actual flames for iterative refinement. If the model is deficient, it is changed and again compared with experiment. The process is repeated until a satisfactory predictive model is obtained.

Incineration, NO_x (NO and NO_2) Formation and Control, Soot Formation and Destruction

Environmental consequences of combustion are still a high priority requiring investigation of the chemistry and process effects on the emissions. Effective means of eliminating the pollutants is also a subject of further research.

Gas Turbines, Diesel Combustion, and Spark Ignition Engines

Application of combustion science to practical power source devices is one of the ultimate aims of developing a fundamental understanding of combustion.

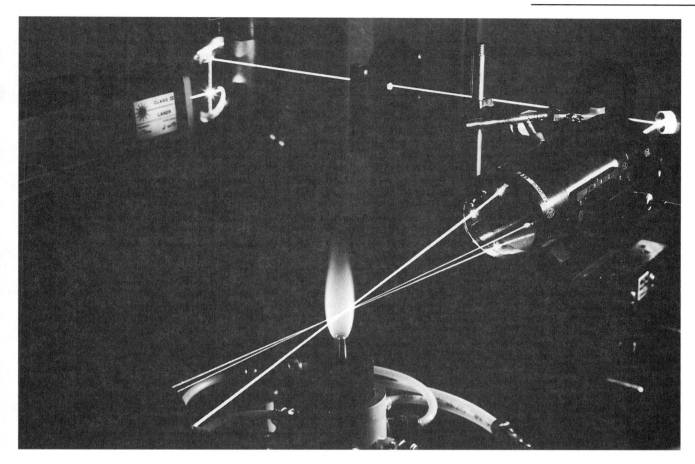

Advanced lasers and computers at the Combustion Research Facility in Livermore, California, are used to study exactly how and why fuels burn. (U.S. Department of Energy)

Using combustion science to improve performance through design changes and engineering techniques is an ongoing research subject.

Droplet and Spray Combustion and Pool Fires

The combustion of liquids is a fertile area for further study. Kowledge of the combustion science of individual droplets as well as groups of droplets helps improve performance of devices that rely on spray burning, particularly diesel engines. Understanding of the science of liquid pool fires potentially effects safety during spills.

High-Speed Combustion, Metals Combustion, and Propellants

Research in these specialized areas is aimed at developing improved and new methods for advanced propulsion. For example, an understanding of high-speed combustion is used in the devel-

opment of supersonic ram jet engines, which are simple alternatives to conventional turbojet engines. Knowledge of metals combustion is relevant to improving the use of metal additives in solid propellants to increase impulse and stability. In general, the study of propellant combustion aids in the development of more stable and longer range rocket engine performance.

Catalytic and Materials Synthesis

These two research areas share the common characteristic of involving inorganic solids in the combustion process. Catalytic combustion research focuses on using the solid to facilitate the oxidation of well-known fuels such as hydrogen and methane. Materials synthesis research focuses on using combustion as a means to react the solids either with each other or a gas, such as nitrogen (which in this case acts as an oxidizer), to make new solid materials.

Microgravity Combustion

Microgravity refers to the environment of extremely low gravity commonly known as a weightless environment. Under microgravity conditions, combustion phenomena that are affected by gravity, such as flames, behave differently than at Earth gravity conditions. Research in this area focuses on using the special microgravity conditions to understand, by contrast, basic combustion processes on Earth.

Fire Safety Research

Research is conducted to increase understanding of the science of combustion specifically as it relates to fires involving homes and plastics, wood, and large-scale spills. This is helpful in the development of fire prevention and extinction techniques.

Detonations

Examining the conditions for the formation and propagation of detonation waves is relevant to special applications of detonations to propulsion as well as safety.

Coal and Char Combustion

The importance of coal as an energy source motivates further research into the combustion characteristics and chemical kinetics of both coal and the material that remains after devolatization, char. Further research will aid in making coal a cleaner and more efficient energy source.

Fluidized Beds, Porous Media Fixed Bed Combustion, and Furnaces.

Specialized practical configurations for combustion have a number of practical applications such as coal burning for energy production. The study of these specialized combustion setups is necessary for better application.

Kenneth Brezinsky

See also: Catalysts; Coal, Production of; Conservation of Energy; Explosives and Propellants; Heat Transfer.

BIBLIOGRAPHY

Borman, G. L., and Ragland, K. W. (1998). *Combustion Engineering*. Boston: McGraw-Hill.

Chomiak, J. (1990). *Combustion: A Study in Theory, Fact, and Applic*ation. New York: Abacus Press.

Combustion and Flame (series). New York: Elsevier.

Combustion Science and Technology (series). New York: Gordon and Breach.

Glassman, I. (1996). *Combustion*, 3rd ed. San Diego, CA: Academic Press.

International Symposium on Combustion (series). Pittsburgh: The Combustion Institute.

Progress in Energy Combustion Science (series). New York: Pergamon.

Turns, S. R. (2000). *An Introduction to Combustion*, 2nd ed. Boston: McGraw-Hill.

Williams, F. A. (1985). *Combustion Theory*, 2nd ed. Menlo Park, CA: Benjamin/Cummings.

COMMUNICATIONS AND ENERGY

To propel or move anything requires energy. Light, electrical waves, or sound waves used in communication are no exception. What is different about communication is the energy used to transmit data.

BASIC SIGNALING

The history of communications and energy shows a relationship that can be expressed by the popular adage, "What goes around, comes around." In ancient times humans communicated through the use of torches, fire, and smoke signals. All three methods required a great deal of energy for the amount of information generated and the short transmission distance. While fire was more than likely first used by prehistoric peoples to cook food and as a source of warmth, it also provided a mechanism for performing a basic signaling method. This method evolved into the foundation for modern communications systems. If we fast-forward to the new millennium, the opposite is true: Communication over great distances with little energy is accomplished through the use of microprocessors and microelectronics. Microprocessors and microelectronics provide global terrestrial and satellite communications, whose use also facilitates the exploration, recovery, and distribution of different types of energy.

Although unknown at the time, some of the earliest uses of fire represented a binary signaling system that was even used during the American Revolution. On the evening of April 18, 1775, Joseph Warren sent for Paul Revere and instructed him to ride from

Charlestown to Lexington, Massachusetts, to warn Samuel Adams and John Hancock that British troops were coming to arrest them. The prearranged signal, which began his historic ride, was based on the placement of lanterns in the bell tower of Christ Church in Boston. Two lanterns were hung to indicate that the redcoats would come by sea rather than by land; the latter would have been signaled by one lantern.

Another popular use of fire for communications during human evolution involved the use of torches. In many locations throughout the world underground deposits of oil seeped to the surface, providing fuel for torches. They could only be used when it was dark and required a line of sight between originator and recipient. Torches were supplemented by smoke signals during daylight. It wasn't until the invention of electricity that communications over relatively long distances were only marginally affected by the elements.

Although many people credit American Indians with first using smoke signals to communicate information, African tribes and ancient Greeks and Romans also used smoke signals. Burning brush or wood and placing cloth or animal hides over the fire enabled people to generate puffs of smoke.

With the discovery of electricity, the evolution of communications occurred at a rapid rate. Smoke signals that were used for hundreds of years to convey information at a word or two per minute were first replaced by the use of copper-based conductors, such as the telegraph, that transmitted twenty to forty words per minute. By the end of the twentieth century, lasers with fiber optic wires transmitted an entire book around the world in under a second.

THE TELEGRAPH

The first notable development in the field of data communications occurred in 1832 when an American, Samuel F. Morse, invented the electric telegraph. When an operator at one location depressed a key, an electrical path was established that allowed current to flow through the path connecting the two locations. At the distant location, electricity flowed through a wire coil, forming a magnet that caused a metal plate (key) to click as it was attracted to the coil. When the key was released it opened the circuit; the distant key was released and it struck a stop bar with a slightly different sounding click. This resulted in two distinct clicks that defined the duration of the operator's key depression and enabled Morse to develop his well-known

code. That code used the time between clicks to represent a dot (short time) or dash (long time). Morse demonstrated the practicality of the telegraph in 1844 when he transmitted the now famous message, "What hath God wrought!," over a wire routed from Washington, D.C., to Baltimore. Similar to torches, lanterns, and smoke signals, the telegraph represents a two-state system, with current either flowing or not. As communications progressed from the telegraph to radio, microwave, satellite, light on fiber, and other systems, the two-state communications system continued to be used. It is the simplest to construct and eventually was developed into the binary system used by all modern communications methods.

In addition to the telegraph, the invention of electricity resulted in the development of other communications systems that paved the way for the modern communications infrastructure. The telegraph was followed by the laying of the transatlantic cable (1858), the invention of the telephone (1876), Marconi's wireless telegraph (1897), television (1927), and satellite communications (1957). Although early implementation of each technology was restricted to a minimum level of practical communication, each technology eventually evolved into a mass market for use by businesses and residential customers.

THE COMPUTER REVOLUTION

The development of the computer fostered the growth in communications. The first electronic computer, developed at the University of Pennsylvania's Moore School of Engineering during World War II, quickly moved from high-energy consumption vacuum tube technology to low-energy consumption semiconductor-based technology by the 1960s. As computers became smaller and used less power, they became more practical for use in telephone company offices. Computerized switches gave automated operator-dependent functions, which enhanced the ability of consumers to rapidly and easily communicate with people around the globe. The invention of the laser and microprocessor during the 1960s and 1970s eventually resulted in communications systems applications. Microprocessors enabled low-cost voice-digitization methods to be implemented that enabled telephone companies to digitize analog conversations at their central offices. Use of lasers made it possible to transmit digitized data over fiber cables using light energy instead of electrical energy, elimi-

A large microwave antenna transmits communications for the Kennedy Space Center in Florida. (Corbis-Bettmann)

optic cable. The use of fiber optic cable not only provided immunity to electromagnetic interference but, in addition, provided a bandwidth several orders of magnitude greater than that obtainable on conventional copper-based media. This enabled a single strand of glass or plastic to transport tens of thousands of simultaneous voice conversations.

As transmission systems were developed to take advantage of the evolution of technology it became possible to communicate further using less energy. While communications carriers were initially the prime beneficiary of the evolution of products that expanded their capability to transport voice, data, and video, the resulting efficiencies were eventually passed along to the consumer in the form of new offerings as well as lower communications costs. By 2000 it was rare to pay more than a dime a minute to make a coast-to-coast long-distance telephone call that cost more than $2 per minute during the 1950s. In addition, the modern telephone call is highly automated and only requires operator intervention for special services, such as setting up an international conference call or reversing charges. In comparison, many long-distance telephone calls made during the 1950s required operator intervention.

Today the majority of energy in the telephone system occurs on the legacy local loop routed from the telephone company central office to the subscriber. The red colored wire in the home or office carried negative voltage (48 Vdc) relative to ground. When a phone is on-hook, it presents a dc resistance near infinity. When a phone is off-hook, it presents approximately 600 ohms to the central office and the current flow rises to 25 mA. As a result of going off-hook, the voltage will drop to between 12 and 15 Vdc. While 25 mA represents a low current, when you multiply this by hundreds of millions of subscribers the energy used by the phone system becomes considerable. Home computers and phone lines that surf the Internet account for approximately 5 percent of electricity consumed in the United States.

THE EVOLUTION OF TELEVISION

Although television dates back to 1927 and was highly publicized during the 1936 Olympics, it was not until post–World War II prosperity that vacuum-tube-based television products reached the mass consumer market. The initial series of televisions manufactured during the 1950s had screen displays that made the viewer

nating the adverse effect of electromagnetic interference generated by lightning and machinery on transmission via copper circuits and microwave.

The use of digital technology and the evolution of the communications infrastructure of carriers to lasers and fiber optic media reduced the energy required to transmit information and increased transmission capacity. Digital transmission is based on the use of microprocessors and digital signal processors (DSPs) that consume less energy than filters and modulators used by analog transmission systems. The use of lasers to transmit information through fiber cables was also more energy efficient because it eliminated the 5 to 10 percent energy loss of copper circuits.

Beginning in the late 1970s most telephone companies in the world began to convert trunk circuits that interconnected central offices from microwave and copper-based transmission to fiber-optic-based transmission. By the mid-1990s almost all of Western European and North American long-distance telephone transmissions resulted in analog voice conversations being digitized and transmitted over fiber squint and the vacuum tubes used so much electricity that they dimmed the lights in the home. Semiconductors also had a profound effect on the television industry because they made televisions smaller and significantly reduced their power consumption. Television became popular home entertainment and it also allowed networks to provide viewers with news broadcasts, presidential press conferences, and special reports. Unfortunately, hills, mountains, and the canyons formed by buildings within cities caused broadcast reflections, echos, and periodically the inability to receive signals. Such problems resulted in the development of the cable television (CATV) industry that evolved into several national multimedia conglomerates. The initial development of the CATV industry was focused on one-way communications. However, by the late 1990s many CATV operators began to replace their one-way amplifier infrastructure with a bidirectional amplifier infrastructure. This upgrade enabled the support of cable modems that allow subscribers to access the Internet at data rates up to 10 Mbps compared to the maximum transmission rate of 56 kbps obtainable over the public switched telephone network (PSTN). CATV modems, as well as those on the PSTN, represent considerable advances in the use of solid state technology over modems developed during the 1960s through 1980s. Packing circuitry built with microprocessors more powerful than mainframe computers that reached the market during the 1980s, today's PSTN and CATV modems consume only a fraction of the electricity of the prior generation of modems, yet have a signal processing capability as great as those of mainframe computers manufactured just a decade earlier.

SATELLITE-BASED SYSTEMS

No discussion of the history of energy and communications would be complete without covering the role of satellites. Although Sputnik communications during 1957 was limited to a series of beeps, today there are hundreds of communications satellites in geostationary and low Earth orbit. Geostationary communications satellites support communications over a predefined arc, while low Earth orbit satellites commonly operate in groups that circle the earth and provide communications coverage over a much wider area. Applications from satellite television broadcasts to Internet access, and global positioning communications to marine transmission to ships at sea are now a reality via the use of satellites. The evolution of satellite communications is closely linked to advances in semiconductor technology that enable the design of miniaturized circuitry requiring less power. This in turn enables communications satellites to be designed employing more transponders (transmitters/receivers) while using the same or a similar solar panel used on earlier orbited satellites.

Through the use of satellite communications, the adage "What goes around, comes around" is a reality today with respect to the relationship of energy and communications. As the search for energy to include petroleum reserves expanded to less populated, more rugged terrain, the need to communicate with field workers increased. Satellite communications is now commonly used as a standard way to facilitate communications with energy exploration workers in the foothills of Montana, at the Arctic Circle, and drilling for gas and oil at various offshore locations throughout the world. A second related area between energy and communications involves the distribution of production. To facilitate communications to mobile locations such as ships delivering oil, an international cooperative known as Inmarsat began service in 1979. Today Inmarsat operates a series of geostationary satellites that provide worldwide communications coverage for telephone, fax, email, and data services. Operating forty land stations in thirty-one countries that transmit and receive communications through Inmarsat satellites, maritime customers are no longer dependent on bouncing radio frequency communications off the ionosphere in an attempt to connect to land.

The role of energy in providing a communications capability is indispensable to maintaining our living standards. Whether it is cell phones powered by batteries, checking our bank account via a toll free telephone number, or surfing the Internet, each of these activities depends on the use of energy. In the history of communications and energy, advances in technology made possible the ability to transmit further and faster using less energy, which has improved productivity and living standards. Within a few years, low power and low cost semiconductor charged coupled devices (CCDs) will more than likely enable the Picturephone to replace the telephone, while other advances in semiconductors may make the Dick Tracy watch telephone a reality. As communications technology progresses, we can continue to expect to do more using less energy.

Gilbert Held

See also: Electricity, History of; Energy Economics; Transportation, Evolution of Energy Use and.

BIBLIOGRAPHY

Held, G. (1997). *The Complete Modem Reference*, 3rd ed. New York: John Wiley & Sons.

Held, G. (1999). *Cisco Security Architectures*. New York: McGraw-Hill.

Held, G. (1999). *Data Communications Networking Devices*, 4th ed. New York: John Wiley & Sons.

Held, G. (1999). *Enhancing LAN Performance: Issues and Answers*, 3rd ed. New York: John Wiley & Sons.

Held, G. (1999). *Understanding Data Communications*, 6th ed. Indianapolis: New Riders Publication.

Held, G. (1999). *Voice & Data Internetworking*. New York: McGraw-Hill.

COMPRESSED AIR STORAGE

See: Storage Technology

CONDUCTION

See: Heat Transfer

CONSERVATION OF ENERGY

Conservation of energy has two very different meanings. In the popular sense, "conserve" means to "save" or "preserve." Electric energy for lights likely had its origin in burning coal, so turning off lights tends to preserve coal, a valuable natural resource. In the scientific sense, conservation alludes to constancy. Succinctly stated, the energy of the universe is constant. Energy can be converted from one form to another, but ultimately there are as many joules of energy after the conversion as before. Every second,

an operating 100-watt lightbulb converts 100 joules of energy. If 10 joules are in the form of light, then 90 joules are in some other form, notably heat to the room. Even though the total amount of energy following the conversion is unchanged, the energy may not be available for some desired purpose. For example, the heat produced by a lightbulb is in the surroundings and is no longer available for other uses. In fact, when energy runs through all possible conversions, the energy ends up as thermal energy in the environment. Addition of thermal energy to the environment can produce local increases in temperature, leading to what is called thermal pollution. This effect is not to be confused with a possible global increase in temperature due to accumulation of carbon dioxide and other gases in the atmosphere.

CONSERVATIVE FORCES

The illustration in Figure 1 depicts a person pushing a box up a ramp. In the process, the person works against the gravitational force on the box and a frictional force between the box and the ramp. The person, the gravitational force, and the frictional force all do work on the box. The same would be said if the ramp were made longer. But interestingly, from the way work is defined, the work done by the gravitational force depends only on the vertical height through which it moves. The work is the same no matter how long or how short the ramp, as long as the vertical height is the same. If the work done by a force depends only on where it started and where it ended up, the force is said to be conservative. Unlike the gravitational force, the frictional force is nonconservative because the work done by it does depend on the path of the movement, that is, the length of the ramp. Potential energy is associated with the work done by a conservative force. For a mass (m) a height (h) above its lowest level the potential energy is $U = mgh$ where g is the acceleration due to the gravitational force (9.80 m/s^2).

MECHANICAL ENERGY

Kinetic energy (the energy of motion) and potential energy (the energy based on position) added together are called mechanical energy. Mechanical energy equals kinetic energy plus potential energy.

$$E = K + U, \text{ with } K = \frac{1}{2} mv^2$$

In an isolated system—one devoid of friction—the mechanical energy does not change. Although friction can never be totally eliminated, there are situations where it is small enough to be ignored. For example, when you hold a book in your outstretched hands, it has potential energy but no kinetic energy because its speed is zero. When dropped, the book acquires speed and kinetic energy. As its height above the floor decreases, its potential energy decreases. The gain in kinetic energy is balanced by a loss in potential energy, and the sum of kinetic energy and potential energy does not change. This is the idea of the conservation principle. Of course, this assumes that there is no friction (a nonconservative force), which is rarely the case in the real world. Frictional forces will always extract energy from a system and produce heat that ends up in the environment. The falling book will come to rest on the floor and have neither kinetic energy nor potential energy. All the energy it had before being dropped will have been converted to heat (and a very small amount of sound energy from its impact with the floor).

Figure 1.

A person does work on a box when pushing it up a ramp and the potential energy of the box increases.

CONSERVATION OF ENERGY WITH A SIMPLE PENDULUM

A simple pendulum with ignorable friction illustrates the conservation of mechanical energy. Pulling the bob (the mass) from its lowest position and holding it, the pendulum has only potential energy; the mechanical energy is all potential. When released, the pendulum bob gains kinetic energy and loses potential energy, but at any instant the sum never differs from the sum at the beginning. At the lowest point of the movement the potential energy is zero and the mechanical energy is all kinetic. As the pendulum bob moves to higher levels, the potential energy increases, and the kinetic energy decreases. Throughout the motion, kinetic energy and potential energy change continually. But at any moment the sum, the mechanical energy, stays constant.

A simple pendulum isolated from nonconservative forces would oscillate forever. Complete isolation can never be achieved, and the pendulum will eventually stop because nonconservative forces such as air resistance and surface friction always remove mechanical energy from a system. Unless there is a mechanism for putting the energy back, the mechan-

ical energy eventually drains and the motion stops. A child's swing is a pendulum of sorts. If you release a swing from some elevated position it will oscillate for a while but eventually will stop. You can push the swing regularly and keep it going, but in doing so you do work and put energy back into the system.

CONSERVATION OF ENERGY IN A SPRING-MASS SYSTEM

When a spring is stretched by pulling on one end, the spring pulls back on whatever is pulling it. Like the gravitational force, the spring force is conservative. Accordingly, there is potential energy associated with the spring that is given by $U = kx^2$, where k, the spring constant, reflects the strength of the spring and x is the amount of stretching from the relaxed position.

A horizontal spring with a mass attached to one end is a form of oscillator—that is, something that periodically returns to the starting position. To the extent that the spring-mass system can be isolated, the mechanical energy is conserved. When stretched but not yet in motion, the system has only potential energy. When released, the mass gains kinetic energy, the spring loses potential energy, but the sum does not change; it is conserved. Much like the oscillating pendulum, both kinetic energy and potential energy change continually, but the sum is constant. Most people do not think of atoms in a molecule as being connected by springs, yet the forces that bind them together behave like springs, and the atoms vibrate. The mechanical energy of the spring-like atomic system is an important energy attribute. A better understanding of the spring-like characteristics of atoms in materials has made possible many advances in sporting goods equipment, from graphite composite vaulting poles to titanium drivers for golfers.

Even though mechanical energy is rigorously conserved only when a system is isolated, the principle is elegant and useful. Water flowing over the top of a dam has both kinetic energy and potential energy. As it plummets toward the bottom of the dam, it loses potential energy and gains kinetic energy. Impinging on the blades of a paddle wheel, the water loses kinetic energy, which is transferred to rotational kinetic energy of the wheel. The principle provides an accounting procedure for energy.

THE FIRST LAW OF THERMODYNAMICS

The first law of thermodynamics is a statement of the principle of conservation of energy involving work, thermal energy, and heat. Engines converting heat to useful work are widely employed in our society, and the first law is vital for understanding their operation. The first law of thermodynamics accounts for joules of energy in somewhat the way a person accounts for money. A person uses a bank to receive and disperse money. If money comes into a bank and nothing is dispersed, the bank account increases. If money is dispersed and nothing comes in, the bank account decreases. If we call U the money in the bank, Q the money coming in or out, and W the money dispersed then the change in the account (ΔU) can be summarized as

$$\Delta U = Q - W$$

Algebraically, Q is positive for money coming in, negative for money going out. W is positive for money going out. If a person put \$100 into the bank and took \$200 out, the change in the account would be

$$\Delta U = \$100 - \$200 = -\$100$$

The account has decreased \$100.

The energy content of a gas is called internal energy, symbol U. The gas can be put in contact with something at a higher temperature, and heat will flow in. If that something is at a lower temperature, heat will flow out. If the gas is contained in a cylinder containing a movable piston, then the gas can expand and push against the piston doing work. In principle, the

work is equivalent to a person pushing and moving a box on a floor. Some agent can push on the piston doing work on the gas. If heat enters the gas and no work is done, the internal energy increases. If no heat enters or leaves the gas and if the gas expands doing work, the internal energy decreases. All of this is summarized in the equation

$$\Delta U = Q - W$$

If 100 J of energy (heat) entered the gas and the gas expanded and did 200 J of work, then

$$\Delta U = 100\,J - 200\,J = -100\,J$$

The internal energy decreased 100 J.

The internal energy of all gases depends on the temperature of the gas. For an ideal gas, the internal energy depends only on the temperature. The temperature is most appropriately measured on the Kelvin scale. The contribution to the internal energy from the random kinetic energy of the molecules in the gas is called thermal energy.

In a monetary bank, money is added and withdrawn, but if in the end the bank is in exactly the same state, the net amount of money in the bank has not changed. Similarly, a gas may expand or be compressed and its temperature may undergo changes, but if at the end it is in exactly the same thermodynamic condition, the internal energy has not changed.

CONSERVATION OF ENERGY AND HEAT ENGINES

Heat engines work in cycles. During each cycle, heat is absorbed, work is done, and heat is rejected. At the beginning of the next cycle the gas is in exactly the same state as at the beginning of the previous cycle; the change in internal energy is zero. To illustrate, consider a conventional automobile engine. The cycle (Figure 2) starts with the piston moving down and pulling a mixture of gasoline vapor and air through the open intake valve and into the cylinder. The intake valve closes at the bottom of the downward motion, the piston moves up, and the gaseous mixture is compressed. At the top of the stroke the mixture is ignited; the gas expands doing work. At the end of the stroke the exhaust valve opens, the cooler gas is forced out the automobile's exhaust, and a new cycle begins.

Because the change in internal energy in a cycle is

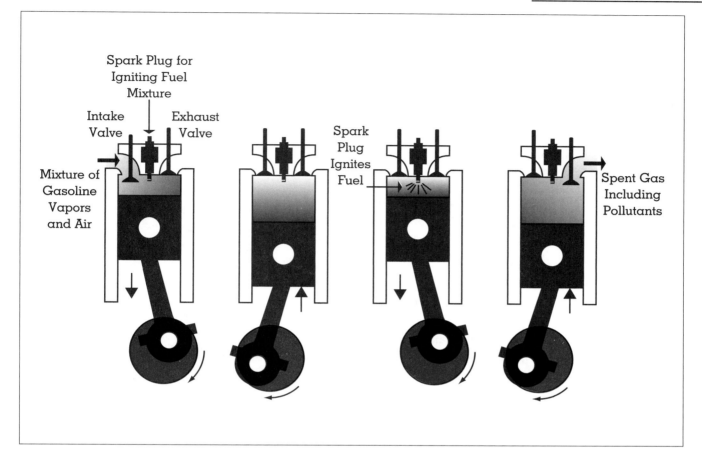

Figure 2.
The cycle of a conventional internal combustion engine.

zero, the first law of thermodynamics requires Q=W. In words the net heat exchanged equals the work done in the cycle. The difference between the heat absorbed and the heat rejected has gone into useful work. The energy absorbed by a gas always takes place at a temperature higher than the temperature at which energy is exhausted. An automobile engine exhausts energy into the environment at a temperature of about 300 K, which is much lower than the temperature of the gasoline vapor-air mixture at the moment of ignition, about 1,000 K.

Engines are used to do work, and a quantitative measure of the performance of an engine is efficiency:

efficiency = work done ÷ energy absorbed

For a given amount of energy absorbed at a high temperature, the more work obtained in a cycle, the more efficient the engine. In symbols,

$$e = W \div Q_{high}.$$

The first law of thermodynamics requires $W = Q_{high} - Q_{low}$ so the efficiency may be written

$$e = (Q_{high} - Q_{low}) \div Q_{high}$$
$$= 1 - (Q_{low} \div Q_{high})$$

If all the heat absorbed were converted into work, the efficiency would be 1, or 100 percent. If none of the heat absorbed was converted into work, the efficiency would be 0. The first law of thermodynamics limits the efficiency of any heat engine to 1 but does not prevent an efficiency of 1. The efficiency of practical heat engines is always less than 1. For example, the efficiency of a large steam turbine in an electric power plant is about 0.5, which is considerably more efficient than the typical 0.35 efficiency of an auto engine.

When two objects at different temperatures are in

283

contact, heat always flows from the hotter one to the cooler one. There is nothing in the first law of thermodynamics that prevents the opposite. The first law only requires that energy be conserved. A heat engine relies on heat flowing from some reservoir to a reservoir at a lower temperature. It is somewhat like a hydroelectric system relying on water flowing from a higher level to a lower level. The efficiency of a hydroelectric system increases as the difference in heights of the water levels increases. The efficiency of a heat engine increases as the difference in temperatures of the two reservoirs increases. In 1824, at age twenty-eight, the French engineer Sadi Carnot reasoned that the maximum efficiency of a heat engine depends *only* on the Kelvin temperatures of the two reservoirs. Formally,

$$\text{maximum efficiency} = e_{max} = 1 - (T_{low} \div T_{high})$$

For a heat engine like a steam turbine in an electric power plant the low temperature is determined by the outdoor environment. This temperature is about 300 K. Engineering considerations limit the high temperature to about 800 K. The maximum efficiency according to Carnot is 0.63 or 63 percent. No matter how skilled the builders of a steam turbine, if the temperatures are 300 K and 800 K, the efficiency will never exceed 63 percent. When you realize that the efficiency can never be larger than about 63 percent, a realizable efficiency of 50 percent looks quite good.

An important message in this discussion is that all the thermal energy extracted from a reservoir is not available to do work. Some will always be lost, never to be recoverable. The principle of conservation of energy guarantees that thermal energy exhausted to the environment is not lost. But the principle does not say that the energy can be recovered. To recover the energy there must be a reservoir at a lower temperature for the heat to flow into. When everything in the environment comes to the same temperature, there is no reservoir at a lower temperature for heat to flow into.

CONSERVATION OF ENERGY AND REFRIGERATORS

Water never flows spontaneously from the bottom of a dam to the top. Water can be forced to flow to a higher level, but it requires a pump doing work. Similarly, heat never flows spontaneously from a lower temperature to a higher temperature. Heat can be forced to flow from a lower temperature to a higher temperature, but it requires work. A household refrigerator is a good example. The noise emanating from a refrigerator is due to an electric motor doing work, resulting in heat flowing from the cool interior to the warmer surroundings. In principle, a refrigerator is a heat engine running backward. The refrigerator operates in cycles and subscribes to the first law of thermodynamics. Work must be done on the working substance in order for heat (Q_{low}) to flow from the lower temperature. At the end of each cycle this heat as well as the work done (W) is rejected at the higher temperature. Conserving energy the first law requires

$$Q_{high} = Q_{low} + W$$

Work is invested to force heat to flow from the interior of the refrigerator. A measure of the performance, called coefficient of performance (COP), is

$$COP = Q_{low} \div W$$

If for every joule of work done, 2 joules of heat flowed out of the refrigerator then the performance would be 2. Using $Q_{high} = Q_{low} + W$, the performance equation can be written

$$COP = Q_{low} \div (Q_{high} - Q_{low})$$

It is quite appropriate to think of a refrigerator as a heat pump. It pumps energy from one region and dumps it into another region at a higher temperature. A commercial heat pump does just this to warm a building during the heating season. There is a lot of energy outside of a building, even though the temperature may be 30°F (16.7°C) lower. This is because there are a lot of molecules in the outdoors, and each molecule contributes to the total energy. The performance of a heat pump decreases as the outside temperature decreases, but if the temperature remains above the freezing point of water (0°C or 32°F) commercial heat pumps can achieve performances between 2 and 4. If it were 2, this means that 2 joules of energy are deposited in the building for every joule of work. This is a significant gain. The temperature below the surface of the ground is several degrees higher than at the surface. Heat pumps drawing energy from the subsurface tend to be more efficient than an above-ground heat pump. This is because it is easier to extract heat from the solid sub-surface than it is from air, and the subsurface temperature is higher.

CONSERVATION OF ENERGY IN CHEMICAL REACTIONS

Conservation of energy is mandatory in chemical and nuclear reactions. To illustrate, consider the combustion of gasoline. One type of gasoline, isooctane, has the molecular form C_8H_{18}. Thermal energy is released when isooctane combines with oxygen according to

$$2C_8H_{18} + 25O_2 \rightarrow 16XO_2 + 18H_2O + \varepsilon\nu\varepsilon\rho\gamma\psi$$

From an energy standpoint the arrow is an equals sign. The total energy on the left side of the equation must equal the total energy on the right side. The sixteen carbon dioxide (CO_2) molecules and eighteen water (H_2O) molecules have lower total energy than the two isooctane (C_8H_{18}) molecules and twenty-five oxygen (O_2) molecules from which they were formed. The difference in energy is liberated as heat. It is possible to have chemical reactions in which the total energy of the molecules on the left side of the equation is less than the total energy of the molecules on the right. But for the reaction to proceed, energy must be added on the left side of the equation.

CONSERVATION OF ENERGY IN NUCLEAR REACTIONS

No name is more universally known in science than that of Albert Einstein, and no equation of physics is more recognizable than $E = mc^2$. In this equation E, m, and c stand for energy, mass, and the speed of light (300,000,000 m/s). Taken literally, the equation suggests that anything having mass has energy. It does not mean the energy of a mass moving with the speed of light. The energy, mc^2, is intrinsic to any mass whether or not it is moving. Because the speed of light is such a large number, the energy of anything is huge. For example, the mass-energy of a penny having a mass of 0.003 kg (3 grams) is 270 trillion joules. This is roughly 10,000 times the energy liberated from burning a ton of coal. As strange as it may seem, mass and energy, are equivalent and it is demonstrated routinely in nuclear reactions that liberate energy in a nuclear power plant.

The nucleus of an atom consists of protons and neutrons that are bound together by a nuclear force. Neutrons and protons are rearranged in a nuclear reaction in a manner somewhat akin to rearranging atoms in a chemical reaction. The nuclear reaction liberating energy in a nuclear power plant is called nuclear fission. The word "fission" is derived from "fissure," which means a crack or a separation. A nucleus is separated (fissioned) into two major parts by bombardment with a neutron.

Uranium in the fuel of a nuclear power plant is designated ^{235}U. The 92 protons and 143 neutrons in a ^{235}U nucleus sum to 235, the number in the ^{235}U notation. Through interaction with a neutron the 92 protons and 144 neutrons involved are rearranged into other nuclei. Typically, this rearrangement is depicted as

$$^{235}U + n \rightarrow {}^{143}B\alpha + {}^{90}K\rho + 3n + \varepsilon\nu\varepsilon\rho\gamma\psi$$

A barium (Ba) nucleus has 56 protons and 87 neutrons; a krypton (Kr) nucleus has 36 protons and 54 neutrons. The 92 protons and 144 neutrons being rearranged are accounted for after the rearrangement. But the mass of a ^{235}U nucleus, for example, differs from the sum of the masses of 92 free protons and 143 neutrons. When you account for the actual masses involved in the reaction, the total mass on the left side of the arrow is less than the total mass on the right. The energy released on the right side of the equation is about 32 trillionths of a joule. The energy equivalent of the mass difference using $E = mc^2$ accounts precisely for the energy released. In the reaction described by the equation above, one-thousandth of the mass of the ^{235}U has been converted into energy. It is energy from reactions like this that ultimately is converted into electric energy in a nuclear power plant. The illustration of a nuclear fission reaction using an arrow becomes an equality for energy when the equivalence of mass and energy is taken into account.

The energy liberated in nuclear reactions is of such magnitude that mass differences are relatively easy to detect. The same mass-energy considerations pertain to chemical reactions. However, the energies involved are millions of times smaller, and mass differences are virtually impossible to detect. Discussions of energies involved in chemical reactions do not include mass energy. Nevertheless, there is every reason to believe that mass energy is involved.

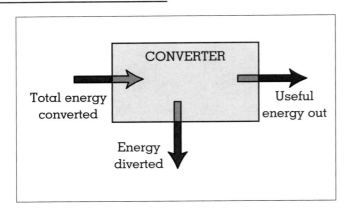

Figure 3.
A descriptive model for an energy converter.

A MODEL FOR CONSERVATION OF ENERGY

Big or small, simple or complex, energy converters must all subscribe to the principle of conservation of energy. Each one converts energy into some form regarded as useful, and each one diverts energy that is not immediately useful and may never be useful. Because energy is diverted, the efficiency defined as

$$\text{efficiency} = \text{useful energy} \div \text{total energy converted}$$

can never be 100 percent. Generally, several energy conversions are involved in producing the desired form. The food we convert to energy in our bodies involves several energy conversions prior to the one a person performs. Energy conversions are involved in our sun to produce light. Photosynthesis producing carbohydrates for the food entails energy conversions. Even the carbon dioxide and water require energy conversions for their formation. Tracking energy conversions is facilitated with the descriptive model shown in Figure 3.

An energy analysis of the production of electric energy in a coal-burning power plant provides an opportunity to illustrate the model. Generating electricity requires burning coal for heat to vaporize water to steam, a turbine driven by steam to drive an electric generator, and an electric generator to produce electric energy.

The diagram in Figure 4, with typical efficiencies for the three converters, describes the fate of 1 J of energy extracted from burning coal. It is important to note that only 0.39 J of electric energy was derived

from the 1 J. More energy in the form of heat (0.61 J) was rejected to the environment than was delivered as electric energy (0.39 J). Thinking of the three converters as one unit converting the 1 J into 0.39 J of electric energy the efficiency of the converter is 0.39. This is just the product of the efficiencies of the three converters:

$$\text{efficiency} = 0.88 \times 0.45 \times 0.99 = 0.39$$

from burning coal into electric energy.

The overall efficiency is smaller than the lowest efficiency in the chain. No matter how efficient all converters in a chain are, the efficiency will always be smaller than the lowest efficiency. As long as the steam turbine is used in the commercial production of electricity, the overall efficiency will be relatively low. Electric power plants using nuclear reactors also use steam turbines, and their efficiency is essentially the same as for a coal-burning plant. Energy in the form of heat is lost in the transmission of electricity to consumers, which reduces the overall efficiency to about 0.33. This means that every joule of electric energy paid for by a consumer requires 3 joules of energy at the input of the chain of energy converters. It also means that every joule of energy saved by turning off lights when not needed saves 3 joules at the input.

The scientific principle of conservation of energy is imbedded in the model for evaluating the performance of a chain of energy converters. Applying the model, we find that the overall efficiency of a chain of energy converters is always less than the smallest efficiency in the chain. Realizing this, strong arguments can be made for conserving energy in the sense of saving. If the overall efficiency of conversion is 10 percent, as it is for an incandescent lightbulb, then saving one unit of energy at the end of the chain saves ten units of energy at the beginning of the chain.

Joseph Priest

See also: Carnot, Nicolas Leonard Sadi; Climatic Effects; Engines; Matter and Energy; Nuclear Energy; Nuclear Fission; Refrigerators and Freezers; Thermal Energy.

BIBLIOGRAPHY

Aubrecht, J. G. (1995). *Energy.* Upper Saddle River, NJ: Prentice-Hall.

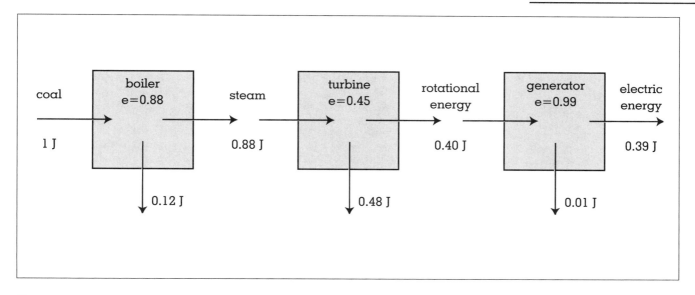

Figure 4.
A model for evaluating the conversion of energy.

Bodansky, D. (1996). *Nuclear Energy: Principles, Practices, and Prospects*. Woodbury, NY: American Institute of Physics.

Brescia, F.; Arents, J.; Meislich, H.; and Turk, A. (1998). *General Chemistry*, 5th ed. San Diego: Harcourt Brace Jovanovich, Inc.

Ebbing, D. D., and Wrighton, M. S. (1987). *General Chemistry*. Boston: Houghton Mifflin Company.

Energy. (1978). San Francisco: W. H. Freeman.

Hecht, E. (1980). *Physics in Perspective*. Reading, MA: Addison-Wesley Publishing Company.

Hecht, E. (1996). *Physics*. Pacific Grove, CA: Brooks/Cole Publishing Company.

Hill, J. W. (1999). *General Chemistry*. Upper Saddle River, NJ: Prentice-Hall.

Hobson, A. (1998). *Physics: Concepts and Connections*. Englewood Cliffs, NJ: Prentice-Hall.

Kondepudi, D. K., and Prigogine, I. (1998). *Modern Thermodynamics: From Heat Engines to Dissipative Structures*. Chichester, NY: John Wiley and Sons, Ltd.

Krame, K. S. (1995). *Modern Physics*. New York: John Wiley and Sons, Inc.

Krauskopf, K. B., and Beiser, A. (2000). *The Physical Universe*. New York: McGraw Hill Higher Education.

Priest, J. (2000). *Energy: Principles, Problems, Alternatives*, 5th ed. Dubuque, IA: Kendall/Hunt Publishing Company.

Ristinen, R. A., and Kraushaar, J. J. (1999). *Energy and the Environment*. New York: John Wiley and Sons, Inc.

Sandfort, J. F. (1979). *Heat Engines*. Westport, CT: Greenwood Publishing Group.

Serway, R. A. (1998). *Principles of Physics*, 2nd ed. Fort Worth, TX: Saunders College Publishing.

CONSERVATION SUPPLY CURVES

The cost of conserved energy (CCE) and its extension, supply curves of conserved energy, are useful tools for investigating the technical potential and economics of energy conservation measures. The CCE is an investment metric that is well suited for analysis of energy conservation investments, and the supply curve approach provides a bookkeeping framework that is ideal for diverse conservation investments. Several people, including Amory Lovins, John Sawhill, and Arthur Rosenfeld, independently developed the general approach in the late 1970s. However, Alan Meier, along with Janice Wright and Arthur Rosenfeld, systematized the concepts and procedures in the early 1980s. This article introduces the cost of conserved energy and supply curves of conserved energy, and explains their application to energy-efficiency issues.

THE COST OF CONSERVED ENERGY

Energy conservation typically involves making an investment that results in lower energy running costs. An investor (or policymaker) is often confronted with a list of possible conservation measures. The investor needs a way to rank the measures and then decide which are worth undertaking. He or she ranks the measures with the help of an investment metric, such as the simple payback time, the benefit-cost ratio, or return on investment. The investment metric provides a means of ranking the opportunities, and then separates the attractive investments from those in which the money would be better invested elsewhere.

Each investment metric has strengths and limitations. For example, the simple payback time indicates the time required to recover the investment, but it ignores any benefits that may occur after the payback time, so measures offering many years of benefits appear no better than short-lived ones. A common drawback of these investment metrics is that the price of energy must be assumed. If the energy price changes, then the payback time must be recalculated.

The CCE spreads the investment over the lifetime of the measure into equal annual payments with the familiar capital recovery factor. The annual payment is then divided by the annual energy savings to yield a cost of saving a unit of energy. It is calculated using the following formula:

$$CCE = (I/\Delta E) \times [d/(1 - (1 + d)^{-n})]$$

where I is the investment or cost of the measure; ΔE is the energy savings (per year); d is the real discount rate; n is lifetime of measure.

The CCE is expressed in the same units as the cost and the energy savings. For example, if the investment is entered in dollars and the savings are in gigajoules (GJ), then the CCE will have the units \$/GJ.

For example a consumer wishes to buy a new refrigerator. The high-efficiency model (offering services identical to the standard model) costs \$60 more but uses 400 kWh/year less electricity. The consumer expects to keep the refrigerator for ten years and has a discount rate of 5 percent. The cost of conserved energy in this case is calculated as follows:

$$\begin{aligned} CCE &= (\$60/400\text{kWh/y}) \times [(0.05/\text{y}) \div (1 - (1 + 0.05)^{-10})] \\ &= \$0.02/\text{kWh} \end{aligned}$$

Here of the cost of conserving a kilowatt hour is much less than the typical residential electricity price \$0.08/kWh.

A collection of conservation measures can be ranked by increasing CCE. The measures with the lowest CCE are the most economically attractive. A measure is cost-effective if its cost of conserved energy is less than the price of the energy it displaces. For example, if a lighting retrofit has a CCE of 3 cents/kWh, then it will be worth doing wherever the electricity tariffs are above 3 cents/kWh. Note that the price of energy does not enter into the CCE calculation, only the decision about economic worthiness.

SUPPLY CURVES OF CONSERVED ENERGY

A supply curve of conserved energy is a devices for displaying the cumulative impact of a sequence of conservation measures. It shows the potential energy savings and CCE of each measure. Figure 1 is an example of a supply curve of conserved electricity for a commercial refrigerator. Each step represents a conservation measure. The step's width is its energy savings and the height is its cost of conserved energy.

The supply curve is useful because it shows which measures should be selected first—the ones on the left—and the cumulative energy savings. Measures with CCEs less than the price of the saved energy are cost-effective. In the example, an energy price line has been drawn to show the cut off point; those measures below the energy price line are cost-effective.

Behind the supply curve approach is a consistent bookkeeping framework. The same data for each conservation measure must be collected and the same CCE calculation performed. This encourages comparison among measures and is important when trying to assess the overall impact of many small measures. Consistent treatment also permits generalizations about the impact of alternative sequences of measures and errors in estimates of energy savings, and minimizes double-counting of energy savings. For example, if a measure is implemented before its position in the sequence shown on the curve, then the energy savings will equal or exceed those indicated, and the CCE will be lower than in the original calculation. These features make the overall approach and results more robust even when some numbers are not accurately known.

The supply curve of conserved energy is useful when trading off the benefits of additional supply against reduced demand through energy conservation, such as homes operating on photovoltaic power systems. There, the costs of supplying additional electricity can easily be compared to the costs of reducing electricity demand because both are

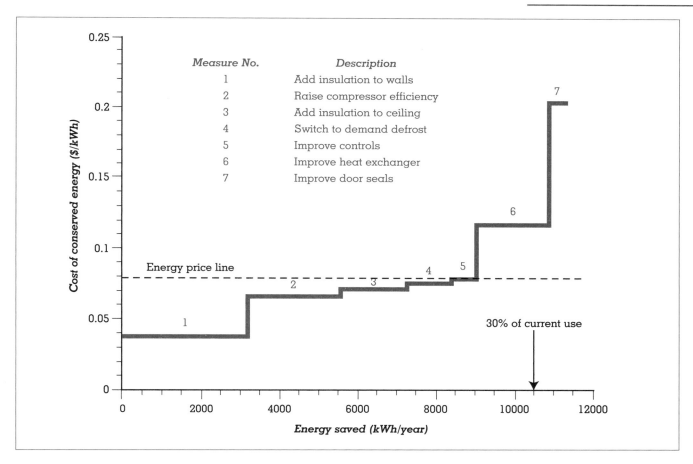

Figure 1.

A micro supply curve of conserved energy for a large commercial refrigerator. Each step represents a conservation measure. The numbers above the steps are keyed to the measure descriptions in the legend. Note that measures 1–4 are cost-effective because the cost of conserved energy is less than the energy price. If energy prices rise, then measures 6 and 7 may also become cost-effective.

expressed in same units, that is, cents/kWh. The economically optimal system will occur when the costs of supply and conservation are equal.

MACRO SUPPLY CURVES OF CONSERVED ENERGY

Figure 1 depicts a micro supply curve of conserved electricity for a single device; however, it is also useful to make macro supply curves of conserved energy, showing the potential cost and energy savings from widespread installation of conservation measures. Figure 2 shows an example of a supply curve of conserved electricity for the U.S. residential sector. Again, each step represents a conservation measure, but here the savings apply to the entire

stock rather than to a single unit. This figure shows the technical potential and should not be confused with a forecast.

These aggregated, or macro, curves are especially useful for policymakers because they show the potential tradeoffs between energy conservation policies and investments in new supplies. This is otherwise difficult because most energy conservation measures are small, highly dispersed, and cannot be instantly undertaken, while energy supplies (such as power plants) typically appear in a few large units.

Consistent bookkeeping is also an important feature of the macro supply curve of conserved energy. Each measure requires, in addition to the data used to calculate the CCE data on the stocks of equipment, turnover rates, etc. The consistent inputs encourage

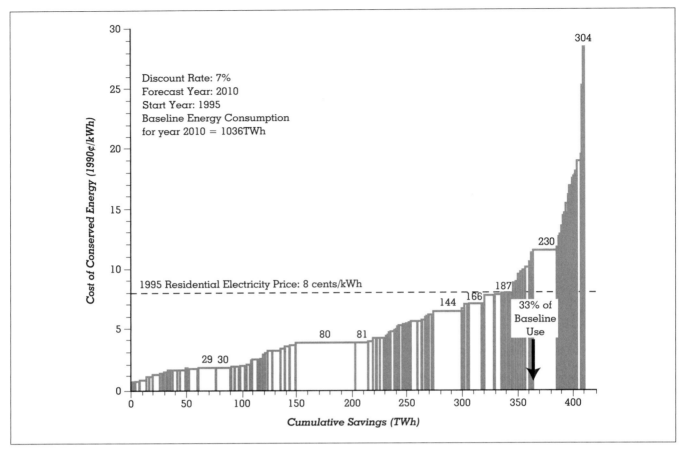

Figure 2.
A macro supply curve of conserved electricity for the U.S. residential sector. Key assumptions are given inside the chart. This supply curve shows estimated savings potentials from 304 different measures. The associated table describing the measures is too long to present here, but certain measures with numbers on top of them are noteworthy from a policy perspective. For example, measure 80 is conversion from conventional water heaters to heat pump water heaters.

confidence in comparisons among measures and in their cumulative impacts.

LIMITATIONS OF THE CCE AND SUPPLY CURVES

Some conservation measures do not easily fit into the form of an initial investment followed by a stream of energy savings because there will be other costs and benefits occurring during the measure's operating life. Furthermore, it is difficult to incorporate peak power benefits into the CCE approach. In these situations, a more precise analysis will be necessary. However, the CCE and supply curve approaches provide first-order identifications of cost-effective conservation, those

measures that should be implemented first, and the overall size of the conservation resource.

Alan K. Meier
Arthur H. Rosenfeld

BIBLIOGRAPHY

American Council for an Energy-Efficient Economy. (1986). *Residential Conservation Power Plant Study.* Washington, DC: ACEEE.

Hunn, B. D.; Baughman. M. L.; Silver, S. C.; et al. (1986). *Technical Potential for Electrical Energy Conservation and Peak Demand Reduction in Texas Buildings.* Austin, TX: Center for Energy Studies, University of Texas.

Interlaboratory Working Group. (1997). *Scenarios of U.S. Carbon Reductions: Potential Impacts of Energy-Efficient* and *Low-Carbon Technologies by 2010 and Beyond.* ORNL-444. Oak Ridge, TN: Oak Ridge National Laboratory.

Meier, A. K. (1982). "Supply Curves of Conserved Energy." Ph.D. diss. University of California, Berkeley.

Meier, A. K. (1983). "What is the Cost to You of Conserving Energy?" *Harvard Business Review* 61(1):36–38.

Meier, A. K.; Wright, J.; and Rosenfeld, A. H. (1983). *Supplying Energy Through Greater Efficiency.* Berkeley, CA: University of California Press.

Panel on Policy Implications of Greenhouse Warming of the National Academy of Sciences. (1992). *Policy Implications of Greenhouse Warming: Mitigation, Adaptation, and the Science Base.* Washington, DC: National Academy Press.

Ruegg, R. T., and Marshall, H. E. (1990). *Building Economics: Theory and Practice.* New York: Van Nostrand Reinhold.

CONSUMPTION

Energy is a commodity, and like any commodity, its level of consumption is largely a reflection of its price. However, unlike most other commodities, energy is not valued as a good unto itself, but as a means to achieve an end—the power behind the technology that makes it possible to do more and improve standards of living. For this reason, huge sums are spent by governments, utilities, and businesses to gather the statistics to forecast consumption patterns. They want to know not only how much energy can be found, extracted, transported, and converted to useful forms, but also how that energy powers today's technology and the emerging technologies of tomorrow.

CONSUMPTION IN THE UNITED STATES

From 1967 to 1973, U.S. energy consumption dramatically increased, from 57.57 quadrillion Btus to 74.28 quadrillion Btus (see Figure 1). In these years leading up to the 1973 Arab oil embargo, energy was inexpensive and growth in consumption was closely linked to population and economic growth. Forecasters could look at population and economic growth trends and accurately project a similar growth in energy consumption. The oil price shock changed this dynamic by spurring energy-efficiency improvements. Overall energy consumption declined to 70.52 quadrillion Btus by 1983 thanks, in large part,

to dramatic improvements in energy efficiency. People in the United States, as well as much of the developed world, learned how to do more with less. More fuel-efficient cars, new energy-efficient homes and office buildings, the retrofitting of existing structures, and more energy-stingy appliances and lighting all played a part in lowering energy consumption from the mid-1970s to the mid-1980s.

The years 1983 through 2000 were marked by stable or dropping energy prices. Adjusting the price of energy for inflation, it was about the same in 1998 as it was in 1967. In the case of gasoline, the inflation adjusted price was actually lower in 1999 than it was before the oil embargo. As a result, people put less effort into conserving energy and became more open to incorporating new energy-hungry devices into their lives. Products that were virtually nonexistent in the 1960—dishwashers, microwaves, telephone answering machines, and personal computers to name a fe—are now considered necessities by many. Americans felt an urge to splurge as consumption gradually reached 94.21 quadrillion Btus by 1997.

Energy-efficiency improvements have offset much of the increased demand from these new appliances. But demand has increased, and will continue to increase, as the combination of continually increasing disposable income and cheap energy accelerates ever more novel inventions and innovations requiring energy.

The U.S. Department of Energy (DOE) projects that the prices of oil and natural gas will rise modestly until the year 2020, and the price of coal and electricity will fall approximately 20 percent by then, which will help push energy consumption to 119.4 quadrillion Btus by 2020, more than double the level of consumption in 1967.

THE DUAL ROLE OF EFFICIENCY

Efficiency has a dual role. While improvements were being made on the use side, major efficiency improvements were happening on the conversion side—the converting of heat energy to electricity. From 1970 to 2000, the efficiency of combined-cycle natural gas plants jumped from about 38 percent to 58 percent. The average efficiency of coal-burning steam turbine plants also improved, with the average for better units attaining 34 to 38 percent efficiency, and a few new units reaching the 45 to 48 percent range.

Efficiency improvements benefit a nation because

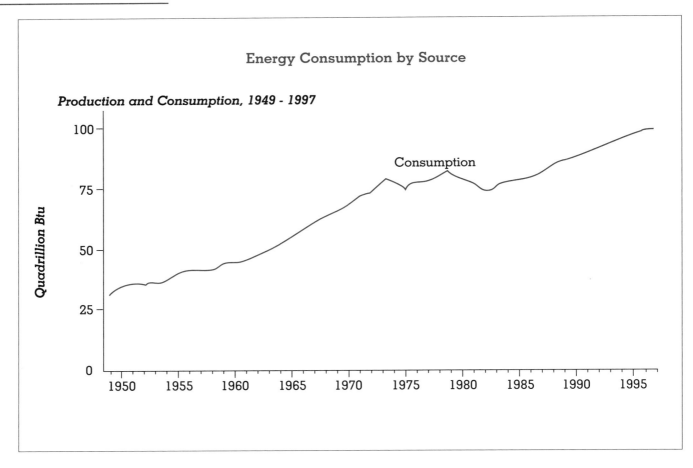

Figure 1.

NOTE: 1×10^{15} British thermal units = One quadrillion Btus

SOURCE: Department of Energy, *Annual Energy Review 1997*.

it takes half as much coal to produce a given amount of electricity with a 40 percent efficient coal-fired steam plant in 2000 as it did with a 20 percent efficient plant in 1970. Even though electricity consumption is expected to increase by more than 20 percent from 2000 to 2020, inefficient plants will continue to be retired, retrofitted, or replaced by efficient plants, so that it will take far less than 20 percent more fossil fuel to meet this greater consumption.

A secondary benefit is that efficiency gains in fossil fuel generation also reduce all types of harmful emissions, even carbon dioxide—the greenhouse gas suspected by many as a major culprit of climate change. A 45 percent efficient plant releases approximately 40 percent less CO_2 per megawatt-hours of electricity produced than a 25 percent efficient plant that it might be replacing.

ENERGY CONSUMPTION BY SOURCE

The U.S. government has tracked consumption by energy source since 1949, the year when petroleum overtook coal as the major source of energy (see Figure 2). Petroleum consumption continued to increase throughout the 1950s and 1960s due to increases in transportation and industrial demand, and as a coal replacement for heating and electric power generation. Natural gas consumption also increased during this period as it became the fuel of choice for home heating.

Following the Arab oil embargo of 1973, natural gas consumption declined until the mid-1980s primarily due to the assumption that the nation was running out of natural gas and because of legislation outlawing the use of natural gas for "low priority" uses. Energy conservation efforts in the industrial, commercial, and residential sectors, primarily the improved energy efficiency of new furnaces and boilers, also were instrumental in this decline. Petroleum consumption peaked later, in 1978, and then began to fall as older vehicles were replaced by more fuel-efficient models, and because of the effort of utilities to switch from petroleum as a fuel for generating electricity.

In the early 1970s, coal consumption once again equaled its earlier peak in the early 1950s and continued to grab a larger share of the electricity-generation market due to the price and supply problems of petroleum and natural gas.

Beginning in 1986 and through the 1990s, natural gas consumption rose again as the Federal Energy Regulatory Commission began deregulating natural gas, and natural gas electricity generation became the choice due to innovations improving the efficiency of generating technology. These new plants were not only more efficient than coal-fired plants, but also less expensive and time-consuming to construct. By 1998, natural gas consumption equaled its 1972 peak of 22.6 quadrillion Btus.

Many nuclear power plants were ordered in the 1960s and early 1970s, but construction slowed in the mid-1970s and halted in the early 1980s because of the high cost of construction, problems with radioactive waste disposal, and political obstacles. Despite no new power plants being built, better management and technology resulted in more energy generation during the late 1980s and 1990s. In 1996, nuclear facility efficiency (the amount of power generated divided by the maximum possible generation) reached an all-time high of 76.4 percent. Although the amount of nuclear-generated electricity more than doubled between 1980 and 1996 (2.74 to 7.17 quadrillion Btus), the future contribution is certain to fall through 2020 for three important reasons: limited potential for further gains in efficiency, many nuclear facilities are scheduled for retirement, and no new facilities are planned.

Out of the 7 quadrillion Btus contributed by renewable energy, more than 95 percent comes from hydroelectric power and biofuels (waste energy, wood energy, and alcohol). Geothermal, solar, and wind are all very minor contributors. Renewable energy's share is unlikely to grow because hydroelectric power faces political and environmental concerns about dams, no new geothermal sites are planned, and biofuel potential is limited. Another major factor hindering growth in renewables was the much lower

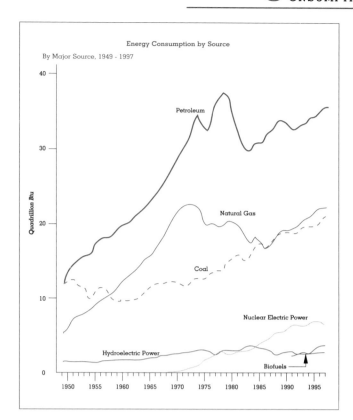

Figure 2.
Note: Because vertical scales differ, graphs should not be compared.
SOURCE: Department of Energy, *Annual Energy Review 1997*.

than expected electric generating cost for coal and natural gas. The projections made in about 1980 for the year 2000 were way off target. In terms of dollars per million Btus of energy, coal was widely projected to reach $3 to $5, not $1, and natural gas was projected to reach $4 to $8, not $2.

In the future, if the criterion for selecting new generating capacity was solely fuel cost, coal will be the number one choice. But the much greater costs of coal-fired plants (primarily to meet local and federal emission standards), as well as the potential of tighter standards, will make gas more attractive in many cases. And although natural gas prices may rise, the fuel costs per kilowatt-hour for gas-fired power plants should remain unchanged as efficiency gains offset the rise in fuel prices.

CONSUMPTION BY SECTOR

Among the four major sectors of the economy—residential, commercial, industrial, and transportation—

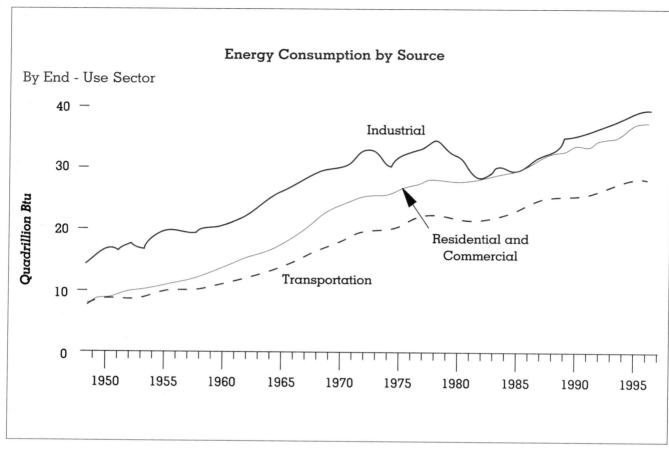

Energy Consumption by Source

By End - Use Sector

Figure 3.

SOURCE: Department of Energy, *Annual Energy Review 1997.*

the industrial sector has historically been the greatest user of energy (see Figure 3). The industrial sector includes energy-intensive industries such as those for paper, metals, chemicals, and petroleum products as well as light-energy industries such as textiles and furniture manufacturers.

In 1951, industrial consumption was 17.13 quadrillion Btus, nearly as much as that for residential and commercial and transportation combined. From the 1950s through the early 1970s all three sectors continued to grow, with the residential and commercial sector showing the greatest increase, a reflection of the growth in office and housing stock.

The period 1973 to 1983 was one in which industrial consumption declined significantly, falling from 31.53 quadrillion Btus to 25.75 quadrillion Btus. There were a number of factors responsible for this steep drop. Foremost among them were the increase

in the price of energy, a drop in output in many energy-intensive industries, and vast improvements in efficiency. Per thousand dollars of manufacturing sales, energy use declined by more than 25 percent between 1977 and 1997. The concerted energy-efficiency efforts of the industrial sector resulted in consumption in the residential and commercial sector briefly surpassing the industrial sector by 1986.

From 1983 to 1998, all three sectors of the economy experienced an increase in consumption largely because the price of all energy sources dropped. In the residential and commercial and the industrial sectors, the greatest growth was in natural gas and electricity consumption. Although new homes were more energy-efficient, they were on average much larger and filled with more energy-hungry housing features that either did not exist or were considered luxuries—air conditioning, Jacuzzis, microwaves, dishwashers, and security systems, to name a few—for home buyers in the 1970s. Some of the consumption increase can be attributed to the need to space-condition more area, as the average home size expanded from 1,600 square feet in 1970 to 2,100 square feet in 1998. The amount of commercial sector space also expanded, which required more energy for air conditioning, heating, and lighting.

Consumption in the transportation sector historically has been less than either the industrial sector or the residential and commercial sector, but it is of greater concern because consumption is almost entirely petroleum as opposed to a mix of petroleum, natural gas, and electricity. And because electricity can be generated from a number of different sources, and because there is greater opportunity to substitute one source of energy for another in generating electricity, the price and supply security that exists in other sectors does not apply to the transportation sector.

Another problem with the dependence of transportation on petroleum is the enormous military, strategic, economic, and political costs of securing the petroleum supply. Estimates of the military costs alone associated with the Mideast region for the period from 1980 to 1993 ranged from $100 billion to more that $1 trillion.

Transportation is also the emissions leader. About 75 percent of carbon dioxide emissions and 45 percent of nitrogen oxide emissions come from the transportation sector. If rising levels of CO_2 are found to be responsible for global warming, and measures are put in place to severely curtail CO_2 emissions, the measures will have the greatest impact on the transportation sector.

As the other sectors have shifted from petroleum beginning in the mid-1970s, the transportation sector has not because of the formidable petroleum-consuming internal-combustion engine. Although competing technologies (electric, fuel cells, and hybrid vehicles) create less emissions, internal-combustion-engine technology continues to make dramatic steps in limiting emissions. It is likely to remain the dominant means of transportation propulsion through 2020, which is why petroleum consumption in the transportation sector will remain high. In 1997, a total of 24.04 quadrillion Btus out of the 36.31 quadrillion Btus of petroleum consumption found use in the transportation sector, consuming nearly all the 24.96 quadrillion Btus that had to be imported.

Close to 65 percent of the petroleum consumed for transportation is gasoline, followed by 18 percent for diesel fuel and 13 percent for jet fuel. The DOE expects that the rise of "cleaner burning" alternative fuels should lower the share of gasoline and diesel, and expects greater air travel to increase the share of jet fuel.

Vehicle technology exists to improve fuel economy to 45 miles per gallon (if 45 miles per gallon was the average, the United States could be a net exporter of petroleum), yet the average consumer of the 1990s values performance and other features far more than fuel economy. This shifting priority away from fuel economy is why the steady improvements in gas mileage through the late 1970s and 1980s halted in 1990s. The trends of more people, more vehicles per person, and more miles driven per vehicle all point toward transportation petroleum consumption increasing significantly in the future.

THE WORLD PICTURE

Despite the Asian economic crisis of the late 1990s, the world is projected to consume more than twice as much energy in 2000 as in 1970, and more than three times that amount by 2020 (see Figure 4). In the world at large, energy consumption is greatest in North America, followed by Europe and the Pacific Rim countries. In the early 1970s, approximately 60 percent of all energy consumption occurred in North America. Consumption growth of the Asian Pacific economies led the world in the 1980s and 1990s, which brought the North American share of energy consumption down, to about 50 percent. Still, looked at from a consumption per capita basis, the 4 percent of the world population who reside in the United States consume 50 percent of all energy. Many feel that this fact could have dire consequences. If the developing world follows the energy-intensive U.S. model, the world will have to deal with the environmental problems associated with much greater energy use.

The United States is by far the most energy-intensive user. Americans consume more than twice as much energy per capita as in Europe or Japan, and are the largest emitter of CO_2. Of the 6 billion metric tons of carbon emissions worldwide in 1996, the United States emitted 1.466 billion metric tons, followed by China at 805 million, Russia at 401 million, and Japan at 291 million. Although many countries consume more energy than China, China consumes

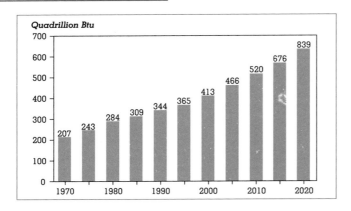

Figure 4.

World Energy Consumption 1970-2020.

SOURCE: U.S. Energy Information Administration.

vast amounts of coal—the worst fossil fuel for CO_2 emissions.

As long as the source for consumption is mainly fossil fuels, who is doing the consuming and how much is being consumed promise to be two of the most contentious aspects of international relations for the foreseeable future. Energy consumption presents a troubling paradox: Greater energy consumption is the result of a higher standard of living, but this greater consumption also portends greater, and often unacceptable, environmental costs.

John Zumerchik

See also: Biofuels; Coal, Consumption of; Economically Efficient Energy Choices; Economic Growth and Energy Consumption; Emission Control, Vehicle; Fossil Fuels; Gasoline Engines; Hydroelectric Energy; Natural Gas, Consumption of; Petroleum Consumption; Population Growth and Energy Consumption; Supply and Demand and Energy Prices; Transportation, Evolution of Energy Use and.

BIBLIOGRAPHY

BP Amoco. (1998). *BP Statistical Review of World Energy.* London: Author.

Green, D. L. (1996). *Transportation and Energy.* Washington, DC: Eno Transportation Foundation.

International Energy Agency. (1998). *Energy Statistics of OECD Countries.* Paris: Author.

Nye, D. E. (1997). *Consuming Power: A Social History of American Energies.* Cambridge, MA: MIT Press.

U.S. Energy Information Administration. (1998). *Annual Energy Review 1997.* Washington, DC: Author.

U.S. Energy Information Administration. (1998). *Annual Energy Outlook 1999.* Washington, DC: Author.

U.S. Energy Information Administration. (1998). *International Energy Outlook 1998, with Projections through 2000.* Washington, DC: Author.

CONTROL SYSTEMS, HISTORY OF

INTRODUCTION

The control of energy in its various forms was always a necessity that became more relevant with the increasing performance requirements of the twentieth century. The control of energy conversion contributes to the optimization in performance and energy efficiency for all processes, machines, and devices.

Technology developments from 1960 to 2000 in the areas of microelectronics and power electronics made possible the development of more complex, efficient and reliable energy controls. In this century there were significant technology mutations, with controls going from mechanical to electromechanical devices, evolving gradually to full electronic controls without moving parts. Since the 1970s electronic controls have been implemented more and more with programmable systems through the use of microcomputers.

The complexity and reduced time constants of modern processes imply the adoption of high performance programmable controllers. This requires not only higher processing speed but also more advanced control algorithms that can optimize the process operation in real time.

Today, with the pressing need to achieve sustainable development, the reduction of the energy losses and the optimization of all processes has promoted the continuous development and implementation of advanced energy control systems in all sectors.

CONTROL SYSTEMS

Basic Concept of the Control

There is a need to have some measurements or observations made on the relevant variables of the controlled system. The data is compared to a reference, and that will cause some feedback on the process to be controlled, in order that value of the controlled variables approaches the desired reference value.

Types of Controls

Manual Controls. The first methods used in energy control involved human intervention. The operator was the sensor (i.e., using his eyes, ears, and hands or using additional devices to quantify the values of the controlled variables), and he was also the actuator controller. The control of the processes was slow and very ineffective. For example, in an old steam engine control the human operator sees the instantaneous pressure and then manually regulates the power of the device (e.g., by adding fuel to a boiler). But in today's industrial reality, this control is not only ineffective but in most cases is not possible.

Another example is the electric generators' excitation control. Early systems were manually controlled (i.e., an operator manually adjusted the excitation system current with a rheostat to obtain a desired voltage). Research and development in the 1930s and 1940s showed that applying a continuously acting proportional control in the voltage regulator significantly increased generator performance. Beginning in the 1950s most of the new generating units were equipped with continuously acting electronic automatic voltage regulators (AVRs).

Mechanical Controls. Mechanical controls have been widely used in steam and internal combustion engines. Although they are low-cost, they can only implement simple control strategies.

One of the oldest energy control systems is the steam engine speed control device, developed by James Watt, consisting in the regulation of motor speed through input steam flow. This device is purely mechanical, and its physical principle is shown in Figure 1.

The angular velocity (ω_{engine}) control of the steam engine load shaft is controlled by a steam valve. As the angular velocity rises, the centrifugal force ($F_{centrifugal}$) increases in the spheres, becoming higher than spring force. The result is a displacement (D_{linear}) of the shaft that partially closes the steam valve. If, due to a load perturbation, the velocity is higher than desired, the increasing speed will force the valve to partially close, decreasing the steam flow. This is an example of a simple negative feedback closed loop

mechanical control. The human operator defines an input (desired engine angular velocity), with the regulation of the shaft length $l_{regulation}$ and the vertical position of the spheres is the angular speed sensor.

Electromechanical Controls. Electro-mechanical control devices are typically used for load control (lighting, ventilation, and heating) in buildings with no feedback signal. The most common device is the electromechanical timer, in which a small motor coupled to a gearbox is able to switch electrical contacts according to a predefined time schedule. They are still in use today, applied to loads with simple scheduling requirements.

Mechanical switches provide a simple manual interface to operate all sorts of loads, but they suffer from all the drawbacks of manual control, namely in terms of speed of response, and also requiring permanent operator awareness.

The first power control devices, with electric insulation between the low power input stage and high power output stage, were electromechanical relays, introduced in the late nineteenth century. They are still widely used today. Applying voltage to the magnetic field winding, an attraction force will be generated between the mobile and fixed iron core that will switch the mobile terminal COM (common terminal) between the normally closed (NC) and normally open (NO) position. Their main advantages are high electric input/output insulation, a high input/output power ratio, high efficiency, and low cost. The disadvantages are the limitations in terms of commutation speed and the limited number of operations (10^5–10^7 cycles). With the invention of the transistor, this type of device evolved into the solid-state relay, which has the advantage of smaller size, higher reliability, and lower input power. The operation principle is based on a light emitting diode (LED), fed by a control signal, which with the emitted light will excite the fototransistor. The zero-cross detector module will then control the firing of the solid-state switches (thyristors), closing the load power circuit. In the late 1990s, the integrated silicon electrostatic relay was introduced, available in an integrated circuit package that has the advantages of small size and very low power consumption. However, the output power is very limited, when compared to electromagnetic and solid-state relays. Basically, the input voltage between a conductive fixed plate and a

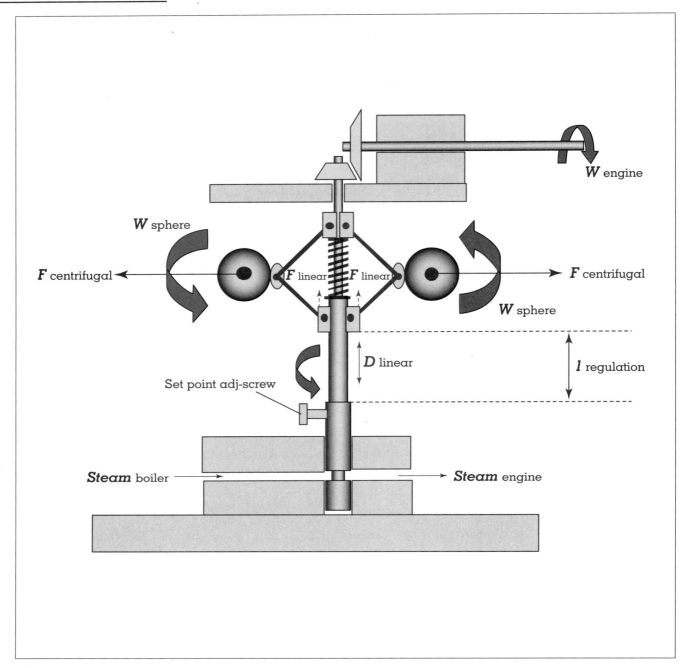

Figure 1.
Example of a pressure control device for a steam engine.

flexible plate creates an electric field between them, which will generate an attraction force. The flexible plate will be deflected and its conductive terminal will switch the contacts.

Electronic Closed Loop Controls. After its development, the concept of closed loop control has become one of the most common tools for systems control. Initially, automatic closed loop controls were widely implemented with electronic analog circuits. Electric power systems were one of the many applications that successfully used these types of controls in power plants.

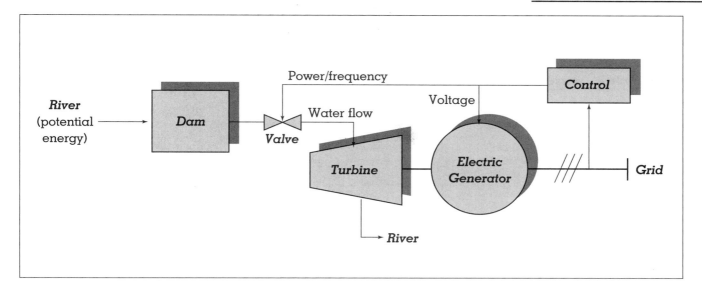

Figure 2.
Simplification of a hydroelectric power plant control system.

In electric power systems, it is essential to have permanent control of the power in electricity production, transportation, and consumption. Because of speed and reliability requirements, electric power systems were the first large systems to use a variety of automatic control devices for the protection of different parts of the system.

A variety of electromechanical relays was used for the detection of abnormal operating conditions (e.g., overload, short-circuit, etc.), leading to the isolation of the faulty components. In the second half of the twentieth century electromechanical relays were progressively replaced by electronic controls that are capable of faster response and higher reliability.

Power generation also makes use of electronic energy controls. A typical hydroelectric power plant using river water flow as its energy source is organized into three main subsystems, corresponding to the three basic energy conversions taking place in the process (Figure 2): the dam, the turbine, and the electric generator (or alternator). The dam stores the water and, using a valve, controls the output water flow. The turbine transforms the kinetic energy of water flow into turbine mechanical power and makes it available at the shaft for final conversion into electrical power in the electric generator. In the generators the electric power and frequency variables are related to the flow of water (or steam in the case of

thermal power plants) in the turbines, and the basic closed loop control involves the reading of the two variables: power and frequency. The controller regulates the function of the valves and thus the turbine valve opening. The first power plants used a speed governor in which the valves were controlled by a mechanism similar to the one described in Figure 1. To guarantee constant voltage in the terminals, an automatic voltage regulator (AVR) is necessary, controlling the generator excitation current as a function of the generator output voltage.

A typical excitation system includes a voltage regulator, or exciter, protection circuits, and measurements transducers. If the terminal voltage decreases below rated value, the control system increases the rotor current and thus the magnetic field; as shown in Faraday's law, the generated voltage is thus forced up to the desired value.

Microcomputer Based Electronic Systems. The introduction of microelectronics and computers in energy control systems enabled the implementation of complex closed-loop control algorithms. A typical example of this type of control is an oven temperature control. Figure 3 depicts a diagram of temperature control in an electric oven. The oven temperature is measured by a sensor that gives an analog output. The sensor output signal is converted to a digital signal using an analog to digital converter and then transmitted to a computer that has a control program. The input temperature is compared with a

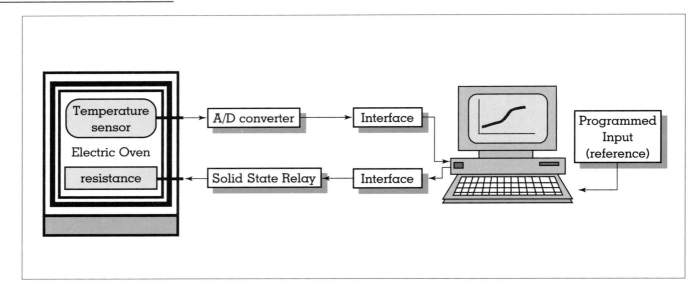

Figure 3.
Temperature control system for an electric oven.

programmed reference temperature, and the computer generates a correcting output to the heater using an interface and a solid-state relay to force the oven temperature to the desired value. A common control algorithm uses a correction signal which is a combination of proportional plus integral plus differential (PID) terms based on the error (difference between desired and actual temperatures).

One of the most recent trends in energy control is the use of fuzzy logic. Fuzzy logic is a multivalued logic that allows intermediate values to be defined between conventional evaluations such as yes/no, true/false, and black/white. Notions such as "rather warm" or "pretty cold" can be formulated mathematically and processed by computers. In this way an attempt is made to apply a more human-like way of thinking in the programming of computers. The employment of fuzzy control is commendable for very complex processes, when there is no simple mathematical model, in highly nonlinear processes, and if the processing of expert knowledge (linguistically formulated) is to be performed. Some applications for fuzzy logic control in the energy field are: automated control of dam gates for hydroelectric power plants, prevention of unwanted temperature fluctuations in air-conditioning systems, improved efficiency and optimized function of industrial control applications, control of machinery speed and temperature for steel works, improved fuel con-

sumption for automobiles, and improved sensitivity and efficiency for elevator control. Fuzzy logic (control) is a way of interfacing inherently analog processes that move through a continuous range to a digital computer that likes to see variables as well-defined numeric values.

Let us consider for example the system to control the temperature in a building is directed by a microcontroller that has to make decisions based on indoor temperature, outdoor temperature, and other variables in the system. The variable temperature in this system can be divided into a range of "states": "cold," "cool," "nominal," "warm," and "hot." However, the transition from one state to the next is hard to pin down. An arbitrary threshold might be set to divide "warm" from "hot," but this would result in a discontinuous change when the input value passed over that threshold. The microcontroller should be able to do better than that. The way around this is to make the states "fuzzy," that is, allowing them to change gradually from one state to the next. The input temperature states can be represented using "membership functions."

The input variables' state now no longer jumps abruptly from one state to the next, but loses value in one membership function while gaining value in the next. At any one time, the "truth value" of the indoor or outdoor temperature will almost always be in some degree part of two membership functions:

0.6 nominal and 0.4 warm, or 0.7 nominal and 0.3 cool, and so on. Given "mappings" of input variables into membership functions and truth values, the microcontroller then makes decisions as to which actions to take based on a set of "rules" that take the form:

IF *indoor temperature* IS *warm*
AND *outdoor temperature* IS *nominal*
THEN *heater power* IS *slightly decreased.*

Traditional control systems are in general based on mathematical models that describe the control system using one or more differential equations that define the system response to its inputs. In many cases, the mathematical model of the control process may not exist or may be too "expensive" in terms of computer processing power and memory. In these cases a system based on empirical rules may be more effective. In many cases, fuzzy control can be used to improve existing controller systems by adding an extra layer of intelligence to the current control method.

Analog Displays

Analog displays are simple devices like mercury thermometers and pressure gauges in which the variation of a physical variable causes a visible change in the device display. This information can be used for monitoring purposes and for manual process control. Although simple and low-cost, most analog displays suffer from poor accuracy. Analog displays have been progressively replaced by electronic sensors.

Sensors

Both electronic and microcomputer-based controls require information about the state of the controlled system. Sensors convert different physical variables into an electric signal that is conditioned and typically converted to a digital signal to be used in microcontrollers. The trend in the construction techniques of modern sensors is the use of silicon microstructures because of the good performance and the low cost of this type of device. In the energy control scope the main quantities to be measured are the temperature, pressure, flow, light intensity, humidity (RH), and the electric quantities of voltage and current.

Temperature. The simplest temperature sensor/ control systems typically use a bimetallic thermostat that integrates two superimposed metal plates with different expansion coefficients. Thus, the deflection of the beam caused by a temperature increase will cause the opening of the contacts. This device is typically used to control electrical heater equipment operation. The most common used devices in temperature measurement are thermocouples, thermistors, semiconductor devices, platinum resistance thermometers, and infrared radiometers.

One widely used temperature sensor is the integrated circuit AD590 introduced by Analog Devices. It generates a current whose value in μA (microamperes) is equal to the temperature in degrees Kelvin (K).

Humidity. Any instrument capable of measuring the humidity or psychrometric state of air is a hygrometer, and the most common types used are psychrometers, dew-point hygrometers, mechanical hygrometers, electric impedance and capacitance hygrometers, electrolytic hygrometers, piezoelectric sorption, spectroscopic (radiation absorption) hygrometers, gravimetric hygrometers, and calibration. The sensor response is related to factors such as wet-bulb temperature, relative humidity, humidity (mixing) ratio, dew point, and frost point.

Pressure. Pressure so defined is sometimes called absolute pressure. The differential pressure is the difference between two absolute pressures. The most common types of pressure-measuring sensors are silicon pressure sensors, mechanical strain gauges, and electromechanical transducers.

Fluid Velocity. The flow of air is usually measured at or near atmospheric pressure. The typical instruments to measure fluid velocity are airborne tracer techniques, anemometers, pilot-static tubes, measuring flow in ducts.

Flow Rate. The values for volumetric or mass flow rate measurement are often determined by measuring pressure difference across an orifice, nozzle, or venturi tube. Other flow measurement techniques include positive displacement meters, turbine flowmeters, and airflow-measuring hoods.

Light. Light level, or illuminance, is usually measured with a photocell made from a semiconductor such as silicon. Such photocells produce an output current proportional to the illumination on the sensor incident luminous density.

Occupancy. In the scope of energy management in buildings, occupancy is an important parameter.

This type of sensor typically includes infrared (IR) and ultrasound sensors. IRs detect the heat released from humans, and ultrasound sensors detect the movement of the human occupant (i.e., the device compares the reflection in different instants; if they are different, something is moving in the sensor distance range).

EXAMPLES OF ENERGY AND LOAD CONTROL

Lighting Controls

Generally in all sectors, light energy consumption is very significant, and typically, control has been done manually or by electromechanical timers. The most sophisticated systems integrate sensors and programmed microcomputer controls. In less important places or with irregular human presence, it is common to use occupancy sensors that activate the lights. Photosensors packaged in various configurations allow the control ambient lighting levels using building automation strategies for energy conservation. Some examples include ceiling-mounted indoor light sensors that are used to continuously dim the available lights in order to produce the desired light level.

Motor Controls

Most induction ac motors are fixed-speed. However, a large number of motor applications would benefit if the motor speed could be adjusted to match process requirements. Motor speed controls are the devices which, when properly applied, can tap most of the potential energy savings in motor systems. Motor speed controls are particularly attractive in applications where there is variable fluid flow. In many centrifugal pump, fan, and compressor applications mechanical power grows roughly with the cube of the fluid flow. To move 80 percent of the nominal flow only half of the power is required. Centrifugal loads are therefore excellent candidates for motor speed control. Other loads that may benefit from the use of motor speed controls include conveyers, traction drives, winders, machine tools and robotics.

Conventional methods of flow control used inefficient throttling devices such as valves, dampers, and vanes. These devices, although they have a low initial cost, introduce unacceptable running costs due to their inefficiency. Several speed control technologies can be used to improve motor system operation.

Electronic variable speed drives are the dominant motor speed control technology. Figure 4 shows the power consumed by a motor driving a fan, using different flow control methods. The main benefits of adjusting motor speed through the use of VSDs include better process operation, less wear in mechanical equipment, less noise, and significant energy savings (50% or more for some type of applications).

With the development of power electronics, the introduction of variable speed drives (VSDs) to control induction motor speed has become widespread. VSDs produce a variable frequency and voltage output that will regulate the motor speed and torque. In the case of closed loop control, the use of speed sensors (encoders) allows more precise control of the speed. The most common VSD type is the inverter-based VSD, in which the 3-phase supply is converted from ac to dc using a solid-state rectifier. Afterwards the inverter uses this dc supply to produce a 3-phase adjustable frequency, adjustable-voltage output which is applied to the stator windings of the motor. The speed of the motor will then change in proportion to the frequency of the power supply. Usually output voltage waveforms can be synthesised over the frequency range of 0–100 Hz.

HVAC Controls

Heating, ventilation, and air conditioning (HVAC) system controls are the link between varying energy demands on a building's primary and secondary systems and the approximately uniform demands for indoor environmental conditions. Without a properly functioning control system, the most expensive, most thoroughly designed HVAC system will be a failure. It simply will not control indoor conditions to provide comfort. The main controlled variable in buildings is zone temperature. The control of zone temperature involves many other types of control within the primary and secondary HVAC systems, including boiler and chiller control, pump and fan control, liquid and air flow control, humidity control, and auxiliary system control subsystems. There are two fundamentally different control approaches—pneumatic and electronic. Various kinds of sensors, controllers, and actuators are used for principal HVAC applications.

Pneumatic Control Systems. The first widely adopted automatic control systems used compressed air as the operating medium. A transition to electronic controls has been occurring since the 1970s. Pneumatic controls use compressed air for the operation of sensors and actuators. The most common pneumatic sensor is the temperature sensor. A typical method of sensing temperature and producing a control signal is a bellows (or diaphragm): if the temperature rises, it swells and moves the flapper to open or close the flapper/nozzle gap. This flapper could also be a bimetallic strip. Humidity sensors in pneumatic systems are made from materials that change size significantly (1–2%) with humidity (typically synthetic hygroscopic fibers). Because the dimensional change is relatively small, mechanical amplification of the displacement is used. An actuator converts pneumatic energy to motion—either linear or rotary. It creates a change in the controlled variable by operating control devices such as dampers or valves. The opening of a pneumatically operated control is controlled by the pressure in the diaphragm acting against the spring. Pneumatic controllers produce a branch line pressure that is appropriated to produce the needed control action for reaching the set point.

Electronic Control Systems. Electronic controls are being increasingly used in HVAC systems in commercial buildings. The main advantages are precise control, flexibility, compatibility with microcomputers, and reliability. With the continuous decrease in microprocessor cost and associated increase in capabilities, its cost penalty is virtually disappearing, particularly when calculated on a per function basis. Modern electronic systems for buildings are based in direct digital controllers (DDC) that enhance the previous analog-only electronic system with digital features. Modern DDC systems use analog sensors (converted to digital signals within a computer) along with digital computer programs to control HVAC systems. The output of this microprocessor-based system can be used to control electronic, electrical, or pneumatic actuators or a combination of these. DDC systems have the advantage of reliability and flexibility (it is easier to accurately set control constants in computer software than by making adjustments at a controller panel with a screwdriver). DDC systems also offer the option of operating energy management systems.

Anibal T. De Almeida
Hashem Akbari
Fernando J. T. E. Ferreira

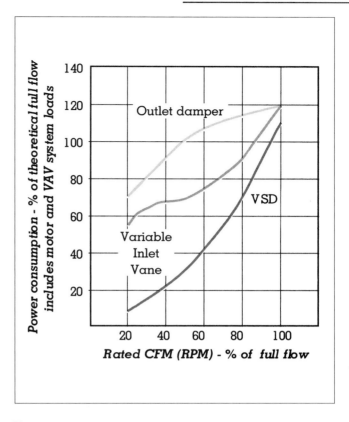

Figure 4.

Power consumption with different flow control methods in a fan system.

See also: Auditing of Energy Use; Behavior; Building Design, Commercial; Building Design, Energy Codes and; Building Design, Residential; Conservation of Energy; District Heating and Cooling; Electric Motor Systems; Heat Transfer.

BIBLIOGRAPHY

de Almeida, A; Bertoldi, P.; and Werner L. (1997). *Energy Efficiency Improvements in Electric Motors and Drives.* Heidelberg, Germany: Springer-Verlag.

Doebelin, E. O. (1990). *Measurement Systems-Application and Design.* New York: McGraw-Hill, Inc.

Eley Associates, Lighting. (1992). *Fundamentals Handbook.* Palo Alto, CA: Author.

Howell, R.; Saner, H.; and Coad, W. (1998). "Principles of Heating, Ventilating and Air Conditioning." Atlanta: ASHRAE, Inc.

Ogata, K. (1990). *Modern Control Engineering.* Englewood Cliffs, NJ: Prentice-Hall.

Weedy, B. M. (1998). *Electric Power Systems.* London: John Wiley & Sons.

CONVECTION

See: Heat Transfer

COOL COMMUNITIES

As modern urban areas have grown, there has been a corresponding growth in darker surfaces and a decline in vegetation, affecting urban climate, energy use, and habitability. Dark roofs heat up more and thus raise the summertime cooling demands of buildings, collectively with reduced vegetation, warming the air over urban areas and creating "heat islands." On a clear summer afternoon, the air temperature in a typical city can be as much as 2.5°C (4.5°F) higher than in surrounding rural areas. (The peak heat island effect occurs during cold winter evenings and is caused primarily by the rapid cooling of the rural areas—the thermal storage of pavements and dark-roofed buildings is much greater than greenery.) Peak urban electric demand rises by 2 to 4 percent for each 1°C (1.8°F) rise in daily maximum temperature above 15 to 20°C (59 to 68°F), so the additional air-conditioning use caused by higher urban air temperature is responsible for 5 to 10 percent of urban peak electric demand, costing U.S. ratepayers several billion dollars annually.

Temperatures in cities are increasing. Figure 1 depicts summertime monthly maximum and minimum temperatures between 1877 and 1997 in downtown Los Angeles, clearly indicating that these maximum temperatures are now about 2.5°C (4.5°F) higher than in 1920. Minimum temperatures are about 4°C (7.2°F) higher than in 1880. In California from 1965 to 1989, the average urban-rural temperature differences, measured at thirty-one pairs of urban and rural stations, have increased by about 1°C (1.8°F). In Washington, D.C., temperatures rose 2°C (3.6°F) between 1871 and 1987. This recent warming trend is typical of most U.S. metropolitan areas and exacerbates demand for energy. In Los Angeles, we estimate a heat-island induced increase in power consumption of 1 to 1.5 GW, costing ratepayers $100 million per year.

Besides increasing systemwide cooling loads, summer heat islands increase smog production. Smog production is a highly temperature-sensitive process. At maximum daily temperatures below 22°C (71.6°F), maximum ozone concentration in Los Angeles is below the California standard of 90 parts per billion. At 35°C (95°F), practically all days in Los Angeles are smoggy (see Figure 2).

HEAT ISLAND MITIGATION

When sunlight hits an opaque surface, some energy is reflected (this fraction is known at the albedo or reflectivity); the rest is absorbed. Use of high-albedo urban surfaces and planting urban trees are inexpensive measures that reduce summertime temperatures. The effects of planting trees and increasing albedo are both direct and indirect. Planting trees around a building or using reflective materials on roofs or walls has a direct effect: altering the energy balance/cooling requirements of that building. Planting trees and modifying albedo throughout the city has an indirect effect: citywide climate modification. By reducing air temperature, the air conditioning requirements of all buildings are reduced. Planting trees also sequesters atmospheric carbon through photosynthesis. Figure 3 depicts the overall process that impacts energy use and air quality within an urban area.

COOL ROOFS

When dark roofs are heated by the sun, they directly raise summertime building cooling demand. For highly absorptive (low-albedo) roofs, the surface/ambient air temperature difference may be 50°C (90°F), while for less absorptive (high-albedo) surfaces with similar insulative properties (e.g., white-coated roofs), the difference is only about 10°C (18°F), which means that "cool" surfaces can effectively reduce cooling-energy use.

For example, a high-albedo roof coating on a house in Sacramento, California, yielded seasonal savings of 2.2 kWh/day (80% of base-case use), and peak-demand reductions of 0.6 kW (about 25% of base-case demand). Field studies of nine homes in Florida before and after applying high-albedo coatings to their roofs yielded an air-conditioning energy-use reduction of 10 to 43 percent, saving on average 7.4 kWh/day (19%). The peak-demand reduction at 5:00 P.M. was 0.2 to 1.0 kW, an average of 0.4 kW (22%).

Researchers have simulated the impact of the urbanwide application of reflective roofs on cooling-energy use and smog in the Los Angeles Basin. They

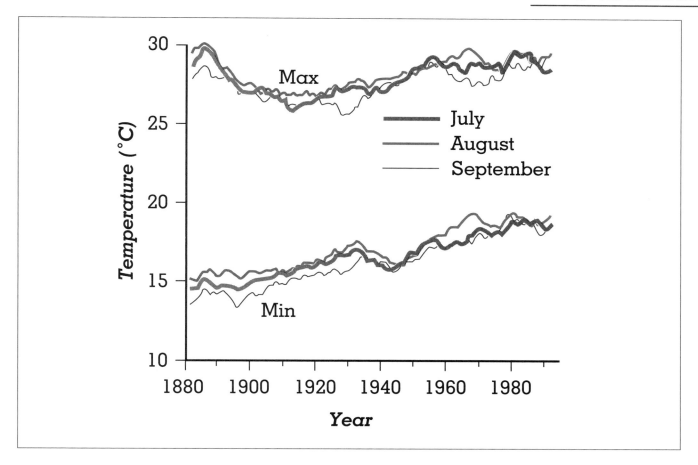

Figure 1.
Ten-year running average summertime monthly maximum and minimum temperatures in Los Angeles, California (1877–1997). The average is calculated as the average temperature of the previous four years, the current year, and the next five years. Note that the maximum temperatures have increased about 2.5°C (4.5°F) since 1920.

estimate that roof albedos can realistically be raised by 0.30 on average, resulting in a 2°C (3.6°F) cooling at 3:00 P.M. on a sunny August day. This temperature reduction significantly reduces building cooling-energy use further. The annual electricity savings in Los Angeles are worth an estimated $21 million. Cooling the air also results in a 10 to 20 percent reduction in population-weighted exposure to smog.

Other benefits of light-colored roofs include a potential increase in their useful life. The daily temperature fluctuation and concomitant expansion/contraction of a light-colored roof is less than that of a dark one. Also, materials degradation because of absorption of ultra-violet light is temperature-dependent. Thus, cooler roofs may last longer than

hot roofs of the same material. Cool roofs incur no additional cost if color changes are incorporated into routine reroofing and resurfacing schedules.

URBAN TREES

The beneficial effects of trees are also both direct and indirect: shading of buildings and ambient cooling. Their shade intercepts sunlight before it warms buildings, and their evapotranspiration cools the air. In winter, trees shield buildings from cold winds. Urban shade trees offer significant benefits by reducing building air conditioning, lowering air temperature, and thus improving urban air quality (reducing smog). Over a tree's life, savings associated with these benefits vary by climate region and can be up to $200

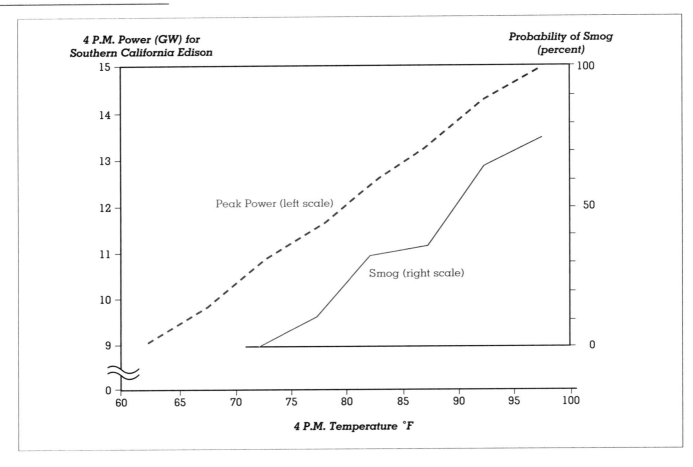

4 P.M. Power (GW) for Southern California Edison

Probability of Smog (percent)

Peak Power (left scale)

Smog (right scale)

4 P.M. Temperature °F

Figure 2.
Ozone levels and peak power for Southern California Edison versus 4 P.M. temperature in Los Angeles as a predictor of smog.

per tree. The cost of planting and maintaining trees can vary from $10 to $500 per tree. Tree planting programs can be low-cost, offering a good return on investments for communities.

Data on energy savings from urban trees are rare but impressive. In one experiment, the cooling-energy consumption of a temporary building in Florida was cut by up to 50 percent after adding trees and shrubs. Cooling-energy savings from shade trees in two houses in Sacramento were about 30 percent, corresponding to average savings of 3.6 to 4.8 kWh/day.

Simulations of the meteorological and energy impact of large-scale tree-planting programs in ten U.S. metropolitan areas show that on average trees can cool cities by about 0.3°C to 1.0°C (0.5°F to 1.8°F) at 2:00 P.M. The corresponding annual air-conditioning savings from ambient cooling by trees in hot climates range from $5 to $10 per 100m^2 of roof area of residential and commercial buildings. Indirect effects are smaller than direct shading, and, moreover, require that the entire city be planted.

There are other benefits associated with urban trees. These include improvement in life quality; increased property values; and decreased rain run-off and hence flood protection. Trees also directly sequester atmospheric CO_2, but the direct sequestration of CO_2 is less than one-fourth of the emission reduction from savings in cooling-energy use.

COOL PAVEMENTS

Urban pavements are made predominantly of asphalt concrete. The advantages of this smooth and all-

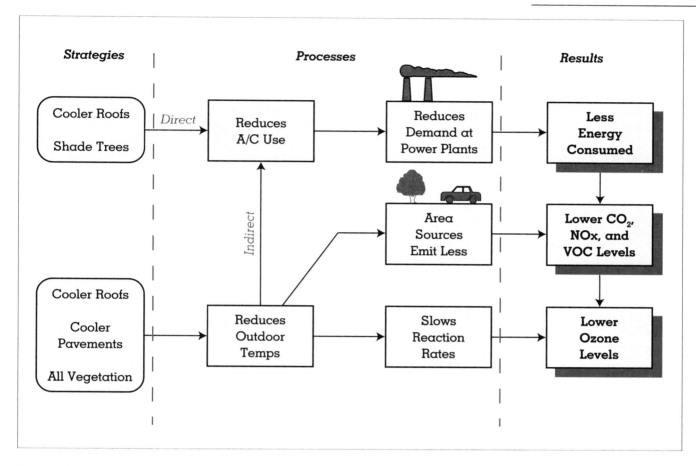

Figure 3.
Methodology to analyze the impact of shade trees, cool roofs, and cool pavements on energy use and air quality (smog).

weather surface for vehicles are obvious, but some associated problems are perhaps not so well appreciated. Dark asphalt surfaces produce increased heating by sunlight. Experimentally, the albedo of a fresh asphalt concrete pavement is only about 0.05. The relatively small amount of black asphalt coats the lighter-colored aggregate. As an asphalt concrete pavement is worn down and the aggregate is revealed, albedo increases to about 0.15 for ordinary aggregate. If a reflective aggregate is used, the longterm albedo can approach that of the aggregate.

The benefits of cool pavements can be estimated by first finding the temperature decrease resulting from resurfacing a city with more reflective paving materials. Cool pavements provide only indirect effects through lowered ambient temperatures. Lower temperature has two effects: (1) reduced demand for electricity for air conditioning and (2) decreased smog production.

Furthermore, the temperature of a pavement affects its performance: cooler pavements last longer. Reflectivity of pavements is also a safety factor in visibility at night and in wet weather, reducing electric street-lighting demand. Street lighting is more effective if pavements are more reflective, increasing safety. In reply to concerns that in time dirt will darken light-colored pavements, experience with cement concrete roads suggests that the light color of the pavement persists after long usage.

SUMMARY

Cool surfaces (cool roofs and cool pavements) and urban trees can reduce urban air temperature and hence can reduce cooling-energy use and smog. A

Benefits	Measures			
	Cooler roofs	Trees	Cooler pvmnts	Totals
1. Direct				
a A/C energy savings (M$/yr)	46	58	0	104
b Δ Peak power (GW)	0.4	0.6	0	1.0
c Present value ($)	153	64	0	
2. Indirect				
a A/C energy savings of 3°C cooler air (M$/yr)	21	35	15	71
b Δ Peak power (GW)	0.2	0.3	0.1	0.6
c Present value ($)	25	24	18	
3. Smog				
a 12% ozone reduction (M$/yr)	104	180	76	360
b Present value ($)	125	123	91	
4. Total				
a All above benefits (M$/yr)	171	273	91	535
b Total Δ peak power (GW)	0.6	0.9	0.1	1.6
c Total present value ($)	303	211	109	

Table 1.
Energy, Ozone Benefits, and Avoided Peak Power of Cooler Roofs, Pavements, and Trees in Los Angeles Basin
NOTE: The present value and surcost data for surfaces are calculated for 100 m^2 of roof or pavement area, and for one tree.
SOURCE: Rosenfeld et al., 1998.

thorough analysis for Los Angeles (see Table 1) showed that a full implementation of heat island mitigation measures can achieve savings of more than $500 million per year. Extrapolating those results, we estimate that national cooling demand can be decreased by 20 percent. This equals 40 TWh/year savings, worth over $4 billion a year by 2015 in cooling-electricity savings alone. If smog reduction benefits are included, savings could total to over $10 billion a year.

Hashem Akbari
Arthur H. Rosenfeld

See also: Air Conditioning; Air Pollution; Atmosphere; Climatic Effects; District Heating and Cooling; Domestic Energy Use; Efficiency of Energy Use; Energy Economics; Environmental Economics; Environmental Problems and Energy Use; Geography and Energy Use.

BIBLIOGRAPHY

Akbari, H., ed. (1998). "Energy Savings of Reflective Roofs." *ASHRAE Technical Data Bulletin* 14(2).

Akbari, H.; Bretz, S.; Taha, H.; Kurn, D.; and Hanford, J. (1997). "Peak Power and Cooling Energy Savings of High-Albedo Roofs." *Energy and Buildings* 25(2):117–126.

Akbari, H.; Davis, S.; Dorsano, S.; Huang, J.; and Winnett, S., eds. (1992). *Cooling Our Communities: A Guidebook on Tree Planting and Light-Colored Surfacing*. Washington, DC: U.S. Environmental Protection Agency.

Akbari, H.; Kurn, D.; Taha, H.; Bretz, S.; and Hanford, J. (1997). "Peak Power and Cooling Energy Savings of Shade Trees." *Energy and Buildings* 25(2):139–148.

Hall, J. V.; Winer, A. M.; Kleinman, M. T.; Lurmann, F. M.; Brajer, V.; and Colome, S. D. (1992). "Valuing the Health Benefits of Clean Air." *Science* 255:812–817.

Heat Island Group. (1999). <http://eetd.lbl.gov/HeatIsland>. Berkeley, California: Lawrence Berkeley National Laboratory.

Pomerantz, M.; Akbari, H.; Chen, A.; Taha, H.; and Rosenfeld, A. H. (1997). "Paving Materials for Heat Island Mitigation." Report LBNL-38074. Berkeley, CA: Lawrence Berkeley National Laboratory.

Rosenfeld, A.; Romm, J.; Akbari, H.; and Pomerantz, M. (1998). "Cool Communities: Strategies for Heat Islands Mitigation and Smog Reduction." *Energy and Buildings* 28(1):51–62.

Rosenfeld, A.; Romm, J.; Lloyd, A.; and Akbari, H. (1997). "White and Green." *MIT's Technology Review* February/March:54–59.

Taha, H. (1996). "Modeling the Impacts of Increased Urban Vegetation on the Ozone Air Quality in the South Coast Air Basin." *Atmospheric Environment* 30(20):3423–3430.

Taha, H. (1997). "Modeling the Impacts of Large-Scale Albedo Changes on Ozone Air Quality in the South Coast Air Basin." *Atmospheric Environment* 31(11):1667–1676.

Taha, H.; Konopacki, S.; and Gabersek, S. (1999). "Impacts of Large-Scale Surface Modifications on Meteorological Conditions and Energy Use: A 10-Region Modeling Study." *Theoretical and Applied Climatology* 62(3–4):175–185.

COOLING TOWERS

See: Heat and Heating

CRACKING

See: Refineries

CRUDE OIL

See: Petroleum

CULTURE AND ENERGY USAGE

INTRODUCTION

The earliest energy theorists were largely physical scientists, some of whom held that the growth and increasing complexity of society were largely synonymous with "progress," construed as a movement toward the higher, the better, and the more desirable. Nobel laureate chemist Wilhelm Ostwald (1907) was of this camp, although other eminent scholars such as the Nobel laureate physicist Fredrich Soddy (1912) and Alfred Lotka (1925), a founder of mathematical biology, also ventured ideas on the relation of energy and evolution yet made no explicit connection.

Energy theorists of cultural evolution are concerned with the whole sweep of cultural evolution, from prehistoric hunters and gatherers to modern industrial societies. This global, secular perspective is useful in assessing the relevance of ideas advanced to account for short periods of time in the history of particular societies. Those who propose an energy theory of cultural evolution emphasize the problem of causality-whether or not the amount of energy a

society uses can be manipulated, and if so, to what extent, by what means, and to what effect (Nader and Beckerman, 1978).

In the social sciences, steps toward an energy theory of cultural evolution were made by British archeologists V. Gordon Childe (1936) and Graham Clarke (1946). However, the first figure in the social sciences who fully developed an energy theory of cultural evolution is anthropologist Leslie White. White (1959) held that culture advances as a consequence of the ability to harness more energy, although we are not to conclude that people, either individually or collectively, can choose to vary their energy harnessing technology and thus vary the rest of their culture. Causality in White's view, runs from materialistic forces like environmental change, population pressure, culture contact, and the like, to "superorganic" technological systems, and thence to superorganic social and ideological systems. Technological systems may determine the rest of the culture, but specific technology in turn can come about and continue in use through forces completely outside the conscious command of the participants in culture.

Sociologist Fred Cottrell's thesis (1955, p. 2) was that "the energy available to man limits what he *can* do and influences what he *will* do." He later added that both material phenomena and choice are involved in any human situation. However, human choice for Cottrell is not directed. To varying degrees choices can be predicted, given information on individual values, the costs to the individuals of making various choices, and the power of the individuals in question to achieve their choices. One assumes that some element of chance is involved, but in a given situation a particular choice may be predicted with a high level of confidence. Although Cottrell is far from White in his rejection of radical determinism, he is equally far from many political philosophers prominent in the history of the West, who assume that society simply represents the ongoing result of innumerable unconstrained individual choices.

Howard T. Odum (1971), an ecologist, is the most diligent in his attempts to reduce all-or nearly all-cultural phenomena to the currency of energy. On first examination his approach to causality is strikingly reminiscent of White. On closer examination, it seems that Odum holds a "possibilistic" position on causality, similar to that of Cottrell. In a section on theories of history, he comments apropos of the rise and fall of civilizations (1971, p. 229) and refers to the expansion of fossil fuel energy use as the "basic cause" of the population explosion. Energy use may cause other social phenomena, as asserted by White, but energy use itself is not seen as the *primum mobile* of all human affairs, else there would be no point in suggesting that humans alter their energy use. Thus, the gross amount of energy harnessed by a society is only a starting point. What matters is the presence and sensitivity of feedback loops by which energy is channeled into "useful work," and not discharged in destructive "short circuits."

As an example, Odum (1971, p. 291) mentions the relations in some states among hunters, the Fish and Game Department, and the game animal populations. Hunters pay a significant amount of money for their hunting licenses. This money is spent by the Fish and Game Department on preserving and augmenting animal populations that are hunted. When the animal populations grow too large to be supported in their natural habitat, more hunting is permitted; when the animal populations fall, hunting is curtailed. He specifically (1971, p. 300) calls for this sort of loop in regard to energy indicating the right to inject fuel into the overheated world economy must be regulated.

Lastly, Richard N. Adams (1975) claims intellectual descent from White. He also ranks the cultural evolution of societies by the amount of energy harnessed, and sees the drive toward the harnessing of increasing amounts of energy by the whole fabric of human cultures as inevitable, as long as energy is available to be harnessed. However, this statement does not amount to a prediction of the course of any particular society. Adams has modified White's determinism so that only global processes are held to be deterministic while local events may manifest a high degree of indeterminism (Lovins, 1976). Just how much room this reexpression allows for the manipulation of a particular society's energetic parameters by national policy decisions is an open question.

All four theorists agree that the amount of energy available constrains possibilities for social change and social action. They also agree on a relationship between energy use and the increase of what is socially desirable. White decouples increasing amounts of harnessed energy from ideas of what is more desirable. Cottrell is also concerned with cultural evolution on a macroscopic scale, focusing on the contrast and transition between "low energy" (unindustrialized, or "third world") societies and "high-energy"

(industrialized) societies. Cottrell also recognizes that there is no necessary coupling, especially in the short term, between cultural evolution (defined again as the harnessing of increasing amounts of energy) and the increase of what is socially desirable. Odum's emphasis on feedback loops is that intricate feedback linkages and a high degree of role specialization are necessary for the realization of individual worth. He places a positive value on systemic stability. This stability is achieved through the use of more energy than that employed by tribal peoples, whom Odum considers to exist at the whims of a fickle natural environment, but less energy than currently employed by industrial countries, which Odum considers to be running, as cancer does, out of control. For him, the position seems to be that quality of life relates not to gross magnitudes of energy but to the complexity and stability of the system of energy production, distribution, and use. There is explicit coupling of growth of energy use without feedback controls with a deterioration of something like the quality of life. On the other hand, Adams argues that a deterioration in the quality of life for some members of a society is an inevitable correlate of increased energy flow.

Energy policy debates having to do with the immediate future in the United States take a more restricted range of time and space than do the global schemes of energy theorists. Further, some of the primitive assumptions of these policy arguments run directly counter to assumptions of global theorists. Policy statements almost always assume, for instance, that energy use is the dependent variable which can be manipulated at will by political decisions (Nader and Milleron, 1979).

Despite general acceptance of some measure of harnessed energy as an index of cultural evolution, it has been more than fifty years since anyone seriously argued that cultural evolutionary "advance" was in itself a movement toward the higher, the better, and the more desirable. Societies may indeed grow larger, more centralized, more internally differentiated, and more powerful as they consume more energy, but as time unrolls these changes have resulted in both improvement and degeneration in the lives of members of those societies, some members experiencing both (Daly, 1974; Duncan, 1975).

The conclusions of the cultural theorists are pertinent because of the clear dangers of putting too much stock in local and short term experience, since technology rarely exists free of people. For example, it is popularly held that energy use per capita versus quality of life in the United States in the period between 1955 and 1975 shows a negative correlation: energy use generally increased while quality of life generally declined. Conversely, the period 1935 to 1955 shows a positive correlation: both energy use and quality of life increased. Neither correlation shows the true relationship, because per capita energy use and quality of life are related to each other in complex ways. While this situation can hardly be illuminated by a simplistic, unidirectional attribution of causality, it does make the point that a mere increase in energy use cannot force a rise in the quality of life (Nader and Beckerman, 1978).

CONTEMPORAY PERSPECTIVE

Many issues in energy policy have been and will continue to be argued on the basis of presumed effects of energy use and energy policy on the quality of life. Some international comparisons of energy use and well-being are useful in their focus on efficiency of energy use (Goen and White, 1975). Others, such as the comparison between the United States and Sweden (Schipper and Lichtenberg, 1976), conclude that far from suggesting an inevitable coupling between level of economic activity and energy use, comparisons actually suggest ways in which more well-being can be wrought from every BTU of fuel and kilowatt-hour of electricity consumed in a given place. Furthermore, the data are persuasive that there is no direct relationship between per capita energy use and standard of living as measured by Gross National Product (GNP) (Weizdäcker, Lovins, and Lovins, 1997).

If the gross per capita energy use in a society cannot be held to correlate with the quality of life therein, what energy measures can? It seems likely that rapid change in the amount of energy used, either up or down (unless driven by zeal), will have a deleterious effect on the quality of life. If we examine American history over the last century or so, and if we compare our experiences with those of other nation-states, it appears that evidence for sharp increases or decreases in energy production and consumption correlates with social downturn. Our own experience of the great depression, as well as the fears that industrial nations share of an oil boycott, lead to the conclusion that a rapid increase in energy use will bring unemployment, shortages, and declining economic indicators of all kinds. On the other hand,

Many non-Western cultures, including the Chinese, tend to use more flexible forms of transportation than automobiles. (Corbis-Bettmann)

experience from the last part of the nineteenth and first part of the twentieth centuries, as well as comparable periods of energy growth from the history of England, Germany, and Japan suggest that rapid increases in energy use are accompanied by fluctuations in the roles of social institutions, unmanageable inequalities in the distribution of power, class, regional, and ethnic conflicts of a serious nature, and a pervasive sense of anomie.

In the social science literature life style has had an accepted meaning-value preference as expressed in consuming behavior. Sociologists have developed methods whereby projections may be made of personal consumption patterns such as a shift in the proportion of consumer dollars spent on services as compared to dollars spent on durable goods. Such research attends to such questions as, "Can consumption patterns change so that less energy is used without altering social preferences?" Other researchers might examine the official statistics on personal consumption expenditures over an extended period of time in order to gain a historical understanding of the nature and scope of changing life styles, up or down. One can observe that Ireland and New Zealand have a very similar consumption of fossil fuel per capita (Cook, 1976). The Republic of Ireland is frequently used as an example of social and economic stagnation and even misery, while New Zealand is typically seen as a society in which everything runs so smoothly and progressively and equitably that its only fault is dullness. Per capita energy use and technologies are parallel. Fads and fashions play a role, as does enthusiasm, zeal, loyalty to tradition, style setting, and being the first on the block.

In Denmark, an intensely ecologically–minded country, moving to wind power made sense because they are located in one of the windiest corners of the world. They moved to wind power development after choosing not to have nuclear power. In 1999, wind energy covers 7 percent of Danish electricity with the goal being 50 percent over the next three decades. Eighty percent of the country's windmills are owned by individuals and cooperatives. Denmark's project is part of a larger voluntary venture for sustainability. Self-sufficiency and the environment are values strongly supportive of conservation and individual initiatives in Denmark, whereas in Japan and France the use of nuclear power has increased from 1987–1997. While long-term results of such choices cannot be easily forecast, energy catastrophies in one country could easily change the picture.

The ethnographic literature is replete with examples that illustrate that within a wide range, the total amount of energy that a society uses per capita is less important than the specific uses made of such energy. For example, the use made of energy and the perception of energy surplus among three precolonial African agricultural groups varied. All three groups

increased the energy at their disposal by obtaining slaves, but they used these for different purposes: the Bemba principally as a trade item, the Tonga as additional labor about the homestead, and the Chokwa as both porters and trade items. All three groups had an energy surplus over and above what they expended. To have enough energy to cope with occasions that require a high expenditure of energy, any society has to live with most of its members operating below full work capacity most of the time. People mobilize their energy resources for short spurts; there are additional food and other resources known to them that could be harvested if needed but which usually are ignored (Cline-Cole et al., 1990). Without long-term storage, the inefficiencies are a necessary condition for survival over time, unlike the assumption that any energy available ought to be used because it is available and not because there is some human purpose to be obtained as a consequence of its deployment.

ENERGY AND SOCIAL ORGANIZATION

There is reason to believe that social organization, the framework within which we operate on a daily basis, may be strongly influenced by energy capability to produce, consume, store, or distribute. In one well-documented example (White, 1962), the pattern of energy control led to drastic change in social organization. The significant technological innovation responsible for the rise of feudalism in Europe was the introduction of the stirrup. The effect of the stirrup was to permit the concentration of greater force on the tip of a spear or the edge of a sword than could be achieved by a rider without stirrups, and to permit the rider to withstand greater force without being unhorsed. The horse had been in common use in Europe for centuries. What changed was not the gross amount of energy available, but the proportion of it that could be concentrated, and the speed and precision with which that concentrated energy could be released. The single consequence of this innovation was that, as arms and armor evolved to the full potential of a cavalry, and the support and forage of horses became requisite for military success, warfare became too expensive for serfs, who could afford neither the equipment nor the land necessary for mounted war. The result was the "flowering" of knighthood, the code of chivalry, the tying of ownership of land to the vassalage of its occupants, the tying of ownership of wealth to public responsibility, and the rest of the distinctive characteristics of feudalism. The control of a particularly important means of locally distributing energy led to a vast change in society itself.

It is anthropological commonplace that those who control scarce but necessary resources control, in large measure, the society that depends on those resources. Around the turn of the nineteenth century and somewhat later, railroads dominated large scale transportation in this country. The owners of railroad companies also dominated political life to an extraordinary degree (Boyer and Morais, 1955). At the grassroots level there was distrust and outright hatred of the railroad companies and their owners which brought us perhaps closer than we have ever been to class warfare in the United States. Nevertheless, the building of railroads led inexorably to capital intensive, highly centralized control.

The power of the railroads was diminished, not only by federal regulation, as history books sometimes argue, but also by the rise of motor vehicles and public roads as viable alternatives. With private trucks and cars the monopoly of a small segment of the population on the scarce "resource" of long range transportation was broken. There is some parallel with nuclear reactor technology since nuclear power is centralized, heavily regulated, subsidized by government, and crucially dependent upon selling prowess quoting scientific and engineering expertise.

Discussions of technologies have not always related to organization and values. Certainly since 1945 the bulk of the discussion has concentrated upon technical issues, such as the adequacy of the emergency core cooling system in light water reactors or the question of nuclear waste. However, the breakthroughs on safety and vulnerability resulted from studies of the culture of nuclear power. Charles Perrow (1999) argues that accidents are "normal" because they are built into the system. Perrow concludes that some complex systems can never be made accident-free because of "interactive complexity" (technological components are too varied for human operators to predict), and "tight-coupling" (small errors escalate too quickly for operators to figure out what is happening). For him, failure is built into a hard-wired system that does not allow for resilient possibilities of recovery.

In addition, discussions have focused on the effect of technology upon individual civil rights and civil liberties, terrorism, financial issues, and decaying techno-carcasses. Such discussions have not utilized in-depth

Americans tend to commute singly in automobiles, which can lead to traffic jams, such as this one near Longfellow Bridge in Boston, Massachusetts. (Corbis-Bettmann)

comparisons of strategies as with solar, nuclear, coal, and conservation; rather, each strategy has been considered separately, often independently of end users and consequences of failures (Kuletz, 1998), as shut downs, health and environmental catastrophies, or wars.

American discussions of nuclear energy first explored civil rights and civil liberties. Dangers of theft or sabotage of a plutonium facility require drastic incursions on individual freedom (Ayers, 1975).

At issue are risks of increased civilian surveillance, the extension of the military clearance system to include civilian workers, and increasing steps towards infringement on privacy such as covert airport services and security of government officials. In sum, safety considerations, so vital in dealing with vulnerable technologies, result in restrictive laws affecting all aspects of behavior, such that there are large increases in the numbers of police (Zonabend, 1993). Safety considerations then justify extraordinary investigation, arrest, and regulatory measures.

The sociopolitical consequences of increased commitment to nuclear technologies which represent only 5 percent of world energy, raises questions of democratic decision-making to safeguard the environment and health and safety of the general public (Holdren, 1976). Some ask if it is worth the price. Research on the social and political implications identifies the crucial contrast between vulnerable and nonvulnerable technologies, and between technological waste and social waste.

Areas of consensus and dissent appear, suggesting that the way energy is used and the purpose to which it is put are important to acceptance if not satisfaction. For example, in Europe, North America, and elsewhere a consensus is forming against wasteful engineering design. Few people would express themselves against improved miles per gallon or improved efficiency of refrigerators. There is more likely to be dissent on social waste; people would be more likely to object to carpooling or trading autos for mass transit. With regard to solar strategies there might be consensus on the democratizing effect of direct solar technology-after all, the sun falls on the rich and the poor, the weak and the powerful, the famous and the anonymous. Particularly the issue of decentralized solar power is symbolic of a greater issue: the preservation of liberty and equity through maintaining some independence from the "big system" (Stanford Research Institute, 1976). Centralized solar energy systems would have few of the dangers associated with highly vulnerable supply technologies, but there is expressed dissent at least among experts. Whatever the disagreements, it is clear that at issue is the value placed on freedom.

A SHIFT IN VALUES

Movement to redirect technological progress has brought about an "efficiency revolution" and the notion of a new industrial revolution incorporating, for example, the production of hypercars, compact fluorescent lamps, water drip systems, desk top computers that give more for less. Such movement is fueled by the realization that if Northern lifestyles spread globally it would take several globes to accommodate such life styles. It is well documented that the world's well-to-do minority uses the most energy, produces the greatest amount of pollution, and contributes greatly to the greenhouse effect. The efficiency revolution is in direct contradiction to supply side wizardry or high tech fantasies of fast breeder reactors, mega-fertilizer factories, gigantic water projects, preferring instead direct solar, hydro power, wind power, biomass, fuel-cell cars and the like (Nader, 1995). In the United States and Europe a coalition of business executives, consumers, environmentalists, labor leaders and legislators are the new energy entrepreneurs in a world where oil and coal fuels are increasingly being viewed as sunset industries (Hawken, Lovins, and Lovins, 1999).

Energy is becoming a multi-disciplinary concern. Aspects which were of interest to physicists, chemists and engineers are now a fixed growing concern for a wide variety of people. All energy research is inextricably interwoven with values, such as those relating to scale, complexity, organization, scientific challenge, and cost. The bulk of energy research that deals with scientific and technological questions is often embedded in deeply held beliefs about the human condition and direction. Technical specialists operating within the limits of their competence produce a clouding of the basic human factors that apply to broad understanding of the human dimensions of energy issues.

Energy can be used to foster different values in society: economic values based on efficiency, political values of democratic decision making, aesthetic values of architectural and environmental beauty, or their opposites. The question arises as to whose values will predominate in relation to the amount of energy produced, the purposes for which energy is used, and the forms and consequences of energy production. The perspective from which these relationships are addressed is important (Nader, 1981). The mode of energy use in industrialized society has been largely determined by producers rather than users. Automobile companies produce cars to sell cars, and Los Angeles's city rail systems were eliminated with this in mind (Mokhiber, 1988, pp. 221–228). Oil companies prefer inefficient cars because they use more gasoline (Sampson, 1975). Central power systems allow utilities to retain control over all productive facilities. Illuminating engineers may be apt to illuminate in a way that adds to profit and convenience of the illuminating engineer.

When quality of life becomes central, the dialogue on energy changes. In such a dialogue the individual user must have equal time. Certain choices might never be made if public welfare was the yardstick. Production and per capita expenditures of energy may then be secondary to the purposes for which energy is used and to the form of energy production. While certain broad correlation of energy use with other social parameters have been examined by social scientists for over a generation, findings are often ignored or misinterpreted. For example, technological progress and increased energy is said to have eliminated the drudgery of women's work while there is ample evidence to the contrary (Bendocci, 1993). Yet energy technologies are still being sold as panaceas for a woman's work life.

Our understanding of the relationship between human energy usage and culture has evolved from the simple paradigm wherein increased gross energy expenditure equaled cultural advancement to complex, non-lineal theories which account for interactions between myriad technological and social forces. The diversity of practical technologies world-wide suggests an enormous variety of solutions on hand and in the making for meeting human energy needs in both developed and developing countries.

Laura Nader

See also: Energy Economics; Ethical and Moral Aspects of Energy Use; Geography and Energy Use.

BIBLIOGRAPHY

Adams, R. N. (1975). *Energy and Structure.* Austin: University of Texas Press.

Ayres, R. W. (1975). "Policing plutonium: the civil liberties fallout." *Civil Liberties Law Review* 10(2):369–443.

Bendocci, C. G. (1993). *Women and Technology: An Annotated Bibliography.* New York: Gareaud Publishing.

Boyer, R. O., and Morais, H. M. (1955). *Labor's Untold Story.* New York: Cameron Assoc.

Childe, V. G. (1936). *Man Makes Himself.* London: Watts.

Clark, G. (1946). *From Savagery to Civilization.* London: Cobbett.

Cline-Cole, R. A., et al. (1990). *Wood Fuel in Kono.* Tokyo: United Nations University Press.

Cook, E. (1976). *Man, Energy, Society.* San Francisco: Freeman.

Cottrell, F. (1955). *Energy and Society.* New York: McGraw-Hill.

Daly, H. E. (1974). "Steady-State Economics Versus Growthmania: A Critique of the Orthodox Conceptions of Growth, Wants, Scarcity, and Efficiency." *Policy Science* 5:149–67.

Duncan, O. D. (1975). "Does money buy satisfaction?" *Social Indicators Research* 2:267–74.

Goen, R., and White, R. (1975). *Comparison of Energy Consumption Between West Germany and the United States.* Washington DC: Government Printing Office.

Hawken, P.; Lovins, A.; and Lovins, L. H. (1999). *Natural Capitalism: Creating the Next Industrial Revolution.* Boston: Little, Brown and Company.

Holdren, J. P. (1976). "Technology, Environment, and Well-Being: Some Critical Choices." In *Growth in America*, ed. C. L. Cooper. Westport, CT: Greenwood.

Kuletz, V. L. (1998). *The Tainted Desert: Environmental Ruin in the American West.* New York: Routledge.

Lotka, A. (1925). *Elements of Physical Biology.* Baltimore: Williams & Wilkins.

Lovins, A. B. (1976). "Energy strategy: the road not taken?" *Foreign Affairs* 54:65–96.

Mokhiber, R. (1988). *Corporate Crime and Violence: Big Business Power and the Abuse of Abuse of the Public Trust.* San Francisco: Sierra Club Books.

Nader, L. (1981). "Barriers to Thinking New About Energy." *Physics Today* 34(3):9, 99–102,

Nader, L. (1995). "Energy Needs for Sustainable Human Development-Anthropological Aspects." In *Energy as an Instrument for Social Change*, eds. Jose Goldenberg and T. B. Johansson. New York: United Nations Development Programme Publications.

Nader, L., and Beckerman, S. (1978). "Energy as it Relates to the Quality and Style of Life." *Annual Review of Energy* 3:1–28.

Nader, L., and Milleron, N. (1979). "Dimensions of the 'People Problem' in Energy Research and the Factual Basis of Dispersed Energy Futures." *Energy* 4(5):953–967.

Odum, H. T. (1971). *Environment, Power and Society.* New York: Wiley (Interscience).

Ostwald, W. (1907). "The Modem Theory of Energetics." *The Monist* 17(l): 511.

Perrow, C. (1999). *Normal Accidents: Living with High-Risk Technologies.* Princeton: Princeton University Press.

Sampson, A. (1975). *The Seven Sisters: The Great Oil Companies and the World They Made.* New York: Viking.

Schipper, L., and Lichtenberg, A. J. (1976). "Efficient Energy Use and Well-Being: The Swedish Example." *Science* 194:1001–13.

Soddy, F. (1912). *Matter and Energy.* London: Oxford University Press.

Stanford Research Institute. (1976). "A Preliminary Social and Environmental Assessment of the ERDA Solar Energy Program 1975-2000." Vol. 1, Main Report. Menlo Park, CA: Stanford Research Institute.

Weizdäker, E. von; Lovins, A. B.; and Lovins, L. H. (1997). *Factor Four: Doubling Wealth—Halving Resource Use.* London Earthscan Publications, LTD.

White, L. T. (1959). *The Evolution of Culture.* New York: McGraw-Hill.

White, L., Jr. (1962). *Medieval Technology and Social Change.* Oxford: Clarendon.

Zonabend, F. (1993). *The Nuclear Peninsula.* Translated from the French by J. A Underwood. New York: Cambridge University Press.

CURIE, MARIE SKLODOWSKA (1867–1934)

Maria Salomee Sklodowska was born to Wladislaw and Bronislawa Sklodowski on November 7, 1867, in Warsaw, Poland. This was a difficult time in Poland's history, since Poland was under Russian control. Nevertheless, Curie excelled in school, and she finished her secondary school education in 1883 with a gold medal. Although her parents were academicians and emphasized the importance of both education and religion, Curie maintained an agnostic view. Her rejection of religion is generally attributed to the loss of her mother and sister while she was still a young child and also may be related to her attraction to science.

Because women were not allowed to pursue higher education in Poland, Curie and her elder sister Bronia decided to earn their advanced degrees in France. While Bronia studied medicine at the Sorbonne, her sister worked as a governess to support her and then joined her in Paris in 1891, enroll in the Faculty of Sciences of the Sorbonne. During this period in France, Marie adopted the French variation of her name. By 1893 Curie had finished first in her class with a degree in physics despite beginning the program with poor French-language skills. In the following year she completed a mathematics degree and met her future husband Pierre a physicist at the School of Industrial Physics and Chemistry. A respected researcher in the fields of magnetism and piezoelectricity, Pierre convinced Marie that they shared an attitude of single-minded devotion to science and they were married in 1895.

The Curies continued to work together in Pierre's

Marie Curie, working in the laboratory with her husband Pierre. (AP/Wide World Photos)

lab after their marriage and in 1897 Marie published her first paper, on magnetism. In September of that same year, her daughter Irène was born. For Marie to continue her career, the Curies hired a house servant, and Pierre's father, Eugène, moved in after the death of his wife. This allowed Marie to continue her search for her doctoral thesis topic. At that time the completion of a doctoral thesis by a woman anywhere in Europe was unprecedented.

Curie chose for her dissertation research the new topic of uranium rays, a phenomenon that had only recently been observed by Henri Becquerel. The mystery was the source of the energy that allowed uranium salts to expose even covered photographic plates. Curie's first efforts in the field were systematic examinations of numerous salts to determine which salts might emit rays similar to those of Becquerel's uranium. After discovering that both thorium and uranium were sources of this radiation, Curie proposed the term "radioactive" to replace "uranium rays." She also discovered that the intensity of the emissions depended not on the chemical identity of the salt but on the amount of uranium in the compound. This key observation eventually led Curie to propose that radioactivity was an atomic property rather than a chemical property. Among the samples that Marie studied were two uranium ores that proved to be more radioactive than pure uranium. This observation led her to propose that the ores contained a new element more radioactive than uranium. Gabriel Lippmann presented the paper describing Curie's work to the Académie des Sciences on April 12, 1898.

Although the Curies noted that one equivalent gram of radium released one hundred calories of heat per hour, they were uninterested in the practical implications of this, as they were both devoted to pure scientific discovery. During their work with pitchblende in 1898, the Curies discovered two new radioactive elements, which they named polonium (in honor of Marie's homeland) and radium. By 1902 they had isolated a pure radium salt and made the first atomic weight determination.

On June 25, 1903, Curie defended her doctoral thesis at the Sorbonne, delivering a review of the research in the area of radioactivity. In December of the same year, the Curies were named joint recipients of the Nobel Prize for Physics along with Becquerel. This award was made for the recognition of the phenomenon they called radioactivity. Due to their poor health, the Curies were not able to travel to Sweden to accept the prize until June 1905. Although they were both clearly suffering from radiation sickness, neither of the Curies realized this at the time. In the meantime, Marie gave birth in 1904 to another daughter, Eve.

In 1906 Pierre was awarded a full professorship and position as chair of physics at the Sorbonne and Marie was promised a position as director of the laboratory which the university planned to create for Pierre. However, in April 1906 Pierre was killed when he stepped into the path of a horse-drawn cart. While this event personally devastated Marie, it was a pivotal point in her professional career. She was offered Pierre's chaired position at the level of assistant professor, making her the first woman in France to obtain a professorship and allowing her to both continue her research and financially support her family.

After Pierre's death, Marie was faced with having to present her work without the support and social skills of her husband. Furthermore, she spent numerous years defending her work from William Thomson (Lord Kelvin) who did not believe that radioactivity

could be an atomic property. Nevertheless, in 1911 Curie became the first person to receive a second Nobel Prize when she was awarded the Nobel Prize in Chemistry for her discovery of polonium and radium and the isolation of radium.

Curie spent much of the remainder of her life raising money for research. Always politically active, she also worked with her daughter Irène (who later earned her own Nobel Prize in Chemistry) during World War I establishing mobile x-ray services and training workers to perform x-rays on the battlefield. It was not until the 1920s that the issue of the health hazards of radium emerged. Despite her own health problems, Curie was reluctant to accept that radiation could be linked to the illnesses and deaths of so many of her colleagues in the field. Eventually Curie was diagnosed with a severe form of pernicious anemia (caused by years of exposure to radiation) and she died on July 4, 1934.

Margaret H. Venable
T. Leon Venable

See also: Nuclear Energy; Thomson, William.

BIBLIOGRAPHY

Curie, E. (1938). *Madame Curie; A Biography,* tr. V. Sheean. Garden City, NY: Doubleday, Doran.

Giroud, F. (1986). *Marie Curie: A Life,* tr. L. Davis. New York: Holmes & Meier.

Ogilvie, M. B. (1988). *Women in Science.* Cambridge, MA: MIT Press.

Rayner-Canham, M., and Rayner-Canham, G. (1997). *A Devotion to Their Science: Pioneer Women of Radioactivity.* Philadelphia: Chemical Heritage Foundation.

Rayner-Canham, M., and Rayner-Canham, G. (1998). *Women in Chemistry: Their Changing Roles from Alchemical Times to the Mid-Twentieth Century.* Philadelphia: Chemical Heritage Foundation.

Reid, R. (1974). *Marie Curie.* New York: Saturday Review Press/E. P. Dutton.

CURRENT, ALTERNATING AND DIRECT

See: Electricity

DAMS

See: Hydroelectric Energy

DEMAND-SIDE MANAGEMENT

WHAT IS DSM?

Demand-side management (DSM) refers to active efforts by electric and gas utilities to assist customers in modifying their use of energy. DSM encompasses a variety of activities designed to change the level or timing of customers' energy consumption. Most discussion of DSM focus on programs that help customers save energy by encouraging them to adopt energy-efficient measures or practices. The most well-known are rebate programs that lower the cost of energy-efficient appliances. Other DSM programs promote changes in the "shape" of a utility's load by shifting demand away from high daytime electricity rates when it is most expensive for utilities to generate electricity.

U.S. utility DSM programs can be divided into seven categories: general information, targeted technical information, financial assistance, direct installation of energy-efficient technologies, performance contracting, load control/load shifting, and innovative tariffs. General information programs aim to increase customers' awareness of their energy use patterns and opportunities to use energy more efficiently. Almost all utilities provide general information, ranging from educational brochures about turning off gas-furnace pilot lights during warm months to bill inserts describing energy-efficient products and services. General information is also distributed through advertisements and by utility representatives.

Targeted technical information programs include audits of customers' current energy use patterns, accompanied by recommendations for ways to use energy more efficiently. Audits are typically offered free of charge although some utilities are experimenting with charging fees.

Financial assistance programs—loans or direct payments—lower the cost of purchasing energy-efficient technologies. Cash payments or rebates have been the most popular type of DSM program. Rebates reduce some or all of the cost of purchasing and installing an energy-efficient technology. Rebates can be fixed payments per unit (e.g., a $100 coupon toward the purchase of an energy-efficient refrigerator) or payments that lower the initial cost of a technology to some predetermined level (e.g., to ensure that the energy saved within three years of purchase will pay for the extra cost of buying an energy-efficient refrigerator). Some utilities offer low-interest loans in place of or in conjunction with rebates. When given a choice, customers generally prefer rebates to loans, so utilities interested in achieving large impacts over a short period of time have devoted a much larger share of DSM budgets to rebate programs than to loans.

Direct installation of energy-efficient technologies programs send utility staff or utility-hired contractors to a customer's premises free of charge to provide energy audits and install pre-selected energy-saving technologies. Because utilities pay the full installation cost, these DSM programs are frequently the most expensive to operate, as measured by the cost of energy saved. Utilities have typically offered installation programs either as a last resort—for example, when there is an imminent threat of supply shortfall—or to serve particular market segments (e.g., low-income residential customers that have proven difficult to reach with other DSM programs.)

Performance contracting are programs in which a third party, often an energy service company (ESCO), contracts with both a utility and a customer to provide a guaranteed level of energy savings. Performance contracting programs involve either competitive bidding, in which ESCOs and customers make proposals to the utility, or "standard offers," in which the utility agrees to pay for energy-saving projects at a fixed price per unit of energy saved. Payment is contingent on verification that the customer actually saves the amount of energy guaranteed in the contract. When an ESCO enters into a performance contract with a utility, the ESCO must recruit utility customers and form a separate contractual relationship with them so that the ESCO can finance and install energy-saving technologies and verify their performance. Performance contracting has mainly involved the public sector (i.e., schools and government buildings), business (i.e., commercial) and industrial, rather than residential, customers.

In load-control/load-shifting programs, utility offers payments or bill reductions in return for the ability to directly control a customer's use of certain energy-consuming devices or to assist the customer in installing a device, which alters the timing of demands on the electric system. In load-control programs, utilities directly control some customer appliances during periods when demand for power is high, such as extremely warm days when increased cooling energy use causes heavy power system loads. Load-control programs rotate groups of appliances (typically water heaters or central air conditioners) on and off for short periods of time, reducing net loads on the power generation system. These programs have usually involved residential customers. Customers can also control loads by adopting load-shifting technologies that allow the timing of the customer's load. Thermal storage, for example, allows a customer to use power when rates are low, such as at night, to generate and store heating or cooling to be used at other times of day when rates are high. "Valley-filling" describes programs that shift (or increase) customer loads to times of day when utility system loads and production costs are low (e.g., during the night).

Innovative tariff programs make it cost effective for customers to reduce or change the timing of energy use. These tariffs include interruptible rates, time-of-use rates, and real-time pricing. An interruptible rate is similar to a load-control program; a customer pays a lower rate in return for agreeing to curtail loads whenever requested by the utility. The customer, rather than the utility, determines which loads to reduce when the request is made. Time-of-use rates set different prices for energy used during different times of day, based on the utility's costs of generating power at those times. Real-time pricing is a sophisticated form of time-of-use rates, in which a utility typically gives customers a forecast of twenty-four hourly energy prices one day in advance. With both time-of-use rates and real-time pricing, customers respond by changing energy use to reduce their costs. Innovative pricing programs have targeted primarily industrial and large commercial customers.

HISTORY OF ELECTRIC UTILITY DSM PROGRAMS IN THE UNITED STATES

The history of DSM in the United States is dominated by the activities of electric utility companies. Historically, electricity service was considered a natural monopoly; it was thought that only one company in a geographic region could efficiently capture the economies of scale offered by electricity generation, transmission, and distribution technologies. In the United States, two primary institutions arose to secure the public benefits associated with increased electrification: publicly-owned municipal utilities were established in some large cities and governed by city councils; privately-owned utilities were also formed, governed by state regulatory authorities. More than 80 percent of the electricity produced and sold in the United States comes from privately-owned utility companies that generate, transport, and distribute power.

For much of their history, electric utilities in the United States promoted new uses of power in order to increase their sales and thus their profits. However, during the 1970s, the dramatic rise in world oil prices and growing concern about the environmental impacts of electricity generation (especially those associated with reliance on nuclear power) led to a new emphasis on conserving energy.

In 1978, the federal government passed the National Energy Conservation Policy Act (NECPA). Among other things, NECPA required utilities to offer onsite energy audits to residential customers. This law acknowledged that saving energy could be cheaper than producing it. We now recognize NECPA as the beginning of modern utility DSM programs. In fact, California and Wisconsin authorized utility DSM programs as early as 1975; these programs were the very first DSM programs, predating NECPA. NECPA encouraged utilities to create, staff, train, and maintain internal organizations devoted to helping customers reduce their electricity use. Prior to this time, utility staff efforts were focused on ensuring reliable service and promoting new uses for electricity.

After only modest increases through the mid-1980s, electric utility DSM programs began to increase dramatically during the late 1980s as a handful of states began to direct their regulated utilities to formally adopt least-cost or integrated resource planning principles. Regulatory efforts to encourage utilities to adopt DSM programs were bolstered by growing evidence of the low cost of technologies that could reduce electricity demand by increasing the efficiency with which energy was used. Energy-efficiency advocates conducted numerous analyses showing that substantial amounts of energy could be saved for much less than the cost of building new power plants, with no change in the level of amenity provided to the customer. Least-cost planning advocates argued that utilities should pursue these demand-side options whenever they were less expensive than the supply-side alternatives they could displace. A number of "market barriers" to consumers choosing cost-effective energy-efficient technologies were identified. DSM programs overcame these barriers by providing information and financial incentives to assist customers in selecting energy-efficient technologies that would lower their energy costs.

THE GROWTH IN UTILITY DSM PROGRAMS

Regulators responded to this evidence by encouraging utilities to increase the size and scope of their DSM programs whenever it could be shown that energy efficiency was a cheaper "resource" alternative than investments in new generating plants. While there had been little utility resistance to the original directives of NECPA in the late 1970s many utilities actively resisted suggestions by their regulators to pursue least-cost planning in general and DSM in particular in the late 1980s.

As political pressure to pursue least-cost planning strategies grew, regulators began to recognize that utility resistance could be traced to longstanding regulatory approaches that had been developed to encourage sales of electricity when the (then) increasing economies of scale meant that increased consumption led to lower costs for all. As a result, utilities had a powerful financial incentive to increase earnings by encouraging customers' consumption and constructing capital-intensive, new power plants to meet growing demand. These incentives meant that most activities to reduce sales were directly contrary to the business interests of the utility.

Two regulatory strategies were developed to overcome utilities' incentive to sell electricity and to insure that investment in programs to reduce customer demand would not be a financial burden on utilities. The first strategy compensates utilities for sales "lost" as a result of cost-effective DSM programs. The second "decouples" revenue from sales by establishing a revenue target that is independent of the utility's sales, and creating a balancing account to compensate for the difference between revenues actually collected and the revenue target. By making total revenues independent of actual sales volumes in the short-run, these approaches took away utilities' incentives to increase loads.

Some states created separate financial incentives for the delivery of superior DSM programs. Three types of incentives have been used. First, utilities earn a profit on money spent on DSM. Second, utilities earn a bonus paid in $/kWh or $/kW based on the energy or capacity saved by a DSM program. Third, utilities earn a percentage of the net resource value of a DSM program. Net resource value is measured as the difference between the electricity system production costs that the utility avoids because of the program(s) and the costs required to run the program(s). These new incentives were instrumental in stimulating growth of DSM energy-efficiency programs.

The effect of these efforts to realign the financial interests of utilities with the public interest was a dramatic increase in utility spending on DSM programs.

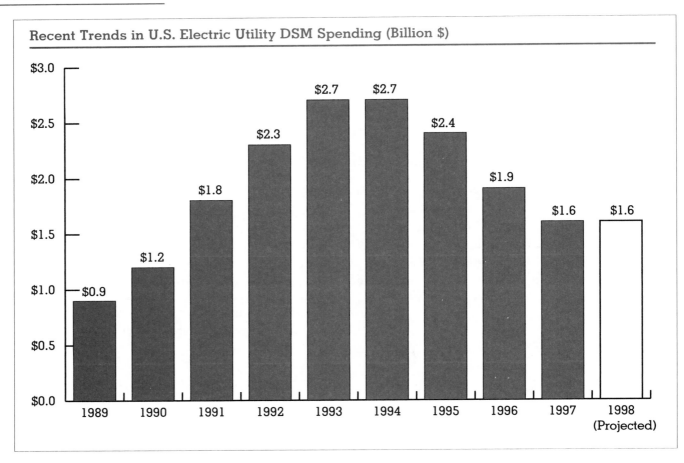

Figure 1.
Trends in U.S. Electric Utility DSM Spending.
SOURCE: U.S. Energy Information Administration

Spending increased so significantly that the U.S. Energy Information Administration began tracking DSM expenditures formally in their annual survey of utility operations starting in 1989. These surveys revealed that DSM spending by electric utilities in the United States had increased from $0.9 billion in 1989 to $2.7 billion in 1994 (see Figure 1). Electric utility DSM programs reached their largest numbers in 1993, accounting for more than $2.7 billion of utility spending or about one percent of U.S. utility revenues.

The record of utility achievements demonstrated that DSM energy efficiency programs were able to save electricity cost effectively. In the most comprehensive analysis to date, an examination of the forty largest energy-efficiency programs targeted to the commercial sector (i.e., office buildings, retail establishments, schools, etc.) showed that these programs

had saved energy at an average cost of 3.2 ¢/kWh and that they were highly cost effective when compared to cost of electricity generation they allowed the sponsoring utilities to avoid. However, not all utilities were equally effective in running energy-efficiency DSM programs. The study also found that some utilities, notably those with large DSM programs, had saved energy at cost of less than 2¢/kWh, while others, mostly making a more modest investment, had saved energy at a cost in excess of 10¢/kWh. Electric utility DSM programs save more than 56 million kWh annually—30 percent more than the total growth of electric sales in the United States between 1996 and 1997. DSM programs have enabled utilities to avoid the construction of approximately forty average size coal-fired power plants.

THE FUTURE OF DSM

Although DSM programs have been successful, the U.S. electric utility industry is in the midst of changes that will continue to affect its future. Starting in 1994, states actively began discussing restructuring their electricity industry to allow customers to choose their supplier of electricity and to increase competition among electricity generators. Utilities responded predictably to this threat of competition. First, they have sought regulatory protection for assets whose cost was in excess of current market prices. They also aggressively began cutting costs in all areas, including DSM. According to the Energy Information Administration, total DSM spending in 1997 had declined to $1.6 billion (from a high of $2.7 billion in 1993). By 1998, several states had restructured their electricity industry and many other states are expected to follow this trend.

In a fully restructured industry, utilities become essentially regulated power distribution companies with only an obligation to connect all customers to the power grid but with no obligation to plan and acquire generation to serve all customers. When utilities are relieved of the obligation to serve certain customers, they are also relieved of the obligation to use least-cost planning principles to acquire resources (including demand-side energy efficiency improvements) on behalf of these customers.

Although electricity industry restructuring renders the traditional utility monopoly franchise obsolete, the public purposes that are served by utility DSM programs remain whenever market barriers prevent cost-effective energy-efficiency decisions from being made. Restructuring means that electricity prices set by the market will better reflect the true value of electricity than prices set by regulatory authorities. If successful, elimination of regulatory mispricing would address an important historic market barrier to energy-efficiency. However, it is unlikely that restructuring will address all the market barriers that prevent cost-effective energy-efficiency actions from being taken. For example, the environmental consequences of electricity generation in particular, which are not currently reflected in the market prices paid for electricity, remain a strong argument for continuing energy-efficiency programs.

In the past, utilities were in a unique position to promote public interest in energy efficiency through DSM programs. Utilities had (1) access to low-cost capital; (2) name recognition among customers and acknowledged technical expertise; (3) lack of direct financial interest in promoting particular energy-efficiency products or services; (4) access to detailed information on customer energy-use patterns; and (5) a system for billing customers for services. Whether utilities will retain these desirable features following restructuring is not known. It will depend on decisions regulators make on the organization and rules governing the firms operating in the market.

Many utilities plan to offer DSM programs as a feature to keep and attract customers. Nonutility energy service providers are concerned that ratepayer funding for DSM programs will be used unfairly to subsidize utility development of business opportunities that will be pursued as unregulated profit-making activities. They argue that utility managers already face conflicts of interest when delivering ratepayer-funded DSM programs that historically served broad public interests, while at the same time attempting to maximize shareholder returns by using these programs to keep utility customers from switching to other suppliers or to increase the utility's dominance in local energy-efficiency service markets.

In a restructured electricity industry, utilities are concerned that including DSM program costs in their regulated rates puts them at a competitive disadvantage to other power providers that are not also required to charge for DSM. A surcharge to recover DSM program costs levied on all electricity users, regardless of their suppliers, eliminates this concern. As of 1998, twelve states had adopted these surcharges to continue funding for DSM programs as well as, in some cases, funding for other "public purpose" activities, such as research and development or promotion of renewable energy.

If the utilities can mitigate conflicts of interest, some states are expected to rely on utilities with good past records to continue to administer DSM programs. If local utilities have had poor past performance with DSM or cannot mitigate conflicts of interest, states may consider (1) administration by an existing or newly-created government agency, and (2) administration by an independent, possibly non-profit entity. Both alternatives raise questions of governance and accountability for the administration of funds.

SUMMARY

Demand-side management programs have been a bold experiment in the active promotion of energy efficiency by electric and gas utilities in the United States. By and large, they have demonstrated that market barriers, which constrain consumers' abilities to lower their energy costs, can be successfully and cost-effectively addressed by well-designed and targeted programs. These programs have represented billions of dollars in utility investments that have allowed utilities to avoid even more costly and environmentally damaging power plants. As the utility industry is restructured, many are hopeful that regulators will continue to enact policies that successfully align utility interests with pursuit of cost-effective energy efficiency opportunities.

Joseph Eto

BIBLIOGRAPHY

Battelle-Columbus Division and Synergic Resources Corporation. 1984. *Demand-Side Management, Evaluation of Alternatives.* Palo Alto, CA: Electric Power Research Institute and Edison Electric Institute.

Baxter, L. 1995. *Assessment of Net Lost Revenue Adjustment Mechanisms for Utility DSM Programs.* ORNL/CON-408. Oak Ridge, TN: Oak Ridge National Laboratory.

Blumstein, C.; Krieg, B.; Schipper, L.; and York, C. (1980). "Overcoming Social and Institutional Barriers to Energy Efficiency." *Energy* 5(4):355–72.

Cavanagh, R. 1988. "Responsible Power Marketing in an Increasingly Competitive Era." *Yale Journal on Regulation* 1:331–366.

Energy Information Administration. 1998. *Electric Power Annual 1997, Volume II.* DOE/EIA-0348(97)/2. Washington, DC: Energy Information Administration. October.

Eto, J.; Kito, M.; Shown, L.; and Sonnenblick, R. 1995. *Where Did the Money Go? The Cost and Performance of the Largest Commercial Sector DSM Programs.* LBL-38021. Berkeley, CA: Lawrence Berkeley National Laboratory.

Eto, J.; Goldman, C.; and Nadel, S. 1998. *Ratepayer-Funded Energy Efficiency Programs in a Restructured Electricity Industry: Issues and Options for Regulators and Legislators.* Washington, DC: American Council for an Energy Efficient Economy.

Eto, J.; Stoft, S.; and Belden, T. 1994. *The Theory and Practice of Decoupling.* LBL-34555. Berkeley, CA: Lawrence Berkeley National Laboratory.

Krause, F., and Eto, J. 1988. *Least-Cost Utility Planning, A Handbook for Public Utility Commissioners, The Demand Side: Conceptual and Methodological Issues.* Washington, DC: National Association of Regulatory Utility Commissioners.

Lovins, A. 1976. "Energy Strategy: The Road Not Taken?" *Foreign Affairs* 55(1):65–96.

Moskovitz, D. 1989. *Profits and Progress Through Least-Cost Planning.* Washington, DC: National Association of Regulatory Utility Commissioners.

Nadel, S. 1992. "Utility Demand-Side Management Experience and Potential—A Critical Review." *Annual Review of Energy and the Environment* 17:507–35.

Solar Energy Research Institute (SERI). 1981. *A New Prosperity: Building a Sustainable Energy Future.* Andover, MA: Brick House Publishing

Stoft, S.; Eto, J.; and Kitom M. 1995. *DSM Shareholder Incentives: Current Designs and Economic Theory.* LBL-38059. Berkeley, CA: Lawrence Berkeley National Laboratory.

DEPARTMENT OF . . .

See: Government Agencies

DEREGULATION

See: Regulation and Rates for Electricity

DESULFURIZATION (OIL AND COAL)

See: Refineries

DIESEL, RUDOLPH (1858–1913)

Rudolph Christian Karl Diesel was a German thermal engineer and inventor of the high-efficiency internal-combustion engine that bears his name.

Much of Diesel's life and brilliant career was tragic—from his business failings, to his struggles to translate theory into practice, to his chronic physical and mental ailments. Yet his research and engine prototypes laid the foundation for one of the world's most efficient and widely used fossil fuel engine technologies.

Diesel was born to German Protestant parents. His father, Theodor, a bookbinder, emigrated from Germany to Paris in 1850, and five years later married Nuremberg-born Elise Strobel, a teacher of German and English. The couple had three children—Louise (b. 1856), Rudolph (b. 1858), and Emma (b. 1860)—and were strict disciplinarians. French was spoken in the Diesel home, and Rudolph's mother also taught him English. Rudolph had few friends, but took an interest in technology, reinforced by frequent visits to the Conservatoire des Arts et Métiers, a technical museum in Paris.

In 1870 the Diesels were forced to leave Paris by a government general expulsion order during the Franco-Prussian War, and in November Rudolph went to live with relatives in Augsburg, Germany. He enrolled in the royal commercial school and then studied at the city's industrial school. In a letter to his parents on his fourteenth birthday, Diesel declared his ambition to become an engineer. Perhaps inspired by a "fire piston" at the school (a device that caused a tinder to glow with compressed air), Diesel became intrigued with compression ignition. He was plagued with chronic health problems, especially headaches and insomnia, yet still excelled at his studies. His final exam grades in 1875 were the highest in the school's history.

Diesel's distinguished academic record drew the attention of Professor Karl Max von Bauernfeind of Munich's Techniche Hochschule, who offered Diesel a two-year scholarship. Ultimately Diesel spent four years in the school's mechanical-technical division, where he studied theoretical machine design under Carl von Linde (a leader in that field and in refrigeration science) Moritz Schroeter, and others. Graduating at the top of his class with a civil engineering degree in 1880, Diesel took a job at the Sultzer Brothers factory in Winterthur, Switzerland. Diesel was then hired by the Linde Refrigeration Company in Paris and soon went to work directly for Linde, traveling regularly to consult with clients.

By this time Diesel had become enamored with the "social question"—the social problems and class conflict fostered by industrialization. Perhaps in reaction to his father's earlier strong embrace of

Rudolph Diesel. (Library of Congress)

magnetic healing, Diesel rejected organized religion in favor of an increasingly popular rational humanism. One of Diesel's key motivations to invent a high-efficiency engine was to help relieve the burdens of the artisan class. Later (in 1903) he published a book called *Solidarismus: Natuürliche wirtschaftliche Erloösung des Menschen (Solidarism: The Natural Economic Salvation of Man)*, in which he called for, among other things, worker-run factories.

In 1882 Diesel met Martha Flasche, a German. The couple married in November 1883 and for the remainder of the decade lived in Paris, where they had three children: Rudolph, Jr. (b. 1884), Hedy (b. 1885), and Eugen (b. 1889).

Diesel began work on an economical engine as early as 1880. The following year he took out his first patents—for machines to make clear ice. Throughout the remainder of the decade he worked on an ammonia vapor engine and (less rigorously) on a solar-powered engine. In 1889 he moved to Berlin to work as Linde's representative there. In

325

1890 or 1891 Diesel began to work out the theoretical basis for a constant temperature (isothermic) engine that would later evolve into the diesel engine. Diesel hoped to create a "universal" (flexible-fuel) engine that would operate on the cycle described by Nicolas Leonard Sadi Carnot and thus waste only about 20 to 30 percent of its energy through heat loss. In his model, the first downward stroke of the piston drew air into the cylinder. That air was then compressed to high pressure and temperature with the return (upward) stroke. With the second downward stroke, fuel was introduced at such a rate that the heat generated by its combustion would counterbalance the natural decline in temperature caused by the expansion of the cylinder space.

Diesel received a patent for his engine design in 1893, the same year he published a book on the subject, *Theorie und Kunstruktion eines rationellen Waärmemotors (Theory and Construction of a Rational Heat Engine)*, which was translated into English in 1894. He then persuaded Machinenfabrik Augsburg (Augsburg Engine Works), led by Heinrich Buz, to form a syndicate with Krupp in April 1893 to manufacture a 2-cylinder, 50-horsepower engine. But Diesel and his backers were unable to produce a smoothly running prototype until 1897. For that to happen, Diesel modified many of the fundamentals of his original theoretical design. The working engine operated at much higher pressures (18 to 33 atmospheres); ran on kerosene (instead of any liquefied or pulverized fuel) and at a new fuel-air mix; used compressed air rather than solid injection; and, most importantly, did not operate at constant pressure.

After announcing the success and imminent commercialization of his engine at the June 1897 meeting of the Society of German Engineers, Diesel began seeking licensees throughout the industrialized world. Three German companies bought patent rights, as did several non-German firms and individuals, including brewing magnate Adolphus Busch in the United States. The Augsburg company managed to produce a reliable 60-horsepower model by 1902. But Diesel's foreign licensees continued to struggle, despite drawings and engineering assistance from Augsburg. Busch's company produced a mere 260 engines between 1902 and 1912, when its license expired, the diesel venture having cost the family millions of dollars.

Meanwhile, Rudolph Diesel's growing fame was haunted by a series of personal and business setbacks. Working to the point of exhaustion, he required months of recuperation in a sanitarium from late 1898 to early 1899 and again in 1901–1902. Some of his critics pointedly challenged the originality of his work, claiming that Diesel's engines operated on principals articulated by others, not on the inventor's original concepts. A poor financial manager, Diesel nevertheless maintained a lavish villa in Munich called Jugendstil. In 1913, when he boarded the steamship *Dresden* at Antwerp, bound for England, Diesel faced financial ruin. Sometime during the evening of September 29–30, he disappeared from the ship's deck. His body was recovered ten days later, and all signs pointed to suicide.

Diesel engines—heavier and more expensive to build per horsepower than gasoline engines, but much more durable and cheaper to operate—made rapid inroads in shipping in the 1920s, heavy-duty trucking and construction equipment in the 1930s, railroads in the 1950s, and began to gain ground in passenger automobiles following the energy crises of the 1970s.

David B. Sicilia

See also: Automobile Performance; Carmot, Nicolas Leonard Sadi; Diesel Cycle Engines; Diesel Fuel; Engines; Gasoline Engines.

BIBLIOGRAPHY

Bryant, L. (1976). "The Development of the Diesel Engine." *Technology & Culture* 17:432–446.

Lytle, R. H. (1968). "Introduction of Diesel Power in the United States, 1897–1912." *Business History Review* 42:115–148.

Thomas, D. E., Jr. (1987). *Diesel: Technology and Society in Industrial Germany.* Tuscaloosa: University of Alabama Press.

DIESEL CYCLE ENGINES

The diesel engine is one of the most widely used global powerplants and can be found in almost every conceivable application. From small single-cylinder models to V20 designs, their horsepower can range from as low as 3.73 kW (5 hp) to as high as 46,625 kW

(62,500 hp). Some important applications of the diesel engine include

Light-duty
- cars
- pickup trucks
- riding lawnmowers

Heavy-duty
- heavy-duty trucks
- buses
- locomotives
- industrial power-generating plants
- oilfilled exploration equipment
- road-building equipment (e.g., backhoes, excavators, crawler tractors, graders, and bottom dumps)
- agricultural, logging, and mining equipment

Marine
- pleasure craft
- sailboat auxiliary engines
- workboats (e.g., tugs)
- oceangoing merchant ships and passenger liners.

In addition, a wide variety of military equipment, including tanks, armored personnel carriers, HUMVEEs and ships, is powered by diesel engines. The governed speed of diesel engines can range from as low as 85 rpm in large-displacement, slow-speed models, to as high as 5,500–6,000 rpm in smaller automotive type models.

Although today's technologically advanced diesel engine is named after the German Rudolph Diesel, it is a direct result of developmental work that began in the late 1700s when the first internal-combustion engine was constructed. This basic concept was further developed in 1824 by a young French engineer named Sadi Carnot. Other individuals added to this knowledge: Lenoir in 1860 with the first commercial internal-combustion engine; Beau De Rochas in 1862; Otto in cooperation with Langen in 1867; Clerk in 1881; Ackroyd-Stuart in 1890, and finally Diesel in 1892. Since its first practical inception in 1895 by Rudolph Diesel, the diesel cycle engine has been a source of reliable, efficient, long-lasting power.

Both gasoline and diesel engines are available in either a two-stroke- or a four-stroke-cycle design. The fundamental difference between the Otto engine cycle (named after Nikolaus Otto, who developed it in 1876) and the diesel engine cycle involves the conditions of the combustion. In the Otto cycle, the almost instantaneous combustion occurs at a constant volume, before the piston can move much. The pressure goes up greatly during combustion. However, in the diesel cycle, combustion occurs under constant pressure, at least for a time, because the piston moves to increase the volume during the burn to hold the pressure constant.

FOUR-STROKE-CYCLE ENGINE

The four-stroke-cycle internal-combustion engine design is widely employed in both gasoline and diesel engines. The high-speed four-stroke-cycle diesel engine produces superior fuel economy, lower noise factors, and ease of meeting exhaust emissions regulations over its two-stroke-cycle counterpart. In the four-stroke-cycle diesel engine, the concept shown in Figure 1 is used. A total of 720 degrees of crankshaft rotation (two complete revolutions) are required to complete the four piston strokes of intake, compression, power, and exhaust. The actual duration in crankshaft degrees for each stroke is controlled by both the opening and closing of the intake and exhaust valves by the camshaft and will vary among makes and models of engines.

In Figure 1 the engine crankshaft is rotating in a clockwise direction when viewed from the front of the engine. During both the intake and the power strokes, the piston moves down the cylinder, while on both the compression and the exhaust strokes the piston moves up the cylinder. Basically during the intake and exhaust strokes the piston acts as a simple air pump by inducting air and expelling burned exhaust gases from the cylinder. On the compression stroke the upward-moving piston raises the air charge to a pressure typically between 30 and 55 bar (441 and 809 psi) based upon the piston compression ratio and whether the engine is naturally aspirated or employs an exhaust-gas-driven turbocharger to boost the air supply pressure. A net loss in energy (waste heat to the cooling, lubrication, and exhaust systems, and to friction and radiation) occurs during the intake, compression, and exhaust strokes, since only during the expansion or power stroke do we return energy (torque, which is a twisting and a turning force) to rotate the crankshaft.

The diesel engine operates with a much higher compression ratio (CR) than does a gasoline engine, and therefore is manufactured with structurally

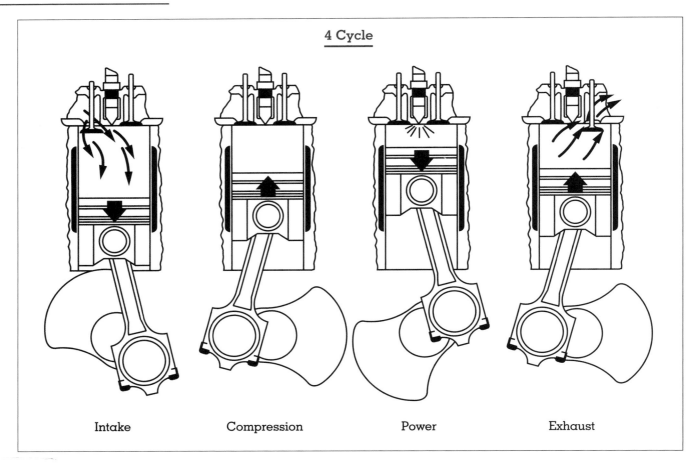

4 Cycle

Intake   Compression   Power   Exhaust

Figure 1.
Sequence of individual piston and valve events for a four-stroke diesel cycle engine.
SOURCE: Detroit Diesel Corporation

stronger components capable of handling this design feature. This higher CR results in a much higher cylinder pressure and temperature; therefore a greater expansion rate occurs when the piston is driven down the cylinder on its power stroke than occurs in a gasoline engine. CR is the difference between the volume of air remaining above the piston while at bottom dead center (BDC), versus that at top dead center (TDC). Typically a gasoline engine will operate with CRs between 9 and 10.5:1, while a DI (direct-injected) diesel CR usually varies between 15 and 17:1. IDI (indirect-injected) engine model CRs usually range between 18 and 23:1. These high CRs result in compressed air temperatures prior to the delivery of fuel from the fuel injector, typically in a range between 649°C to 927°C (1,200°F to 1,700°F).

This hot air converts the injected fuel from an atomized liquid to a vapor to permit self-ignition (it establishes a flame front) without the need for a spark plug.

DIESEL CONSTANT PRESSURE CYCLE

Diesel's original concept was for his slow-speed engine to operate on a constant-pressure design throughout the power stroke, obtained by continually injecting both compressed air and fuel. To increase the efficiency of the diesel cycle, his first engines used no cooling system, with disastrous results. Later engines, with cooling systems, corrected this part of the problem but resulted in cylinder heat losses accompanied by frictional, radiated, and exhaust heat losses. In addition, although com-

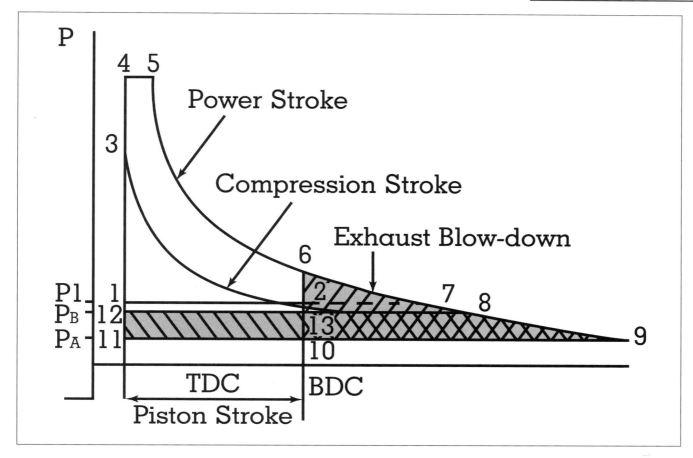

Figure 2.

Pressure-volume curve illustration for a turbocharged and direct-injected high-speed heavy-duty four-stroke diesel cycle engine.

pressed air and fuel were supplied to the cylinder throughout the power stroke, the increasing cylinder volume as the piston moved down on its power stroke was unable to maintain a high enough air temperature and pressure to sustain effective combustion, therefore the air/fuel ratio was not conducive to continued combustion. In today's electronically controlled high-speed diesel engines, fuel is injected for a number of degrees after top dead center (ATDC). This maintains cylinder pressure at a fairly constant level for a given time period, even though there is an increase in clearance volume above the descending piston. Engineers designing and testing engines like to compare the air standard cycles under actual engine performance with corresponding values for highly idealized cycles based on certain simplified assumptions.

PRESSURE-VOLUME CURVE

The energy used and returned to the engine crank-shaft/flywheel is illustrated in Figure 2, showing a PV (pressure-volume) diagram for a turbocharged and direct-injected high-speed heavy-duty four-stroke-cycle diesel engine. This schematic simplifies the internal operation of a piston throughout its four strokes. In Figure 3 we show the actual combustion operating principle in graphic form, with the piston at 90 degrees before TDC and at 90 degrees ATDC.

In Figure 2 you can see that from position 1 to 2 the piston moves down the cylinder on the intake stroke as it is filled with turbocharger boost air higher than atmospheric pressure, as indicated in line Pl. Depending on the valve timing, actual inlet valve closure will control the degree of trapped cylinder air

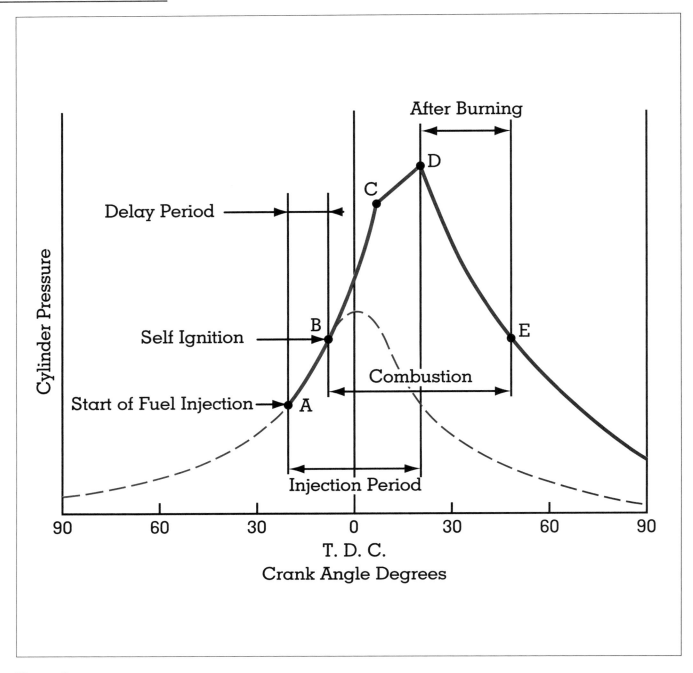

Figure 3.
Example of the sequence of events within the cylinder at 90 degrees BTDC (before top-dead center) and 90 degrees ATDC (after top-dead center).

SOURCE: Zexel USA, Technology and Information Division

pressure. In this example, compression starts at position 2, as the piston moves up the cylinder. Fuel is injected at a number of degrees BTDC (before top dead center), and there is both a pressure and a temperature rise as the fuel starts to burn from positions 3

to 4, As the piston moves away from TDC on its power stroke, positions 4 to 5, the continuous injection of fuel, provide a constant pressure for a short number of crankshaft degrees. From positions 5 to 6, the piston is driven downward by the pressure of the expanding gases. The point at which the exhaust valves open BBDC (before bottom dead center) depends on the make and model of the engine used. The work represented by area 6–7–2 is available to the hot end of the turbocharger (turbine wheel) from the hot, pressurized exhaust gases. Line PA indicates atmospheric pressure along points 9–10–11. The exhaust manifold pressure is shown as line PB and the exhaust gas blowdown energy is represented by points 6–9–10. The exhaust process from the engine cylinder is shown among points 6, 13 and 12, where 6 through 13 is the blow-down period when the exhaust valves open and the high-pressure gases expand into and through the exhaust manifold. From points 13 to 12 the piston moves from BDC to TDC, displacing most of the exhaust gas out of the cylinder. Therefore the potential work of the exhaust gases in this turbocharged engine is represented by the crosshatched areas identified as points 10–11–12–13. The maximum energy to drive the turbine of the turbocharger is that shown in the area identified as points 6–9–10 and 10–11–12 and 13. Ideally during points 6–9–10, if both the cylinder pressure and the turbocharger inlet pressure could both be maintained at equal levels before the piston moves upward from BDC on its exhaust stroke, a system close to ideal would be created when using a pulse turbocharger system.

The higher CR and the fact that diesel fuel contains a higher heat content per gallon or liter (approximately 11%) than does gasoline are two of the reasons why the diesel engine produces better fuel economy and higher torque at the crankshaft and flywheel. Further fuel economy improvements can be attributed to the fact that in the diesel engine, the throttle pedal is used to control the cylinder fueling rate directly. In a gasoline engine, however, manual operation of the throttle (gas pedal) directly controls the volume of air entering the engine intake manifold by restricting the size of the opening.

Diesel engines, however, operate on an unrestricted air flow at all speeds and loads to provide the cylinders with an excess air charge. This results in a very lean air/fuel ratio of approximately 90:1 to 100:1 or higher at an idle speed. At the engine's rated speed (full load maximum power output) the air/fuel ratio will drop to 20:1 to 25:1 but still provide an excess air factor here of 10 to 20 percent. This excess air supply lowers the average specific heat of the cylinder gases, which in turn increases the indicated work obtained

from a given amount of fuel. Compare this to most gasoline electronically controlled engines, where the TBI (throttle body injection) or MPFI (multiport fuel injection) system is designed to operate at a stoichiometric or 14.7:1 air/fuel ratio.

TWO-STROKE-CYCLE DIESEL ENGINE

Figure 4 illustrates the basic operational concept for a two-stroke cycle vee-configuration diesel engine where a power stroke is created every 360 degrees of crankshaft rotation versus the 720 degrees needed in the four-stroke-cycle design. The two-stroke design eliminates the necessity for individual intake and exhaust strokes required in a four-stroke cycle engine; therefore it becomes necessary in the two-stroke-cycle engine to employ a gear-driven blower to supply a large supply of low-pressure fresh air for:

- combustion of the injected fuel;
- cooling (approximately 30% of the engine cooling is performed by air flow, while 70% is performed by coolant flow within the engine radiator or heat exchanger);
- scavenging of exhaust gases from the cylinders;
- positive crankcase ventilation.

In the two-stroke-cycle diesel engine the cylinder-head-located poppet valves are to permit scavenging of the exhaust gases. Fresh air for the four functions listed above is force-fed through a series of ports located midway around the cylinder liner. Each liner receives its air supply from an "air box" cast within the engine block. On high-speed diesel engines this air pressure varies between 27.6 kPa and 48.3 kPa (4 and 7 psi) higher than atmospheric throughout the engine speed range. When an exhaust-driven turbocharger is used in conjunction with the gear-driven blower, air-box boost pressures at engine full load typically range between 172 and 207 kPa (25 and 30 psi) above atmospheric.

With reference to Figure 4, each piston downstroke provides power, while each upstroke provides compression of the blower/turbo-supplied cylinder air. The actual number of degrees of both of these strokes will vary based on the specific engine make and model, and its year of manufacture needed to comply with U.S. Environmental Protection Agency (EPA) exhaust emission limits.

Note in Figure 4 that in a high-speed engine, the

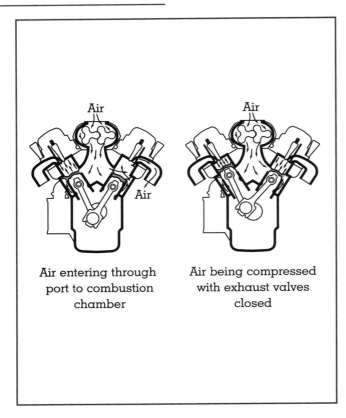

Figure 4.

Two-stroke diesel cycle engine principle of operation.

SOURCE: Detroit Diesel Corporation

power stroke begins at TDC and ends at approximately 90 to 92 degrees ATDC (after top dead center), when the exhaust poppet valves start to open by camshaft action. This allows the pressurized exhaust gases to start flowing from the cylinder through the open exhaust valves. As the piston continues moving down the cylinder, the exhaust gas pressure decreases. At approximately 59 degrees BBDC (before bottom dead center), the piston begins to uncover the cylinder liner ports, permitting the now higher air-box pressure (ABP) to enter the cylinder; this is the start of the "scavenging" stroke. Since the ABP is now higher than the exhaust gas pressure, positive displacement of the exhaust gases out of the cylinder takes place. The scavenging process lasts for approximately 118 degrees of crankshaft rotation (59 degrees BBDC and 59 degrees ABDC (after bottom dead center). It is during this 118-degree period that scavenging; cooling of the piston, liner, valves, and cylinder head; and inducting fresh air for combustion purposes occur.

During the piston upstroke the exhaust valves do not close until after the cylinder liner ports have been covered at approximately 59 degrees ABDC by the upward-moving piston. Typically the exhaust valves close between 3 and 5 degrees after port closure. The piston is now on its compression stroke. The start of fuel injection at all speeds is variable, based on the year of manufacture, the make and model of the engine, and whether it is mechanically or electronically controlled. The fuel injection duration typically lasts for 10 to 14 degrees at an idle speed between 500 and 700 rpm beginning at BTDC and ending just at TDC, or a few degrees ATDC based on the engine make and model and its year of manufacture.

DIESEL COMBUSTION TYPES

The majority of existing diesel engines now in use operate on what is commonly referred to as a direct-injection (DI) design (see Figure 5). This means that the high-pressure injected fuel (as high as 30,000 psi or 207 MPa) enters directly into the combustion chamber formed by the shape of the piston crown. In the indirect-injection (IDI) system the injected fuel is sprayed into a small antechamber within the cylinder head. Combustion begins in this small chamber and forces its way into the main chamber, where it consumes the remaining air required for additional combustion. IDI engine designs require use of an electrical glow plug to initiate satisfactory combustion, something that is not required in a DI engine. Use of a glow plug allows the IDI engine to burn a rougher grade of fuel than the DI engine. Fuel is injected at a lower pressure in the IDI; in addition, the larger combustion surface area of the IDI engine creates greater heat losses. This results in the IDI engine consuming approximately 15 percent more fuel than an equivalent horsepower (kW) rated DI engine. Additionally, using a rougher grade of diesel fuel results in higher exhaust emissions than from a DI engine using high-quality, low-sulfur fuel.

THERMAL EFFICIENCY

Rudolph Diesel's original intent was to produce a low-heat-rejection internal-combustion engine without the need for a cooling system. He believed that this would provide less heat losses from the combustion process and provide him with a superior heat, or thermally efficient (TE), design concept. To his chagrin,

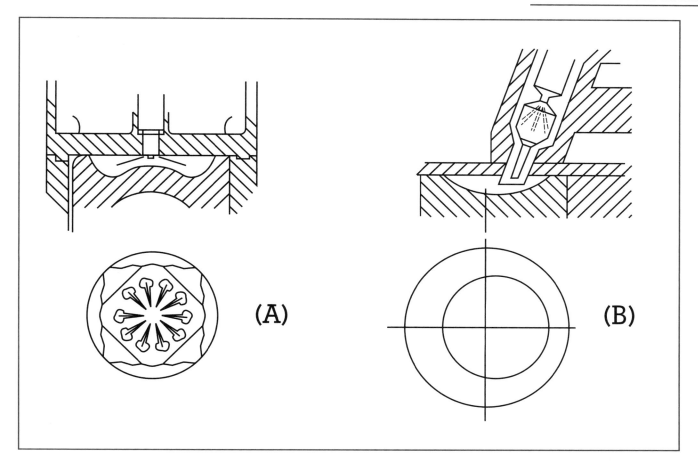

Figure 5.
Principles of (A) DI (direct-injection) and (B) IDI (indirect-injection) combustion chamber designs.

however, he found that this was not a feasible option when his first several test engines failed to perform to plan.

Basic physics involving friction and heat losses prevents the construction of a perfect internal-combustion engine. If friction between moving components could be eliminated, the mechanical efficiency (ME) of the engine would improve. Similarly, if we could eliminate heat losses from the combustion process, we could improve the TE of the combustion process. if no friction or heat losses existed we could design a perfect or ideal engine—one that could provide closer to 100 percent efficiency. A simplified way to consider TE is that for every dollar or hundred cents of fuel consumed by the internal-combustion engine, how much is returned as usable power? Therefore the TE is a comparison of the actual ratio of useful work performed in the engine versus the total energy content of the fuel consumed. Typically gasoline engines today return twenty-eight to thirty-six cents on the dollar. The diesel engine returns approximately forty to forty-three cents on the dollar. It is a measure of how efficiently an internal-combustion engine uses the heat released into the combustion chamber from the fuel to produce mechanical power. Based on the specific make and model of the engine, the cooling and exhaust systems typically account for heat losses of about 23 to 27 percent each; friction losses can range between 7 and 9 percent; while radiated heat from the engine accounts for 3 and 5 percent. Therefore this combination will generally account for a 57 percent heat loss, resulting in a TE of approximately 43 percent in current electronically controlled high-speed diesel engine models. However, many stationary diesel power plants that recapture waste exhaust heat for cogeneration purposes can return TEs in the mid-50-percent and higher ranges. New mechanical and electronic design concepts plus the adoption of ceramic components are just some of the new technologies being adopted not only to improve thermal efficiency but also to drastically reduce exhaust emissions into the atmosphere.

DIESEL COMBUSTION

The diesel engine without the benefit of a spark plug does not generate instantaneous combustion, as occurs within the gasoline engine. Instead, the diesel cycle relies on the high-pressure atomized-injected fuel mixing with the hot compressed air to cause it to vaporize. Once this vaporization occurs, the air/fuel mixture generates a flame front to initiate combustion. This concept creates what is known as "ignition delay" and is one of the characteristics that gives the diesel cycle its unique pinging noise. The longer the ignition delay, the louder the combustion noise (hard combustion) due to the larger volume of injected fuel that collects within the combustion chamber prior to actual ignition. The start of ignition includes the fuel injected prior to this phase and is known as the "premixed flame."

Once the fuel ignites, the remaining fuel being injected has no ignition delay, since it is being sprayed directly into an established flame front. Under full-load conditions, peak cylinder pressures can average between 1,800 and 2,300 psi (12,411 and 15,859 kPa). These tremendous pressures produce the power within the diesel engine during the power stroke, resulting in a higher overall BMEP (brake mean effective pressure—the average pressure exerted on the piston throughout the power stroke), versus the much lower values in a gasoline engine and the higher TE levels of the diesel.

DIESEL EXHAUST EMISSIONS

All internal-combustion engines, due to their inherent design characteristics, are unable to burn the injected fuel to completion. The make, model, year of manufacture, cylinder displacement, speed, and load all affect the percentage of emissions emitted into the atmosphere and the air we breathe. Therefore, major research and development is a continuing effort to clean up the type and quantity of pollutants. The EPA is a government agency charged with setting the limits on all industrial and internal-combustion-engine limits. The European Economic Community as well as Asian and other countries have

similar agencies tasked within these same parameters.

Major exhaust emissions from internal-combustion engines targeted by the EPA can be categorized into the following areas:

1. Carbon dioxide (CO_2), although nonpoisonous, does contribute to "global warming." Complete combustion in an internal combustion engine produces CO_2 and water.

2. Carbon monoxide (CO) is a colurless, odorless, and tasteless gas. Inhalation of as little as 0.3 percent by volume can cause death within thirty minutes. The exhaust gas from spark ignition engines at an idle speed has a high CO content. For this reason NEVER allow an engine to run in an enclosed space such as a closed garage.

3. Oxides of nitrogen (NO_x) have two classes. Nitrogen monoxide (NO) is a colorless, odorless, and tasteless gas that is rapidly converted into nitrogen dioxide (NO_2) in the presence of oxygen. NO_2 is a yellowish-to-reddish-brown poisonous gas with a penetrating odor that can destroy lung tissue. NO and NO_x are customarily treated together and referred to as oxides of nitrogen.

4. Hydrocarbons of many different types are present in exhaust gas. In the presence of nitrogen oxide and sunlight, they form oxidants that irritate the mucous membranes. Some hydrocarbons are considered to be carcinogenic. Incomplete combustion produces unburned hydrocarbons.

5. Particulate matter, in accordance with U.S. legislation, includes all substances (with the exception of unbound water) that under controlled conditions are present as solids (ash, carbon) or liquids in exhaust gases.

Diesel engines, due to the combustion processes described herein, tend to have a rougher time meeting some of the specific exhaust emissions standards. The primary cause of combustion noise and the generation of oxides of nitrogen in the diesel engine can be traced to that portion of the combusted fuel that burns as a very rapid premixed flame. On the other hand, the slower-burning diffusion flame (fuel-injected after the start of ignition) is the primary cause of soot and unburned hydrocarbons.

At this time it is not possible to produce a totally

soot-free diesel engine because heterogeneous combustion always produces soot. Diesel engine operation, due to local concentrations of overly rich mixtures in the diffusion flame, leads to an increase in the emission of black smoke to a moderate extent even with moderate excess air. The relatively low exhaust gas temperatures of diesel engines create a problem for effective catalytic emission control of hydrocarbons, particularly in light-duty diesel engines.

Reduction of exhaust emissions is being tackled in two ways by engineers, including precombustion and postcombustion technology. One of the most effective methods now being researched and adopted includes use of synthetic fuel made from natural gas. This fuel is crystal clear, and just like water, it has no aromatics, contains no sulfur or heavy metals, and when used with a postcombustion device such as a catalytic converter any remaining NO_x or other emissions can be drastically reduced. Estimates currently place the cost of this fuel at $1.50 per gallon, with availability in 2004 to meet the next round of stiff EPA exhaust emission standards.

Some precombustion technology involves improvements in internal engine hardware components, various engine sensors, and electronically controlled common-rail fuel injection equipment. Other systems now on test incorporate the addition of a small quantity of a reducing agent such as urea (sometimes called carbamide) injected into the combustion chamber, resulting in a chemical reaction that releases ammonia. This in turn converts the exhaust gases into nontoxic levels of nitrogen and water.

The second method used to reduce exhaust emissions incorporates postcombustion devices in the form of soot and/or ceramic catalytic converters. Some catalysts currently employ zeolite-based hydrocarbon-trapping materials acting as molecular sieves that can adsorb hydrocarbons at low temperatures and release them at high temperatures, when the catalyst operates with higher efficiency. Advances have been made in soot reduction through adoption of soot filters that chemically convert CO and unburned hydrocarbons into harmless CO_2 and water vapor, while trapping carbon particles in its ceramic honeycomb walls. Both soot filters and diesel catalysts remove more than 80 percent of carbon particulates from the exhaust, and reduce by more than 90 percent emissions of CO and hydrocarbons.

EPA diesel exhaust emissions limits for 1998 on-highway diesel truck and bus engines in g/bhp-hr (grams/brake horsepower/hour) when using existing 0.05 percent low-sulfur diesel fuel were: hydrocarbons, 1.3; CO, 15.5; NO_x, 4.0; and particulate matters, 0.1. Regulations due to come into effect beginning with the 2004 model year represent approximately a 50 percent reduction in emissions of NO_x and particulate matters, as well as reductions in hydrocarbons.

DIESEL AUTOMOTIVE CONSUMER RELUCTANCE

Despite the superior fuel economy of the diesel engine and its longer life to overhaul versus that for most of its equivalent gasoline counterparts, the general automotive consumer has preferred the choice of the gasoline engine in passenger cars and light trucks. However, within the European Community, the better fuel economy of the diesel accounts for up to 35 to 40 percent of all vehicle sales. In North America, diesel pickup trucks from all of the domestic manufacturers are popular options. To capture a greater percentage of the vehicle market, diesel engines require some technological improvements to bring their overall performance closer to that of the gasoline engine.

The following are the nine basic advantages of gasoline over diesel:

1. Quieter operation (no *ignition delay* or diesel knock; lower peak cylinder pressures and temperatures).
2. Easier starting, particularly in cold-weather operation (gasoline vaporizes at a much lower temperature than does diesel fuel; spark plugs provide instant combustion).
3. Less unpleasant odor, particularly in the exhaust gases.
4. Tends to burn visually cleaner at the exhaust tailpipe, since it operates in a closed-loop electronic mode (oxygen sensors interacting with the powertrain control module) to maintain an ideal air/fuel ratio of 14.7:1.
5. Quicker acceleration (no ignition delay) at lower engine speeds.
6. Can operate at higher speeds (rpm). Less inertia forces due to lighter components.

7. Good fuel economy at steady-state highway cruising speeds.

8. Lower weight, resulting in a higher power-to-weight density.

9. Generally lower production costs.

The following are the four basic advantages of diesel over gasoline:

1. Greater mileage between engine overhaul/repair (more robust).

2. Superior fuel economy (more thermally efficient) particularly at low speeds due to lack of restriction of the air flow; air flow restriction occurs in a gasoline engine through the throttling action of the gas pedal.

3. Lower carbon monoxide levels.

4. Higher crankshaft torque-producing capability.

DIESEL TECHNOLOGICAL ADVANCES

Since the 1985 model year, many heavy-duty high-speed diesel engines have been equipped with turbocharged aftercooled engines using electronic fuel injection controls. By the 1990 year all major high-speed engine original equipment manufacturers (OEMs) in North America employed electronic controls, since adopted by other diesel engine OEMs. Metallurgical advances have provided lighter but stronger engine components, and the use of plastics and fiberglass and aluminum alloys has increased in many external engine components

Four-valve cylinder heads, overhead camshafts, ceramic turbocharger components, crossflow cylinder heads, two-piece cross-head pistons, electronically controlled injectors, and hydraulically actuated electronically controlled unit injectors that do not require a pushrod and rocker arm assembly are all in use on existing engines. Future rockerless valve control engines, nonferrous piston and liner components, and turbocompounding will all improve the thermal efficiency of future engines.

Variable valve timing similar to that now in use in gasoline engines will become more common on all internal-combustion engines. Future valveless engines might employ rotating hollow-type shafts in place of the long-used poppet valves, or an electric solenoid will be used to operate both intake and exhaust valves, once again reducing valve-train frictional losses to improve overall thermal efficiency. Turbocompounding is a process whereby the hot, pressurized exhaust gases, after driving the turbocharger, will be directed to a large expansion turbine geared to the engine crankshaft to return additional energy, which would otherwise be wasted by flowing out of the exhaust system. The result will be a substantial increase in thermal efficiency. Turbocompounding is not yet available in a full-scale production engine, but look for this feature on future engines. Low-flow cooling and lube systems that have been in use for some time have reduced parasitic losses to further improve fuel economy. Future cooling systems will employ ceramic components, permitting the engine to run at a higher coolant temperature and providing a further increase in thermal efficiency.

Robert N. Brady

See also: Automobile Performance; Carnot, Nicolas Leonard Sadi; Combustion; Diesel Fuel; Diesel, Rudolph; Engines; Gasoline and Additives; Gasoline Engines; Government Agencies; Otto, Nikolaus August; Thermodynamics.

BIBLIOGRAPHY

Bosch Automotive Handbook, 4th ed. Cambridge, MA: Robert Bentley.

Brady, R. N. (1996). *Modern Diesel Technology*. Englewood Cliffs, NJ: Prentice Hall.

Brady, R. N., and Dagel, J. F. (1998). *Diesel Engine and Fuel System Repair*, 4th ed. Englewood Cliffs, NJ: Prentice Hall.

Lilly, L. R. C., ed. (1984–1985). *Diesel Engine Reference Book*. Stoneham, MA: Butterworth.

Stinson, K. W. (1980) *Diesel Engineering Handbook*, 12th ed. Norwalk, CT: Business Journals

DIESEL FUEL

Liquid fuels for use in internal-combustion engines are extracted and refined from crude oil, with diesel fuels being part of the middle distillate or kerosene fraction. Kerosene was initially derived from coal pyrolysis. The initial main use of this type of distillate was for the kerosene lamp, which had replaced lamps based on whale oil.

In 1859 at Titusville, Pennsylvania, Edwin Drake drilled the first successful oil well, and by the late 1880s most kerosene was made from crude oil. The search for crude accelerated in 1892, when Charles Duryea built the first U.S. automobile powered by a gasoline-fueled internal-combustion engine. Rudolph Diesel patented a compression ignition engine running on middle distillate at about the same time. While this engine was more fuel-efficient, it proved too complex to manufacture, not gaining in popularity until the middle of the twentieth century.

PRODUCTION OF DIESEL FUEL

Within this encyclopedia is an article covering crude oil refining in greater detail. However, as a brief introduction, Figure 1 gives a typical example of the products extracted and refined from a barrel of crude oil. Diesel fuel averages a little more than 18 percent of the total, or approximately 9.2 gallons from each 45-gallon barrel of crude oil.

To extract the products shown in Figure 1, the crude oil is heated in a distillation furnace. The resulting liquids and vapors are discharged into a distillation tower (also called a fractionating unit). This tower is hottest at the bottom, with the temperature dropping gradually toward the top. The distillation (vaporization temperature) separates the crudes into various fractions according to weight (specific gravity) and boiling point. Distillate runs down through the tower over a series of horizontal trays that are perforated to allow the upflow of vapors. Each tray is cooler than the one below it, thus providing a temperature gradient throughout the height of the tower. As different fraction reaches the tray where the temperature is just below its boiling point, its vapors can condense and change back into liquid, where it can be drawn off if desired.

The very lightest fractions that rise in the tower remain in a vaporized state and are used as a fuel in the refinery. Other fractions condense at various points in the tower according to their boiling points, as shown in Figure 2. Medium-weight liquids including kerosene and diesel fuel remain in the tower middle; consequently they are referred to as middle distillates. Components for diesel fuel, which have a higher boiling point than gasoline, boil in the range of 175°C to 355°C (347°F to 671°F). Following distillation, many of the distillate fractions are

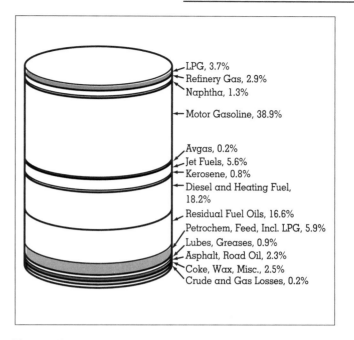

Figure 1.

Typical end-products extracted from a barrel of crude oil.

processed further to purify them or to convert them into lighter or heavier fractions, according to market demands.

Quantities of middle distillates, such as diesel fuel, can be increased by several processes. Mild thermal cracking, a process known as visbreaking (viscosity breaking), breaks heavier molecules down with heat to reduce their viscosity. The heaviest fractions from the distillation towers can also be transformed using more severe conditions to "crack" heavy hydrocarbon molecules into lighter ones. Fluid catalytic cracking ("cat cracking") uses intense heat, low pressure, and a catalytic substance to accelerate these thermal reactions. Hydrocracking employs a different catalyst, slightly lower temperatures, much greater pressure, and a hydrogen atmosphere to convert heavy molecules.

Residues may also be processed by removing carbon. These processes include deasphalting and coking, which produce usable liquid fuel components while rejecting a carbon-rich phase (asphalt or coke).

Diesel fuel makeup can represent various combinations of volatility, ignition quality, viscosity, sulfur level, specific gravity, and other characteristics. Various additives are used to impart special properties

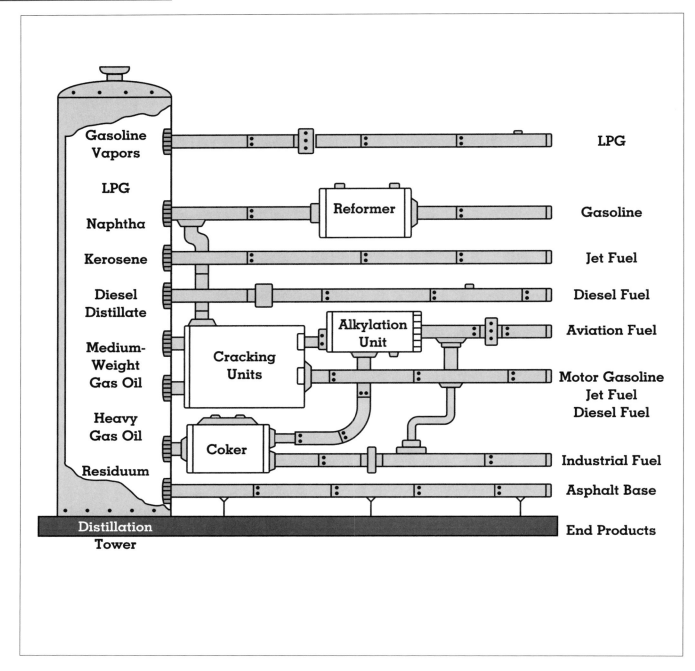

Figure 2.
Simplified arrangement of a crude oil refining system.
ꜱᴏᴜʀᴄᴇ: Chevron Research and Technology Company, a division of Chevron USA Inc.

to the finished diesel fuel. These can include ignition quality improvers, oxidation inhibitors (stability), biocides, rust preventatives (anticorrosion), metal deactivators, pour point depressants, demulsifiers, smoke suppressants, detergent dispersants, conductivity improvers, dyes, and deicers.

DIESEL FUEL CHARACTERISTICS

All liquid fuels and oils manufactured and sold in the United States since 1919 must comply with the engineering standards and quality specifications published by the American Petroleum Institute

(API). Many of API's standards are used around the world.

The main elemental constituents of diesel fuel are carbon (86% by weight) and hydrogen (13% by weight). Diesel fuel hydrocarbons contain molecules with eight to fifteen carbon atoms. The density of diesel fuel will vary based upon the grade, with a range of 0.815 to 0.855 kg/l (6.80 to 7.13 lb/gal) based upon its API gravity rating. Heating value, measured in British thermal units (Btus), depends on the fuel's heat of combustion and also whether the heat of vaporization in the water formed by combustion of the hydrogen in the fuel is usefully recovered. The higher heating value results if the water vapor is condensed. On the other hand, the lower heating value typifies the more common case, where the uncondensed water vapors are lost with the hot flue gases up the chimney or through the exhaust pipe.

Diesel fuel has a specific gravity that is lighter than water (specific gravity-1.00). Therefore a diesel fuel will float on water. Typical boiling points for diesel fuel will vary between 180°C and 360°C (356°F to 680°F), with the actual 100 percent boiling point (vaporization) being based upon the fuel grade. Its latent heat of vaporization is approximately 42.5 MJ/kg. Diesel fuel ignition temperature will vary based upon its grade, cetane number, and distillation range (vaporization temperature), but will average 250°C (482°F).

DIESEL FUEL GRADES

Diesel fuel oil is graded and designated by the American Society for Testing Materials (ASTM). For high-speed heavy-duty diesel engine, basically only two grades are now considered acceptable for transport use so that the engine exhaust emissions can meet Environmental Protection Agency (EPA) air pollution standards for toxic particulates and sulfur. These are the Nos 1-D and 2-D grade fuel oil classification. Fuel classifications below grades Nos 1-D and 2-D, which are designed for use in larger, slow-speed diesel engines, are not considered acceptable for use in high-speed automotive or truck engines. The No. 1-D fuel (typically API gravity 44) is a lighter distillate than a No. 2-D grade (typically API gravity 38 to 39).

Consequently, all things being equal, an engine operating on the same No. 2-D fuel with greater heat content per gallon will yield superior fuel economy than when operating on a grade No. 1-D fuel. This will generally result in increased engine output, and 5 to 10 percent better fuel mileage. In summary, lower-density fuels as well as winter blended fuels have lower volumetric heat content, lower viscosity, and poor lubrication characteristics.

FUEL PROPERTIES

The individual characteristics that particularly affect the performance of diesel fuels stem from the specific requirements of the diesel engine.

In the internal-combustion engine, air at ambient temperature and pressure is used to evaporate a rather volatile fuel (gasoline), and the combustible mixture is valved into the combustion chamber. Here it is compressed by the piston upstroke, and it is then ignited by a spark plug. The expanding hot gases drive the piston into its downstroke, thus performing work. The most important characteristics of the fuel are volatility (ready evaporation in cold weather but no vapor lock in hot weather) and smooth ignition without any engine knock. The latter quality is measured by the octane number.

In the diesel engine, compression ratio is much higher than in an internal-combustion engine. This adiabatic compression raises the air temperature above the ignition temperature of the diesel fuel. The diesel fuel is sprayed into the combustion chamber at the top of the piston upstroke via a high-pressure injection nozzle. The fuel still has to have enough volatility to vaporize quickly, and there has to be enough heat capacity in the compressed air to supply the heat of vaporization to the fine droplets. If these conditions are not met, as in a cold engine or with heavier fuels, the burning droplets will tend to form smoke and soot. The cetane number is analogous to the octane number in internal-combustion engines, and it is the single most important parameter in judging diesel fuel quality.

Distillation Temperature

This is the temperature at which the liquid fuel will vaporize when injected into the combustion chamber; therefore it becomes an important factor in the ignition-delay period of the fuel. Lower-boiling-range fuels such as No. 1-D are more volatile, while No. 2-D has a lower volatility, therefore requiring higher temperature to vaporize or boil. Due to these factors, higher-volatility fuels such as No. 1-D are preferential in applications where the engine will idle for long periods, such as in city bus/coach installations, or when operating in subzero ambient temperatures.

Boiling Point

The temperature at which the fuel is boiled off or vaporized at the refinery is known as the "end point temperature" listed in ASTM test D86, while ASTM spec D975 uses a 90 percent boiling point or distillation temperature to determine its suitability to vaporize. However, a number of major heavy-duty, high-speed diesel engine manufacturers specify that prior to selecting a diesel fuel you should ensure that a 95 percent distillation temperature is considered to ensure better combustion.

Cetane Number

Diesel fuel uses a numbering system whereby the higher the cetane number, the more volatile the fuel. Since a diesel engine requires that the injected fuel must vaporize from its liquid state and self-ignite from the heat of the air compression, a low-cetane-number fuel will result in a longer ignition delay, hard starting, incomplete combustion, smoke in the exhaust, a noisier engine (hard combustion), and lower horsepower. In cold weather operation, noticeable white smoke at the exhaust stack will last until the engine warms up. In engines equipped with charge-air-cooling to lower the turbocharger air inlet temperature, a low-cetane fuel usually will cause white exhaust smoke during light load operation. In high-speed heavy-duty electronically controlled diesel engines throughout the 1990s and into the new millennium, a minimum fuel cetane number of 45 is generally specified by the engine manufacturer for satisfactory performance. Cetane number is determined by putting the engine through a series of performance tests. The fuel marketer may add special additives to the fuel to improve the cetane number.

Sulfur Content

Stringent EPA exhaust emissions regulations introduced in October 1993 mandate that heavy-duty on-highway vehicle engines *must* use a low-sulfur diesel fuel that contains a maximum sulfur content of 0.05 percent. Prior to this time, the sulfur content was typically limited to 0.50 percent. However, most quality No. 2 brands ranged between 0.23 and 0.28 sulfur. Off-highway and marine engines still use cheaper fuels with a higher sulfur content, but EPA plans on all diesel engines eventually employing low-sulfur fuel to lower exhaust emissions. Diesel fuel sulfur content above 0.3 percent mass will cause premature piston ring wear, cylinder wear, and deposit formations, plus an increase in exhaust particulates. High-sulfur fuel for off-highway equipment can be identified by its greenish appearance due to the addition of a blue dye to identify it from the low-sulfur, honey-colored fuel for on-highway engines. Low-sulfur diesel fuel leaves approximately 0.01 g/bhp-hr sulfate in the raw exhaust, or approximately one-fifth or less lower than regular high-sulfur fuel. Because much of the crude oil imported into the United States has a high sulfur content, refining it to produce low-sulfur diesel fuel requires additional and more expensive procedures in the refining process.

Fuel Viscosity

Besides fulfilling the combustion requirement of the engine, diesel fuel also must act as a lubricant and cooling agent for the fuel injection system components. As the engine heats up, the fuel temperature also will increase resulting in a decrease in fuel viscosity. Therefore, to prevent scuffing and scoring of the fuel injection components, a selected fuel must meet minimum viscosity specifications. Additionally, hot fuel will reduce the metered quantity for a given throttle position, resulting in lowered horsepower. Thus, many current electronically controlled diesel fuel systems employ fuel coolers to maintain fuel operating temperatures at 65°C (150°F) or less.

Cloud Point

This condition is of concern only when equipment operates in subzero ambient temperatures. Since diesel fuel extracted from crude oil contains a quantity of paraffin wax, at some low ambient temperatures this paraffin will precipitate and create wax crystals in the fuel. This can result in plugging of the fuel filters, resulting in a hard or no-start condition. Any moisture in the fuel can also form ice crystals. Cloud point temperatures for various grades of diesel and other fuels should be at least 12°C (21.6°F) below the ambient temperature. In cases where cloud point becomes a problem, a fuel water separator and a heater are employed.

Pour Point

The pour point is the temperature at which the diesel fuel will no longer flow and is typically listed as being 5°C to 8°C (9°F to 14.4°F) lower than the cloud point of –13°C to –10°C (9°F to 14°F).

Flash Point

Due to its low vaporization temperature, diesel fuel is a fairly safe transportable fuel. Flash point has little to do with the fuel's combustibility factor or the performance characteristics of the engine. Rather it is a measure of the temperature at which the fuel oil vapors flash when in the presence of an open flame. Safety in handling and storage are the main points warranting consideration for flash point. A No. 1-D fuel has a flash point of 38°C (100°F), while a No. 2-D fuel flash point is 52°C (126°F). The storage of diesel fuel should always take place in clean and contaminant-free storage tanks. Tanks should be inspected regularly for dirt, sludge, and water that can cause microbial growth. Tanks need to be drained and cleaned regularly if signs of contamination are evident.

Carbon Residue

Carbon residue is expressed as a percentage by weight of the original sample of the fuel, with the amount determined by burning a given quantity in a sealed container until all that remains is carbon residue. The amount of carbon residue left within the combustion chamber of the engine has a direct bearing upon the internal deposits and affects the cleanliness of combustion, particularly the smoke emissions at the exhaust stack.

Ash Content

Diesel fuel may contain ash-forming materials in the form of abrasive solids or soluble metallic soaps. These solids cause wear of injection equipment, pistons, piston rings, and liners as well as increasing engine deposits. Ash content is expressed as a percentage of the weight of the original test sample of the fuel when burned to completion in an open container.

Water Content

All diesel fuels tend to contain trace water, expressed in parts per million (ppm). With the very high fuel injection pressures now used in electronically controlled diesel engine, fuel-filter/water separators are widely used, since water allowed to circulate freely through the injection system can result in seizure of components and erosion of injector orifice holes, and in extreme cases the high compressibility factor of water can blow the tip off of the fuel injector.

SYNTHETIC DIESEL FUEL

Engine manufacturers and oil refiners are researching and developing a synthetic blended diesel fuel. The many advantages of diesel power can be greatly improved by reducing the exhaust emission levels to comply with ever stricter EPA-mandated levels.

Substantial benefits can be achieved through the use of synthetic diesel fuels that burn much cleaner. Research has shown that current diesel engines with electronics technology can burn up to twenty times cleaner on this type of fuel versus existing fuel blends. This synthetic fuel is made from natural gas and can run in today's engines without modification. The fuel is crystal clear. It has no aromatics or sulfur, and it contains no heavy metals. With no sulfur, the addition of a catalytic converter to the exhaust system can further clean up the nitrous oxides and other emissions to where the diesel is as clean as its gasoline counterpart. Fuel refiners estimate that they can produce this synthetic diesel fuel for less than $1.50 per gallon. It will assist diesel engine manufacturers in meeting the next round of strict EPA exhaust emissions limits.

SUMMARY

In summary, diesel fuel with very low to no sulfur content is now possible with chemical and technological advances. Along with catalytic converters, electronic fuel systems, and sensors, the diesel engine for the new millennium will be capable of complying with ever more stringent EPA exhaust emissions. The diesel engine will continue to serve as the main global workhorse for all of the many thousands of different applications of its power cycle.

Robert N. Brady

See also: Automobile Performance; Combustion; Diesel, Rudolph; Diesel Cycle Engines; Engines; Fuel Cell Vehicles; Gasoline and Additives; Gasoline Engines; Government Agencies; Hybrid Vehicles; Kerosene; Synthetic Fuel.

BIBLIOGRAPHY

Bosch Automotive Handbook, 4th ed. Cambridge, MA: Robert Bentley

Brady, R. N. (1996). *Modern Diesel Technology.* Englewood Cliffs, NJ: Prentice-Hall.

Brady, R. N., and Dogel, J. F. (1998). *Diesel Engine and Fuel System Repair*, 4th ed., Englewood Cliffs, NJ: Prentice-Hall.

Lilly, L. R. C., ed. (1984–1985). *Diesel Engine Reference Book*. Stoneham, MA: Butterworth.

Stinson, K. W. (1980). *Diesel Engineering Handbook*, 12th ed. Norwalk, CT: Business Journals.

DI-METHYLETHER

See: Alternative Fuels and Vehicles

DIRECT CURRENT

See: Electricity

DIRECT CURRENT MOTOR

See: Electric Motor Systems

DISTRICT HEATING AND COOLING

Thermal energy delivered to a building from an outside source is known as district heating and cooling, which can range in size from small systems serving two or three buildings to networks serving entire cities. District heating and cooling is widely used in developed countries throughout the world and offers numerous advantages over individual building apparatus, including greater safety and reliability, reduced emissions, and greater fuel flexibility, particularly in using alternative fuels such as biomass or waste.

The earliest examples of district heating were Roman hypocausts, a type of hot-air furnace often adapted to warm several buildings in close proximity, such as the three temples at Carnutum (Vienna). The hypocaust and other Roman technologies were reintroduced during the Renaissance, serving primarily as starting points for improvements. Meanwhile, city fathers in Chaudes Aigues, a small town in the volcanic Cantal region of southern France, had by 1322 levied a tax on several houses heated by a natural hot spring channeled through open trenches dug in the rock. The history of this system, which still operates warming 150 residences, includes the introduction of wooden pipe, later replaced by plastic conduits. Accounts of this system appeared in numerous architectural works and may have been the inspiration for the proposed introduction of district heating in London in 1622 by Dutch polymath Cornelius Drebbel. This scheme was primarily intended to distribute heat for cooking and warming and thus reduce air pollution, caused by individual coal stoves. Unfortunately, Drebbel's patron, Prince Charles, was more concerned with wooing the Spanish Infanta, and Drebbel used his talents for other purposes, including building the first working submarine.

Although Roman engineers almost exclusively used hot air for heating, they extensively employed hot water in public baths. This technology was also reborn in the late sixteenth and early seventeenth centuries by Sir Hugh Plat, Solomon DeCaus, and others during what some have called the Rosicrucian Enlightenment.

Steam and hot water were also used to heat extensive horticultural nurseries in England and Sweden. A French engineer, Bonnemain, used hot water to heat several large residences, and in 1785 a Bonnemain-type system was used to heat the three large buildings of Count Potemkin's Taurida Palace in St. Petersburg. A competitor of James Watt's, Matthew Murray, heated his house in 1804 by piping steam from his nearby factory, and in 1808 a Scottish engineer, Robertson Buchanan, wrote that "a number of neighboring buildings might be served with one boiler." In 1826, the religious utopian community of Old Economy, Pennsylvania, used the waste heat from its steam engine to heat several other buildings by means of buried steam pipes.

Despite a strong awareness of the potential advantages of district heating in the early nineteenth century, widespread adoption did not occur until technology was developed to handle the numerous problems associated with heat distribution. The Great

Main thermae of Sbeitla. (Corbis Corporation)

Exhibition of 1851 marked one of the positive turning points, with a separate boiler plant providing steam to power exhibits in the enormous Crystal Palace, a 600-meter-long glass structure. A similar exhibition, complete with separate boiler house, was held in New York two years later.

District heating began to appear more frequently in institutional complexes, which were becoming more numerous in both Europe and America. A new state capital complex in Springfield, Illinois in 1867 was served by a steam system, and the 1876 Croydon Asylum in the outskirts of London used hot water. Another large steam system was built for the American Centennial Exposition of 1876 in Philadelphia.

In March 1877, New York inventor Birdshill Holly introduced the Holly Steam Combination System to provide heat and power using steam distributed through an underground piping network. The first

Holly Steam system was installed in Lockport, and within five years was installed in nearly fifty cities in North America and Europe. Although not all of the early systems were successful, and many were later abandoned for various reasons, the Holly System serving the downtown district of Denver, Colorado began service in November 1880 and has been operating ever since. A variant of Holly's design has operated in New York City since March 1882, providing reliable steam service to a large portion of Manhattan.

Developers in the late nineteenth century also recognized the need for artificial cooling in their customers' buildings. The New York Steam Company offered steam-driven absorption chillers as early as 1886, and four years later a district cooling company began operating in Denver by distributing chilled brine through underground pipes. Several other cities had such service by the 1920s, with systems distributing either brine or ammonia, often serving meat-packing and similar industries. Many electric companies also made ice in their central plants and distributed it to their customers for domestic refrigeration. During the 1930s, large, chilled water refrigeration plants were installed at Rockefeller Center in New York city and at the U.S. Capitol in Washington. In 1940 Southern Methodist University installed a central chiller plant and district cooling system at its campus in Dallas. Many institutional and industrial users installed large district cooling systems, and in 1962 commercial district cooling service was started in downtown Hartford, Connecticut, the first of many such systems.

Despite the early synergies between electric generation and district heating in America, the tremendous growth in generator size made it impractical to locate plants close to urban areas. Coal shortages during World War I (primarily caused by government takeover of railroads) sparked a movement to locate power plants close to coal mines, since it was thought to be more economical and even more environmentally desirable to transport electricity via wires. Small district heating systems that distributed heat from a power plant measured in kilowatts could not compete against plants making tens or even hundreds of megawatts of power. Many systems, however, benefited from managers who recognized the advantages of district heating and located plants where they could still serve heating customers.

At about the time that many American systems were being dismantled, district heating began to be incorporated into European cities such as Paris and Hamburg. The new Soviet Union studied various heating technologies throughout the world and adopted district heating to warm its new cities. In Iceland, the pipe for a planned geothermal system for the capital of Reykjavik was still in Copenhagen harbor when the Nazi invasion stopped the shipment. The United States eventually assumed responsibility for supplying Iceland, and calculated that it was more efficient to provide and transport twenty-four miles of steel pipe rather than deliver oil for Icelandic furnaces.

Many European cities, such as Rotterdam, incorporated district heating while rebuilding after World War II. Investigators from the American Strategic Bombing Survey were surprised to discover that aerial bombs had little effect on high-temperature hot-water piping in German factories. Within a short time such systems appeared on American military bases and, later, on the many colleges and universities

opened in the post-war period. A Danish engineer working at the American air base in Thule, Greenland, invented a better way to prefabricate heating pipe. He returned home to start a company that was able to take advantage of the oil embargo that hit Denmark especially hard in the early 1970s.

Since Denmark imported nearly all of its energy in the form of petroleum, the Danish government adopted an ambitious plan to become energy self-sufficient. This was accomplished over the next twenty years through a combined effort to increase energy efficiency and to make use of alternative fuels, especially renewable resources such as wind and biomass. District heating played a large role in this effort, allowing previously wasted heat from factories and power plants to be used for useful heating. By 2000, more than 55 percent of Danish residences were connected to district heating networks. The environmental advantages of district heating have also become of importance due to global warming concerns. Utilizing the energy wasted in industrial and electric generation plants avoids burning additional fuel in individual buildings.

District energy systems primarily serve commercial and institutional buildings in the United States, although some systems serve multi-unit residential buildings. According to U.S. Department of Energy survey data, about 10 percent of commercial building floor space used district heating and about 4 percent used district cooling as of 1995. Steam was the predominant type of energy produced and distributed, accounting for about three-quarters of distributed energy. The median size system serves about two million square feet of floor space.

Overall, there were about six thousand district energy systems operating in the United states in 2000. They collectively provide over one quad of energy annually—about 1.3 percent of all energy used in the United States. Most systems serve institutions such as colleges, hospital complexes, and military installations, that is, they serve a number of buildings owned by a single organization. Utility systems that sell heating and/or cooling to separately owned buildings account for only around 16 percent of the district energy provided in the United States.

About 10 percent of the district energy systems are part of combined heat and power systems where both electric energy and useful thermal energy are produced. The electric generation capacity of these systems totals about 3,500 MW, about 0.5 percent of total installed electric generation capacity in the United States. However, the use of combined heat and power along with district energy distribution is growing, with new systems installed in recent years in Philadelphia, Penn., Trenton, N.J., St. Paul, Minn., and elsewhere.

Morris A. Pierce

DOMESTIC ENERGY USE

Domestic energy involves the production and consumption of heat, light, and power for a variety of activities. Heat provides space and water heating as well as cooking. Light, both artificial and natural, is necessary for indoor activities. Power is necessary for moving water, transporting people, fuel, and goods, and for numerous other domestic activities, including building and cultivation. For much of human history, energy came from just two sources: the sun and musclepower, both human and animal. At some point, fire became domesticated and over time became the largest and today, including its form of electricity, almost the only source of domestic energy. At the beginning of time, domestic energy was rather crude but generally uniform. Today it remains virtually unchanged in some areas, intensely sophisticated in others, and by no standard uniform. Even advanced developed cultures with similar living standards use vastly different amounts; for instance, on a per capita basis northern Europeans use half the energy as their North American cousins living in a similar climate.

As the philosopher Thomas Hobbes noted, historically human life has often been nasty, brutish, and short. For most of human history, tending the domestic sphere would have been quite sufficient to cause this misery. Communities, and later cities, first appeared during times when agricultural production exceeded the demands of the agricultural population. Most, perhaps all, of these cities were in mild climates, such as the Fertile Crescent, the Mediterranean basin, and their counterparts in America and Asia. The word "climate" itself is derived from the Greek word for latitude, and Aristotle and other Greek philosophers believed that the narrow band of latitude they lived in (along with its undiscovered counterpart in the Southern Hemisphere) were the only areas where humans could survive. Water was moved by muscle, windows let in light during the day, and torches, tapers, oil lamps and the glow of a fire provided it at night. Cooking could be done with wood confined in a ring of stones, but charcoal braziers were quite common, along with a remarkable variety of cooking and baking devices dating from preclassical times, often to bake bread. One example was the tannûr, which was conical and partly sunk into the earth. It is the only form of furnace mentioned in the Old Testament and is precisely detailed in Jewish law. Artificial warming could have been provided by a fire or a charcoal brazier, although Greek architects occasionally incorporated passive solar heating into their designs. In several locales fuels such as chaff, straw, reeds, rushes and dried dung were (and still are) commonly used. Warm bathing was virtually unknown among the Greeks, who considered it effeminate.

The Romans were the first real technologists, and as their empire expanded to the north, their heating systems and gravity-powered water aqueducts followed. The Roman hypocaust—a central heating system with an undeground furnace and tile flues to distribute the heat—was widely used and adapted to local fuels, such as coal in Britain. Charcoal braziers were commonly used for domestic heating, but stoves also appeared in certain areas. Emperor Julian wrote in 363 C.E. that he did not care for the Parisian stoves during a visit there, but was almost suffocated by fumes from a brazier used to warm his sleeping room. The Roman architect Vitruvius wrote the architectural treatise *De Architectura* which contains extensive descriptions of warming techniques. Romans also loved public baths, which along with bread and circuses were used to mollify the masses during hard times. Although there are mixed opinions regarding the fuel efficiency of the hypocaust, the sheer number of them exhausted forests from the Italian peninsula and required fuelwood to be imported from North Africa by ship. After the fall of the Roman Empire, the era known as the Dark Ages swept over Europe. Roman heating technology fell into disuse, but historians of climate generally see this period as being rather warmer than normal. The arrival of the "Little Ice Age" in the fourteenth century led to a rather intensive period of technological innovation, marked most prominently by the widespread adoption of the chimney to replace the communal fire, whose smoke escaped through a hole in the roof. The chimney soon led to improved fire-places, and in an historical instant society changed, as a household could contain and manage multiple fires. In the 1370s William Langland's "The Vision of Piers the Plowman" assessed the chimney bitterly:

> Woe is in the hall each day of the week.
> There the lord and lady like not to sit.
> Now every rich man eats by himself
> In a private parlor to be rid of poor men,
> Or in a chamber with a chimney,
> And leaves the great hall.

This social upheaval was probably not matched until our own era, when families used to bonding around a single television set were wrenched apart by cheap portable sets that allowed individuals to watch in their own rooms. Once started, however, the process of social division was irreversible and indeed accelerated, as technology improved, at least for the upper classes. Chimney historian Leroy Dresbeck argues that chimneys may have promoted the art of love more than troubadors, but the Italian gentleman Octavian seems to have not minded the old way of communal heating, when as an overnight guest he would choose a place near one or another of the women of the house "by whom he was sometimes well received and sometimes got his face scratched."

One notable exception was a small village in the volcanic Cantal region of southern France. By 1322, several residences in Chaudes Aigues were being taxed for using hot water from a geothermal hot spring to heat their houses. The water was channeled to each house through a trench cut in the volcanic rock, which was covered with stone flooring in the living areas of the house. This system is still in operation and today heats 150 houses through a network of plastic piping. Apart from this notable exception, most improvements in domestic energy technology seem to have been aimed at the wealthier classes. For example, Elizabethan lawyer Sir Hugh Plat invented methods to heat greenhouses and generate steam to clear the skin of gentlewomen. He had not forgotten lesser mortals, however, and also invented a better candle, which would burn more than 120 hours, and a "coal ball," which was made from coal mixed with other ingredients to reduce smoke and increase fuel efficiency. His coal balls had the further advantage of being able to be manufactured by disabled veterans and orphaned children, which would reduce the

public dole in addition to saving one-third of fuel consumed in England.

Plat was only one of several inventors active during what one historian has called the "Rosicrucian Enlightenment" in early Stuart England. Another of these was Cornelius Drebbel, a Dutchman who in 1622 proposed a district heating system in London to allow residents to warm their houses and cook without creating smoke. Drebbel also demonstrated an artificial cooling apparatus to the king and others in Whitehall Palace, who were driven out into the summer heat due to the intense cold. Among Dribbel's other inventions was the first working submarine. A contemporary German, Franz Kessler, corresponded with his London comrades and in 1618 published the first heating book, simply titled *Saving Fuel*. This book was illustrated with a number of stoves commonly used throughout Europe and discusses the various economies of the different types. He also predicted, quite accurately, that as heating apparatuses became more efficient, their use would become much more widespread. His book was translated into French the following year, but Kessler is remembered now more for his paintings and optical telegraph.

This early enlightenment came to an end in the chaos of the Thirty Years' War and the English Civil War, but the work of these men was revived later in the seventeenth century as growing national economies and populations required ever greater resources. The search for better heating apparatuses became critical as a bitter cold wave swept over Europe in the early eighteenth century. Technologists responded to the need, and in 1713 Nicholas Gauger published *La Méchanique de Feu*, which described a number of methods to improve heating systems. By 1720 Paris was suddenly warm, and newspapers in both London and Paris advertised firms that could heat rooms of any size "without the least suffocation." Warmth came with a price, however, and the streets of Paris were soon cluttered with thousands of carts laden with wood and sawyers plying their trade.

The great immigration to America began at this time, and English colonists found themselves surrounded by vast forests that could fuel "Christmas fires every day." Benjamin Franklin in the 1730s noted that Pennsylvania Germans burned one-fourth to one-fifth the fuel consumed by English colonists and had fewer children die during epidemics. He made a careful study of fireplace and stove technology, primarily relying on Gauger's book for basic information, and in 1744 advertised details of his new Pennsylvania Fire Place. Franklin's work was the most extensive investigation of heating systems up to that time, examining fuel, comfort, environmental effects, and health in some detail. He chose not to patent his inventions as he did not want to benefit from a product designed for the general good. Perhaps his great contributions were to debunk the English notion that cold houses were both healthy and inevitable, and to advance the idea that properly applied technology could deliver not only better warmth but also better health. Franklin's work had enormous impact on improving domestic heating systems throughout America and Europe.

Although open fireplaces were still common in houses for both heating and cooking, the stove came into widespread use during the nineteenth century. Part of this was due to the increasing number of manufacturers who could produce a quality product at low cost, but probably more important was the disappearance of wood as a domestic fuel over the course of that century, in America, just as it had in England three centuries earlier. This had probably been inevitable. Within two decades of the day the first English landed in Boston, the surrounding fuel supply had been burned up. One of the reasons why Massachusetts acquired Maine in 1677 was to tap its seemingly inexhaustible forests. Even Harvard University in 1800 had its own ship to bring fuel-wood for its students and faculty from its land in Maine. As Americans moved westward across the continent, they were sure that the forests would last forever. In fact they did not last long at all, and only the discovery of coal in large quantities in Wyoming made it possible to build and operate the transcontinental railroad after the American Civil War. One of the most telling stories of this is Laura Ingall Wilder's *The Long Winter*, in which she describes living in a small town in South Dakota during the winter of 1880–1881. There were no trees (her six-year-old sister had never seen one), and the community was entirely dependent on coal delivered by railroad. During a particularly severe blizzard the tracks were blocked for several weeks, and Laura and her family burned straw and furniture to maintain a minimum amount of heat.

Coal could not be burned easily in an open fireplace, and the transition from wood to coal created enormous demand for coal stoves and other heating technology. The most promising was steam, where the boiler could be placed in a separate room (or even building) to allow coal to be kept away from the living area of a house. Steam apparatus was initially very expensive, and its initial markets were quality housing in larger cities. As the technology improved, particularly for safety devices, the prices dropped and they found a wider audience. One of the other advantages of steam was that it could also cook food. Steam kitchens had been thoroughly explored by Count Rumford early in the nineteenth century, although they were primarily limited to institutions, hotels, and large restaurants. The first residential steam heating apparatus used steam from a kitchen boiler and was described to the Royal Society in 1745 by Colonel William Cook.

On a larger scale, Lockport, New York, inventor Birdsill Holly in 1877 resurrected Drebbel's 1622 plan to provide district heating and cooling in London. Although Holly's district steam system became widespread in the commercial, institutional, and industrial marketplaces, his first systems included steam supply to residences for domestic use, including space and water heating and cooking. A variety of domestic steam appliances were invented. Demonstrations of steam-cooked meals were held for reporters in several cities, yet the idea never caught on. One that did was to use gas for these same purposes, which at that time meant gas manufactured in a local gas works by heating coal or coke and capturing the gaseous residue. Manufactured gas had been used since the late eighteenth century for street lighting as well as for interior illumination in a small number of commercial and larger residential buildings, but in the 1850s became more widespread in cities as clean whale oil became scarce and expensive. Kerosene was later to become widely used as a domestic lighting fuel, but gas made significant inroads into the residential market in many communities.

Although manufactured gas was too expensive for space heating, the small quantities necessary for domestic cooking and water heating provided a desirable market, and several local gas companies actively pursued it. Often a second meter was installed for the heating and cooking service, which was charged at a lower price. Manufacturers of coal stoves responded by developing a line of "summer stoves" that were primarily for cooking and water heating when the larger, winter heating stove was not necessary. Before long, salesmen from kerosene companies such as Standard Oil were also selling kerosene stoves for the same purposes, creating an intensely competitive market in some communities.

While this market struggle was going on, Thomas Edison was installing his first electric light plant, on Pearl Street in New York City. It began service in 1882, but it took a long time before electricity showed up in most houses. In 1907 only 7 percent of American households were electrified, and many of these probably retained some gas lighting as well for competitive reasons. Twenty years later, half of American households were electrified, and gas lighting was rapidly becoming only a memory. Much has been written about the struggle between alternating-current and direct-current electrical systems, but the much larger (and still ongoing) struggle has been gas and electricity for domestic cooking, heating, and refrigeration. Lighting was won very quickly by the incandescent lamp, later joined by fluorescent and other types to improve efficiency. Electric appliances such as toasters, irons, vacuum cleaners, and washers also prevailed in the market, often despite the early efforts of many local electric companies to discourage their use. Several local electric companies first became aware that housewives were using electric irons when power was shut off during the day (a normal occurrence in residential areas, since lights were not in use) and numerous complaints poured in.

Domestic refrigeration prior to World War I meant an icebox cooled with ice delivered by wagon and later by truck. Both electric and gas models appeared during the 1920s, and initially gas refrigerators were quite popular, since they were quiet and less expensive to operate. In 1930 only 5 percent of American households had a refrigerator, but within twenty years 80 percent had one. Although gas refrigerators are still used in recreational vehicles, hunting cabins, and other areas where electricity in unavailable, it is worthwhile to remark on their superb reliability. The American Consumer Product Safety Commission in 1998 issued a safety notice on a potential hazard from Servel gas refrigerators dating back to 1933, of which a large but unknown number were still in service. Once electric refrigerators were made quiet and efficient, they easily captured the market.

There still is intense competition between gas and electricity for the large ovens and ranges market. While electricity has completely prevailed in smaller appliances, including the relatively recent microwave oven, clothes dryers and domestic water heaters are also intensely competitive, but household heating by natural gas has become prevalent in most regions of the country where it is available. One of the interesting domestic energy sidenotes of the twentieth century was the "all-electric house," which was promoted as the ultimate in cleanliness and efficiency. Unfortunately such houses included electric space heating, which is not only the most expensive heating fuel but also the reason why many utilities built nuclear power plants. Domestic electric space heating is still prevalent in Norway, Québec, and areas of the northwestern United States with large amounts of hydroelectric power, but it remains to be seen how it will manage in an age of deregulated electric markets. Air conditioning also has become more widespread, not only in southern climates but also in any region with warm summer temperatures. Despite ongoing attempts by the gas industry to market gas residential cooling, this has largely been conceded to the electric utility industry.

Domestic water supply also has changed dramatically since the mid-1850s. Prior to that time water was collected from a source outside the house and delivered in buckets or other containers. A farmhouse without a spring or natural well probably would have a hand pump and later a windmill-powered pump, which became very popular in the American West. In urban areas a household water supply would have become more widely available, and over time transitioned from an intermittent supply, which might have provided water four hours a day, to a constant supply and eventually to a constant-pressure supply. Constant and generous sources of clean water led to indoor flush toilets; lush, green lawns in desert areas; and washing the family car. The availability of new forms of energy also led to other significant domestic improvements, including the telephone and the automobile.

It is easy enough to describe and even quantify domestic energy use as it has changed over the centuries and particularly since 1900, but it is much more difficult to assess how these changes have affected society, culture, and individuals for better or for worse. Ruth Schwartz Cowan argued in *More Work for Mother* (1983) that this new domestic technology resulted in the disappearance of domestic ser-

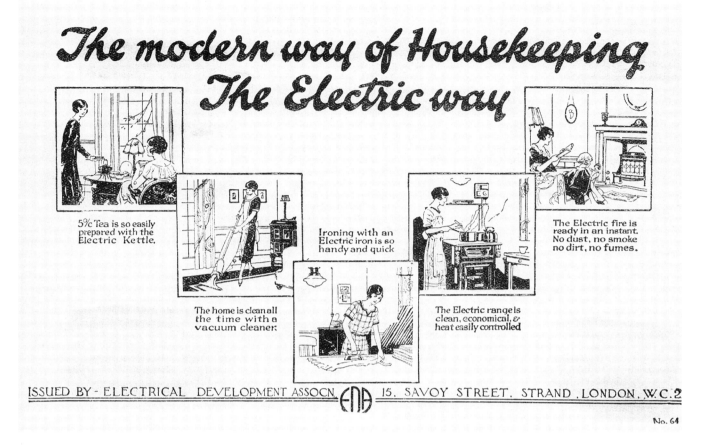

The modern way of Housekeeping The Electric way

5% Tea is so easily prepared with the Electric Kettle.

The home is clean all the time with a vacuum cleaner.

Ironing with an Electric iron is so handy and quick

The Electric range is clean, economical, & heat easily controlled

The Electric fire is ready in an instant. No dust, no smoke no dirt, no fumes.

ISSUED BY - ELECTRICAL DEVELOPMENT ASSOCN. 15, SAVOY STREET, STRAND, LONDON, W.C.2

No. 64

An early twentieth century advertisement from the Electrical Development Association (United Kingdom) promotes the use of electricity for housekeeping. (Corbis Corporation)

vants and higher standards of cleanliness that the lady of the house had to maintain. Yet even in 1950 a single working person would have been hard pressed to prepare his or her own meals and likely would have lived in a boardinghouse that provided one or more meals each day, or eaten in restaurants. Today meals are relatively easy to prepare and without much effort can even be quite elegant. It has, in fact, become so easy that many men now view domestic cooking as a fun thing to do. Probably this statement alone summarizes the changes that energy and technology have wrought in the household.

Morris A. Pierce

See also: Air Conditioning; Air Quality, Indoor; Appliances; Building Design, Residential; Coal Consumption of; Consumption; Edison, Thomas Alva; Electric Power Transmission and Distribution Systems; Heat and Heating; Insulation; Lighting; Natural Gas, Consumption of; Petroleum Consumption; Refrigerators and Freezers; Thompson, Benjamin; Water Heating; Windows.

DRILLING RIGHTS

See: Property Rights

DRIVETRAINS

The principal role of the automotive drivetrain is to transfer power from the engine output shaft to the drive wheels of the vehicle. Among its other functions are to multiply engine torque for improved vehicle performance, to operate the engine at a point offering good fuel economy, to enable operation of the vehicle in reverse, and to allow the engine to continue operation at idle speed while the vehicle is stationary. The transmission makes all of these functions possible. The drivetrain also incorporates a differential, so that when turning the vehicle, the outside drive wheel can rotate faster than the inside wheel. Finally, the drivetrain must accommodate the relative motion that occurs between the engine/transmission and the drive wheels. This requirement is fulfilled by the universal joint.

TRANSMISSION

In the absence of drivetrain losses, the power available to the drive wheels is equal to the power delivered by the engine. Power is proportional to the product of torque (the turning effort around the center of rotation) and rotational speed (typically measured in revolutions per minute, or rpm). The inevitable drivetrain losses are accounted for by a drivetrain efficiency. Losses in the transmission, and also the differential, arise from the friction of gears, bearings, and seals, and from churning of the gear lubricant. Churning losses depend on lubricant viscosity, which for a given lubricant varies with its temperature. In addition, the contemporary automatic transmission experiences flow losses in the torque converter and parasitic losses from the transmission pump used to provide the pressurized oil that controls shifting automatically.

Drivetrain efficiency depends on the rotational speed of individual components in the drivetrain and on the torque being transmitted. With a manual transmission, maximum drivetrain efficiency can be as high as 95 percent, but with an automatic transmission it falls closer to 85 percent. The passenger-car transmission seldom runs at its maximum efficiency point, however. For both transmission types, drivetrain efficiency trends toward zero as the transmitted torque approaches zero. Note the following:

$$\text{wheel torque} \times \text{wheel rpm} = \text{drivetrain efficiency} \times \text{engine torque} \times \text{engine rpm}$$

The wheel torque is equal to the product of the rolling radius of the drive wheels and the propulsive

force acting on the wheel axle to move the vehicle forward. Thus:

$$\text{propulsive force} \times \text{wheel radius} \times \text{wheel rpm} = \text{drivetrain efficiency} \times \text{engine torque} \times \text{engine rpm}$$

This expression indicates that for a given delivered engine power and drivetrain efficiency, represented by the right side of the second equation, the propulsive force can be increased by turning the drive wheels at a slower speed. The transmission does this by altering the speed ratio between the engine and the drive wheels.

This ability of the transmission to change speed ratio is also important for improving fuel economy during normal vehicle driving. Some of the power developed within the cylinders of a traditional gasoline engine never reaches the output shaft. Rather, it is spent in overcoming mechanical friction in the engine and in pumping the fresh air-fuel mixture into the cylinders and the exhaust products out of the cylinders. Generally, the friction losses are decreased by operating the engine at a lower rotational speed. At a given rotational speed, the pumping losses are decreased by operating the engine with a more open inlet throttle valve. The transmission allows the engine to satisfy the road-load power requirement of the vehicle at a combination of rotational speed and throttle opening that offers good fuel economy commensurate with satisfactory engine response to a sudden demand for increased power.

The gearbox of a manual transmission houses a selection of gears of different diameter. The input/output speed ratio of a pair of meshed gears varies inversely with their diameters. The diameter of meshing gears is proportional to the number of teeth on each gear. The driver of the vehicle uses the gear-shift lever to select which gears mesh. The typical manual transmission for a passenger car has three to six forward speeds from which to choose, with five being the most common at the close of the twentieth century.

A three-speed version is represented in Figure 1. As shown, the transmission is in neutral—that is, it is transmitting no torque to the output shaft. The input and output shafts share a common axis and are paralleled by a countershaft. Drive gears on the input and output shafts are always meshed. The output shaft is splined so the gears it carries can be slid axially along it. In low gear, which is selected for starting a vehicle from rest and which gives the highest input/output

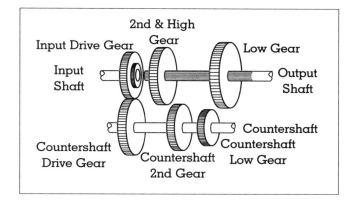

Figure 1.
Gearbox schematic for a three-speed manual transmission.

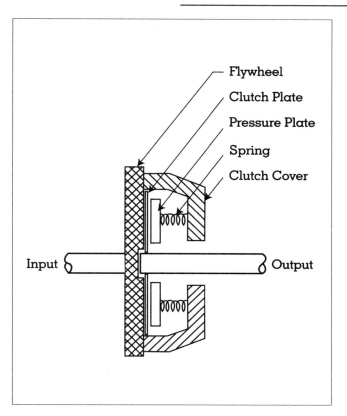

Figure 2.
Schematic of an automotive clutch.

speed ratio (greatest multiplication of engine torque), the low gear on the output shaft is slid axially to the left to mesh with the low gear on the countershaft. In second gear, the meshed low gears are disengaged and second gear on the output shaft is moved to the right to mesh with second gear on the countershaft. The gear at the left end of the output shaft contains teeth that can be engaged with matching teeth added to the input drive gear. In high gear, used for vehicle cruising, high gear on the output shaft slides to the left, locks onto the input drive gear, and engine torque is transmitted directly through the gearbox without multiplication.

When the transmission is shifted into reverse, an idler gear (not shown in Figure 1) is interposed between appropriate gears on the countershaft and output shafts to reverse rotation of the output shaft. When the vehicle is stationary with the engine running, the transmission is shifted into the neutral condition of Figure 1.

During gear shifting, the driver depresses a clutch pedal to disconnect the engine from the transmission input until the newly chosen gears are meshed. A disengaged clutch is shown schematically in Figure 2. Input torque is supplied by the engine flywheel, which is attached to the crankshaft. A clutch plate is mounted parallel and in close proximity to the flywheel. The clutch plate is coated with a friction lining. On the opposite side of the clutch plate is a pressure plate.

When the driver disengages the clutch by pushing down on the clutch pedal, a lever system (not shown) moves the pressure plate to the right against compres-

sion springs, as illustrated. When the driver engages the clutch by releasing the clutch pedal, the springs press the pressure plate and clutch plate against the flywheel, transmitting engine torque to the transmission input shaft. The friction linings on the clutch plate slide until engagement is completed. This prevents jerking of the vehicle during gear shifts.

Most automatic transmissions in production retain a gearbox, but the mechanical clutch is replaced by a rotating fluid unit that eliminates the need for a driver-operated clutch. In the gearboxes of most automatic transmissions for passenger cars, planetary gearing replaces the countershaft arrangement of Figure 2. A planetary gear set, illustrated in Figure 3, is comprised of a sun gear and a ring gear with intervening planet pinions, which are connected together by a planet carrier. The speed ratio across a planetary set depends on which of its three gear elements is prevented by a clutch from rotating, and also on which of the remaining two gear elements is the input. If the planet carrier is fixed, the ring and the sun rotate in opposite directions. If either the ring or the sun is fixed, the remaining elements rotate in the same direction.

The fluid unit in the typical automotive automatic transmission is called a hydrodynamic drive because it transmits power solely by dynamic fluid action in a closed recirculating path. Hydrodynamic drives are often classified into fluid couplings and torque converters. Both were patented in Germany in 1905 by Herrmann Foettinger. The original application was in marine vessels. In the 1930s, hydrodynamic drives began to appear in transit buses, where freedom from manually shifting a transmission was particularly desirable for a driver who had such other duties to perform as fare collection. Hydrodynamic transmissions also saw limited application in military vehicles during World War II.

The first passenger-car automatic transmission to see widespread use was the Hydramatic from General Motors. It used a fluid coupling, an example of which is illustrated in Figure 4. The fluid coupling has a torus-shaped split housing in which vanes are set radially to the axis of rotation. The impeller, on the right, is fastened to the engine. The turbine, on the left, is fastened to the gearbox input shaft. The vanes in the rotating impeller transfer engine torque into the transmission fluid as it is pumped outward. That torque is transferred by the swirling impeller discharge flow to the turbine vanes as the fluid flows radially inward to reenter the impeller.

Although the torque delivered by the turbine is equal to the engine torque absorbed by the impeller, their rotational speeds are not equal. When the engine is operating and the vehicle is stationary, there is 100 percent slip in speed from impeller to turbine, and the efficiency across the coupling is zero. If the vehicle is stationary with the engine idling, this provides the infinite output/input speed ratio of a disengaged clutch in a manual transmission, thus facilitating elimination of the clutch pedal. However, when the driver's foot is removed from the brake, the torque of the idling engine passes through the transmission to cause vehicle creep. During constant-speed cruising at moderate to high road speed, the slip in rotational speed between impeller and turbine is as little as 2 to 3 percent, yielding a fluid-coupling efficiency of 97 to 98 percent.

The torque converter soon replaced the fluid coupling as the hydrodynamic device of choice in the automatic transmission. A schematic of a simple

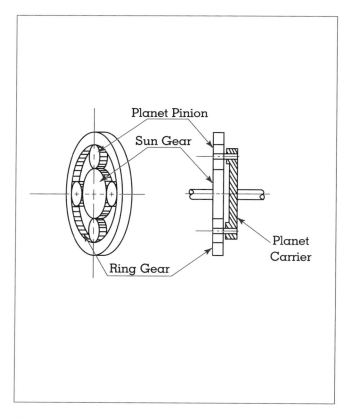

Figure 3.
Planetary gear set.

torque converter appears in Figure 5. As in the fluid coupling, transmission fluid circulates around a torus. However, the torque-converter flow path contains an additional member: a row of reactor vanes. The reactor is mechanically connected to the stationary housing of the torque converter through a one-way clutch. This clutch allows the reactor to rotate around the axis of the torque converter when the fluid forces impose on it a torque in one direction, but to lock it against rotation when the torque acting on it is in the opposite direction. In contrast to the vanes in the fluid coupling, torque-converter vanes in the impeller, turbine, and stator are all carefully curved out of the plane of the rotational axis.

The distinguishing performance characteristic of the torque converter, in contrast to the fluid coupling, is that it is capable of multiplying torque. Torque multiplication is made possible by vane curvature and the presence of the reactor. When the converter is stalled—that is, the turbine and the reactor are stationary—the torque delivered to the gearbox is typically 2

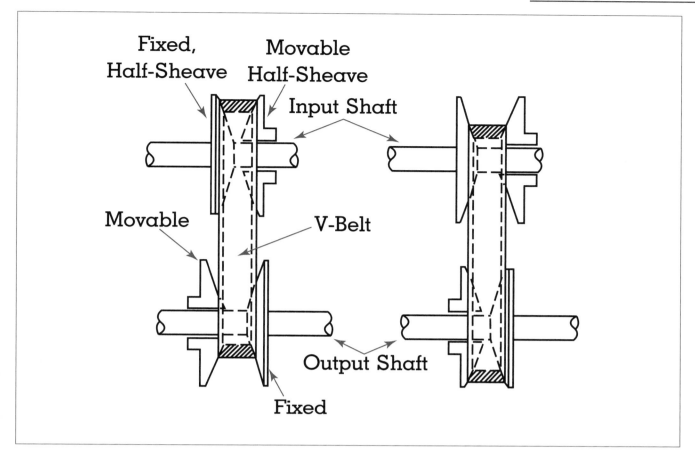

Figure 4.
Fluid coupling.

to 2.5 times the input torque from the engine. This torque multiplication normally allows a given level of vehicle acceleration performance to be achieved with one less step in the gearbox than if either a mechanical clutch or a fluid coupling were used. Since 1950, passenger-car torque-converter transmissions of the type described have usually employed gearboxes with two to five forward gears, with four most common at the end of the twentieth century.

As a vehicle is accelerated from rest, the initially stationary turbine accelerates toward impeller (engine) speed, and torque multiplication falls steadily. In a typical automotive converter, by the time the turbine has reached about 85 percent of the impeller speed, the torque ratio across the converter has dropped to unity and the one-way clutch allows the reactor to rotate. The speed ratio at which this occurs is known as the coupling point. Beyond the coupling point, the rotating reactor can no longer redirect the flow of torque-converter fluid and so cannot multiply input torque. The torque converter then acts like a fluid coupling, with efficiency peaking as high as 97 to 98 percent as a result of slip between impeller and turbine. In modern automatic transmissions, a torque-converter clutch is added that mechanically links the turbine to the impeller at this condition to prevent the slip responsible for such efficiency losses. Actually, slight slip is often allowed in the torque-converter clutch to minimize transmission of engine torque pulses to the vehicle drive wheels.

For many years, gear shifting in the automatic transmission has been controlled hydraulically, typically in response to vehicle speed and the position of the engine inlet throttle. In recent times that control has been reassigned to an electronic computer, which facilitates greater smoothness and flexibility in shift control.

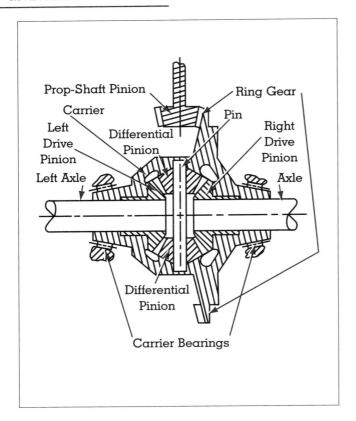

Figure 5.
Torque converter.

The substantial forces required to actuate gearbox clutches are still developed hydraulically, however.

Since 1985, more than three-quarters of new cars sold in the United States have had automatic transmissions. In fact, many licensed drivers in the United States have never learned to shift a manual transmission. In Europe, the high cost of gasoline has helped to retain the manual transmission because it is generally conceded to offer somewhat better fuel economy, although interest in automatic shifting there is growing. The congested driving encountered in Japan provides a greater incentive for the automatic transmission there than in Europe.

Practically all of the automatic transmissions sold in recent years have been of the torque-converter/gearbox type described, but there is growing interest in other approaches. Generally these have been of the continuously variable transmission (CVT) type. It is expected that by eliminating step changes in engine speed during shifting, as occurs with a gearbox, driving will be more pleasant. Also,

eliminating the discrete speed steps of the gearbox with a CVT is expected, on average, to allow the engine to operate more efficiently. Countering those positions, the gear shifts of a traditional torque-converter/gearbox automatic have been smoothed until nearly imperceptible by evolutionary improvements that have included electronic shift control. Further, the increasing number of speeds in the gearbox has helped to diminish the margin of fuel-economy improvement attributable to a continuously variable ratio. It is noteworthy that Daimler and Benz equipped their first mass-produced gasoline-engine-powered vehicle with a CVT in 1886.

One type of CVT that has recently seen limited production use is the belt and variable sheave arrangement. Modern research on this approach started during the 1960s in the Netherlands. In that country, van Doorne makes a transmission of this type that uses a belt comprised of steel blocks joined by a steel band. On the left, spacing between the sheave halves has been decreased on the input shaft and increased on the output shaft to decrease the rotational speed of the output shaft. On the right, sheave-half spacings have been reversed to drive the output shaft faster than the input shaft. Because of load limitations on the belt, this CVT is suitable only with small engines.

A second type of CVT relies on rolling contact between two metal surfaces. The surfaces require lubrication by a special type of oil. General Motors worked on a toroidal transmission of this type in the 1920s and 1930s, but it lost out to the hydrodynamic type. The toroidal traction drive was actively pursued by Perbury in England, as well as by others, during the 1970s and 1980s. As of the start of the twenty-first century, it is receiving renewed attention by several developers striving for improved fuel economy.

In a toroidal traction drive, toroidal input and output disks face one another, separated by a number of rollers that contact the toroid surfaces. The rollers are mounted so that they can be tilted to vary the radius from the centerline of the disks where they contact the toroids, and therefore determine the input/output speed ratio of the rotating disks. A substantial axial force must be applied to the disks to prevent the rollers from slipping on the disk surfaces. To avoid excessive losses when the torque transmitted is low, this force needs to be modulated in proportion to the torque transmitted.

DIFFERENTIAL

As a vehicle turns a corner, the wheel on the outside of the turn must rotate faster because it travels farther, and the inside wheel must rotate slower than if the vehicle were traveling straight ahead. When the two wheels are only there to support the vehicle and not to deliver a propulsive force, this is not a problem because the wheels need not be mechanically connected. However, if they are drive wheels producing propulsive force from an engine, a way must be found to deliver torque to both wheels as they turn at different speeds, even though the input torque from the engine is supplied at a single speed.

In France in 1769, Nicholas Cugnot designed a self-propelled three-wheeled artillery gun tractor powered by a steam engine. He avoided the problem of propulsion in turns by delivering torque only to the single front wheel, as in a modern child's tricycle. However, instability of the tricycle on rough terrain necessitated transition to a four-wheel configuration. To manage its problem of propulsion in turns, only one of the two rear wheels was powered. With only the one wheel driving, though, traction proved inadequate in some situations. To improve traction, both rear wheels were driven, but then in sharp turns one of the wheels had to be manually disconnected from the power source. Finally, in about 1827, the Pecqueur idea of differential gearing was conceived to deliver power to both driving wheels while allowing them to rotate at different speeds in turns. Today the differential is so common that the early problem of propelling a four-wheeled vehicle in turns is forgotten by most.

In a rear-wheel-drive automotive differential, the differential is preceded by a "final drive," which consists of a pair of gears—a small pinion gear fixed to the prop shaft and meshing with a larger ring gear. These gears are beveled so the rotational axis is changed from the fore-and-aft orientation of the prop shaft to the transverse orientation of the rear axles. The difference in gear diameters causes the axle speed to be reduced while the delivered torque is correspondingly multiplied, as suggested by the first equation above. The input/output speed ratio typically falls between 2.5 and 4.5. Some such form of final drive is necessary because the overall speed ratio required from engine to drive wheels cannot normally be provided by the transmission alone.

Fixed to the ring gear and therefore rotating with it about the axle centerline is a carrier. Two identical beveled pinion gears are free to rotate about a pin fixed in the carrier. Each of these differential pinions engages both the right and the left drive pinions, which are fixed to the left and the right axles, respectively, but are free to rotate within the carrier.

When the vehicle drives straight ahead, the differential pinions do not rotate about the pin on which they are mounted. However, they tumble with the pin, end over end, about the axle centerline because the pin is fastened to the carrier and its ring gear. In this condition, the drive pinions, and with them their respective axles, rotate with the carrier and drive the vehicle forward.

In a turn, the carrier continues to tumble the differential pinions and their mounting pin about the axle centerline. However, if the vehicle turns to the left, the right drive pinion rotates about its centerline in the carrier in the forward direction. This increases the rotational speed of the right drive wheel, which is equal to the rpm of the carrier *plus* the rpm of the right drive pinion in the carrier. The rotating right drive pinion also makes the differential pinions turn on their mounting pin in opposite directions. This causes the left drive pinion to rotate about its centerline in the carrier at a speed equal and opposite to that of the right drive pinion. Now the rotational speed of the left drive wheel equals the rpm of the carrier *minus* the rpm of the left drive pinion in the carrier, and the difference in drive wheel speeds necessary to avoid wheel slip during the turn is achieved.

The differential can introduce a problem when driving on a slippery road because the torques on the two differential pinions are always equal, thus delivering equal torques to the drive wheels. In the extreme, if the tire of one drive wheel rests on ice and therefore lacks traction, the differential allows it to spin freely while the opposite drive wheel, and the vehicle itself, remain at rest. Special differential designs have been devised to overcome this problem. Traction is also improved in some vehicles through the application of four-wheel drive, whereby additional shafts and gears are employed to distribute engine power to all four wheels of the vehicle.

UNIVERSAL JOINT

The drive wheels of an automobile follow the irregularities of the roadbed. Although deflection of pneumatic tires absorbs some of these irregularities, the suspension system, incorporating springs and shock absorbers, is the primary means for smoothing out the ride for the passengers. The suspension system connects the drive-wheel axles to an assembly including the body, engine, and transmission. Thus as the vehicle moves forward, the road-induced vertical oscillations of this assembly, and therefore the transmission output shaft, differ from those experienced by the drive wheels.

Using a rear-wheel-drive vehicle as an example, a prop shaft delivers power from the transmission output, near the middle of the vehicle, to the differential that drives the wheels at the rear. If this prop shaft is designed to be stiff, which is normally the case, it could not be rigidly attached to the transmission output at one end and the differential at the other because of the differences in vertical movement between the drive wheels and the chassis. Typically, two universal joints are inserted into the drivetrain to accommodate this situation.

Following the operating principle of the most commonly used universal joint, the input shaft and the output shaft both terminate in yokes that are oriented in mutually perpendicular planes. The branches of each yoke are pinned to a cross connector so that each yoke can pivot about its beam of the cross. This mechanism was employed in the sixteenth century by Italian mathematician Geronimo Cardano, who used it to maintain a shipboard compass in a horizontal plane, regardless of the movement of the ship. Consequently it is often called a Cardan joint.

The seventeenth-century British mathematician Robert Hooke showed that if the input and output shafts do not share the same rotational axis, a uniform input rotational speed is transformed into a nonuni-form rotational speed on the output shaft. He further showed how to avoid this characteristic by correctly phasing two joints in series. Because he acquired a patent on this mechanism, the Cardan joint is also known as the Hooke joint.

In a front-wheel-drive car, the drive wheels experience not only the road-induced vertical motion of the rear wheels but also must rotate about a vertical axis to accommodate steering. Several different configurations of constant-velocity universal joints have been developed to manage such motion. These constant-velocity joints are larger and more expensive than the joint described above.

Charles A. Amann

See also: Automobile Performance; Electric Vehicle; Fuel Cell Vehicles; Gasoline Engine; Hybrid Vehicles.

BIBLIOGRAPHY

Beachley, N. H., and Frank, A. A. (1980). "Principles and Definitions for Continuously Variable Transmissions, with Emphasis on Automotive Applications." ASME paper 80-C2/DET-95.

Society of Automotive Engineers. (1979). *Universal Joint and Driveshaft Design Manual*. Warrendale, PA: Society of Automotive Engineers.

Zingsheim, E. W., and Schall, M. M. (1955). "Progress of Torque Converters." *SAE Transactions* 63:84–97.

DYNAMITE

See: Explosives and Propellants

ECONOMICALLY EFFICIENT ENERGY CHOICES

If a person has the choice of installing oil, gas, or electric systems to heat a house and believes that any one of the three would perform equally well, the system that is cheapest is the efficient choice. If, however, the individual compares heating with an oil furnace to heating with a wood-burning stove, monetary cost may not be the only consideration.

Wood may be cheaper and oil more convenient. If two people are confronted by the same information and one chooses wood while the other chooses oil, both decisions may be economically efficient in the sense of maximizing the utility, or satisfaction, of the decisionmaker.

People differ in their preferences and in the value they put on their time. For consumer choices in particular, the entire list of qualities of services provided by different energy sources can be important. For example, both kerosene lamps and electric lights can be used to illuminate a home. Most households shifted quickly to electricity once it became available, and few would shift to kerosene today even if kerosene for lamp fuel were free. The lighting provided differs in more ways than cost.

Inputs and outputs usually can be valued according to market price. The opportunity cost of any choice is the best opportunity that has to be given up to make that choice. For the firm, the opportunity cost of oil or natural gas is measured by the price the firm must pay for it. That does not imply that the firm will always choose the cheapest fuel, because the costs of using the fuel must also be considered. Even in a sit-uation where electricity is many times as expensive per British thermal unit (Btu) as coal, the firm may choose to buy electricity to operate lights, motors, and computers.

FINDING THE OPTIMAL MIX OF INPUTS

Energy can almost always be replaced in part by other inputs. For example, a steam pipe can be insulated more heavily or an industrial process can be modified to use more labor and less energy. Economic efficiency does not imply minimizing the use of energy or any other input, but rather finding the appropriate mix of inputs. The economically efficient level of inputs is reached when the last dollar spent on energy yields the same amount of benefits as the last dollar spent on labor or materials or any other input.

In discussions of economic efficiency, the concept of decreasing marginal rate of substitution plays a crucial role. This simply means that for a great many different activities, it becomes increasingly difficult to substitute one input for another as one continues to make substitutions. For example, a household with access to both electricity and natural gas probably will use gas for heating the house, electricity for lighting and refrigeration, and might choose to use either one for cooking, heating water, and drying clothes. It is possible to shift all of those activities to either energy source, although electric heat is expensive in most applications and gas refrigerators and lighting are rarely used when electricity is available. The substitutions become increasingly expensive as one moves to one extreme or the other. If the quality of the service offered by either fuel is identical, then the most efficient mix of energy sources would be the cheapest mix, which would depend on the relative prices of natural gas and electricity.

Firms face similar substitution possibilities in many of their activities and especially in industrial processes. Large commercial and industrial firms can even substitute electricity that they generate themselves for some or all of their purchased electricity. In the choice between purchasing and generating electricity, the two are perfect technical substitutes—that is, one can be substituted for the other without encountering a diminishing marginal rate of substitution. However, the cost of generation will increase as the firm tries to cover its own peak loads.

An investment in new equipment generally must be made before one energy source can be substituted for another. If new investment is required, the decision to switch fuels is not lightly made in response to fuel price fluctuations, particulary if they are viewed as temporary. In such cases, energy choices are most easily made when the activity is in the planning stage. For example, in some places new homes constructed in the late 1970s were not allowed to have connections to natural gas lines because the policy of the federal government was based on the assumption that gas reserves would soon be exhausted. New houses that were built with electric resistance heating systems required major investments before they could be converted to natural gas when the misguided policy was abandoned.

ADJUSTING FOR THE TIMING OF COSTS AND BENEFITS

In comparing the economic performance of energy alternatives, it is essential to take account of the times at which costs are incurred and benefits received. For example, in the case of a railroad line that crosses a range of mountains, it is possible to save fuel on every trip by tunneling under the mountains instead of traveling over them. Does this mean that the mountain crossings should be replaced by a tunnel? Suppose that the tunnel is on a lightly traveled route, costs $1 billion to construct, and will last indefinitely. If it saves $1 million per year in fuel and other operating costs on an ongoing basis, is the project economically efficient? Ignoring the opportunity cost of capital, the sum of the annual savings would eventually (after one thousand years) equal the cost of construction. But the resources devoted to its construction have an opportunity cost and the project is not cost-effective when the opportunity cost is taken into account.

Discounting makes it possible to compare costs incurred at one time with costs and benefits received at another taking into account the opportunity cost of capital. In the absence of such a procedure, one cannot compare alternatives that differ in the timing of their costs and benefits.

A dollar that will be received a year from today has a "present value" of $1 divided by $(1+r)$, where r is the discount rate, which is equal to the opportunity cost of capital; and a dollar that will be received two years from today has a present value of $1 divided by $(1+r)(1+r)$ or $(1+r)^2$. A payment that is to be received t years from today must be divided by $(1+r)^t$. If the opportunity cost of capital is fairly high, savings that will be realized many years from today will be heavily discounted. For example, if r is 10 percent, the present value of a dollar that will be received seven years from today is about 51 cents. If a dollar will be received twenty-five years from today, its present value is not even a dime. The total value of the tunnel that saves $1 million per year indefinitely is only $1 million divided by r. If the opportunity cost of capital is 10 percent, the tunnel is worth only $10 million. The economy will not prosper if it sinks $1 billion in building a tunnel that will generate only $10 million of benefits.

While this example is constructed to be an extreme case, it illustrates the importance of not being misled that a long-lasting stream of returns necessarily means that a capital investment will be profitable. Returns from energy savings to be received far in the future will have a low present value unless some mechanism works persistently to raise future energy prices at a rate that is commensurate with, or exceeds, the discount rate.

After the average crude oil price increased from $3.18 per barrel in 1970 to $21.59 in 1980, many analysts forecast skyrocketing energy prices for the remainder of the century. The "middle price path" of the U.S. Energy Information Administration in 1979 projected a nominal price of $117.50 per barrel in 1995! Such forecasts seemed to be soundly based not only in recent experience but also in the economic theory of exhaustible resources. As a consequence, U.S. industries invested heavily in energy conservation measures, with the result that industrial consumption of energy decreased from 31.5 quads in 1973 to 27.2 in 1985. Some of this investment was probably not warranted on economic efficiency grounds because prices ceased to rise after 1981, and even plummeted to $10 per barrel in 1986.

Whether the most energy-efficient equipment is also the most economically efficient depends on the circumstances. For example, a truck operator may be able to cut the fuel costs in half by spending $100,000 to replace an old truck with a new, energy-efficient model. But if a long-haul trucker spends $30,000 per year on fuel, the half saved is a significant amount. But if the truck is used mainly to shuttle containers between a port and nearby warehouses, the annual fuel bill might be $5,000 and the half saved ($2,500 per year) not enough to justify spending $100,000 for the new truck.

Similarly, an electric motor can use electricity that costs more than the motor during a year of continuous operation. Even if the motor is in perfect condition, it may be cost effective to replace it with a new motor that is a few percentage points more efficient at converting electricity into work. In many applications, however, an electric motor operates only a few hours per year. In such cases, the cost of the electricity is negligible relative to the cost of a new motor, so that even a large gain in energy efficiency is not worth the cost.

TAKING INTO ACCOUNT ALL INPUTS

As can be seen from the above examples, one characteristic of the concept of economic efficiency is that it takes account of all inputs, not just energy. Even if one were interested only in conserving energy, the economic approach would guide one to use labor, capital, and other inputs to conserve the greatest possible amount of energy for the budget. Of course, the economic approach is not usually associated with minimization of any one input or maximization of any one output, but rather with the minimization of costs for a given level of benefit or maximization of net benefits.

PRICES GUIDE DECISIONS

Because economic efficiency is simply a description of the rules by which individuals and firms can gain the most of what they want, another characteristic of economic efficiency is that firms and individuals do not need orders or special incentives to induce them to pursue it. They are simply acting in their own interest as they see it.

All of the information required for firms and individuals to pursue economic efficiency is conveyed by the price system. The price of natural gas relative to coal conveys information to all potential users of the

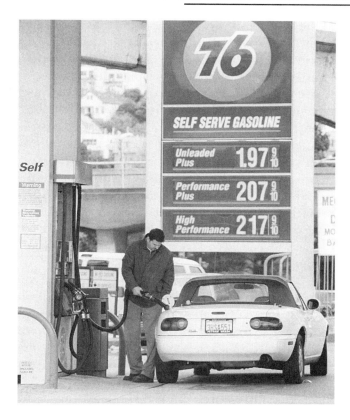

A motorist fills his tank at a Union 76 service station where two out of three grades of gasoline cost more than $2 per gallon (March 7, 2000). West Coast fuel prices on the spot market shot up in reaction to the highest oil prices in nearly a decade. (Corbis-Bettmann)

two fuels about their relative scarcity. The price of the output relative to the inputs conveys information to potential producers about whether an activity will be profitable to expand. Decisionmakers can then pursue activities to the point where the benefit of expanding any activity or of any input in any activity is equal to the cost of that expansion. The price system is especially valuable because it conveys subtle information that is otherwise very difficult to ascertain or to factor into the analysis. Attempts to allocate particular inputs by rules or bureaucratic orders in wartime or in other controlled economies have invariably proved extremely inefficient, at best, and often disastrous.

PRICES MAY FAIL TO REFLECT EXTERNAL COSTS

This powerful effect of prices in conveying information throughout the economic system naturally leads to the question of whether the information conveyed about inputs and outputs generally, and about energy specifically, is accurate. Critics have indicated various ways in which energy prices can be misleading. One classic problem is that of external costs. For example, unregulated coal mining pollutes streams with acid drainage from underground mines and silt from unreclaimed surface mines. In such a world, the market price of coal, to which firms and individuals react in making their decisions, is too low because it does not include such damages. One possible solution to this problem is to assign ownership of the stream to someone (anyone), who would then charge polluters for the damage done. The cost and market price of coal would then incorporate the formerly external costs.

Generally, the United States has not followed this approach. Instead, regulations have been promulgated to specify either the production techniques that must be used to eliminate or lessen the external costs, or the permissible levels of emissions of pollutants. These regulations have helped to clean the environment and also have increased the cost of energy, but few economists would claim that the existing set of regulations leads to the same behavior and the same efficiency that a perfected set of prices would. One particular difficulty is that regulations rapidly become obsolete as technology and markets change, whereas prices adjust to changed circumstances and exert pressure for behavior to adjust accordingly.

Another characteristic of the economic-efficiency concept is that it does not require arbitrary decisions by the analyst about, for example, how coal should be evaluated compared with natural gas. The question of whether 1 Btu of coal is equal to 1, or perhaps 1/2, Btu of natural gas is answered directly by the market. The weightings of the marketplace, revealed in relative prices, vary with scarcity, cost of production, technology, and human preferences. Decisionmakers do not need to think about the underlying reasons, however. They need to know only current prices (and make their best guesses about future prices).

Prices can fail to reflect true social cost for reasons other than externalities. Factors such as taxes, subsidies, monopolies, and fear of expropriation also can cause prices to diverge from marginal social cost.

NET ENERGY ANALYSIS

The characteristics of economic efficiency noted above are considered advantages by most economists and disadvantages by a small group of critics, many with an ecological orientation. The most politically influential challenge to the concept of economic efficiency comes from "net energy analysis" (NEA). This type of analysis attempts to convert all inputs and outputs into weighted energy equivalents in the hope that the resulting project appraisals will be more stable and consistent than those provided by economic analysis. Arriving at weights for different forms of energy and energy equivalents for labor and capital has proved to be difficult and controversial. Moreover, some analysts question what use can be made of the results of NEA if they differ from those of economic analysis.

AREAS OF AGREEMENT AND DISAGREEMENT AMONG ECONOMISTS

Within the mainstream of economics, no serious challenges to the conventional analysis of economic efficiency have been sustained. Judgments differ on the extent to which market prices may need to be adjusted to compensate for externalities or other imperfections. For example, does global warming (presumed to result from emission of carbon dioxide and other greenhouse gases) pose a serious enough threat that the prices of all fuels containing carbon should be raised by imposing a tax equal to the amount of the damage done? If so, how much is that amount? Note that the disagreements about this issue reflect real gaps in knowledge about the effects of carbon dioxide emissions, not disagreement about the concepts and analysis.

Similarly, economists generally agree that analyses involving time require discounting according to the standard formulas, but disagree regarding which discount rate should be used. The basic issue is that using a high discount rate is equivalent to saying that benefits or costs that are expected far in the future do not receive much weight in decisions made today. Why worry about the costs of global warming if they will not be felt for a century or more? At any reasonable discount rate, the value of a dollar received a century from today is negligible. Some critics have argued that we owe something to future generations and therefore should value their preferences as highly as our own.

The supply of capital is limited, however. If an investment that yields a 2 percent return is adopted because it yields benefits to future generations, but if it consumes capital that could have yielded a 20 percent rate of return to an investor, the capital stock will grow more slowly. To maximize growth of physical capital and personal income, the highest-yielding investments should be chosen.

William S. Peirce

See also: Economic Externalities; Efficiency of Energy Use; Efficiency of Energy Use, Economic Concerns and; Efficiency of Energy Use, Labeling of; Energy Economics; Environmental Economics; Green Energy; Industry and Business, Energy as a Factor of Production in; Industry and Business, Productivity and Energy Efficiency in; Risk Assessment and Management; Subsidies and Energy Costs; Supply and Demand and Energy Prices; Taxation of Energy; True Energy Costs; Utility Planning.

BIBLIOGRAPHY

Gilliland, M. W. (1975). "Energy Analysis and Public Policy." *Science* 189:1051–1056.

Hayek, F. A. (1945). "The Use of Knowledge in Society." *American Economic Review* 35:519–530.

Huettner, D. A. (1976). "Net Energy Analysis: An Economic Assessment." *Science* 192:101–104.

Mikesell, R. F. (1977). *The Rate of Discount for Evaluating Public Projects*. AEI Studies 184. Washington, DC: American Enterprise Institute for Public Policy Research.

Peirce, W. S. (1996). *Economics of the Energy Industries*. Westport, CT: Praeger.

Pindyck, R. S., and Rubinfeld, D. L. (1998). *Microeconomics*, 4th ed. Upper Saddle River, NJ: Prentice-Hall.

U.S. Department of Energy, Energy Information Administration (1979). *Annual Report to Congress*. Washington, DC: U.S. Government Printing Office.

ECONOMIC EXTERNALTIES

An economic externality exists whenever the well-being of some individual is affected by the economic activities of others without particular attention to the welfare of that individual. For example, smog-related illnesses such as bronchitis and exacerbated cases of childhood asthma have been blamed, to some extent, on the emissions of nitrogen oxides from automobiles and large fossil-fuel-burning power plants. These illnesses have high treatment costs that are not incorporated in the related electricity-production and oil-consuming activities of the power plant and transportation industries, and must therefore be borne by the affected third parties. Air pollution sort is a classic example of an economic externality, and is called a negative externality because it has external costs.

Many environmental problems arise from externalities of energy exploration, production, refining, distribution, and consumption. This is especially so for fossil fuels. Air pollution, global warming and climate change, and acid rain are due mainly to emissions of carbon, sulfur, and nitrogen oxides associated with the burning of fossil fuels. Coastal and marine degradation, wildlife habitat destruction, and the availability and quality of fresh water can be blamed to some extent on oil spills, drilling for oil and gas, coal mining, and the underground storage of oil and gasoline. The nuclear power industry deals constantly with toxic-chemical and hazardous-waste issues. Some of these important environmental problems and others (e.g., deforestation and desertification) can also be attributed to the changing patterns of, and increases in, population, land use, transportation, and industry. Energy plays a significant role, and energy-environment externalities often have strong socioeconomic and environmental welfare effects.

Market forces determine much of energy production and use. Associated externalities are often beyond the capacity of the market to resolve. To understand how energy externalities impose costs on society, one must first understand how markets allocate resources (including energy) efficiently. Figure 1a shows a typical demand and supply diagram for a commodity (e.g., coal) or service. For many goods, the demand curve reflects marginal private benefit (MPB) and the supply curve reflects marginal private cost (MPC), since commodities usually are produced and consumed privately. The demand (or marginal benefit) curve is downward-sloping to reflect the fact that people will pay less for additional units of a good as they consume more of it. The demand curve also shows people's willingness to pay for a good, and so the downward slope means that as the price of the good decreases, people are willing to buy more of it. Thus the demand curve shows the amount of a good that is demanded at each price. Similarly, the supply (or marginal cost) curve shows the amount that is produced at each price. The upward slope of the supply curve reflects increasing costs of production, and also that producers are willing to supply more at higher prices.

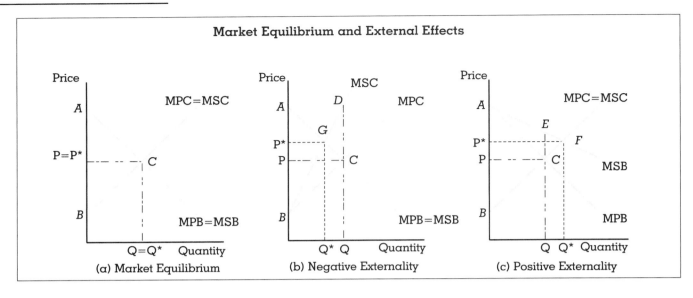

Market Equilibrium and External Effects

(a) Market Equilibrium (b) Negative Externality (c) Positive Externality

Figure 1.

The areas under the curves represent benefits from consuming, and costs of producing, the commodity. These benefits and costs increase as more of the good is consumed or produced. Benefits are higher than costs up to the point where MPB equals MPC, and thereafter costs are higher. Therefore net private benefits are maximized when MPB equals MPC, and Q units of the good are demanded and supplied at a price of P. The area enclosed by triangle ABC in Figure 1a represents maximum net private benefits. Social net benefits are maximized when MPB and MPC are identical to marginal social benefit (MSB) and marginal social cost (MSC), respectively. Markets efficiently allocate resources to achieve this outcome, and there is market failure whenever divergence exists between MPC and MSC, and/or between MPB and MSB. Market failure is caused by many factors, including:

- externalities;
- imperfect markets—when markets are not competitive;
- incomplete markets—when property rights are not well defined to enable exchange;
- public goods—goods that are indivisible and may be free to some consumers;
- imperfect information—when costs and benefits are not fully known by all;
- nonconvexities—when MSC is shaped so that it crosses MSB at several points.

Social net benefits must be used in considering how energy externalities impose costs on society. In Figure 1, private market forces promote production and consumption of Q units at a price of P, and social net benefits are maximized when MSB equals MSC with production of Q★ units at a price of P★. In Figures 1b and 1c, triangles ABG and ABF represent maximum attainable social net benefits, respectively. Market production and consumption of Q units provide social net benefits equal to area ABG less area GCD in Figure 1b, and area ABF less area ECF in Figure 1c. Thus externalities impose costs on society by making it impossible to gain maximum social net benefits. In Figure 1b we can see that a negative externality such as energy-related pollution implies private market production of too much energy and pollution. Similarly, Figure 1c shows that a positive externality such as plant growth enhancement by carbon dioxide emissions from the burning of fossil fuels implies that too little energy and positive externalities are produced. In both cases, the market price for energy is too low. In reality, Figure 1b best represents the case of energy externalities, since the pollution effects outweigh the plant growth enhancement effects (i.e., the resultant external effect of energy production and use is negative).

The solution to the problems posed by energy externalities is to internalize the externality, so that the external costs and/or benefits are included in the transactions and other activities involved in the production and consumption of energy. For example, public policy that uses a tax to raise energy prices and/or restrict energy production to socially desirable levels would solve energy externality problems. Energy is an important factor of production, and any policy that affects energy price or quantity ultimately affects the entire economy.

The problems that energy externalities present are further complicated by the nature of energy pollution, other market failure issues, and inappropriate government intervention. Internalizing the externalities of air pollution and other emissions is sound in theory, but in practice, quantifying pollutants and their impacts and equitably dealing with the problem is very difficult. Energy pollutants are emitted from both stationary and mobile sources and may accumulate in the environment. Stationary sources (e.g., power plants) are generally large and few, are run by professional managers, and provide simple local pollution patterns. Mobile sources (e.g., cars) are abundant, are and run mainly by individuals, and complicate local pollution patterns. Pollutants that accumulate in the environment can affect future generations, and introduce intergenerational equity problems and issues. For example, carbon dioxide emissions in excess of the absorptive capacity of the environment can accumulate. This is because carbon dioxide is not a true pollutant because it is essential for plant life and is absorbed by plants and the oceans. Although carbon dioxide is the main greenhouse gas blamed for global warming, there is considerable scientific uncertainty regarding global warming and climate change, and there are critical measurement issues in determining both market and external costs and benefits.

The impact of energy pollutants on the environment can be local, regional, or global. As the zone of influence of pollutants extends beyond local boundaries, the political difficulties of adopting and implementing control measures are compounded. For example, sulfur oxides are regional pollutants, while excess carbon dioxide is a global pollutant. Carbon dioxide pollution policy thus requires international cooperation, but sulfur oxide policies may require only national policy. The United States has a program for trading sulfur emissions, and Japan taxes sulfur oxides, but there seems to be little progress with international attempts to control carbon dioxide. Developing nations fear that participating in carbon dioxide emission reduction programs will retard their economic development. This introduces international income distribution issues, but in many cases intranational income distribution issues also must be addressed. In short, the nature of energy pollutants, uncertainty and measurement issues, income distribution effects, intergenerational equity, economic development, and political difficulties make public policy regarding energy externalities very challenging.

Samuel N. Addy

See also: Air Pollution; Climatic Effects; Environmental Economics; Environmental Problems and Energy Use; Ethical and Moral Aspects of Energy Use; Government and the Energy Marketplace; Historical Perspectives and Social Consequences; Industry and Business, Energy as a Factor of Production in; Market Imperfections; Subsidies and Energy Costs; Supply and Demand and Energy Prices; True Energy Costs.

BIBLIOGRAPHY

Baumol, W. J., and Oates, W. E. (1988). *The Theory of Environmental Policy.* Cambridge, Eng.: Cambridge University Press.

Folmer, H.; Gabel, L. H.; and Opschoor, H., eds. (1995). *Principles of Environmental and Resource Economics.* Cheltenham, Eng.: Edward Elgar.

Goldemberg, J. (1996). *Energy, Environment, and Development.* London: Earthscan.

Kahn, J. R. (1998). *The Economic Approach to Environmental and Natural Resources,* 2nd ed. Orlando, FL: Dryden Press.

Landsberg, H. H., ed. (1993). *Making National Energy Policy.* Washington, DC: Resources for the Future.

Portney, P. R., ed. (1990). *Public Policies for Environmental Protection.* Washington, DC: Resources for the Future.

Tietenberg, T. H. (1996). *Environmental and Natural Resource Economics,* 4th ed. New York: HarperCollins.

ECONOMIC GROWTH AND ENERGY CONSUMPTION

Energy is a vital ingredient to economic growth. This has been recognized at least as long as economic statistics have been compiled by government, and probably for much longer than that. Perhaps the best example of the fundamental role that energy plays in large, complex national economies is found in the

1973–1974 oil embargo, when oil-producing nations of the Middle East restricted supply and prices rose fourfold in a space of a few months. The resulting chaos in the oil-consuming economies of the industrialized West was widely considered to be a direct result of the embargo. In the United States alone, Gross Domestic Product—an accepted measure of economic activity—fell in 1974, after two decades of steady growth. The high cost and scarcity of oil was seen as the primary cause.

ENERGY AS AN INPUT TO PRODUCTION

What makes energy and economic growth go hand-in-hand? Traditionally, economists since Adam Smith have discussed the major inputs to economic activity as being land, labor, and capital. While very descriptive of the agrarian economies of the seventeenth and eighteenth centuries, the growth of industrial nations in the nineteenth century can be seen in retrospect to have been the result of a fourth major input, energy. Energy can be seen simply as the ability to multiply the work of laborers exponentially. Where the agrarian society had to make use of horses and mules for transportation services, the industrial economy could take advantage of the miracle of the internal combustion engine, which, when powered by gasoline, could lower the costs and increase the availability of transportation by orders of magnitude. Where once laborers did their jobs with scythes, shovels, and other tools, energy enabled them to increase their outputs tremendously by powering great machines such as tractors, cranes, and pile drivers. Power for illumination allowed the growth of multiple "shifts," greatly increasing the output that could be produced over a given period of time.

ENERGY INTENSITY

The ratio of energy consumption to economic activity is referred to as the "energy intensity" of an economy. Energy intensity may also be measured at lower levels of aggregation, such as at the industrial or transportation sectors of an economy. In general, as nations move into a more industrialized state, they find that their energy intensity greatly increases, as the demands of a more complex economy require a greater amount of energy per unit of output. Another way to think about the role of energy is that the ability to harness it technologically allows an economy to

1997 Rank	Country	Energy Intensity (Thousand Btu per 1990 U.S. Dollar of Gross Domestic Product)	
		1980	*1997*
1	United States	17.77	13.84
2	Japan	7.78	6.36
3	Germany	9.37	7.60
4	France	8.94	7.42
5	Italy	7.02	6.48
6	United Kingdom	11.70	9.07
7	China	109.63	45.53
8	Canada	22.05	18.59
9	Brazil	10.73	13.26
10	Spain	8.82	8.06

Table 1.
Ten Largest World Economies' Energy Intensity, 1980–1997

greatly increase its economic potential (e.g., to expand its production possibilities frontier). An example of such growth might be that of the United States during the late nineteenth and early twentieth centuries, as it began the transformation that would make it the world's leading industrial power, powered by its then-abundant supplies of petroleum. Another example would be China during the two decades starting in the late 1970s, as it moved away from a totally state-controlled economy to a partially market-driven system. During this period, China experienced one of the highest rates of growth in the world, driven largely by its ability to harness its huge coal reserves in the production of electricity and for transportation services.

THE EFFECT OF ECONOMIC MATURITY ON ENERGY USE

As nations become more economically mature, two effects are typically seen. One, the rate of economic growth necessarily slows, as the base of economic activity expands and opportunities for easy expansion become more scarce. Two, the use of energy becomes more efficient as consumers and manufacturers become more knowledgeable about its use, and technological progress enables economic output to be produced with less energy input.

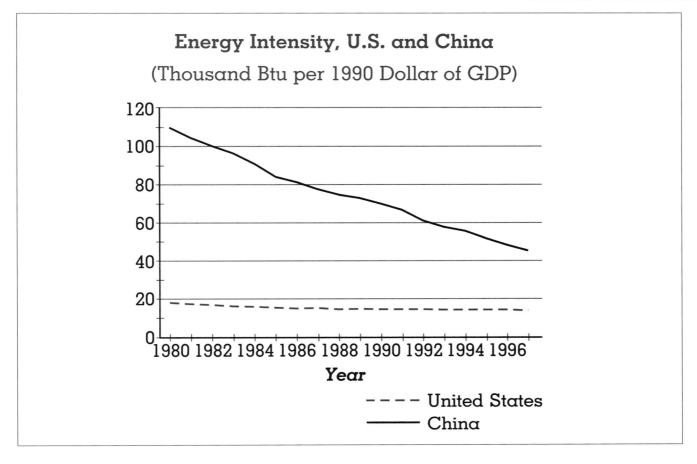

Energy Intensity, U.S. and China
(Thousand Btu per 1990 Dollar of GDP)

- - - - United States
——— China

Figure 1.

The post–World War II experience of North America and Western Europe provides an excellent example of both of these effects. Between 1949 and 1973, the year of the oil embargo, the United States expanded its output of goods and services at an annual rate of 4.1 percent; Great Britain showed a growth rate of 3.0 percent. The oil embargo caused a major structural shift in these energy-consuming nations. It took more than a decade for these countries to return to the economic growth rates they had enjoyed before 1973. Not until after 1983 did Great Britain return to consistent economic growth of 3 percent or more. In the United States, the post-embargo period was characterized by economic growth rates that were more moderate than the pre-embargo period, especially after a second round of oil price increases in 1979–1980. But the second effect also became evident, as energy efficiency, stimulated by the extraordinary rise in energy prices, became a hallmark of the industrialized nations. In the United States, energy intensity fell more than 30 percent between 1970 and 1986; in France, the decline was 21 percent between 1970 and 1990.

FACTORS DECREASING ENERGY INTENSITY

More efficient automobiles were large contributors to the decline in overall energy intensity, as both consumers and government regulators took steps to increase the efficiency of cars by making them lighter, smaller, and equipped with more efficient engines. Homeowners also contributed to the reduction in energy intensity, by lowering thermostats, using more efficient lighting, and purchasing more efficient furnaces, air conditioners, and electric appliances. Some of these behaviors were mandated by laws and regulations forcing manufacturers to market appliances that consumed less energy, or to label their

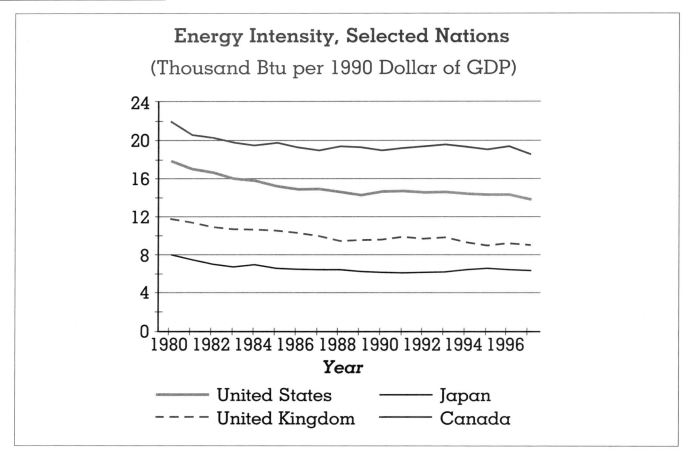

Figure 2.

average energy consumption. But most were the result simply of supply and demand stimulating long-term changes in habits that reduced the consumption of a high-priced economic input, energy, in favor of other, lower-priced inputs.

THE WORLD'S LARGEST ENERGY USERS

Past patterns of economic growth shape current patterns of energy consumption. The largest energy-consuming nations tend to be those whose economies are also the largest, such as the United States, Japan, and Western Europe. In 1997, the world's ten largest economies—ranging from the United States to Spain, and representing more than three quarters of the world's gross domestic output—also accounted for 58 percent of the world's energy consumption (Table 1). The energy intensities of these nations varied widely, however, ranging from less than 6,400 British thermal units (Btu) per

1990 U.S. dollar of Gross Domestic Product (GDP) in Japan, the most energy efficient of the world's largest economies, to more than 45,000 Btu per GDP dollar in China (Figures 1 and 2). In large measure, these variations reflected the progress of each nation in reaching economic maturity. It also reflected the relative abundance of energy resources within these countries. Japan, for instance, has very little indigenous energy supplies, and is highly dependent on energy imports. China, on the other hand, is both a developing country and one with large domestic reserves of coal. Japan, along with many of the world's consuming nations, has also discouraged consumption by taxation; in 1998, regular leaded gasoline in Japan cost approximately $2.80 per gallon, of which about $1.70 was taxes. By contrast, the same gallon cost about $1.05 in the United States, only about $0.35 of which was due to taxes.

THE ENVIRONMENTAL IMPLICATIONS OF ECONOMIC GROWTH AND INCREASING ENERGY CONSUMPTION

Economic growth and increasing energy consumption are not always considered an unalloyed benefit. There are significant environmental consequences to energy consumption, including increased concentration of carbon gases in the atmosphere, emissions of sulfur dioxide (that cause acid rain) and nitrogen oxides (precursors to smog), water pollution caused by oil spills, and land issues related to coal mining and other energy production. Debates about how to ameliorate these effects inevitably include discussions of the economic impacts of such amelioration, and the effect on economic well-being of higher energy taxes or outright bans on consumption or production of certain kinds of energy. Non-polluting energy sources such as hydroelectricity, solar and wind energy, or, more controversially, nuclear-based electricity-for which there are considerable concerns about safety and waste disposal—have been discussed as long-term alternatives to the more traditional fossil-based fuels. While coal, oil, and natural gas have clearly been indispensable to the growth of today's modern industrial economies, it is also a certainty that the supplies of these energy forms are ultimately limited, and that future economic growth will depend on a long-term transition to other energy sources.

OUTLOOK

Some analysts have posited that, given the wide variation in energy intensities of the world's economies, we can look forward to a time when economic growth and energy consumption are effectively uncoupled. Moreover, this uncoupling can make it possible to reduce the environmental consequences of economic growth at a relatively low cost, because such growth will not mean an inevitable rise in the consumption of fossil fuels. Certainly the success of Japan and Western Europe in achieving low energy intensities is evidence that further reduction is possible in other parts of the world.

On the other hand, some of the differences in energy intensities reflects infrastructure that may be difficult to change. Some nations, such as the United States and Canada, are large energy consumers because their climate requires above-average heating and cooling, and their vast size makes transportation a more difficult and energy-intensive activity. China, which has similar characteristics with respect to climate and size, has the additional burden of being a largely agrarian, labor-intensive economy that still uses energy in relatively inefficient ways because it has not yet achieved the technological progress characteristic of the developed countries. Until more efficient means of heating, cooling, and transportation are found, and greater progress is made toward use of more efficient technologies in the developing countries, economic growth will likely continue to require additional energy consumption, but perhaps less so than in the past.

The possibilities of new sources of energy, such as energy from hydrogen, may also some day become economical, and help to uncouple fossil fuel consumption from economic growth

Scott B. Sitzer

See also: Energy Intensity Trends; Environmental Problems and Energy Use; Industry and Business, Energy as a Factor of Production in.

BIBLIOGRAPHY

Clark, J. G. (1991). *The Political Economy of World Energy.* Chapel Hill and London: University of North Carolina Press.

International Energy Agency. (1994). *Energy in Developing Countries.* Paris: OECD Publications.

International Energy Agency. (1999). *Energy Prices and Taxes: Quarterly Statistics, Second Quarter 1999.* Paris: OECD Publications.

Jorgenson, D. W. (1978). "The Role of Energy in the U.S. Economy." *National Tax Journal* 31(3):209–220.

Jorgenson, D. W. (1998). *Energy, the Environment, and Economic Growth.* Cambridge, MA: MIT Press.

U.S. Department of Energy, Energy Information Administration. (1998). *International Energy Annual 1996.* Washington, DC: Author.

EDISON, THOMAS ALVA (1947–1931)

Thomas Alva Edison, is the archetype of American ingenuity and inventiveness. He played a critical role in the early commercialization of electric power. He designed the first commercial incandescent electric light and power system and his laboratory produced

Thomas Edison, holding an Edison Effect Lamp at the West Orange, New Jersey, laboratory. (Corbis Corporation)

the phonograph, a practical incandescent lamp, a revolutionary electric generator, key elements of motion-picture apparatus, and many other devices. He was owner or co-owner of a record 1,093 U.S. patents

"Tom" Edison was born in the small town of Milan, Ohio, the son of middle-class parents. He was educated at home by his parents rather than at the local school, where he was thought to be of low intelligence. As a boy he showed an early proclivity for chemistry experiments and for turning a profit, first peddling vegetables, then newspapers. When the Civil War began, Edison was exempted from service because of deafness in one ear; he became a telegrapher, and one of the fastest operators in the corps of generally brash, swaggering men who ran the railroad telegraph systems. He drifted around the country, winding up in New York in 1869, searching for a job. During these years he tinkered with some contrivances, but these had led nowhere.

Then, hanging around looking for a job in a New York brokerage office during one frenzied day of trading, he jumped in to fix the telegraphic stock ticker, which had broken down. He was hired on the spot at a large salary but stayed only a few months, leaving to form a company devoted to the business of invention. The company's first product, an improved stock ticker, was sold to his previous employer for the astounding sum of forty thousand dollars, and Edison set up operation with a staff of fifty men in Newark, New Jersey.

During its first years, Edison's company devoted itself to the manufacture of stock tickers and to improvements in telegraphic equipment but then spread into other areas, offering to provide inventions as ordered. In 1876 the laboratory moved to Menlo Park, New Jersey, and the staff eventually grew to about one hundred. Over the years, a number of men who worked with Edison at Menlo Park achieved fame as inventors and scientists in their own right. This group included Nikola Tesla, the inventor of the alternating current electric induction motor; John Fleming, the inventor of the vacuum tube diode; William Dickson, the inventor of the first sound movie; Arthur Kennelly, a discoverer of the ionosphere; and Edward Acheson, the inventor of carborundum.

During the following years a brilliant series of inventions at his laboratory earned Edison the appellation "Wizard of Menlo Park." As early as 1878, the mere announcement that Edison intended to produce a practical electric light was sufficient to cause the price of gas illumination stock to fall sharply.

Some of Edison's commercial inventions were produced solely to break the monopolies of patents already granted. Many others represented improvements or changes of known devices; these included Edison's electric light and dynamo and his quadraplex telegraph and improved telephone transmitter. This does not detract from the importance of his work, because in the cases of the electric light and dynamo, in particular, his work led to commercially practical devices that were widely adopted. Although some inventions, such his motion picture apparatuses, were not the result of his work alone, but the result of the joint efforts of the staff of the laboratory, Edison's contribution as leader in these projects cannot be ignored.

One case in particular vividly delineates Edison's own individual original genius. In 1877 he accidentally discovered that he could obtain an audible sound from a mechanical arm touching a rapidly rotating tin disk he had inscribed with a spiral series of dots and dashes representing telegraph signals. He then sketched and gave his machinist for construction a drawing of the first phonograph. The device worked at first try. There had never before been even a description of a machine for recording and replaying sound.

Edison's greatest contribution was the design of an electric power distribution system to provide power to factories and homes. He challenged the commonly held but mistaken idea that an unavoidable loss of half the power would occur in the generator and designed a power station, a revolutionary electric generator, a system of radiating power lines to consumers and electric meters to measure consumption. In 1882 he opened the first electric power station, on Pearl Street in lower Manhattan. He correctly calculated that he would have to manufacture bulbs for less than forty cents to make a profit; he achieved this goal in the fourth year of operation. After he had recouped his initial losses, he sold out of the business to support his other activities.

Edison often portrayed himself as a tough businessman whose sole interest was the profit to be made from an invention. However, he was not an outstanding businessman, and he made a number of obvious blunders during his career. He led a bitter, losing fight against the adoption of ac power distribution; he did little to keep some brilliant assistants; he made financial miscalculations; and he seemingly did not understand marketing and failed to meet his customers' desires. Nor was Edison a man of science; as a matter of fact he showed little interest in scientific matters. Edison was interested in the creation of new technology, and the close connection of technology and science was not as evident in the nineteenth century as it is now. He later admitted that while conducting his experiments he had had no understanding at all of the nature of electric current. Indeed, his own reminiscences make it clear that his real motivation had not come from the desire for profit or knowledge but came instead from the creative urge to invent. That urge drove Edison for the rest of his life. Although he produced relatively little after the turn of the century, and lived to be an old man, he never completely stopped working.

Leonard S. Taylor

See also: Electricity; Electric Power, Generation of; Electric Power Substations; Electric Power Transmission and Distribution Systems; Lighting; Tesla, Nikola.

BIBLIOGRAPHY

Brittain, J. E. (1977). *Turning Points in American Electrical History*. New York: IEEE Press.

Israel, P. (1998). *Edison: A Life of Invention*. New York: John Wiley & Sons.

Norman, D. A. (1999). "Backing the Right Technical Horse." *IEEE Spectrum* 36(2):5–62.

Wilson, M. (1954). *American Science and Invention*. New York: Bonanza Press.

EFFICIENCY OF ENERGY USE

Although energy efficiency was already heavily emphasized in the 1970s as a key strategy for energy security, more recently it has also been recognized as one of the most cost-effective strategies for reducing environmentally harmful emissions. Energy efficiency is more than just a resource option such as choosing between coal, oil, or natural gas. It curbs demand rather than increasing supply, and thus provides additional economic value by preserving the resource base and reducing pollution.

For specific applications, we can calculate the ratio of the measure of the goods or services provided to the energy input required. For example, in the transportation sector, energy efficiency is based on miles per gallon for personal vehicles, seat-miles per gallon for mass transit, and ton-miles per gallon for freight transportation.

For the entire economy, with its countless services and inputs, economists usually define the "service" or economics efficiency as the entire GDP (Gross Domestic Product) divided by E, the annual total primary energy used: Economic Efficiency = GDP/E.

Economists also track the reciprocal, E/GDP, which is called energy intensity. For example, the energy intensity of the United States in 1998 was 91 quads/$8.5 trillion in 1996 dollars (1996$), which divides out to be 10,700 Btu/$. (Note: 1 "quad" = 1 Q = 1 quadrillion Btu = 10^{15} Btu.)

Measuring energy efficiency gains for the entire economy is not a precise science since the population continues to expand, new technologies continue to be introduced, and there is great variability in the behavior of individuals using technology. Nevertheless, trends in economic effiency and energy intensity best reflect the impact of energy efficiency improvements.

ORIGINS OF ENERGY EFFICIENCY

The increased availability of energy fueled the Industrial Revolution. The United States became the world's largest oil producer, and the new fossil fuels were abundant and modestly priced. A technology's energy efficiency was not a key part of capital investment decisions. Energy-efficient technology as a priority ranked well behind improved performance.

Energy intensity declined from 60,000 Btu/$ (1992$) in the 1850s to 13,000 units in 1995. There were rapid drops in the 1860s as the switch was made from wood to more efficient coal, and, starting in the 1920s, as the switch was made to even more efficient oil and gasoline. The 1973 OPEC oil embargo and the next eleven years of rising energy prices triggered the final dip. The overall drop by a factor of 4.6 in 145 years corresponds to a steady annualized efficiency gain of 1.1 percent.

Before the OPEC embargo, there was no Department of Energy, and energy efficiency was not considered to be a government responsibility. Other aspects of energy were understood to be appropriate for government support. For example, research and development (R&D) on futuristic power supply technologies such as fission nd fusion was funded by the Atomic Energy Commission. From fixed year (FY)1948 through FY1972, in 1999 constant dollars, the federal government spent about $22.4 billion for nuclear (fission and fusion) energy R&D and about $5.1 billion for fossil energy R&D. The government also had a role in electrification as an economic development strategy. The entire rural electrification effort, including the federally subsidized Power Marketing Administration is still a major government program today. But it took an OPEC embargo to convince Americans to create a Department of Energy (DOE) in 1974 and to use public funds for efficiency research and development.

RECENT TRENDS IN U.S. ENERGY INTENSITY

The relative lack of importance of energy prices changed dramatically with the OPEC oil embargo. Even though energy prices were still a small fraction of total costs, people and businesses began to make energy-efficient capital-investment decisions in expectation of higher prices.

From 1974 through 1992, Congress established several complementary energy-efficiency and energy-conservation programs. By the 1980s, this concern over finite resources had dissipated as higher prices encouraged greater innovation in efficiency and in resource recovery. Energy efficiency had become a cost-saving, "demand-side management" tool that helped to avoid expensive power plant construction. The DOE's 1995 report, *Energy Conservation Trends,* states that energy efficiency and conservation activities from 1973 through 1991 curbed the pre-1973 growth trend in primary energy use by about 18 Q, an 18 percent reduction.

By the late 1980s, concerns over air pollution began to play a role in the government rationale for energy efficiency, which in the 1990s was followed by concern over global warming. Neither concern had a strong impact on energy use. Since 1985, national energy use has climbed about 20 Q, reaching a record high of 92 Q in 1999. From FY1980 through FY1999, the DOE spent $7 billion on energy efficiency R&D, which accounted for about 10 percent of all energy supply R&D.

E/GDP experienced a steep decline during the eleven OPEC years (1974 through 1985) and a recent equally steep decline starting in 1997. The latter drops may be associated with the rapid growth of the U.S. economy and with the explosive growth in information technology and the Internet. From 1960 through 1973, energy prices were low and there was almost no improvement in E/GDP. Similarly, after the collapse of OPEC in 1985, prices were again low, energy policy wavered, and E/GDP leveled off. The overall drop from 18 in 1973 (the year before the embargo) to 10.5 in 1998 is a drop of 57 percent, and corresponds to a steady gain of 2.2 percent/year for 26 successive years.

Thus, improved energy efficiency can be credited with energy savings of $232 billion in 1999. The arithmetic is as follows. A drop to 57 percent corresponds to a savings of 43 percent; but roughly one-third of the gain (about 15%) came from structural change as we switched from a smokestack to a service economy, and only two-thirds (about 30%) came from a true increase in efficiency. So pure efficiency has reduced our energy intensity only to 70 percent (not 57%). Had our efficiency stayed frozen at its 1973 value, we would now use more energy by the factor $1/0.7 = 1.43$, that is, we would use 43 percent more energy for every dollar of GNP. Our 1999 energy bill was $540 billion, and would have been larger by 43 percent, which is $232 billion. The fraction that can be attributed to government intervention in the marketplace is highly debatable. Clearly, the marketplace would have made the energy efficiency improvements anyway; yet, a considerable portion would not have taken place without government intervention.

It is interesting to compare this huge annual saving of $232 billion with two other 1999 expenditures. The total non-military discretionary federal budget was $300 billion; therefore, efficiency savings pay for three-fourths of our entire civilian discretionary budget. Efficiency savings also equate to a large percentage of the U.S. Social Security budget, which in 1999 was $392 billion.

As technology develops steadily, there follows a corresponding decline of E/GDP, averaging about 1 percent/year. This can be accelerated in the marketplace by new fuels, new technologies, and innovations in existing technology. Government intervention—such as efficiency labels, performance standards for buildings and equipment, tax incentives, utility policy, and voluntary agreements with industry—which is usually implemented during periods of rising energy prices—can further accelerate the decline in E/GDP.

INTERNATIONAL COMPARISONS OF ENERGY INTENSITY

The E/GDP for the United States has sloped steadily downward from 18,000 to 11,000 Btu/$. Europe and Japan are typically only half as energy-intensive as the United States. An explanation is that, during their development, Western Europe and Japan were petroleum-poor compared to the United States, so energy use was perceived to imply imports (and risk of supply disruption) and trade deficits. Thus, they adopted tax policies to conserve energy. The United States took the opposite path; to stimulate economic growth, domestic oil and gas production was subsidized.

Among E/GDPs for developing countries or regions, the most notable is that of China, reaching 110,000 to 120,000 Btu/$ until 1976, but then declining steadily 5.2 percent/year for 21 years to 40,000 in 1997. This two-thirds drop shows the striking potential savings for other developing economies. The former Soviet Union (FSU) had a steady but inefficient

economy until 1989. After that the FSU's rise in E/GDP is mainly because of the collapse of GDP. Eastern Europe comes next. It started off indistinguishably from the Soviet Union, made small improvements until 1989, and then rapid improvements as it adopted market economies. It is expected that the FSU curve will also soon turn down as its GDP picks up. Although India's efficiency trends are not currently in the right direction, developing countries, and particularly the FSU, have a high potential for cost-effective efficiency gains. Well below India come the industrialized countries, with the United States at the top and Japan at the bottom.

ENERGY SAVINGS IN THE BUILDINGS SECTOR

Energy is used in buildings to provide a variety of services such as lighting, space conditioning, refrigeration, hot water, and electronics. In the United States, building energy consumption accounts for slightly more than one-third of total primary energy consumption. Percentages reported for energy consumption and related carbon emissions in all four sectors are based on the Energy Information Administration's *Annual Energy Outlook 2000,* DOE/EIA-0383(2000), December 1999. By 2010, significant changes are expected to occur that will affect how buildings are constructed, the materials and systems used to build them, and the way in which buildings are maintained and used. A wide array of technologies can reduce energy use in residential and commercial buildings. Using sensors and controls to better manage building energy use, and improving building design and construction materials to maximize the thermal resistance of the building shell can also significantly reduce building energy requirements. For example, a cool white roof can reduce air-conditioning energy use by 20 percent.

Appliances have shown very dramatic improvements in energy efficiency, and perhaps the most impressive efficiency gains have come in improving refrigerators. In what follows, we show that refrigerators' efficiency gains are due to the interplay of regulatory and technological advancement. Two energy regulatory innovations (appliance labels, soon followed by standards) and a major technological innovation (blown-in foam insulation) led to the change from an annual energy use growth of 7 percent/year to a drop of 5 percent/year.

In the 27 years between the 1974 peak annual usage of 1,800 kWh and the 2001 federal standard of 450 kWh, refrigerator energy use dropped to one-quarter of its former use, even as the average volume grew from 18 cu. ft. to 20 cu. ft. This corresponds to a compound annual efficiency gain of 5.1 percent. As for economic savings, by the time 150 million refrigerators have reached year-2001 efficiency, compared to 1974, they will save 200 billion kWh/year, which corresponds to the output of forty huge (1 GW) power plants, and to one-third of the nuclear electricity supplied last year in the United States. Consumers will save annually $16 billion annually in electric bills, but their net savings will be only $10–11 billion, because there is a cost premium for the improved refrigerator (typically repaid by bill savings in three years). This $16 billion annual electricity saving from refrigerators alone roughly matches the entire $17 billion wholesale annual value of all United States nuclear electricity.

This surprising equality arises because an efficient appliance saves "expensive" electricity at the meter, at an average retail price of 8 cents/kWh; whereas one kWh of new wholesale supply is worth only 2–3 cents at the power plant. Thus, even if electricity from some future new remote power plant is "too cheap to meter," it still must be transmitted, distributed, and managed for 5–6 cents/kWh. It is impossible to disentangle the contribution of standards and of accelerated improvement in technology, but clearly the combination has served society well.

ENERGY SAVINGS IN THE INDUSTRIAL SECTOR

The industrial sector is extraordinarily complex and heterogeneous. It includes all manufacturing, as well as agriculture, mining, and construction. In the United States, industrial energy consumption accounts for slightly more than one-third of total primary energy consumption. Recent data show nonmanufacturing industries such as agriculture and construction have maintained their energy use growth rate while that for manufacturing has dropped.

Still, the manufacturing sub-sector accounts for about 70 percent of industrial-sector energy consumption. Nonenergy-intensive manufacturing accounts for an increasing share of energy use; for example, electronic equipment is expected to have a growth rate twice that of the manufacturing sector as

a whole. The most energy-intensive (in terms of energy used per dollar of output) manufacturers are iron and steel, pulp and paper, petroleum refining, chemicals, and cement; together, these industries account for about half of the primary energy consumed in the industrial sector.

Of the end-use sectors, the industrial sector—especially in its more energy intensive industries—has shown the greatest and the fastest energy efficiency improvements. For example, over the past quarter century, the U.S. steel industry has reduced its energy intensity by nearly 50 percent; the cement industry has improved its fuel efficiency nearly 30 percent since 1975 although, since 1986, the energy intensity improvements have slowed somewhat. The energy use per pound of product in the chemicals industry has fallen at an average of 2 percent per year, and its energy efficiency continued to increase during periods when energy price was stable or falling (though less steeply when the price was falling). By using more of its former waste products for energy, the pulp and paper industry increased its purchased fuel efficiency by nearly 45 percent from 1972 to 1994.

Clearly, the more recent industrial energy efficiency gains are due more to technological progress than to energy prices. Unlike the buildings and transportation sectors, industry has adopted some supply-side energy-efficient technologies that reduce emissions without necessarily reducing energy demand. These include more efficient use of by-product fuels and retrofitting boilers for combined heat and power. On the electricity demand side, some generic improvements, such as high-efficiency motors and advanced motor system drives and controls, have applications in almost all types of industry.

Other energy-efficient technologies are more sector-specific. For example, in the future, the steel industry, even with increases in recycling, will need to make some steel from ore. A new cokeless steel-making process could cut energy use 30 percent relative to a blast furnace by going directly from solid ore to steel. This "smelt reduction" technique could also increase the industry's productivity, as its investment costs and operating costs are much lower. There are many reduction opportunities of this order of magnitude in various industries, but the single biggest "bang for the buck" is in more efficient heat and power systems.

Combined Heat and Power (CHP) systems, also called "cogeneration" systems, generate electricity (or

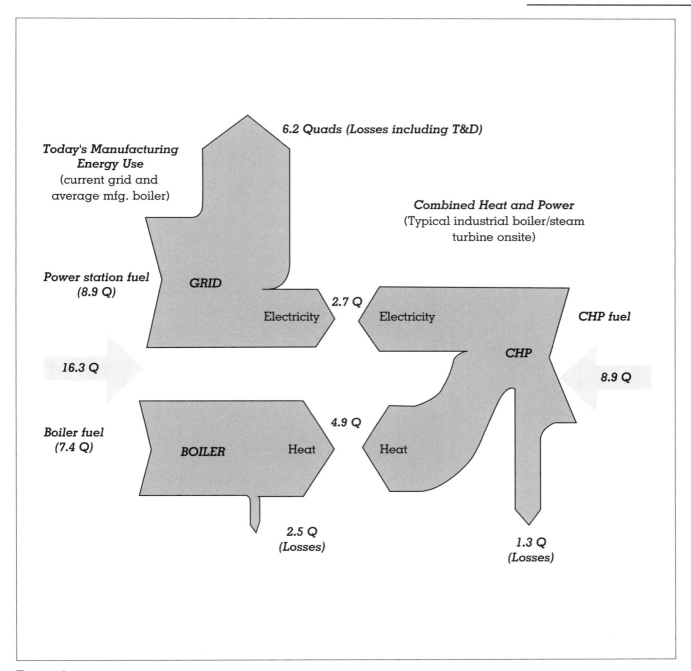

Figure 1.

GRID + Boiler SHP vs. Industrial Gas Turbine CHP, which provides manufacturers with electricity and
steam using 45 percent less fuel than SHP.

mechanical energy) and heat simultaneously at the point of use. Figure 1 shows that in 1994, manufacturers used 7.4 quads to generate electricity and, together with the on-site steam produced from separate boilers, required 16.3 quads of fuel, for a system thermal efficiency of 46.5 percent. If produced jointly as CHP at 85 percent efficiency (clearly achievable based on the previous figure), the total fuel requirements would be only 8.9 quads, nearly 50 percent less. Replacing much of industrial Separate Heat and Power (SHP) with CHP by 2010 is not so far-fetched. According to one source (Kaarsberg and Roop, 1999),

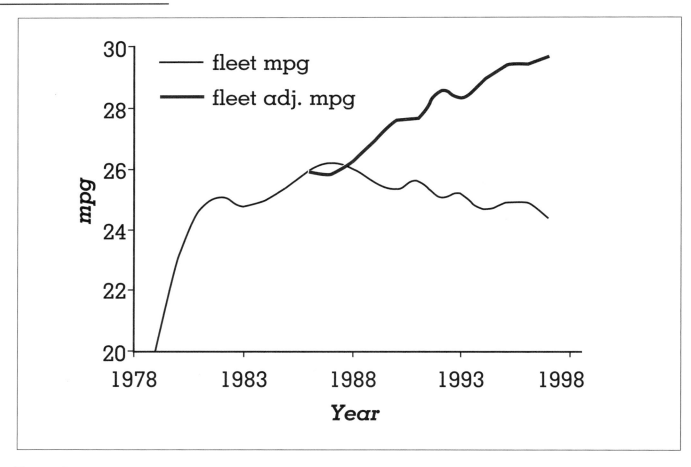

Figure 2.
Fuel economy for the new vehicle "fleet," including light trucks (actual mpg and mpg adjusted for mass, power, and with light truck fraction frozen at 29 percent (1986 value).

ENERGY SAVINGS IN THE TRANSPORTATION SECTOR

Transportation accounts for about one-quarter of total U.S. primary energy consumption. Since 1986, average new car horsepower has increased nearly 40 percent. The average miles per gallon (mpg) of new light-duty vehicles, new cars, and light trucks combined has not changed significantly since 1982. In the absence of new efficiency standards, carmakers' technical improvements respond to consumer demands for roomy, powerful vehicles. Between 1986 and 1997, the average fuel economy for new passenger cars increased by less than 2 percent (from 28.2 to 28.6 mpg), while the horsepower (hp) per weight increased by 27 percent (from 3.89 to 4.95 hp/100 lb), and weight further grew by 9 percent. The fuel economy of the entire fleet (including a growing fraction of light trucks) decreased of 6 percent (0.5% /year) over this period. Figure 2 shows our calculation of what the fleet fuel economy could have been if carmakers had focused on reducing fuel use rather than increasing power, weight, and size.

PARTNERSHIP FOR A NEW GENERATION OF VEHICLES (PNGV)

The PNGV is a government-industry (General Motors, Ford, DaimlerChrysler) research partnership. One of the most highly publicized PNGV goals is to triple the fuel efficiency of a car (with a prototype by 2004) while preserving safety, performance, amenities, recyclability, and holding down costs. As a result of PNGV, federal government R&D in advanced automotive technologies has been reorganized and redirected toward this ambitious goal. There are PNGV programs in advanced materials; electric drives, including power electronics; high-power energy storage devices; fuel cells; and high efficiency low-emission diesel engines.

A key element of the PNGV is the development of an advanced hybrid-electric vehicle. Energy losses from conventional engine idling or running at part load are eliminated in hybrid vehicles-which are available today. Efficiency is doubled with such hybrid propulsion systems. One-hundred units of fuel needed in today's new car (averaging 28.6-mpg) produces the same amount of drive power as 50 units needed by the electric hybrid. Even today's relatively efficient gasoline spark-ignition internal combustion (IC) engine loses 84 units of energy per 100 units of fuel in. The most easily reduced of these losses are the "standby" losses, which account for 11 percent on average (or up to 20% under increasingly typical congested conditions). These losses occur when the engine is either idling or running at far less than 100 percent load. Standby losses are so high because the engine is oversized to allow for acceleration and therefore runs almost entirely at less-efficient part load. Other losses, in descending order, are exhaust, radiator, engine friction and pumping losses, and accessories (electrical system, pumps, etc., but not including major seasonal loads such as air conditioning). Once the power is delivered to the drive train, it is finally completely dissipated in braking and in rolling and wind resistance.

The 2x electric hybrid suffers no standby engine losses. It features a less powerful, more fuel-efficient IC engine/generator running at full load (or off) and uses a battery-powered electric motor to boost acceleration. Thus, the IC engine is not oversized. It has only two modes: maximally efficient full load, or "off." Engine power typically is delivered not to a drive shaft, but to a generator, on to a storage battery, and then to electric motors on the wheels. While the engine is off, the battery powers these motors. When the engine is on, the battery is either being recharged or is boosting acceleration. The hybrid configuration also helps to reduce driving losses. Instead of friction-braking, an electric hybrid uses its electric motor as a generator to recapture braking energy and charge the battery. Thus, braking losses are reduced by 70 percent. They would only be eliminated if the batteries and generator were 100 percent efficient and all four wheels could recapture the braking energy.

Energy losses also occur as the car is propelled. The 16 units of power are dissipated in rolling resistance and aerodynamic drag (and in today's cars, in friction-braking). The next challenge to reach 3x is to reduce the weight and rolling losses with lighter materials and reduce drag with sleeker designs. A major obstacle to reaching the PNGV affordability goal is the high cost of advanced, lightweight body and tire materials. In such a path, about half the savings are due to the propulsion system and half to the improved (lighter-weight, more aerodynamic) envelope.

This is only one of the many "paths" to achieve 3x efficiency. Other ways to eliminate standby losses being investigated by the PNGV program include advanced diesels, direct injection stratified charge gasoline engines, and fuel cells. With these options, the engine's efficiency could double without going to a hybrid configuration.

It is too soon to judge the PNGV program. Surprisingly, Honda and Toyota were the first companies to introduce 3x efficiency hybrids into the U.S. marketplace in 1999 and 2000, respectively. These two companies were not part of the PNGV program, and thus the early introduction shows that the marketplace could efficiently develop and market efficient vehicle technology without the aid of government. Whether the PNGV participants can catch up and market far superior hybrid vehicles is yet to be seen.

AN OPTIMISTIC CONCLUSION

Above, we said that (apart from crises like the eleven OPEC years) energy intensity E/GDP falls about 1 percent year (E is inversely proportional to energy efficiency, η). Here we are concerned with trends in United States, and later, world energy use, so we write:

$$E = E/GDP \times GDP$$
$$= Const/\eta \times GDP$$

Then to stabilize E, η must rise annually not by merely 1 percent, but fast enough to cancel our desired annual growth in GDP (or gross world product GWP), that is, about 3 percent. We have discussed six examples of why this is achievable, given sufficient motivation to do so:

1. For the last three years, in the United States, the annual growth in η has averaged not 1 percent,

but 3.5 percent; that is, while GDP has surged nearly 4 percent/year, energy use has leveled off. It is still unclear how much of this gain is a real trend from increased productivity and the explosive growth of information technology and the Internet (particularly business-to-business e-commerce). But if even part of this gain continues, it is very good news for reducing carbon emissions.

2. During the eleven OPEC "crisis" years, it is well known that η grew 3 percent/year, of which 1 percent was "structural" (moving from a "smokestack" to a service economy) and 2 percent was a pure efficiency gain. This annual 2 percent is measured for our whole stock of energy using equipment, most of which has a service life longer than the 11-year "experiment." Thus, cars last 12 years, refrigerators 15, buildings 50, and so on. So this gain in the stock must lag the gain in new products, or (re-worded) the rate of improvement of new products must lead that of the stock. Arthur Rosenfeld and David Bassett (1999) have crudely estimated this lead/lag correction, and find that new products improved 5 percent/year.

3. Our refrigerator discussion showed that under appliance standards, refrigerator energy use has been dropping more than 5 percent/year for 27 successive years. Yet the payback time for the improving technology has stayed at two to five years. This suggests that significant steady gains can be kept up for a very long time.

4. During the OPEC years, auto fuel economy improved 7 percent/year. After correcting for the 40 percent increase in power, we see an adjusted gain from 1975 through 1997 of 4 percent/year. If the PNGV 3x car at 80 mpg is a significant fraction of the new car fleet by 2010, this rate of improvement will have been sustained for 35 years.

5. Combined Heat and Power (CHP) is generally 1.5-2 times more efficient than separate heat and power. It grew nearly five-fold in the United States in the years between the enactment of CHP incentives with the passage of the Public Utilities Regulatory Policy Act of 1978 (PURPA) and today. During some of that period it grew at more than 15 percent per year and now accounts for 9 percent of electricity generation.

THE WORLD'S NEED FOR ENERGY IN 2100

Next we estimate the world's need for energy in 2100 (E_w), under three scenarios, where the symbol α will denote the annual gain in energy efficiency η:

1. "BAU" (Business as Usual): $\alpha = 1$ percent/year, its historic non-crisis rate.
2. "No Regrets": $\alpha = 2$ percent/year; that is, the world runs scared of climate change.
3. "De-Materialization": $\alpha = 3$ percent/year; a bit less than the actual rate for the United States for 1997 to 2000.

For this brief discussion, we factor world energy in 2100 as

$$E_w (2100) = Pop. \times e^{-100\alpha} \times \frac{Watts_P}{Cap.} \bigg|_{2000} \quad (1)$$

where $Pop._{2100} \approx 10$ billion and $\alpha = -(d/dt)(E/GDP) = 1$ to 3 percent per year. Population by 2100 will probably level off at slightly under 10 billion, and $Watts_p(2000)$/capita is the rate of primary energy use today, considered a satisfactory goal by the majority of people in developing countries today. For $Watts_p(2000)$/capita we propose 5 kW (that is, 5 kW-years of energy for the year 2000), corresponding to that of Western Europe today. Thus, we assert that a poor African or Indian today would happily aspire to a year-2100 standard of living, health, transportation, and so on equal to that of Germany or Scandinavia today (even if they have few SUVs).

We also note a 1985 study (by Goldemberg et al.) showing that the best then-available technology could yield a Western European lifestyle at only 1 watt of primary energy per capita.

The exponential factor arises because, in 100 years, this 5 kW/capita will drop:

$$E_w (2100) = 10^9 \times (5kW_p/Cap.) \times e^{-100\alpha}$$
$$= 50 TW_p \times e^{-100\alpha} \quad (2)$$

Next, we note that world primary power today is slightly over 10 TW, so we can write:

$$E_w (2100) = 4 E_w (2000) \times e^{-100\alpha} \quad (3)$$

Table 1 shows how many 2000 Worlds of power

Scenario	α(%/year)	e‑$_{100\alpha}$	$E_w(2100)$
BAU	1%	1/e	1.5E$_w$(2000)
No Regrets	2%	(1/e)^2	0.5E$_w$(2000)
De-Materializing	3%	(1/e)^3	0.2E$_w$(2000)

Table 1.
Number of Today's Worlds of Additional Energy Needed in 2100

we must construct by 2100. Many authors ignore the sensitive dependence on α and discuss the need for a huge program of technology development and construction. Our view is that in any given year it is cheaper and easier to improve efficiency by 2 percent than to add 2 percent to the world's entire energy supply. Since this can be done purely by investments that produce net (life-cycle) savings, we call it a "No Regrets" scenario. The third line of Table 1 corresponds roughly to the current α for the United States. If this were to continue, and spread globally, gross world product can grow at a healthy annual 3 percent, and energy use could level off at today's rate. If we raised α only to 2 percent we would need to grow supply by only 1 percent/year.

Of course, we in the present developed world will not give up "our" present 7–8 TW, but Equation 3 and Table 1 show that under our "No Regrets" scenario, 10 billion people in developing countries need add only 0.5 new "worlds" of energy supply to provide them with an attractive current Western European standard of living. This is what we labeled above as an "Optimistic Conclusion," but it is a continuous and enduring challenge to the inventiveness of technologists and policy-makers of the twenty-first century.

Arthur H. Rosenfeld
Tina M. Kaarsberg
Joseph J. Romm

See also: Cogeneration; Hybrid Vehicles.

BIBLIOGRAPHY

Goldemberg, J.; Johansson, T. B.; Reddy, A. K. N.; and Williams, R. H. (1985). "Basic Needs and Much More in One Kilowatt per Capita." *Ambio* 14(4–5):190–200.
Kaarsberg, T. M., and Roop, J. M. (1999). "Combined Heat and Power in Industry: How Much Carbon and Energy can Manufacturers Save?" *IEEE Aerospace and Electronic Systems Magazine*, pp. 7-11, January
Rosenfeld, A. H., and Bassett, D. (1999). "The Dependence of Annual Energy Efficiency Improvement on Price and Policy." International Workshop on Technologies to Reduce Greenhouse Gas Emissions. Crystal City, VA, May 1999, Plenary IV. Paris, France: International Energy Agency. <http://www.IEA.org/workshop/engecon/>.
U.S. Department of Energy, Office of Energy Efficiency and Renewable Energy. (1997). *Scenarios of U.S. Carbon Reductions Potential Impacts of Energy Technologies by 2010 and Beyond.* <http://www.ornl.gov/ORNL/Energy_Eff/labweb.htm>.

EFFICIENCY OF ENERGY USE, ECONOMIC CONCERNS AND

The large increases in energy prices during the 1970s encouraged extensive interest in reducing energy use and in using energy resources efficiently. The common belief resulting from the energy crisis was that private markets would not provide an adequate supply of energy and would not conserve the use of energy resources. A particular concern was that households and businesses have insufficient incentives to invest in energy-saving technologies. A simplified investment model illustrates this point. The characteristic of any investment is that an initial commitment of funds is made with the expectation of a future payoff. Funds received in the future are of less value than identical funds today; hence investors must discount future cash flows.

Textbooks on investment present a simple model where the net present value (NPV) of an investment equals annual future revenues (*R*) summed and discounted at the rate *r*, minus the initial investment cost, *I*. Using t as a time subscript to denote different years, the equation is

$$\text{NPV} = -I + \sum R_t / (1 + r)^t$$

Revenues are summed from an initial period throughout the economic lifetime of the investment. The business decision rule is that an investment is profitable when its net present value is positive. If revenues accrue at a constant rate, continuously and forever, the equation becomes simpler. In the inequality $I < R/r$, an investment is profitable if annual revenue, *R*, divided by the discount rate, *r*, exceeds the initial investment cost. For instance, assume an initial investment of $100 yields annual net revenues of $11 and the discount rate is set equal to the interest rate for obtaining credit of 10 percent. The net present value of the investment (R/r) is $11/.10, or, $110, which exceeds the initial investment. The investment is profitable, and businesses and households who find 10 percent an acceptable annual rate of return are inclined to make such an investment in an efficient market.

In the above equation, r can indicate the internal rate of return on an investment. Suppose that an investment in an energy-saving technology cost $100 and reduces energy costs by $20 per year indefinitely. The reduction in costs is comparable to net revenues received. The above equation can be modified as follows: I equals R/r, where the values of I and R are specified and the value of r is computed. Hence $100 equals $20/$r$, and r equals 0.20, or 20 percent. The internal rate of return on the $100 investment is 20 percent per year. An investment is generally profitable when its internal rate of return exceeds the (interest rate) cost of obtaining credit. The investment is attractive when its internal rate of return exceeds the investor's hurdle rate, which may vary depending on the riskiness of the investment, and on the rate that can be earned from alternative uses of the investment funds.

Another investment performance measure is the payback period, which, in its simplest form, is the number of years until the net revenue from an investment equals the initial cost. In the above example of a $100 investment cost and a $20 annual payoff, the investment pays off in five years. The payback investment criterion, a crude rule of thumb, is to accept the investment if it continues to return revenue well after the payback time is reached, without any offsetting costs. This criterion has less appeal than investment decision rules based on net present value and rates of return. Investments with a short payout period may yield a high net value or internal rate of returns, but this result is not inevitable. Investments with a low return for the first few years but a very high return in later years can have a long payout period but still offer a high net return. The payoff period has some merit in considering highly risky investments where risk is a function of time. Where the entire initial investment is at risk—such as in a country with a politically unstable government—recovering an initial investment as soon as possible is

important. Note, however, that payback is essentially a break-even measure, not a measure of profitability.

Households and businesses use energy jointly with technologies to produce energy services. In almost every application, consumers have a choice between highly efficient technologies that cost more initially but have lower energy costs, and less efficient technologies that have lower initial costs but higher operating costs. For instance, electric heat pumps use differing amounts of electricity to produce space heating and cooling, depending on their efficiency. The standard investment model indicates that rational consumers will invest in the more efficient heat pump if the present value of energy saving exceeds the higher initial investment cost of the more efficient unit, other factors being equal.

Data obtained from the U.S. Energy Information Administration illustrate the trade-off between efficient heat pumps and currently purchased models. In Table 1, the current popular model heat pump uses 6,973 kilowatt-hours (kWh) per year on average to produce heating and cooling, whereas the new and efficient model uses 5,279 kWh per year. Energy efficiency is sometimes interpreted as a simple technical coefficient, such as the amount of energy required to perform a unit of work. Using this technical definition, the efficient heat pump is necessarily more energy-efficient than the current model. Whether this energy efficiency investment makes good economic sense is another matter.

The efficient heat pump reduces energy use by 1,676 kWh per year on average. Is the efficient model heat pump a good investment? Suppose the incremental cost of the efficient unit, as compared with the less efficient unit, is $1,000, and electricity cost 10 cents per kWh. With this price of electricity, the efficient heat pump reduces electricity costs by $167.60 per year. Taking a simplified approach for purposes of illustration and assuming that each unit lasts indefinitely and has no repair, maintenance, or replacement costs, and ignoring possible tax effects, the internal rate of return may be calculated as $1,000 = $167.60/$r$, which is 16.76 percent per year. If the household can borrow money at, say, 10 percent per year and earn 16.76 percent, the investment makes economic sense. If we assume a 10 percent discount rate, the present value of the investment is $1,676, which exceeds the initial investment cost. The net present value is $676, which indicates that the investment is feasible.

	Current Model		Efficient Model	
	Annual kWh	Efficiency	Annual kWh	Efficiency
Heating	4742	7.5	3984	9.4
Cooling	2297	12	1802	15.3
Annual kWh	6973	—	5297	

Efficiency data are seasonal energy efficiency ratings (SEER) for cooling and heating; seasonal performance factors (HSPF) for heating.

Table 1.
Illustrative Electricity Use In Current and Efficient Residential Electric Heat Pumps
SOURCE: Arthur D. Little, September, 1998. *EIA Residential and commercial Building Technologies—Reference Case.*

The technical literature on the economic return to energy-efficient investments is vast and yields two main conclusions: First, energy-efficient investments frequently offer a positive net present value, or alternatively, a high internal rate of return; second, energy-efficient technologies often fail to achieve a significant market share, at least for the first few years after introduction. For instance, J.G. Koomey, A. H. Sanstad, and L. J. Shown (1996) conclude that consumers fail to purchase energy efficient light bulbs, even though such purchases offer very high internal rates of return. Paul Ballonoff (1999) challenges their estimates and argues that their estimated internal rates of return are the result of erroneous calculations.

There is an enormous controversy about whether consumers and businesses undertake an efficient level of investment in energy-saving technologies. The view of energy conservationists is that energy-efficient investments offer consumers abnormally high rates of return, but consumers still refuse to make such investments. Furthermore, government efforts are required to encourage energy-efficient purchases in these inefficient markets. The alternative view, often associated with mainstream economists, is that consumers make reasonably efficient choices. Taking this view, additional government regulations are more likely to impose market inefficiencies than reduce them.

In the illustration of the electric heat pump, the simple internal rate of return is 16.76 percent. We observe that some households purchase the efficient model, but others purchase the current model. Are these households irrational, or are there simple explanations? Many economists are concerned whether energy-efficient investments make good economic sense. There are numerous explanations for the refusal of households to buy the more efficient heat pump. The cost of borrowing money is not 10 percent, as assumed above, but for some households the cost of borrowing could be a credit card rate, such as 18 percent per year. At this rate, the NPV of the investment is negative. In the above illustration, the price of electricity is 10 cents per kilowatt-hour, but this is an average cost, not the marginal cost of incremental electricity use, which may be much lower. If the incremental cost of electricity is only 5 cents per kilowatt-hour, the annual energy cost saving is $83.80. The internal rate of return is 8.38 percent. Some households find this rate attractive and purchase the efficient heat pump; others do not. Furthermore, this example ignores other system costs, or assumes them to be the same as for the current system, which may not be realistic if the efficiency gains are achieved through newer, less proven, technology.

The alternative estimate of net present value assumes that households are identical; but households have important differences. The energy saved by the efficient heat pump depends on climate and household conditions. Heat pumps have a lower payoff in very cold regions than in regions with moderate climates. Efficient heat pumps are an attractive investment in some regions; but in other regions, the current model is a better choice. Residential structures that are "tight" offer lower benefits to the efficient heat pump. The net present value of the efficient heat pump may be positive for some households but negative for others.

Technologies that offer increased efficiency tend to be the newest technologies that are entering the market. The adoption rate for many new technologies and products may increase very slowly over time, even when such purchases appear to offer high internal rates of return. The adoption rate of new technologies and products tends to be low at first, but as information becomes more widespread over time, the adoption rate of superior technologies increases. Information about a technology or product is a broad concept. Consumers must first become aware of the new technology. Consumers must also become aware of the performance and cost features of the new technology. Acquiring this information is time-consuming and expensive, but is required if a new technology is to capture a large share of the market. New energy-efficient technologies do not immediately achieve a large market share even when their estimated NPV is positive. Moreover, most new technologies and products that are not energy-related also do not immediately achieve a large market share when their NPV is positive.

The energy-efficient heat pump is likely to achieve a small market share for at least its first few years after introduction. Similarly, other energy-efficient technologies, as well as nonenergy related technologies, are likely to achieve a small market share after they enter the market.

The initial rejection of the efficient heat pump by many consumers may be well founded or not. In addition to the sound economic rationales for rejection, there may be market-impediment explanations. Consumers may have imperfect information and be unaware of the energy savings of new, efficient technologies. The transaction cost of acquiring information and making an efficient choice may be just too high. Because of these impediments, households often fail to make investments that would actually save them money over time.

In summary, the main points in this controversy are first, the contention by energy conservation proponents that numerous investment opportunities currently exist to reduce energy costs. Further, these investment opportunities would meet the investment criteria noted above. Although there are few challenges to this contention, it is not the center of controversy. Conservationists contend that the failure of private markets to make these investments is indicative of serious inefficiencies. Some economists dispute this contention and provide explanations of why normal, well-functioning markets may defer on energy-efficient investments.

Ronald J. Sutherland

See also: Auditing of Energy Use; Capital Investment Decisions; Conservation Supply Curves; Economically Efficient Energy Choices; Economic Growth and Energy Consumption; Efficiency of Energy Use; Efficiency of Energy Use, Labeling of; Energy Economics; Environmental Economics; Government and the Energy Marketplace; Industry and Business, Energy as a Factor of Production in; Industry and Business, Productivity and Energy Efficiency in; Supply and Demand and Energy Prices.

BIBLIOGRAPHY

Arthur D. Little, Inc. (1998). *EIA-Technology Forecast Updates-Residential and Commercial Building Technologies-Advanced Adoption Case.* Washington, DC: Energy Information Administration.

Ballonoff, P. (1999). "On the Failure of Market Failures." *Regulation: The Cato Review of Business and Government* 22(2):17–19.

Koomey, J. G.; Sanstad, A. H.; and Shown, L. J. (1996). "Energy-Efficient Lighting: Market Data, Market Imperfections, and Policy Success." *Contemporary Economic Policy* 14(3):98–111.

EFFICIENCY OF ENERGY USE, LABELING OF

The U.S. Federal Trade Commission (FTC) issued the Appliance Labeling Rule in 1979 in response to a directive from Congress in the Energy Policy and Conservation Act of 1975 (EPCA). In addition to mandating that the FTC promulgate this labeling Rule, EPCA directed the U.S. Department of Energy (DOE) to establish energy conservation standards for residential household appliances and to develop and maintain test procedures by which members of the appliance industry could measure the efficiency or energy use of these products.

The Rule requires manufacturers of most major household appliances to show energy information about their products on labels so consumers purchasing the appliances can compare the energy use or efficiency of competing models. Without this energy use information on EnergyGuide labels, purchasers would have no way to assess the energy efficiency of appliance products and thus would not be able to include the information as a criterion in their purchasing decisions. At the time it was published, the Rule applied to refrigerators, freezers, dishwashers, clothes washers, water heaters, window air conditioners, furnaces, and boilers. In 1987 the FTC included central air conditioners and heat pumps, and in 1989 the FTC added a requirement for a simple disclosure for fluorescent lamp ballasts. In 1993 and 1994, the

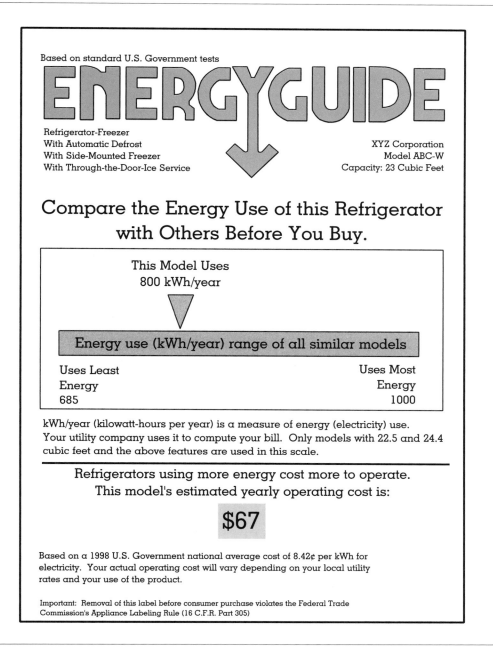

Figure 1.
Sample label for refrigerator-freezer.
NOTE: The required dimensions of a full-sized label are 5½ inches (width) by 7⅜ inches (height).

FTC added requirements for the disclosure of water use for certain plumbing products, the disclosure of energy-related information for light bulbs and fluorescent lighting products, and the disclosure of energy-efficiency information for pool heaters. The FTC exempted other products listed in EPCA from labeling requirements—such as stovetops and ovens, clothes dryers, television sets, space heaters, humidifiers, and dehumidifiers—because it did not believe that the energy use among different models was great enough to be a potentially significant factor for consumers considering purchasing the products.

To comply with the Rule, manufacturers of all covered major household appliances must place a black-and-yellow "EnergyGuide" label on each of their products that shows the energy consumption or efficiency of the labeled product (Figure 1). The labels also must show a "range of comparability" bar or scale (published regularly by the FTC) that shows the highest and lowest energy consumption or efficiencies for all appliance models that have features similar to those of the labeled one. In addition to the energy consumption or efficiency disclosure, the labels must show the products's estimated annual operating cost, based on a specified national average cost for the fuel the appliances use. Manufacturers must derive the efficiency, energy use, and operating cost information from the standardized tests that EPCA directed DOE to develop. The information on the EnergyGuide label also must appear in catalogs from which the products can be ordered.

Appliance manufacturers must attach the labels to the exterior surface of their products or use hang tag labels, which must be attached in such as way as to be easily viewed by a consumer examining the product. Manufacturers of furnaces, central air conditioners, and heat pumps also must attach EnergyGuide labels to their products showing the products's efficiency and the applicable range of comparability, but the labels need not contain operating cost information. As with appliances, the information on the EnergyGuide also must appear in catalogs from which the products can be ordered.

The requirements for products other than major home appliances vary depending upon the product. Manufacturers of fluorescent lamp ballasts must disclose an encircled "E" on ballasts and on luminaires containing ballasts, as well as on packaging for both. The "E" signifies compliance with DOE's energy conservation standards for those products. Manufacturers of showerheads, faucets, toilets, and urinals must disclose, on the products and their packaging and labeling, the water usage of their products in terms of gallons and liters per flush, per minute, or per cycle. Manufacturers of certain incandescent bulbs, spot and flood bulbs, and screw-base compact fluorescent bulbs must disclose, on packaging, the light output in lumens, energy used in watts, voltage, average life, and number of bulbs. They also must explain how purchasers can select the most energy-efficient bulb for their needs. Manufacturers of certain tube-type fluorescent bulbs must disclose on

packages an encircled letter "E" and a statement that the "E" logo means the bulb meets U.S. federal minimum efficiency standards.

The FTC can assess penalties under the Rule against manufacturers for violations of the above requirements. The Rule also states that energy-use-related representations regarding covered products, including print and broadcast advertisements, must be based on the DOE test procedures.

James G. Miller

See also: Economically Efficient Energy Choices

BIBLIOGRAPHY

Appliance Labeling Rule, Code of Federal Regulations, 1979. Vol. 16, part 305.
Energy Policy and Conservation Act of 1975, U.S. Code. Title 42, secs. 6291–6309.

EINSTEIN, ALBERT (1897–1955)

Albert Einstein the twentieth century's most renowned scientist, was born in Ulm, in the kingdom of Warttemberg, now part of Germany, the son of Hermann Einstein, a small businessman, never very successful, and Pauline Einstein (née) Koch. In 1881 Maria, his only sibling, was born. In 1880 the family moved to Munich, where Einstein attended public school and high school, always doing well. (The story that he was a poor pupil is a myth, probably caused by his dislike of formal education.) In those years he also received private violin lessons and, to comply with legal requirements, instruction in the elements of Judaism. As a result of this inculcation, Einstein went through an intense religious phase at about age eleven, following religious precepts in detail and (he later told a friend) composing songs in honor of God. A year later, this phase ended abruptly and forever as a result of his exposure to popular books on science, to "the holy geometry book" (as he called it) on Euclidean geometry, to writings of Kant, and more.

In 1895 Einstein took the entrance examination at the Federal Institute of Technology (ETH) in Zurich but failed because of poor grades in literary and political history. In 1896, after a year of study at a high school in Aaratu Switzerland, he did gain admission, however. In that year he gave up his German citizenship and became stateless, in 1901 he became a Swiss citizen.

During his next four years as an ETH student, Einstein did not excel in regular course attendance, relying far more on self-study. In 1900 he passed his final examinations with good grades, which qualified him as a high school teacher in mathematics and physics. For the next two years he had to be satisfied with temporary teaching positions until in June 1902 he was appointed technical expert third class at the Patent Office in Berne.

In January 1903 Einstein married Mileva Maric, (of Greek-Catholic Serbian descent), a fellow student at the ETH. In 1902 the couple had a daughter out of wedlock, Lieserl, whose fate remains unknown, and after marriage they had two sons, Hans Albert (1904), who became a distinguished professor of hydraulic engineering in Berkeley, California, and Eduard (1910), a gifted child who became a student of medicine in Zurich but who then turned severely schizophrenic and died in a psychiatric hospital.

In 1914 the Einsteins separated, and they divorced in 1919. Thereafter Einstein married his cousin Elsa Einstein, who brought him two stepdaughters. He had several extramarital affairs during this second marriage.

None of Einstein's first four papers published between 1901 and 1904 foreshadowed his explosive creativity of 1905, his *annus mirabilis*, in which he produced: in March, his proposal of the existence of light quanta and the photoelectric effect, work for which in 1922 he received the Nobel Prize; in April, a paper on the determination of molecular dimensions, which earned him his Ph.D. in Zurich; in May, his theory of special relativity; in September, a sequel to the preceding paper containing the relation $E = mc^2$. Any one of these papers would have made him greatly renowned; their totality made him immortal.

Only after all these publications did Einstein's academic career begin: *privatdozent* in Berne, 1908; associate professor at the University of Zurich, 1909, the year of his first honorary degree (Geneva); full professor at Karl Ferdinand University, Prague, 1911; professor at the ETH, 1912; professor and member of the Prussian Academy of Sciences, Berlin, 1914–1932, where he arrived four months before the outbreak of World War I.

In 1915 Einstein cosigned his first political docu-

Albert Einstein. (AP/Wide World Photos)

ment, "Manifesto to Europeans," in which all those who cherish European culture were urged to join in a League of Europeans (never realized). Far more important, in that year Einstein completed his masterpiece, perhaps the most profound contribution to physics of the twentieth century: his general relativity theory, on which he had been brooding for the previous eight years. In the special theory all laws of physics have the same form for any two observers moving relatively to each other in a straight line and with constant, time-independent, velocity. In the general theory the same is true for all kinds of relative motion. This demands a revision of Isaac Newton's theory of gravitation. Space is curved, Einstein now asserted, the amount of curvature depending on how dense matter is at that place; matter determines by its gravitational action "what shape space is in."

The superiority of Einstein's over Newton's theory became manifest in 1915, when Einstein could for the first time explain an anomaly in the motion of the planet Mercury (advance of the perihelion), known observationally since 1859. He also predicted that light grazing the sun bends by a factor of two larger than predicted by Newton's theory.

In 1916 Einstein completed his most widely known book "on the special and the general theory of relativity, popularly explained," wrote the first paper on gravitational waves, and became president of the *Deutsche Physikalische Gesellschaft*. In 1917 he became ill, suffering successively from a liver ailment, a stomach ulcer, jaundice, and general weakness, but nevertheless he managed to complete the first paper on relativistic cosmology. He did not fully recover until 1920.

In November 1919 Einstein became the mythical figure he is to this day. In May of that year two solar eclipse expeditions had (in the words of the astronomer Eddington) "confirm[ed] Einstein's weird theory of non-Euclidean space." On November 6 the president of the Royal Society declared in London that this was "the most remarkable scientific event since the discovery [in 1846] of the predicted existence of the planet Neptune."

The next day the *Times of London* carried an article headlined "Revolution in Science/New Theory of the Universe/Newtonian Ideas Overthrown." Einstein had triumphed over Newton (who, of course, remains a stellar figure in science). The drama of that moment was enhanced by the contrast with the recently concluded World War I, which had caused millions to die, empires to fall, and the future to be uncertain. At that time Einstein emerges, bringing newlaw and order. From that time on the world press made him into an icon, the divine man, of the twentieth century.

At about that time one begins to perceive changes in the activities of Einstein, now in mid life. He began writing nonscientific articles. In 1920 he was exposed to anti-Semitic demonstrations during a lecture he gave in Berlin. At the same time, Jews fleeing from the East came literally knocking at his door for help. All that awakened in Einstein a deepened awareness of the Jewish predicament and caused him to speak up and write in favor of Jewish self-expression by means of settling in Palestine, creating there a peaceful center where Jews could live in dignity and without persecution. Thus he became an advocate of what may be called moral Zionism, though he never was a member of any Zionist organization.

The 1920s also was the period of Einstein most extensive travels. In 1921 he paid his first visit to the United States, for to raise funds for the planned Hebrew University, being honored on the way, including being received by President Warren G. Harding. In 1922 his visit to Paris contributed to the normalization of Franco-German relations. Also in that year he accepted membership in the League of Nations' Committee on Intellectual Cooperation. In June Walter Rathenau, foreign minister of Germany, a Jew and an acquaintance of Einstein, was assassinated. After being warned that he, too, might be in danger, Einstein left with his wife for a five-month trip abroad. After short visits to Colombo, Singapore, Hong Kong, and Shanghai, they arrived in Japan for a five-week stay. The press reported that, at a reception, the center of attention was not the empress, everything turned on Einstein.

On the way back, they visited Palestine. In introducing Einstein at a lecture, the president of the Zionist Executive said: "Mount the platform that has been awaiting you for two thousand years." Thereafter Einstein spent three weeks in Spain. In 1925 he journeyed to South America, lecturing in Buenos Aires, Montevideo, and Rio de Janeiro. Apart from three later trips to the United States, this was the last major voyage in Einstein's life.

All these multifarious activities took a lot of Einstein's energies but did not keep him from his physics research. In 1922 he published his first paper on unified field theory, an attempt at incorporating not only gravitation but also electromagnetism into a new world geometry, a subject that was his main concern until the end of his life. He tried many approaches; none of them have worked out. In 1924 he published three papers on quantum statistical mechanics, which include his discovery of so-called Bose-Einstein condensation. This was his last contribution to physics that may be called seminal. He did continue to publish all through his later years, however.

In 1925 quantum mechanics arrived, a new theory with which Einstein never found peace. His celebrated dialogue with Niels Bohr on this topic started at the 1927 Solvay Conference. They were to argue almost until Einstein's death without agreeing.

In 1928 Einstein suffered a temporary physical collapse due to an enlargement of the heart. He had to stay in bed for four months and keep to a salt-free diet. He fully recuperated, but he stayed weak for a year. The year 1929 witnessed his first visit with the Belgian royal family, leading to a life-long correspondence with Queen Elizabeth.

Einstein had been a pacifist since his young years, but in the 1920s his position became more radical in this respect. For example, in 1925 he, Gandhi, and others signed a manifesto against obligatory military service, and in 1930 another, supporting world government. In that year and again in 1931 Einstein visited the United States. In 1932 he accepted an appointment as professor at the Institute for Advanced Study in Princeton, originally intending to divide his time between Princeton and Berlin. However, after he and his wife left Germany on December 10 of that year, they would never set foot in Germany again—in January 1933 the Nazis came to power. Though remaining pacifist at heart, Einstein was deeply convinced that they could be defeated only by force of arms.

Because of the new political situation, Einstein changed his plans, arriving on October 17, 1933, in the United States to settle permanently in Princeton. Whereafter he left that country only once, in 1935, to travel to Bermuda to make from there application for permanent residency in the United States. In 1940 he became a U.S. citizen.

Also in his new country Einstein remained a prominent figure. In 1934 he and his wife were invited by Franklin and Eleanor Roosevelt and spent a night at the White House. Einstein remained scientifically active—he wrote, in fact, some good papers, but nothing as memorable as in his European days.

In 1939 Einstein wrote to Roosevelt to draw his attention to possible military use of atomic energy. His influence on these later developments was marginal, however. In 1943 he became consultant to the U.S. Navy Bureau of Ordnance but was never involved in atomic bomb work. In 1944 a copy of his 1905 paper on special relativity, handwritten by him for this purpose, was auctioned for six million dollars as a contribution to the war effort. (It is now in the Library of Congress.)

After the war he continued to speak out on political issues, such as his open letter to the United Nations urging the formation of a world government, and his frequent condemnations in the press of Senator Joseph McCarthy's activities. After the death of Chaim Weizmann, the first president of Israel, Einstein was invited but declined to be his successor.

In 1948 Einstein was found to have a large intact aneurysm of the abdominal aorta. In 1950 he wrote his testament, willing his papers and manuscripts to Hebrew University (where they are now). On April 11, 1955, he wrote his last letter, to Bertrand Russell, in which he agreed to sign a manifesto urging all nations to renounce nuclear weapons. On April 13 Einstein wrote an incomplete draft for a radio address that ends: "Political passions, aroused everywhere, demand their victims." On the afternoon of that day his aneurysm ruptured. On April 15 he entered Princeton Hospital, where he died on April 18 at 1:15 A.M. His body was cremated that same day. The ashes were scattered at an undisclosed place. The following November his first great-grandson was born.

Abraham Pais

See also: Nuclear Energy; Nuclear Energy, Historical Evolution of the Use of; Nuclear Fission; Nuclear Fusion.

BIBLIOGRAPHY

Pais, A. (1982). *Subtle Is the Lord: The Science and the Life of Albert Einstein.* Oxford, Eng.: Oxford University Press.
Pais, A. (1997). *Einstein Lived Here.* New York: Oxford University Press.

ELASTIC ENERGY

A material is said to be elastic if it returns to its original shape after being deformed. Elastic energy is energy contained by an object as a result of deforming it from its relaxed position. A rubber band used to hold a stack of papers together and a trampoline are elastic. Deforming something requires application of a force. A person pulls on a rubber band to stretch it; a gymnast pushes down on a springy trampoline to deform it. The act of doing work to deform an elastic material produces elastic energy in the material. This elastic energy can then be used to do work as the deformed material returns to its original shape.

Springs made from coils of wire are used to store elastic energy which is a form of potential energy. The most common type of spring is cylindrically shaped, with coils evenly spaced and of the same diameter. A ballpoint pen uses a coil spring to hold the point in place for writing and to return it to the case for protection.

The energy stored in a spring depends on the strength of the spring and the deformation that may be either an extension or compression from its relaxed

Multiple exposure photography is used to illustrate the movement of a pole vaulter. (Corbis-Bettmann)

condition. For many springs, the deforming force is directly proportional to the deformation; doubling the force doubles the deformation. For example, if a 10 N force stretches a spring 0.01 m then a 20 N force will stretch it 0.02 m. Such a spring is said to obey Hooke's law, and the relation between force and deformation is written $F = kx$, where F is the deforming force in newtons (N), x is the deformation in meters (m), and k is the spring constant in newtons/meter (N/m). The stiffer the spring, the larger the spring constant. The elastic energy of a spring obeying Hooke's law is given by $E = \frac{1}{2}kx^2$. The energy depends strongly on the deformation because doubling the deformation (x) quadruples the energy. In other words, a spring stretched 0.1 m will have four times the elastic energy as when it was stretched 0.05 m.

Some watches and clocks, not powered by batteries, use springs to store energy for their mechanical mechanism. Usually, the springs are made by winding wire in a flat spiral. Rather than storing energy by

stretching or compressing, the spirals wound tighter. Rewinding a watch by turning a knob amounts to tightening the spirals of the spring inside the watch. Energy released when the spring unwinds is used to run the clock mechanism. The design is such that it takes only a few seconds to store the energy that is then released over twenty-four hours or more.

Elastic energy is not limited to springs. In fact, it is an important property that material scientists consider in selecting materials for products. A bent diving board, a drawn bow, and a planted vaulting pole are just a few of the many examples of elastic materials. It is not easy to calculate the elastic energy for these materials applications, but it is easy to see how energy is involved. A diver does work on the board by jumping on it, a hunter does work on the bow by pulling on the strings, and a vaulter does work on the pole by bending it. Elastic energy is stored as a result. The diver recovers the elastic energy of the board and rises upward when the board relaxes. The arrow recovers the elastic energy of the bow when the hunter releases the string. And the vaulter recovers the elastic energy of the pole and rises upward.

Joseph Priest

See also: Potential Energy.

BIBLIOGRAPHY

Hobson, A. (1995). *Physics: Concepts and Connections.* Upper Saddle River, NJ: Prentice-Hall.

Serway, R. A., and Faughn, J. S. (1995). *College Physics.* Philadelphia, PA: Saunders College Publishing.

ELECTRIC CIRCUIT

See: Electricity

ELECTRICITY

Electrical and electronic devices and machines have become an integral part of contemporary life, ranging from household appliances and computers to huge industrial machines. When home and business owners pay the monthly bill from the electrical power company for the use of all of these items, they are paying for energy very conveniently delivered over electrical wires from the power company.

Although the delivery of electricity to homes and businesses has been possible only during the past century, static electricity was observed by the ancient Greeks over 2,000 years ago. They also noticed natural magnets, called lodestones, found near the town of Magnesia. These were important discoveries because scientists now know that electricity and magnetism are intimately related. Magnetism is used by power companies to produce the electricity used every day.

ELECTRIC CHARGES AND ELECTRIC FORCE

If the humidity is low, it is common to experience a shock when touching a metal doorknob after walking across a carpeted floor. This is static electricity, and it can be studied by rubbing a hard rubber rod on some fur, and then touching the rod to a small metal ball that is suspended on the end of a silk thread. The ball quickly bounces away from the rod and is repelled by it, as shown in Figure 1a. The rod and the ball are electrified or charged with the same type of (net) charge. Further, when a glass rod is rubbed with silk and touched to a second similar ball, the second ball is repelled by the glass rod, shown in Figure 1b. But when the two balls are brought near each other (without touching), they are clearly attracted to each other as shown in Figure 1c.

In the mid-1700s Benjamin Franklin proposed that there are only two kinds of electrical charge, which he called plus and minus. He defined the net charge on the rubber rod to be negative, and the charge on the glass rod to be positive. Further, charges of the same kind are repelled from each other, while opposite charges are attracted to each other. The amount of electrical charge (often represented by the letter q or Q) is measured in coulombs (abbreviated as C) in the Standard International system of units, called S.I. units.

CHARGES IN THE STRUCTURE OF AN ATOM

Electrons particles in matter with a negative charge were discovered around 1900. All matter is made of tiny atoms packed closely together. The structure of the atom was proposed to be like a tiny solar system, with negative electrons revolving in orbits around a very tiny positive nucleus (see Figure 2). The charge ("e") of the electron is a certain fixed amount: only a millionth of a trillionth of 0.16 coulombs (or 0.16×10^{-18} C). The total negative charge of all of the electrons in an atom is exactly the same (but of opposite sign) as the charge of the positive nucleus; thus an atom taken as a whole is normally uncharged or "neutral."

Since a piece of matter (e.g., a piece of rubber or copper) is made up of a great many neutral atoms, the piece is itself normally neutral or uncharged. When a rubber rod is rubbed with fur, some electrons are pulled from the fur onto the rod, giving it extra negative charge (the fur is then deficient of electrons so it is positively charged).

ELECTRIC CIRCUITS AND CURRENT FLOW

There were few applications of electricity (or of magnetism) before the invention of the battery by Alessandro Volta in 1800. A battery can cause charges to move for long periods of time. The movement or

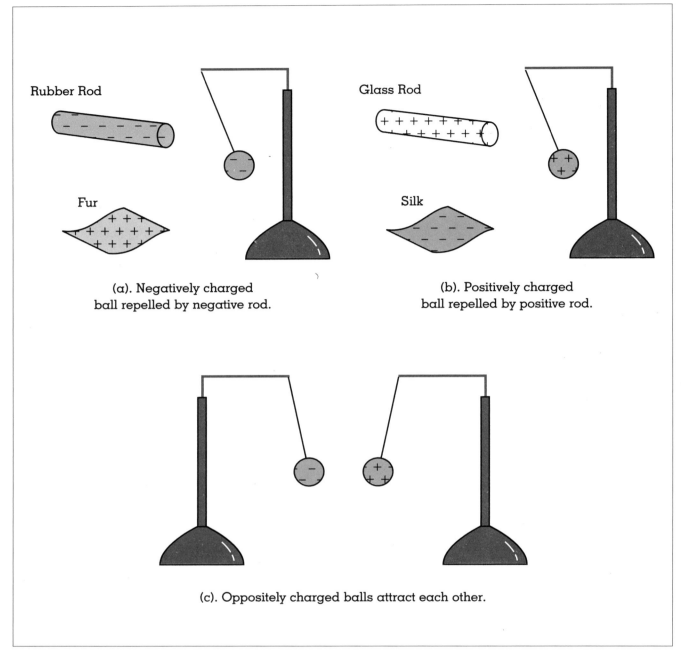

Figure 1.
Static electricity experiments.

"flow" of electrical charge is called a current. A basic battery cell consists of two different metals (called electrodes) immersed in an acid or salt solution (called an electrolyte). Through chemical interactions, one electrode develops extra electrons and becomes negatively charged, while the other develops a deficiency of electrons and becomes positively charged; these are respectively labeled the negative (-) and positive (+) battery cell terminals.

Work (W) is done by a battery whenever it pushes a positive charge (+q) away from the (+) terminal (through space outside the battery) to the (-) terminal. The "potential" or "emf" or "voltage" (V) of a battery is defined as the work done in this process divided by

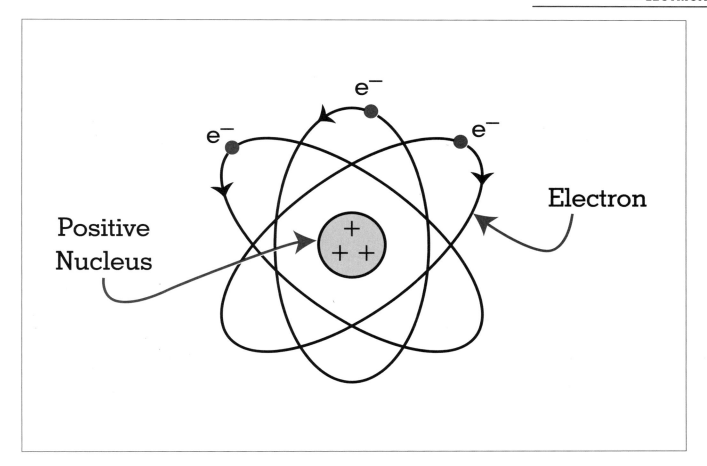

Figure 2.
Model of an atom with three electrons.

the charge: V= W/q. In S.I. units the voltage V is in volts V to honor Volta. A battery cell often used in flashlights is the "carbon-zinc dry cell." If unused and fresh, the carbon-zinc cell has a voltage of about 1.5 V. Many flashlights use two carbon-zinc cells placed end-to-end or in "series." Strictly speaking, the two cells together constitute a "battery," and the voltage of this battery is 3.0 V. The common 12 V "lead-acid" battery used in automobiles has six cells connected in series with it; each lead-acid cell has a voltage of 2.0 V.

Suppose a long thin metal wire is connected by a pair of thick wires between the terminals of a battery. This is a basic "electric circuit" as shown in Figure 3a. In all metals, each atom permits roughly one of the outer electrons to move quite freely in the material; these are called the "free electrons." In contrast, all electrons of the atoms of good electrical "insulators," such as glass, rubber, and air, are tightly bound to the atoms and are not free to move through the body of

the material. The free electrons are repelled from the negative battery terminal and attracted toward the positive terminal, so that a continuous movement of charge (an electrical current) results around this complete path or "circuit." But there are frictional effects (electrons bumping into atoms as they move along) that resist the movement or flow of electrons through the wires, especially in the thin wire. The frictional effects result in the electrical "resistance" (R) of the wire, and cause the wires to heat up, especially the long thin wire.

An important property of this or any electrical circuit is the rate that charge moves past a place in the circuit (e.g., out from or into a battery terminal). The electrical current (I) is defined to be the charge (Q) that flows, divided by the time (t) required for the flow: I = Q/t. In S.I. units the current (I) is in amperes (A).

In the early 1800s, Georg Ohm studied the effect

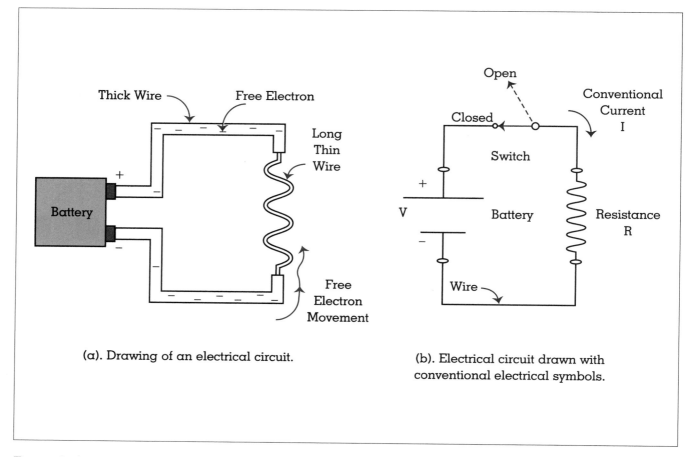

(a). Drawing of an electrical circuit.

(b). Electrical circuit drawn with conventional electrical symbols.

Figure 3.
Basic d.c. electrical circuit.

on current of changing the battery voltage as well as the length, cross-sectional area, and material of the wire in an electrical circuit like Figure 3a. He found that the current flow (I) was proportional to the voltage (V) and inversely proportional to the resistance (R) of the wire to this flow. The resistance (R) is quite constant for a given piece of metal wire; Ohm's law states that we may write that $I = V/R$. The S.I. units of resistance are ohms (Ω). The resistance to current flow of a wire is larger if the wire is longer, but smaller if the wire is thicker (larger cross-sectional area).

The smallest diameter copper wire generally permitted within the walls for U.S. household wiring is called no. 14 wiring; such wire has a diameter of about 1/16 inch (1.63 mm). A 10 foot length (about 3 meters) of no. 14 copper wire has a resistance of about 0.025 ohms; note that a 100 foot length (about 30 meters) has a resistance of 0.25 ohms.

Figure 3b shows the basic electrical circuit of Figure 3a as it is drawn using conventional electrical symbols. Note the symbol for a battery. The connecting wires with negligible resistance are drawn as lines, while an element with significant resistance is drawn as a zigzag line. An electrical switch is a length of metal that can be moved so that a space of air appears in the circuit; the air space has a large resistance (perhaps billions of ohms) so that virtually no current flows through it when it is "open" (as represented by the dashed line in Figure 3b. When the switch is closed the circuit path is complete and current flows around the "closed circuit." It is conventional to draw the current flow direction as from the (+) battery terminal to the (-) terminal. This is the direction that (+) charges would move if they were moving within the wire, and this was assumed to be the case from about 1800 to 1900. It is now known electrons move in the opposite direction, but the old conventional direction is almost universally used. In any case, the current flows in only one direction in this circuit; this situation is called "direct current" or "d.c."

ELECTRICAL POWER AND ELECTRICAL ENERGY

As electrical current flows through a wire of significant resistance, the wire is heated and the wire gives off heat energy. The wire in an incandescent light bulb is heated so much that it glows and gives off light energy as well as heat energy. This energy is supplied by a battery or other source of emf. Further, "power" is the rate that energy is being supplied, or that work is being done: power = energy/time, or power = (work done)/time. In S.I. units power is measured in watts. Using the definition of voltage and of current one can show that when a current flows from a battery (or any source of emf), the battery is delivering power at the rate of the voltage times the current: $P = VI$. For example, suppose a 115 V battery is supplying 0.87 amperes of current to a light bulb connected to the battery; then the power being used by the light bulb is equal to $(115V) \times (0.87 A) = 100$ watts. This particular bulb is a "100 watt light bulb." Note that because 1,000 watts equals one kilowatt, this bulb is using a tenth of a kilowatt or 0.10 kilowatts (0.10 kW) of power.

Since power = energy/time, it follows that energy = power × time; that is, energy supplied is equal to the power multiplied by the time-interval. The unit of electrical energy unit used by the electrical utility company is (kilowatts) × (hours), or kWh. Suppose a 100 W bulb is turned on for one day or 24 hours. The energy used by the bulb is $(0.10 \text{ kW}) \times (24 \text{ h}) = 2.4$ kWh. A typical electrical energy cost in the United States might be 8 cents per kWh, so the cost for using this bulb for an entire 24 hour day would be $(2.4 \text{ kWh}) \times (8 \text{ cents/kWh}) = 19.2$ cents.

Note that electrical energy used (and paid for) depends not only on the power consumption by a device or appliance, but also the length of time that it is used in a month. Therefore people are sometimes surprised to discover that there is a modest electric bill even after they have been gone from home on vacation and "everything" had been turned off. However, a number of devices around the home were probably operating, such as electric clocks, the electric motor on the furnace, the refrigerator, and the freezer. There are also many applicances operating on standby mode, such as televisions, answering machines, and cordless appliance. These are sometimes called "phantom power" devices.

MAGNETS, MAGNETIC FIELD, AND ELECTROMAGNETS

Magnets also repel or attract each other. A magnet always has both a north pole (N) and a south pole (S). Again, "opposites attract and likes repel" so that two magnets with north poles close to each other are repelled from each other, even without touching. The influence in the region near a magnet is often pictured as a "magnetic field"; invisible "magnetic field lines" are imagined, and often drawn in a diagram of a magnet. Figure 4a illustrates a steel bar "permanent magnet" with its magnetic field lines.

In the early 1800s H. C. Oersted discovered that an electric current flowing through a coil of wire produces a magnetic field, as shown in Figure 4b. A long coil of wire, often wound on an iron core to enhance the magnetic field, is called a "solenoid" or "electromagnet" and attracts iron or steel objects, just like a permanent magnet.

ELECTRIC GENERATOR

Shortly after Oersted's discovery, Michael Faraday found that changing the number of magnetic field lines within a coil of wire induces an emf or voltage in the coil so that a current flows if there is a closed circuit. Thus not only can a current produce a magnetic field, but a changing magnetic field can generate a current. Figure 5a shows the basic idea of an electric generator. A coil consisting of a single turn of wire is pictured; this is twirled (rotated) in front of a bar magnet so that the magnetic field lines pass through the coil, first from one side and then the other side. The changing magnetic field within the coil results in an alternating emf or voltage that is measured between the terminals A and B, and a current that reverses directions (i.e., an alternating current or a.c.) flows through the attached "load" resistance.

The alternating voltage V_{AB} measured between terminals A and B is graphed in Figure 5b. If the coil is turned at 60 cycles per second, then the frequency of alternation is 60 cycles per second (called 60 Hz in S.I. units). This is the standard frequency for commercial a.c. power in the United States (in Europe the standard is 50 Hz). By increasing the number of turns (or loops) in the coil, or by increasing the magnetic field strength, it is possible to increase the amount of emf or voltage generated (e.g., the peak voltage V_{peak} illustrated in Figure 5b). A typical power company generator might generate 10,000 V_{rms}. This

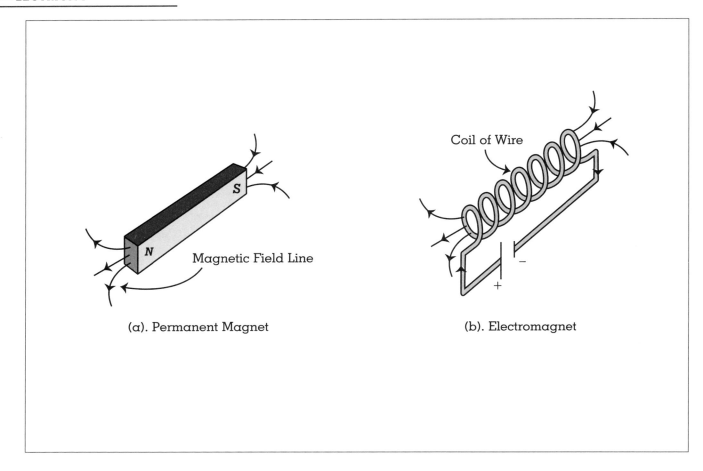

(a). Permanent Magnet

(b). Electromagnet

Figure 4.
Magnets with associated magnetic field lines.

is much too large to be safe for household use, so the generated voltage must be "transformed" down to about 115 V_{rms} that is present at a regular household outlet. (But first it is actually transformed up to much higher voltage to reduce energy lost in transporting the electrical energy.)

Note that a.c. voltages (also a.c. currents) are usually measured and quoted as "r.m.s." The r.m.s. value is 70.7 percent of the peak value. This is done so that the formula to calculate electrical power in the a.c. case is the same as for the d.c. case stated earlier: power is the voltage times the current (provided both are rms values).

It is important to realize that it requires effort to turn the coil of an electrical generator if current is being supplied by the generator; that is, work must be done to turn the generator coil. Conservation of energy requires that the energy used to turned the coil (i.e., mechanical energy input to the generator) is at least as

much as the electrical energy produced by the generator in a given period of time. Electrical generators have relatively high efficiency, so that the electrical power output is perhaps 90 percent of the mechanical energy input, with the remaining 10 percent lost in various heating effects in the generator. Most commercial (utility company) electrical generator coils in the United States are turned by steam engines that burn coal to obtain the energy to generate the electricity.

TRANSFORMERS AND DELIVERY OF ELECTRIC POWER

Alternating voltage (called a.c. voltage) can be quite readily changed to a different value through the use of an electrical transformer. Note that an electrical transformer will *not* transform a d.c. voltage to another d.c. voltage value; this is a principal reason why commercial electricity in the United States (and

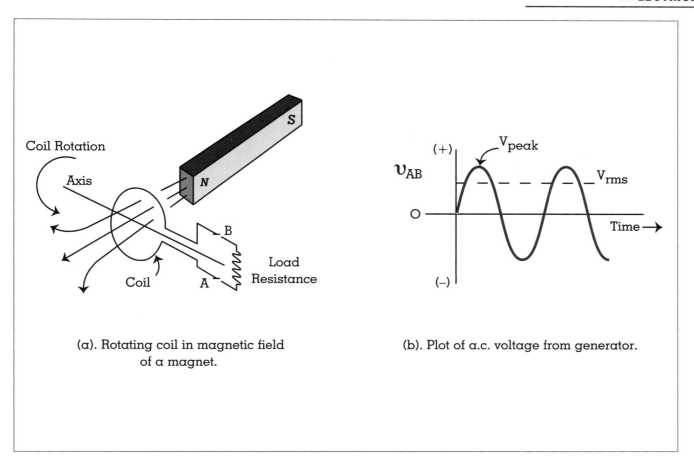

(a). Rotating coil in magnetic field of a magnet.

(b). Plot of a.c. voltage from generator.

Figure 5.
Basic a.c. generator and voltage.

most of the world) is almost always a.c. rather than d.c. A transformer consists basically of two coils of wire, with one coil often wound on top of the other. Electrical voltage is supplied to the coil called the "primary" coil, while voltage is output from the "secondary" coil. Then the basic transformer relationship is that the ratio of the secondary voltage to primary voltage is the same as the ratio of secondary-coil-turns to primary-coil-turns.

Suppose 10,000 V_{rms} (e.g., from an electrical power plant generator) is input to a transformer where the number of secondary turns is 75 times more than the number of primary turns. Then the voltage from the secondary will be 75 × 10,000 V_{rms} = 750,000 V_{rms}. This very high voltage level is actually produced by the secondary of the transformers at a modern commercial electrical power plant; the 750,000 V_{rms} is connected to the high voltage transmission lines (thick wire cables) that are used to transport the elec-

trical power over long distances (perhaps hundreds of miles). The high voltage is used to reduce the energy lost to heating of the transmission wire resistance by the electrical current. The power lost to heating of the transmission wire is proportional to the square of the current flowing, so that the losses mount rapidly as the current flowing increases. But the power transmitted is equal to the transmission line voltage times the current flowing. For a given power, the current flowing will be much less (and energy lost in the transmission process is much less) if the transmission line voltage is very large. At the output end of the transmission line, a transformer is used again to step the voltage down to a level appropriate for local distribution (e.g., 2,400 V_{rms}), and a neighborhood transformer is finally used to step the voltage down to 220 V_{rms} and 110 V_{rms} for use in homes and businesses.

Dennis Barnaal

393

See also: Electric Power, Generation of; Electric Power Transmission and Distribution Systems; Electricity, History of; Transformers.

BIBLIOGRAPHY

Aubrecht, G. (1995). *Energy*, 2nd ed. Englewood Cliffs, NJ: Prentice-Hall.

Faughn, J.; Chang, R.; and Turk, J. (1995). *Physical Science*, 2nd ed. Fort Worth, TX: Harcourt College Publishers.

Giancoli, D. (1998). *Physics*, 5th ed. Upper Saddle River, NJ: Prentice-Hall.

Priest, J. (1991). *Energy: Principles, Problems, Alternatives*, 4th ed. Reading, MA: Addison-Wesley Publishing Co.

ELECTRICITY, HISTORY OF

Electrical effects were known to the ancients through the attraction that amber, when rubbed, had for lightweight objects. They were also aware of the seemingly unrelated phenomenon of lightning. There is the further possibility that a form of electric battery was used for electroplating in Mesopotamia, but the evidence is meager and, even if true, there were no long-term consequences.

The term "electric" comes from the Greek word for amber and was coined by William Gilbert in his book on magnetism, published in 1600. Gilbert showed that other materials had this same attractive property and made the important observation that it was quite different from the attractive property of magnetism.

Although the subject was not abandoned, serious experimentation arguably began with the work of Francis Hauksbee, Stephen Gray, and Charles Dufay in the early decades of the eighteenth century. By rubbing glass rods to generate electric charge, and using silk threads as conductors, they were able to develop a basic understanding of conduction and to stimulate thinking about the nature of electricity.

Two inventions in the 1740s changed the electrical scene dramatically. One was the frictional machine, which made it possible to generate continuous streams of electricity relatively easily; the other was the condenser, or Leyden jar, which made possible the storage and sudden discharge of substantial quantities of electric charge.

Rubbing a simple glass rod can produce on the order of 50,000 volts, but the capacitance (the ability to store electric charge) of the surface is low and the energy available in a discharge has been estimated as less than 0.001 joule. Electrostatic machines could generate somewhat higher voltages, the ultimate being the large plate machine constructed by Martinus van Marum in Haarlem in 1785. It produced sparks up to 60 centimeters long, which (based on the size of the electrodes) translates into about 330,000 volts. Since van Marum's machine also had a substantial prime conductor (an arrangement of brass tubes and spheres that could store the charge), the energy available in a discharge could be as much as 10 joules. With 100 large Leyden jars this increased to 30,000 joules (about 8 watt-hours).

The apparatus available to a typical experimenter would of course be more modest, but with even one or two Leyden jars, energy levels in the tens of joules are easily possible.

Benjamin Franklin was in the fortunate position of beginning his electrical experiments in 1747, just as these new devices were becoming available. He was quick to discover that the energy they delivered was capable of a substantial physiological effect, but they also made it possible to study electricity in a more systematic matter. Franklin conceived of electricity as a fluid made up of particles that were attracted to ordinary matter in a fashion similar to gravity. He also explained the action of the Leyden jar as due to electrical fluid on one side of the glass being held in place by attraction to matter that had lost its electrical fluid on the other side. This was a serviceable theory (especially when the concept of repulsion was added a few years later by Franz Aepinus) that was the basis for the mainstream of experimental work that followed.

During the remainder of the eighteenth century, electricity was a popular subject for public demonstrations, but the scope of experimental work was limited. Franklin suggested a means for testing the hypothesis that lightning was electrical by using pointed conductors to draw charge from the atmosphere. This was done successfully in France (using a long iron rod) in 1752 and possibly by Franklin (using a kite) somewhat later. Much inconclusive work was done on biological and medical effects of electricity. There were some chemical experiments, including the decomposition of water. In 1785 Charles Coulomb demonstrated that the electrostatic force varied inversely as the square of the distance.

An article by Alessandro Volta in the British journal *Philosophical Transactions*, based on investigations of the previous year, stimulated a whole new range of electrical experiments and discoveries that would be important to energy history. He found that a "pile" of pairs of zinc and silver discs, separated by moistened pads, produced a continuous current of electricity. In an alternative arrangement, the liquid was contained in a series of cups. Plates of silver and zinc were placed in the cups and linked together in series from one cup to the next. Volta's announcement immediately let to a host of extensions and variations on his basic scheme, the ultimate being the Great Battery of the Royal Institution in 1810. This battery consisted of two hundred troughs, each with ten pairs of four-inch square copper and zinc electrodes. Although the energy content of these was not large, the fact that it could be discharged in a controlled fashion over time led to a number of new applications. Most important in the first few years were those in the area of chemistry. Numerous compounds were broken up and several new elements were identified. Furthermore, the basis was laid for electrical theories of the chemical bond.

These early experimenters quickly found that battery discharge was hindered by a process called polarization, or the buildup of deposits on electrodes. Thus began a long period—which continues to the present—to find a combination of materials that will efficiently convert chemical to electrical energy. It would be even better, of course, if the process were reversible, so that electrical energy could be stored in chemical form and then released when needed. A practical form of storage battery was devised by Gustav Planté in 1859, using lead and lead-oxide electrodes in sulfuric acid.

The energy available from nineteenth-century batteries varied depending on the materials used and the form of construction. The voltage from a single cell ranged between one and two volts, currents were typically a few amps, and time of discharge (more difficult to estimate) could be measured in hours. The implication is therefore that a single cell might be good for a few watt hours. This was not enough for economical "power" applications, but it was sufficient for the telegraph in the 1840s and the telephone in the 1870s.

In the 1880s there was great interest in trying to develop storage cells efficient enough and with

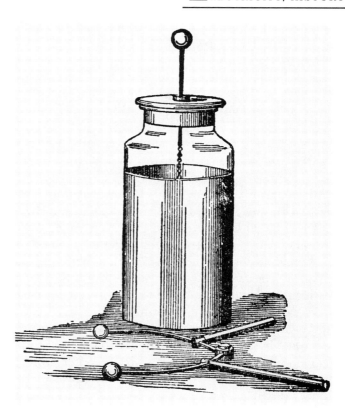

The Leyden jar and discharger used to collect electric fluid in a bottle half filled with water (undated engraving). (Corbis Corporation)

enough capacity to act as load equalizers for the emerging power industry, and at the end of the century there was further pressure for a design that would power an electric car. This was a problem that remained intractable not only through the nineteenth century but through the twentieth as well.

Several electrical scientists in the early part of the nineteenth century, influenced at least in part by their understanding of German *naturphilosophie*, expected forces of nature to be intimately connected to each other, and some of them spent extraordinary amounts of time looking for the relationship. One of these was a Dane, Hans Christian Oersted, who, after an exhaustive series of experiments, in 1820 found that electricity could indeed produce a magnetic effect. Further experiments by Michael Faraday demonstrated, in 1821, that by proper orientation of an electric current and a magnetic field it was possible to produce continuous motion in what soon would be called a motor. It took an additional ten frustrating years for him to prove what he instinctively felt to be true, that, in a fashion inverse to what

Oersted had discovered, magnetism somehow could produce electricity. He called his device an electromagnetic generator.

For the next thirty-five years motors and generators were little more than classroom demonstration devices, limited as they were by the magnetic fields of permanent magnets. Generators were employed in a few specialized cases, for instance in some isolated lighthouses where large magneto machines were used to power arc lamps, notably by the Société l'Alliance in France, but this was the exception. The best of the machines were bulky, weighing on the order of 2,000 kilograms, and produced a little over two kilowatts at an efficiency of under 20 percent.

A breakthrough came in 1867 with publications by Charles Wheatstone, Werner Siemens and S. Alfred Varley that they had, independently, invented generators in which both the armature and the field magnets were electromagnets. Similar constructions were described by others but did not lead to practical consequences. These generators were, in other words, "self-excited." This was accomplished because the core of the field coils kept a small amount of residual magnetism that produced a small current in the moving armature, part of which was fed back into the field coils, which very rapidly reached their full magnetic strength. Improvements in armature and field design led to further efficiencies. These were successfully applied by the Belgian Zenobe Gramme, who by 1872 was manufacturing commercial dynamos (self-excited generators). His first machines weighed about half as much as the Alliance magnetos and produced more electricity at higher efficiency. In 1876, direct comparisons of magneto generators (Alliance) and dynamos (Gramme and Siemens) showed that a dynamo of as little as one tenth the weight, and little more than one tenth the cost, for the same input horsepower had more than twice the efficiency of the magneto.

The new dynamos made arc lighting economically feasible, especially for street lights. Initial installations were made in Paris in 1877, London at the end of 1878, San Francisco in 1879, and New York in 1880. The other principal commercial application in those early years was for electroplating.

The success of arc lighting in turn stimulated the search for a less intense incandescent lamp. It was Thomas Edison who, after more than a year of exhaustive effort, won this race. But for our purposes it is important to note that at the same time he was seeking improvements in the dynamo. Others reasoned that the best system was one in which the internal resistance of the dynamo equaled the external resistance, since this would maximize the amount of energy provided to the external load. Edison, however, was quick to point out that this was hardly a bargain because it limited the efficiency to 50 percent. He therefore designed generators with low internal resistance and was able to achieve efficiencies in excess of 90 percent. His first successful dynamos, with which he lit his first lamps, produced about five kilowatts. In 1881, the "Jumbo" dynamo, which he shipped to the Paris electric exhibition, produced approximately 120 kilowatts.

The 1880s saw a phenomenal growth in the number of (mainly incandescent) electric lighting systems, especially in the United States. After 1886, street railways also came into use. Electric lights could run on either alternating (ac) or direct current (dc), but motors for street cars or any other purpose needed dc. There was also the possibility that truly practical storage batteries would be developed that could serve as load equalizers for dc systems. Finally, there was evidence that for the same voltage, alternating current was more dangerous.

No effective storage battery appeared, however, and the safety argument was for most people not a determining factor. Therefore, when Nikola Tesla invented a practical induction motor in the late 1880s, the advantages of dc became inconsequential. Alternating current could easily be transformed from low to high voltage and back, and at high voltage it could be transmitted over long distance with relatively low loss. The "battle of the systems" could be said to have been officially over with the merger of the Edison company and Thomson-Houston in 1892, and especially with the adoption of ac for the planned Niagara Falls generating plant in 1893. Westinghouse built ten two-phase, 25-Hz alternating current dynamos based on several Tesla patents for the initial Niagara power station; two were installed in 1895, the final one in 1900. Each was rated at 5000 horsepower. Electricity was transmitted at the generated 2200 volts to local industrial plants and at 11,000 volts (converted to three phase) to Buffalo.

In the United States the early decades of the new century were dominated by ever-increasing size: steam-turbine generators with capacities that reached over a thousand megawatts, and transmission voltages (the highest ones now often dc) above 700,000 volts. But after mid-century these "economies of

Members of the Royal Institution attend a lecture given by Michael Faraday on magnetism and light (London, England, 1846). (Corbis-Bettmann)

scale" were proving not to be so economical after all, and there was a decided move toward smaller gas-turbine co-generation plants.

Another legacy of the late nineteenth century was identification of the electron by an appropriate interpretation of the Pieter Zeeman effect in 1896, and more especially by J. J. Thomson's experiments the following year. Starting in the 1920s, physicists devised machines that could accelerate electrons (or protons) to increasingly higher energies and then cause them to collide with themselves or with atoms of various elements. This proved to be a very effective means for studying the fundamental properties of matter, with the result that higher and higher energies

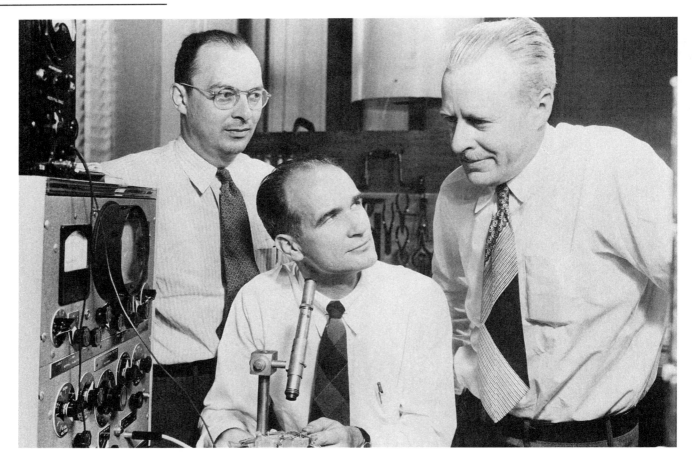

William Shockley (seated), John Bardeen (standing, left), and Walter H. Brattain doing transistor research at Bell Telephone Laboratories (New York, 1948). (Corbis Corporation)

were sought. By the end of the century, bunches of electrons in the Stanford Linear Accelerator were being raised to the level of 50 billion electron volts. On the average, in continuous operation, this means a power level of tens of kilowatts. Individual bunches, operating over very short periods of time, attain power levels approaching a million billion kilowatts.

On a more practical level, Edison's observation in 1881 that the filaments of his lamps emitted negatively charged particles of carbon led to John Ambrose Fleming's invention of a two-element "valve" in 1904 and Lee de Forest's three-element "audion" in 1906. De Forest thought that gas in his audion tube was essential to its operation, but by 1912 several people realized that, to the contrary, a critical feature was a really good vacuum. This made possible the rapid development of effective vacuum tube amplifiers for wireless sets, for long-distance telephone lines (transcontinental service opened in

1915), for radio receivers (after 1920), and eventually for television, radar and microwave systems. As rapid switches vacuum tubes were also critical to the development of early computers.

Of special interest is the fact that with a good vacuum, these tubes, in properly arranged feedback circuits, could "oscillate" and produce radio waves. By the mid-twenties they were competing effectively with the popular high-frequency alternating dynamos, the largest of which generated 200 kilowatts. A typical high-power radio station was, however, operated at 50 kilowatts.

In the 1940s, higher-frequency devices—notably the magnetron and the klystron—produced radiation in the microwave region, which was especially useful for radar. By the early postwar period, these could easily generate continuous signals in the range of several kilowatts, with pulses in the megawatt region.

Even higher frequencies were achieved by the

maser, conceived by Charles Townes in 1951, and by the laser, first demonstrated by Theodore Maiman in 1960. The maser, operating in the microwave region, has been used mainly in amplifiers. Its optical- (and near optical-) frequency sibling, the laser, has found a variety of uses, ranging from eye surgery to weapons, where the amount of energy that can be released is important. Some versions have operated in continuous fashion at one kilowatt and above, or with very short pulses up to 100 megawatts.

The study of electrons trapped in matter (commonly termed "solid state") led eventually to the invention of the transistor in 1947 by Walter Brattain, John Bardeen, and William Shockley at Bell Laboratories, and then to the integrated circuit by Robert Noyce and Jack Kilby a decade later. Use of these devices dominated the second half of the twentieth century, most notably through computers, with a significant stimulus to development being given by military expenditures.

Another way of considering the changing "face" of electricity is to look at the professional organizations it has spawned. Through much of the nineteenth century electricity was produced at low energy levels (mostly by batteries) and employed in operations (mainly the telegraph) that were much more complicated mechanically than electrically. The self-excited dynamo and the incandescent lamp changed all that, and by the mid-1880s schools of electrical engineering began to appear in American colleges. This shift can be seen in Britain where the Society of Telegraph Engineers, founded in 1871, became the Institution of Electrical Engineers (AIEE) in 1888. In the United States, where there was no prior engineering society (although there were journals), some like-minded practitioners joined together to form the American Institute of Electrical Engineers in 1884. The significance of the new wireless-electronic technology became evident with formation of the Institute of Radio Engineers (IRE) in 1912.

Merger of the AIEE and the IRE in 1963 into the Institution of Electrical and Electronics Engineers (IEEE) preserved in its name a sense of the earlier bifurcation of the profession, but in fact the lines had become quite blurred and the numerous constituent societies drew and continue to draw from both traditions. At the end of the century, the IEEE was easily the largest professional organization in the world with membership over 350,000. Its thirty-six constituent societies provided an indication of the degree to which electricity has pervaded the modern world, ranging from "Communications" and "Lasers & Electron

Optics" to "Neural Networks" and "Engineering in Medicine & Biology," from "Power Electronics" and "Vehicular Technology" to "Aerospace and Electronic Systems" and "Robotics & Automation." Significantly, the largest constituent group was the Computer Society.

Bernard S. Finn

See also: Cogeneration Technologies; Edison, Thomas Alva; Electricity; Electric Motor Systems; Electric Power Transmission and Distribution Systems; Matter and Energy; Regulation and Rates for Electricity; Siemens, Ernst Werner von; Tesla, Nikola; Thomson, Joseph John; Townes, Charles Hard; Turbines, Gas; Turbines, Steam; Volta, Alessadro; Wheatstone, Charles.

BIBLIOGRAPHY

Adams, E. D. (1927). *Niagara Power: History of the Niagara Falls Power Company, 1886–1918*. Niagara Falls, NY: The Niagara Falls Power Company.

Bowers, B. (1982). *A History of Electric Light and Power*. Stevenage, UK: Peter Peergrinus, Ltd.

Finn, B. (1991). *The History of Electrical Technology: An Annotated Bibliography*. New York: Garland Publishing, Inc.

Heilbron, J. L. (1979). *Electricity in the 17th and 18th Centuries: A Study of Early Modern Physics*. Berkeley: University of California Press.

Hughes, T. P. (1983). *Networks of Power: Electrification in Western Society, 1880–1930*. Baltimore: Johns Hopkins University Press.

King, W. J. (1961–1962). "The Development of Electrical Technology in the 19th Century." *U.S. National Museum Bulletin* 228. Papers 28–30:231–407.

McMahon, A. M. (1984). *The Making of a Profession: A Century of Electrical Engineering in America*. New York: IEEE Press.

Riorden, M., and Hoddeson, L. (1997). *Crystal Fire: The Birth of the Information Age*. New York: W. W. Norton & Company.

ELECTRICITY REGULATION AND RATES

See: Regulation and Rates for Electricity

ELECTRIC LEAKAGE

See: Electricity

ELECTRIC MOTOR SYSTEMS

Electric motors are everywhere. These ubiquitous devices come in a wide variety of sizes and power outputs, ranging from a fraction of a watt to huge multikilowatt applications. Tiny ones operate computer disk drives, power windows/mirrors, and windshield wipers; moderate-sized ones run appliances such as fans, blenders, electric shavers, and vacuum cleaners; a large drive pumps, elevators, sawmills, and electric trains and vehicles.

There has been not only growth in the total number of electric motors (more standard appliances in use), but also a proliferation in their use for new, novel applications. Both trends will continue to increase demand for the electricity to run electric motors. In the United States, electric motors are responsible for consuming more than half of all electricity, and for the industrial sector alone, close to two-thirds. Since the cost of the electricity to power these motors is enormous (estimated at more than $90 billion a year), research is focused on finding ways to increase the energy efficiency of motors and motor systems.

EARLY DEVELOPMENT

Electric motors are devices that convert electrical energy into mechanical energy. Devices that do the opposite—convert mechanical energy into electrical energy—are called generators. An important example is power plants, where large gas, steam, and hydroelectric turbines drive generators to provide electricity. Since generators and motors work under the same principles, and because construction differences are minimal, often generators can function as motors and motors, as generators, with only minor changes.

The early development of electric motors and generators can be traced to the 1820 discovery by Hans Christian Oersted that electricity in motion generates a magnetic field. Oersted proved the long-suspected promise that there is indeed a relationship between electricity and magnetism. Shortly thereafter, Michael Faraday built a primitive electric motor, which showed that Oersted's effect could be used to produce continuous motion.

Electric motors consist of two main parts: the rotor, which is free to rotate, and the stator, which is stationary (Figure 1). Usually each produces a magnetic field, either through the use of permanent magnets, or by electric current flowing through the electromagnetic windings (electromagnets produce magnetism by an electric current rather than by permanent magnets). It is the attraction and repulsion between poles on the rotor and the stator that cause the electromagnet to rotate. Repulsion is at a maximum when the current is perpendicular to the magnetic field. When the plane of the rotator is parallel to the magnetic field (as in Figure 1) there is no force on the sides of the rotator. The left side of the rotator receives an upward push and the right side receives a downward push; thus rotation occurs in a clockwise direction. When the rotator is perpendicular to the magnetic field, forces are exerted on all its sides equally, canceling out. Therefore, to produce continuous motion, the current must be reversed when reaching a vertical direction. For alternating-current motors this occurs automatically, since the alternating magnetic attraction and repulsion change directions 120 times each second; for direct current-motors, this usually is accomplished with a device called a commutator.

Before useful electric motors could be developed, it was first necessary to develop practical electromagnets, which was primarily the result of work done by Englishman William Sturgeon as well as Joseph Henry and Thomas Davenport of the United States. In 1873 Zénobe-Théophile Gramme, a Belgian-born electrical engineer, demonstrated the first commercial electric motor. A decade later Nikola Tesla, a Serbian-American engineer, invented the first alternating-current induction motor, the prototype for the majority of modern electric motors that followed. Magnetic fields today are measured in units of teslas (symbol T), to acknowledge Tesla's contribution to the field.

MOTOR TYPES

The electrical power supplied to electrical motors can be from a direct-current (dc) source or an alternating-current (ac) source. Because dc motors are more

Direct-Current Electric Motor

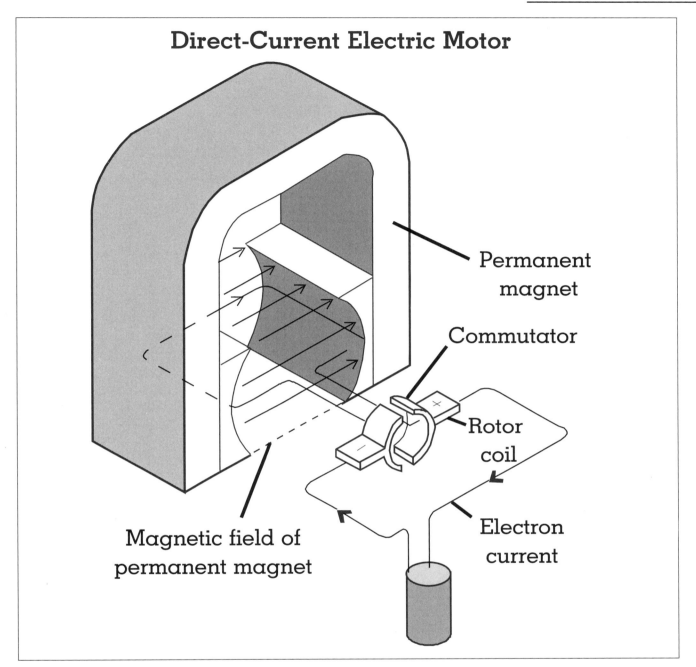

Figure 1.
The attraction and repulsion between the poles on the rotor and stator cause the electromagnet to rotate. Direct-current motors usually need commutators to achieve continuous motion.

expensive to produce and less reliable, and because standard household current is ac, ac motors are far more prevalent. However, the growing demand for portable appliances such as laptop computers, cordless power drills, and vacuums ensures a future for dc motors. For these portable applications, the alternating current charges the battery, and the direct current of the battery powers the dc motor.

Based on the way magnetic fields are generated and controlled in the rotor and the stator, there are several additional subclassifications of direct- and alternating-current motors.

Direct Current Motors

Direct-current motors are classified as separately excited motors, series motors, shunt motors, and compound motors. The field winding of a separately excited motor is in a circuit that is energized by a separate dc source; the field winding is not physically connected to the armature circuit (containing the armature winding).

For series, shunt, and compound motors, only one power supply is needed. In a series motor the field winding is connected in series with the power supply and the armature circuit. In a shunt motor the field winding is connected across, or in parallel with, the dc supply, which energies the armature winding. In a compound motor, also known as a cumulative compound motor, the field winding is physically in parallel with the armature winding circuit and is magnetically coupled to a coil or a winding in series with the armature winding.

Direct current motors are most appropriately used in applications where a dc power supply is available or where a simple method of speed control is desired. The fans used in automobile heating and air conditioning systems are driven by direct-current motors.

Alternating-Current Motors

Alternating-current motors are classified as induction motors or synchronous motors. Faraday found that a stationary wire in a magnetic field produced no current. However, when the wire continues to move across magnetic lines of force, it produces a continual current. When the motion stops, so does the current. Thus Faraday proved that electric current is only produced from relative motion between the wire and magnetic field. It is called an induced current—an electromagnetic induction effect.

Induction motors usually entail insulated wiring windings for both the rotor and the stator, with the stator connected to an external electric power source. Between the narrow gap of the stator and the rotor, a revolving magnetic field is established. A current can be established only when the waves of the rotor and stator windings are not in phase—not at a maximum simultaneously.

The induction motor is the most common motor in industrial applications, and are also very prevalent for smaller applications because of their simple construction, reliability, efficiency, and low cost. Historically they have been found in applications that call for a constant speed drive, since the alternating-current power supply is of constant voltage and frequency; however, the continual development of more powerful and less inexpensive solid-state electronic devices has allowed for electronic inverters (which control voltage and frequency) to more accurately control the speed and torque of induction motors, thereby matching the control performance of a motor. Since induction motors are less expensive, more compact, more efficient, and more reliable (better voltage overload capabilities), it is likely that induction motors will continue to replace motors in most applications.

Synchronous motors operate like induction motors in that they rely on the principle of a rotating magnetic field, usually produced by the stator. Synchronous motors differ in that the rotor generates a constant unidirectional field from a direct-current winding powered by a direct-current source. This field interacts with the rotating field. To get around the need for direct current-power, the stator can be constructed from permanent magnets so that the permanent magnetic field and rotating field can be synchronized. Synchronous motors are most useful for low-load applications where constant speed control is crucial, such as in phonographs, tape recorders, and electric clocks.

ADVANCES IN PERFORMANCE, EFFICIENCY, AND RELIABILITY

Innovations in designs and materials led to continual advances in the performance, efficiency and reliability, of electric motors throughout the Twentieth century. The best results for motor designers occur when starting with the function at hand and working back toward the power source, optimizing each element along the way, foremost being the improvement of the end use of mechanical energy. If end use mechanical energy is curtailed, the demand for power generated from an electric motor declines, and consequently the demand for electricity to run the motor. For a factory or warehouse conveyer belt system, a streamlined design, better bearings, and lighter components can yield far greater energy savings than replacing a standard-efficiency motor with an energy-efficient model.

Performance

There are few shortcuts. If you want a powerful electric motor, it is going to have to be large and entail an extensive amount of copper windings. That is why it is much more cost-effective to rebuild many large (more than 100 horsepower) industrial motors than to replace them with new motors.

Efficiency

In theory, electric motors can be more than 95 percent efficient. Since electric motors can to convert almost all the electrical energy into mechanical energy, it partly explains the continued growth of electrical technology at the expense of competing technologies. It is widely believed that the electric motor will eventually replace even the internal-combustion engine (in which only about 25 percent of the heat energy is converted to mechanical energy) once the costs of better-performing battery and fuel cell technologies decline.

In practice most electric motors operate in the 75 to 90 percent range, primarily because of core magnetic losses (heat losses and electric current loss), copper resistive losses, and mechanical losses (friction in the bearings, and windings, and aerodynamic drag). At low speeds core magnetic losses are greatest, but as speeds get higher, core magnetic losses decline, and copper resistive losses become more dominant. The mechanical losses do not vary much, remaining fairly constant at both low and high speeds.

There were no standards for energy-efficient motors until the National Electrical Manufacturers Association (NEMA) developed design classifications for energy-efficient, three- phase induction motors in 1989. This standard was made the national minimum efficiency level by the Energy Policy Act of 1992, which went into effect in October 1997. Manufacturers responded to this higher efficiency level by reducing losses through the use of better materials, improved designs, and precision manufacturing.

Aside from the efficiency of the motor itself, energy efficiency is very dependent upon proper sizing. While the efficiency of a motor is fairly constant from full load down to half load, when a motor operates at less than 40 percent of its full load, efficiency drops considerably, since magnetic, friction, and windage losses remain fairly constant regardless of the load. Moreover, the power factor drops continuously as the load drops. The problem is most discernible in small motors.

The obvious answer is to properly size the motor for an application. However, properly sizing a motor is difficult when motors are required to run at a wide spectrum of loads. In many cases a decrease in the motor's speed would reduce the load while maintaining the efficiency.

A major hurdle to greater efficiency is the constant-speed nature of induction and synchronous motors. Nevertheless, considerable advances have been made in improving motor speed controls that essentially better optimize the motor speed to the task at hand, resulting in substantial energy savings, decreased wear of the mechanical components, and usually increased productivity from the user.

Usually the lowest first cost solution is to use multispeed motors with a variety of torque and speed characteristics to match the different types of loads encountered. The more costly and more energy-efficient choice is to use electronic adjustable speed drives that continuously change the speed of AC motors by controlling the voltage supplied to the motor through semiconductor switches. Energy efficiency is also achieved by converting the 60-hertz supply frequency to some lower frequency, thereby enabling induction motors to operate at slower speeds, and thus consuming less energy. Slower speeds may be desirable for many applications such as fans and conveyor belts. There are several different types of electronic adjustable-speed drives, yet no one technology has emerged as superior to all others. That is because the multitude of different motors, different sizes (horsepower) and speeds, and control requirements make it difficult for one control system technology to be superior for all applications.

The cost premium for a motor equipped with speed control can be substantial, sometimes costing twice that of a single-speed motor. But the energy savings from speed control can be substantial, especially for fan and pump systems. Electronic adjustable speed drives continue to become more attractive because the costs of microelectronics and power electronics technologies continue to fall as performance and energy efficiency improve.

Reliability

Electric motors have proven to be very reliable and continue to become more reliable because of better materials and designs. However, because of the excellent reliability record, long life cycles, and the lower first cost of rebuilding motors instead of purchasing new, more energy-efficient models, it will be decades before energy-efficient motors significantly penetrate the market.

Even when the time comes to make a purchasing decision, an energy-efficient motor purchase is not a certainty. Sometimes an energy-efficient motor will be the economically efficient choice; at other times, not. The capital investment decision is based on the cost in relation to performance, efficiency and reliability. Moreover, the decision depends on the application and the amount of time the motor is in operation. It can be the major component of a product (drill or mixer), or a minor component (computer disk drive); it can be the major component cost of a product (fan), or it can be a minor component cost (stereo tape deck); it can run almost constantly (fan, pump, and machinery), or only a few minutes a day (vacuums and power tools). For example, contractors purchase circular saws almost solely based on performance and reliability. Time is money, and since the saw is operating only a few minutes a day and the contractor is often not responsible for the electricity costs to run the motor, energy efficiency is not a consideration; performance and reliability are what matter most. On the other hand, an industrial user, who runs huge electric motors twenty-four hours a day to work pumps, machinery, and ventilation equipment, is very concerned with energy efficiency as well as performance and reliability.

John Zumerchik

See also: Batteries; Capital Investment Decisions; Consumption; Economically Efficient Energy Choices; Electricity; Electric Power, Generation of; Faraday, Michael; Fuel Cells; Fuel Cell Vehicles; Magnetism and Magnets; Oersted, Hans Christian; Tesla, Nikola.

BIBLIOGRAPHY

de Almeida, A. T., and Greenburg, S. (1997). "Energy Efficient Technologies: Electric Motor Systems Efficiency." In *CRC Handbook of Energy Efficiency*, ed. F. Kreith and R. E. West. Boca Raton, FL: CRC Press.

Beaty, H., and Kirtley, J. (1998). *Electric Motor Handbook*. New York: McGraw-Hill.

Electric Power Research Institute. (1992). "Electric Motors: Markets, Trends and Applications." EPRI Report TR-100423. Palo Alto, CA: Author.

Nadel, S.; Shepard, M.; Greenburg, S.; Katz, G.; and Almeida, A. (1992). *Energy-Efficient Motor Systems: A Handbook on Technologies, Programs and Policy Opportunities*. Washington, DC: American Council for an Energy Efficient Economy.

ELECTRIC POWER, GENERATION OF

INTRODUCTION

Electric power systems can be thought of as being comprised of three important sectors: generation, transmission, and distribution. For most utilities, generation capital equipment costs account for approximately 50 percent of total plant in costs. Generation also accounts for close to 75 percent of total operation and maintenance expense.

Generation is the production process center of the power industry. This production process is multifaceted and starts with the conversion of primary energy, such as fossil fuels, uranium, and the kinetic energy of water, to electrical energy. The process by which this primary energy is converted to electricity varies depending upon the prime mover, or technology, of the power generator. Mainstream generation technologies include hydroelectric facilities, internal combustion or combustion turbine facilities, and steam generation facilities. Alternative electric generation can include prime movers powered by the wind, sun, or some other renewable fuel such as biomass or solid waste.

Hydroelectric facilities use the kinetic energy of falling water to turn a water turbine to create electricity. These facilities usually have limited technical applicability and are located in geographic regions that meet certain elevation, water level, and stream flow requirements. The advantage of hydroelectric facilities is that they are virtually free of fuel costs. Their disadvantage, in addition to their limited geographic applicability, is that they have relatively high capital costs.

Historically, internal combustion engines and combustion turbines have been considered unique and limited types of generation facilities. These are similar in many respects to engines used in the auto-

Interior of electrical power plant. (JLM Visuals)

motive and aeronautics industries. Both technologies burn one or a combination of various fossil fuels (oil, diesel, propane, or natural gas) to create mechanical energy to turn electric generators. The advantage of these technologies is that they have relatively low capital costs. Since these technologies are run on fossil fuels, their disadvantage rests with their relatively high and potentially volatile operating costs.

The traditional power generation work horse has been the steam generator. This technology uses fossil fuels (coal, natural gas, and oil) to heat water in a boiler to create steam. The steam, in turn, drives an electricity generator. Nuclear power is a special case of the steam generator, using uranium and nuclear fission to create steam. While these facilities have high capital costs, their operating costs are lower than their nearest competitor, combustion turbines.

On the fringe of electricity generation technolo-gies are alternate-fuel generators, including solar photovoltiac and thermal applications, wind powered applications, and the use of waste agricultural by-products such as rice hulls or bagasse, and even garbage. These technologies have traditionally played a small role in the overall generation portfolio of util-ities in the United States, given their relatively high cost and limited capacity. Most of these technologies have been promoted under the auspices of research and development or within the context of relatively unique niche applications.

GENERATION PLANNING AND OPERATION

In the past, generation planning consisted of develop-ing and maintaining a portfolio of facilities to meet the various types of electricity loads that occur in any given hour, across any given day, in any given season. Reliability tended to be the most important planning

Power plant exterior. (JLM Visuals)

consideration, followed closely by cost. Thus, generation planning strategies consisted of constructing and operating enough power plants to meet demand on a cost effective basis. In many instances, having the ability to meet sudden surges in demand entailed constructing and maintaining large capacity reserve margins that remained idle during large parts of the year.

Since load varies considerably across hour, day, and year, utilities have traditionally segmented their generation facilities into three classifications: baseload generation, intermediate or cycling generation, and peaking generation. Baseload generators are typically steam generation facilities used to service minimum system load, and as such are run at a continuous rate. While these units are the most efficient to operate, they are costly to start up from a cold shut down (designed for continuous combustion); therefore, they are usually run at a near-constant rate.

Intermediate load plants are typically older steam units or combustion turbines brought on line during periods of forced or planned outage of baseload units. Intermediate units can also be thought of as units that bridge the dispatch of baseload and peaking units during periods of unusually high demand. These units can be older and less efficient than baseload units. Peaking units are typically combustion turbines that have the ability to generate electricity immediately, and serve temporary spikes in demand such as during a heat wave when residential and commercial air conditioning demands begin to surge.

In the past, electric utilities dispatched generating units to meet demand on a lowest-to-highest cost basis. This form of dispatch is commonly referred to as "economic dispatch." The marginal or incremental cost of dispatching units is traditionally the benchmark used to rank order available generators. These marginal costs, in the very short run, are typically

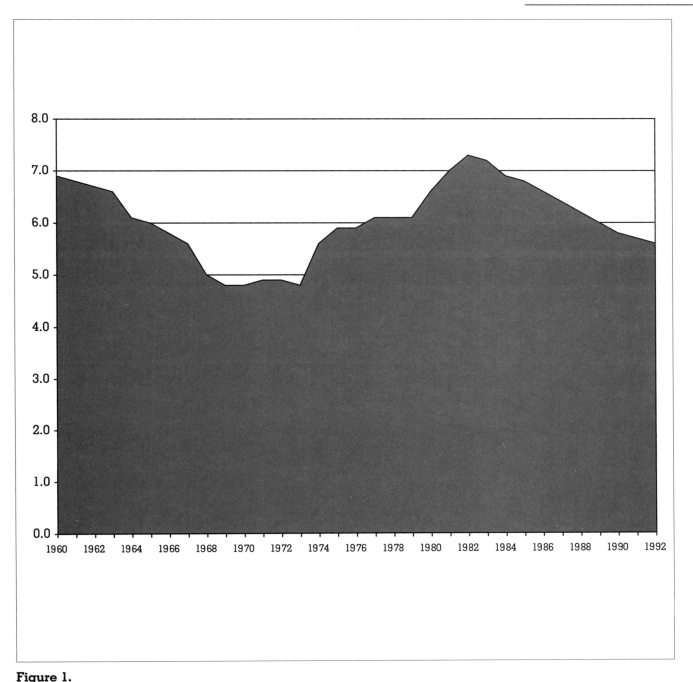

Figure 1.
Real Electricity Prices (1995 cents per kWh)

SOURCE: Energy Information Agency. (1995). *Electric Power Annual*. Washington, DC: U.S. Department of Energy.

associated with changes in fuel costs and other variable operating and maintenance (O&M) costs. Historically, baseload units, which are almost always large coal, hydro, or nuclear units, had the lowest incremental costs and were dispatched first to meet load. As load increased during the day, or across seasons, less efficient intermediate or cycling units, which generate electricity at slightly higher costs, were brought on line. Higher cost peaking units would be the last types of units brought on line, for example, during a heat wave with a resulting large demand for air conditioning. The cost of the last dis-

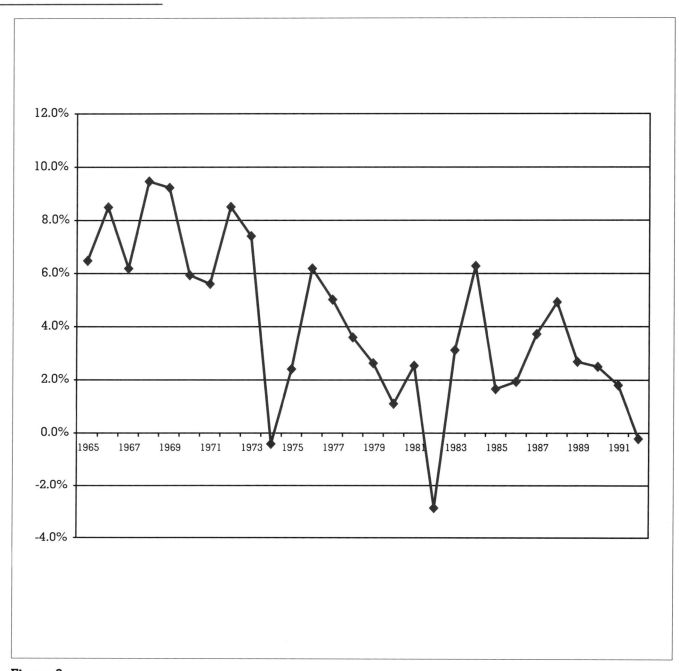

Figure 2.

Annual Rates of Electricity Demand Growth

SOURCE: Energy Information Agency. (1995). *Electric Power Annual*. Washington, DC: U.S. Department of Energy.

patched unit would therefore define the system marginal costs, often referred to as the system "lambda."

Given the importance and relative size of the electricity generation sector, shifts in the costs of constructing and operating electric power plants can have considerable influence on final rates of electricity for end users. Figure 1 shows the historic trend of real, or inflation-adjusted, electricity prices over time. Noticeable in the graph are the spikes that occurred in the early 1970s and again in the early 1980s, when utilities in the industry were faced with a number of almost insurmountable challenges leading to the undermining of their unique natural monopoly cost advantages.

The history of electricity generation planning can be broken into two distinct periods: one period proceeding, and one period following the energy crisis of 1973. Prior to 1973, electricity generation planning was a relatively straightforward endeavor. During this period, forecasted increases in load were met with the construction of new generation facilities. Utilities typically tried to meet this load with the most cost effective generation technology available at the time. As shown in Figure 2, the annual rate of electricity demand prior to 1973 grew at an annual average rate between 6 and 10 percent. This constant, significant growth placed many utilities in the position of having multiple construction projects ongoing at any given time. The period following the energy crisis of 1973 dramatically changed the generation planning process for utilities. During this time the industry was plagued by high inflation and interest rates, high fuel prices, financial risk, and regulatory uncertainty. These uncertainties increased costs, which in turn had a deleterious affect on electricity demand. Dramatic electricity price increases, resulting from the volatile operating environment of the post–1973 environment, stifled the growth of electricity demand and set strong incentives to end users to conserve electricity. As a result, utilities found themselves with considerable excess capacity that quickly became technologically and economically obsolete. This excess and uneconomic capacity, in combination with eventual emergence of new technologies and increased competition, resulted in an undermining of the natural monopoly justification for electric utility regulation.

THE ECONOMIC REGULATION OF POWER GENERATION

Historically, the electric power industry was characterized as being a natural monopoly. Natural monopolies typically occur in industries with very large fixed capital costs and relatively low operating costs. The cost characteristics of these industries tend to make them the most efficient producers in a given regional market. However, since these natural monopolies face no competition, they have the ability, if left unchecked, to charge prices that could be considerably above costs. Industries with large infrastructure requirements, such as telecommunications, water and wastewater, natural gas, and electric power, have historically been considered natural monopolies.

Many industrialized nations grapple with the unchecked power that infrastructure industries can have in any given market. In these instances, government has two public policy options. First, government can expropriate, or nationalize, these industries. Here the government takes over power generation ownership and operates the industry in the public interest by providing service at a reasonable (government determined) price.

Under the second policy option, the government can maintain private ownership and regulate firms operating in the public trust. This has been the unique policy option exercised within the United States for a greater part of the past century. Despite some municipal and federal government ownership of electricity generation facilities, much of electric generation capacity is investor-owned, privately controlled electric companies. Figure 3 shows the electricity generation capacity ownership percentages by type of entity.

Beginning in the 1920s, an extensive set of electric power industry regulation arose based upon the notion that this industry, like others, is a natural monopoly. An additional rationale for power industry regulation has been that electricity, like so many other regulated utility industries, is imbued with the public interest. Perhaps one of the greatest influences on power generation over the past half century has been the role of government and its public policies.

Since electric power moves within and between states, this regulation has roots in both federal and state jurisdictions. Federal intervention in electric power markets has its origins in the Federal Power Act, the Public Utility Holding Company Act, and the Rural Electrification Act. State regulation has evolved from state statutes, constitutions, and other legal precedents. At the federal level, electric power sales are regulated by the Federal Energy Regulatory Commission (FERC), while at the state level, regulation is directed by state Public Utility Commissions (PUCs).

Regulatory bodies at both the federal and state level attempt to ensure that electric power is provided economically, and in a safe and reliable manner. The primary method of electricity regulation has been rate of return, or cost-based, regulation. Here regulators set the rates utilities are allowed to charge their customers. This cost-based regulation allows

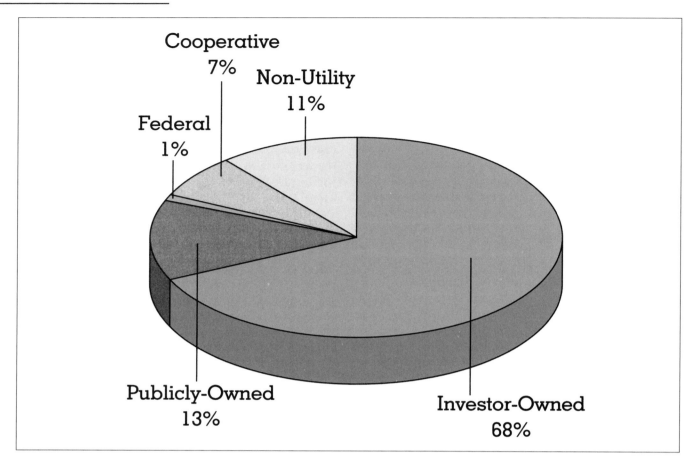

Figure 3.

Ownership Percentages of Total U.S. Generating Capacity

SOURCE: Energy Information Agency. (1995). *Electric Power Annual*. Washington, DC: U.S. Department of Energy.

utilities to recover their prudently incurred costs and also earn a reasonable rate of return on their investments. In return, utilities are granted exclusive franchises and monopoly service privileges

In the period from the 1930s to the early 1970s, regulation in the power industry was relatively uneventful. The energy crisis of the 1970s and early 1980s, however, dramatically changed the regulatory environment for electric power generation. During this period, state and federal government became exceptionally proactive in both generation planning and operation.

At the federal level, Congress passed the National Energy Act of 1978, which was composed of five different statutes: (1) the Public Utilities Regulatory Policy Act (PURPA), (2) the National Energy Tax Act, (3) the National Energy Conservation Policy Act, (4) the Power Plant and Industrial Fuels Act (PPIFA), and (5) the Natural Gas Policy Act. The general purpose of the National Energy Act was to ensure sustained economic growth during a period in which the availability and price of future energy resources was becoming increasingly uncertain. The two major themes of the legislation were to: (1) promote the use of conservation and renewable/alternative energy and (2) reduce the country's dependence on foreign oil.

While all aspects of the National Energy Act affected the electric power industry, PURPA was probably the most significant. PURPA was designed to encourage more efficient use of energy through industrial cogeneration. These cogenerators produce electricity through the capture of waste steam from their production processes. As a result of

PURPA, a whole new class of electricity generation facilities emerged, commonly referred to as "qualifying facilities" or QFs. PURPA required utilities to interconnect and purchase power from any QF at a rate not to exceed the utility's avoided cost of generation. This legislation opened the door to competition in the generation portion of the industry by legitimizing and creating a market for non-utility generation.

State regulation during this period continued the trend of promoting nonutility sources of generation. While PURPA was passed by the federal government, it was the responsibility of state regulators to set the rates at which utilities were required to purchase cogenerated or QF power. In response to the energy crisis of the period, state regulators began to set overly generous rates for QF power to stimulate conservation and alternative sources of electricity in a period of uncertainty. As a result of the generous rates and guaranteed market for nonutility generated power, generating capacity from nonutilities increased from close to 2 percent in 1978 to over 8 percent a decade later.

Regulators also began to require utilities to subject themselves to competitive bidding when they had a need for additional capacity. In addition, regulators began to require utilities to investigate other alternatives to the construction of new generation facilities, including the evaluation of demand-side management, or energy conservation measures, as a means of meeting future load growth. As a result of both of these policies, the fundamental premise of utility regulation and generation planning came under fire as more and more cost-effective, reliable, and alternative means of meeting electricity needs began to emerge.

ELECTRICITY GENERATION AND POWER INDUSTRY RESTRUCTURING

The structural and institutional environment for electric generation began to change dramatically in the late 1980s and throughout the 1990s as more and more competitive providers of electricity began to emerge. By the early 1990s, policy makers were actively discussing the possibility of restructuring the industry by introducing competition into the generation portion of the business. Throughout the 1990s, the terms "restructuring," "deregulation," and "competition" became virtually synonymous.

The passage of the Energy Policy Act of 1992 (EPAct) is considered the watershed federal legislation opening the door to complete power generation competition. This legislation allowed the Federal Energy Regulatory Commission (FERC) to order utilities to "wheel" or transport power over their transmission lines on behalf of third parties on an open access and nondiscriminatory basis. In subsequent years, FERC passed Order 888 and Order 889, which established the rules and institutions under which interstate or wholesale competition would be allowed. This wholesale competition was restricted to customers that were bulk power customers buying on behalf of other customers such as municipal utilities, rural cooperatives, and other IOUs. Retail competition, that is, competition for residential, commercial, and industrial customers, soon followed.

The origin of retail competition has run almost parallel to wholesale restructuring initiatives. State restructuring initiatives began initially in California and were soon adopted in New England. Both regions of the country were suffering from exceptionally high retail rates that were, in some cases, double the national average. Ratepayers, typically industrial ratepayers, appealed to regulators to allow competitive forces, rather than continued regulation, to discipline electric power generation and power markets.

The advent of competition has virtually transformed the industry in every aspect, including its name. In the not too recent past, the industry was referred to as the "electric utility industry." Today, given its significantly wide and numerous participants, it is more appropriate to refer to the industry as the "electric power industry." This new power industry has new power generation and sales participants with names such as qualifying facilities, exempt wholesale generators, merchant facilities, small power production facilities, power marketers, and sales aggregators.

The Mechanics of the Restructuring Process

Restructuring is the process of completely reorganizing the electric power industry. The generation portion of the industry will become more competitive, while the transmission and distribution portion of the industry will remain under regulation. While many specific aspects of restructuring differ between different states, and between federal and state jurisdictions, there are three common transition procedures.

The first transition procedure requires vertically integrated, former electric utility companies to unbundle, or separate, their electric power generation and energy sales operations from other utility operations. This separation is required to prevent former utilities from using their monopoly transmission and distribution assets in an anticompetitive manner to benefit their generation and sales operations. This divestiture, or separation, can be either physical or functional. Under physical divestiture, deregulated utilities are required to sell either all or a portion of their generating assets; the assets are "physically" removed from the former utility's control. Under functional divestiture, utilities are simply required to establish separate corporate affiliates, with stringent rules of conduct between regulated and unregulated companies. Most states opt for functional divestiture.

The second transition procedure establishes independence for the transmission system. This procedure is also required to ensure that a monopoly asset, in this case transmission, is not used in an anticompetitive manner. Two institutional structures are currently being debated for this transmission independence: an Independent System Operator (ISO), or a Transmission Company (Transco). The ISO transmission governance structure is typically associated with a multiutility, nonprofit association. Transcos are typically associated with a single-utility, for-profit, governing board. At the end of the 1990s, the ISO was the more prevalent of the two governance structures, with the Transco proposals gaining in number and popularity.

The third transition procedure defines the rules under which competitive suppliers of electricity can compete for end users. There are two polar models that are often debated for power market organization: the direct access (or bilateral contracts) regime, and the Poolco regime. Under direct access, consumers enter into direct contracts with competitive suppliers of electricity, and competitive providers of electricity enter into contracts with, and pay an access fee to, the local (regulated) distribution company for the use of local power lines.

A Poolco regime is a centralized market structure consisting of an ISO and a competitive Power Exchange (PX), where the ISO handles the physical deliveries and coordination of power flows within a regional power system, and the PX handles all the transactional issues associated with system power sales. In the Poolco regime, regional power market competitors submit bid prices and capacity offers into the competitive PX. Load from local distribution companies, representing all electricity end users, are then aggregated by the Poolco. Hour-ahead bid prices are used to construct a least cost dispatched and an hourly supply curve, and an hourly market equilibrium price is determined at the point at which the PX-determined supply curve intersects total regional aggregated demand. Least cost dispatch information is then transmitted from the PX to the ISO that controls all system coordination and security issues.

Many states have debated the efficiency and equity of both the direct access and Poolco market structures. Like many other restructuring transition issues, final policy decisions tend to be some hybrid or amalgamation of both approaches. Alternatively, some states have moved forward with a more centralized process (i.e., Poolco), with gradual implementation of more disaggregated trading regimes (i.e., direct access) at a later date. Since restructuring rules and laws are promulgated at the state level, it is very likely that market structures will be evolving and moving targets well into the early part of the twenty-first century.

THE SHIFT IN TECHNOLOGICAL PARADIGMS

One of the most dynamic factors underlying changes in the power industry has been technology. From the early days of the industry, designing, constructing, and operating more efficient generating units has been a priority. For a good part of the early to mid-twentieth century, the electric power industry, like other major capital intensive manufacturing industries, was one of the leading sectors of the economy in terms of technical innovation and productivity growth.

The amount of heat input, measured in British thermal units (Btu's), needed to generate a kilowatt hour of electricity with steam turbines decreased by almost 40 percent between 1925 and 1945, and by 35 percent during the period 1945 to 1965. During this period, scale became an important factor in power generation planning. Bigger was clearly better, and remained the premier planning paradigm for the utility sector of the industry until the mid to late 1980s. Larger plants usually entailed larger thermal efficiencies, which in turn reduced costs. However, gains in thermal efficiencies tapered throughout the 1970s. As the gains disappeared, so too did the ability to offset the exogenous economic changes in costs that occurred during the energy crisis. The only nonfossil technology of promise during this period, nuclear power, fizzled under the pressure of cost acceleration and inflation, rapidly increasing safety regulations, imprudent management, and regulatory and financial uncertainty. The accident at Three Mile Island in 1979 all but assured the industry that it had run out of large-scale technological innovations in power generation.

However, out of the ashes of the technological failures of the 1970s and early 1980s came a new technological innovation that dramatically changed the nature of the power industry. The experiences of the decade showed that the industry needed a technology that was flexible, modular, could be constructed quickly, and had minimal environmental impacts. Advances in the aerospace industry made it possible to deliver combustion turbine technologies that met the requirements of a new power generation environment.

Ironically, throughout the early 1980s, it was the nonutility generation portion of the industry that began to aggressively adopt the new efficient combustion turbine and combined cycle applications of the new natural-gas fired technologies. Widespread nonutility deployment of these technologies was the direct result of PURPA and the guaranteed market for nonutility generated power. Combined cycle plants, in particular, were rapidly preferred technologies for onsite generation at large nonutility generation facilities throughout the United States. The rapid deployment of these small, modular, and highly efficient facilities was an underlying technological rationale for introducing competition into generation markets.

Combined-cycle plants were in many ways an extension to the idea of cogeneration. These plants were effectively natural gas-fired combustion turbines with additional waste heat recycling unit—thus, a combined cycle of electric generation. The first stage generates gas-fired electricity from a turbine, while the second stage captures the waste heat to run a second-stage electric generator. Clearly, market participants with small scale power generation construction experience, like industrial cogener-

ators, can develop and operate projects of this nature. With these technologies, utilities need not be the only party participating in power generation construction and operation.

The popularity of combined-cycle units has increased dramatically over the past several years in both the utility and nonutility generation of electricity. As shown in Figure 4, in the year 2000, 3 percent of total generating capacity consisted of combined cycle technology, while traditional steam generating capacity comprised 69 percent of total. By the year 2020, however, these percentages shift dramatically in favor of combined cycle technologies with over 20 percent of total generating capacity invested in this technology.

The widespread adoption of these combined cycle and combustion turbine units represents a technology paradigm shift from large central station generation to more modular, flexible generating units. Under the new planning paradigm, size is less important than flexibility, fuel availability (natural gas), and location to load center. While some scale is still presumed to have benefits under this new paradigm, it is not the foremost consideration that it was a decade before.

The newest paradigm in power industry is known as distributed generation (DG) or, more generally, distributed energy resources (DER). Here, small scale power generation and storage equipment is located at the distribution—not transmission—level of interconnection. DER/DG includes such technologies as reciprocating engines, micro-turbines, fuel cells, and small solar photovoltaic (PV) arrays. While many of these technologies are relatively expensive now, future deployment, as well as changes in competitive generation market conditions, can make a number of applications cost-effective. DER could usher in a new level of competition much like its predecessor, the combined cycle technology, did a decade earlier.

CONCLUSIONS

At the beginning of the 1990s, the power industry was considered an old and tired industry in the United States and global economy. However, changes stimulated by the forces of new technology, environmental consciousness, and public policies promoting competition have brought about a renaissance in the power generation portion of the electric power industry. Like other large-scale manufacturing industries, the power

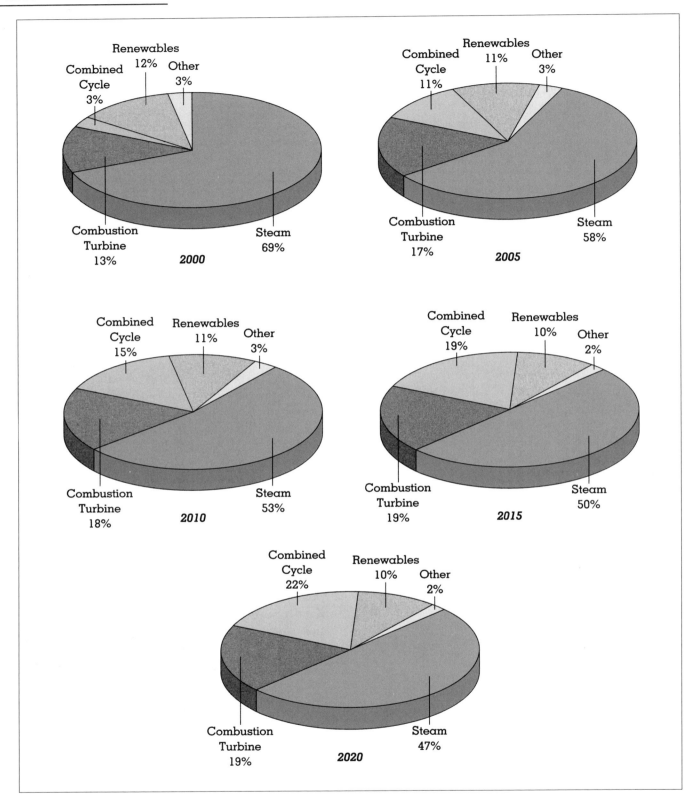

Figure 4.

Projected Changes in Total Capacity Percentages by Prime Mover

SOURCE: Energy Information Agency. (1995). *Electric Power Annual.* Washington, DC: U.S. Department of Energy.

industry has restructured and retooled, taking advantage of informational, technological, and managerial innovations. Competition, choice, and changes in the way power is generated, delivered, and sold to end users should continue this trend well into the next century.

David E. Dismukes

See also: Cogeneration; Demand-Side Management; Engines; Hydroelectric Energy; Market Transformation; Supply and Demand and Energy Prices; Turbines, Gas; Turbines, Steam.

BIBLIOGRAPHY

Energy Information Administration. (1993). *The Changing Structure of the Electric Power Industry*, 1970–1991. Washington, DC: U.S. Department of Energy.

Utility Data Institute. (1994). *Electric Utility Power Plant Construction Costs*. Washington: DC: Utility Data Institute.

ELECTRIC POWER, SYSTEM PROTECTION, CONTROL, AND MONITORING OF

Protection is the branch of electric power engineering concerned with the principles of design and operation of equipment (called "relays" or "protective relays") which detect abnormal power system conditions and initiate corrective action as quickly as possible in order to return the power system to its normal state. The quickness of response is an essential element of protective relaying systems—response times of the order of a few milliseconds are often required. Consequently, human intervention in the protection of system operation is not possible. The response must be automatic, quick, and should cause a minimum amount of disruption to the power system.

THE NATURE OF PROTECTION

In general, relays do not prevent damage to equipment; they operate after some detectable damage has already occurred. Their purpose is to limit, to the extent possible, further damage to equipment, to minimize danger to people, to reduce stress on other equipment, and above all, to remove the faulted equipment from the power system as quickly as possible so the integrity and stability of the remaining system is maintained. There is a control aspect inherent in relaying systems which complements the detection of faults and helps return the power system to an acceptable configuration as soon as possible so that service to customers can be restored. There is also a vital need to constantly monitor the power and the protective systems to analyze operations for correct performance and to rectify errors in design, application, or settings.

Reliability, Dependability, and Security

Reliability is generally understood to measure the degree of certainty that a piece of equipment will perform as intended. Relays, in contrast with most other equipment, have two alternative ways in which they can be unreliable. They may fail to operate when they are expected to, or they may operate when they are not expected to. This leads to the two-pronged definition of "dependability," the measure of certainty that the relays will operate correctly for all faults for which they are designed to operate and "security," the measure of certainty that the relays will not operate incorrectly for any fault.

Zones of Protection

Relays have inputs from several current transformers (CTs) and the zone of protection is bounded by these CTs. While the CTs provide the ability to detect a fault inside the zone, circuit breakers (CBs) provide the ability to isolate the fault by disconnecting all of the power equipment within the zone. Thus, a zone boundary is usually defined by a CT and a CB. When the CT is part of the CB it becomes a natural zone boundary. When the CT is not an integral part of the CB, special attention must be paid to the fault detection and fault interruption logic. The CT still defines the zone of protection, but communication channels must be used to implement the tripping function. Figure 1 shows the zones of protection in a typical system.

Relay Speed

It is, of course, desirable to remove a fault from the power system as quickly as possible. However, the relay must make its decision based upon voltage

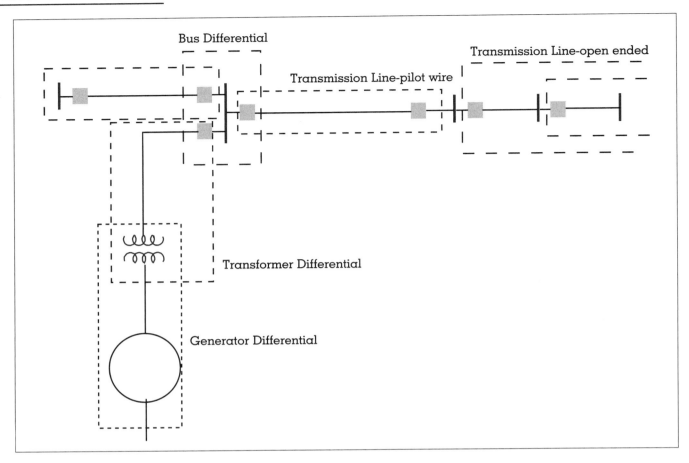

Figure 1.
Zones of protection.

and current waveforms which are severely distorted due to transient phenomena which follow the occurrence of a fault. The relay must separate the meaningful and significant information contained in these waveforms upon which a secure relaying decision must be based. These considerations demand that the relay take a certain amount of time to arrive at a decision with the necessary degree of certainty. The relationship between the relay response time and its degree of certainty is an inverse one and is one of the most basic properties of all protection systems.

Although the operating time of relays often varies between wide limits, relays are generally classified by their speed of operation as follows:

Instantaneous—These relays operate as soon as a secure decision is made. No intentional time delay is introduced to slow down the relay response.

Time-delay—An intentional time delay is inserted between the relay decision time and the initiation of the trip action.

High-speed—A relay that operates in less than a specified time. The specified time in present practice is 50 milliseconds (3 cycles on a 60 Hz system)

Ultra high-speed—This term is not included in the present relay standards but is commonly considered to be operation in 4 milliseconds or less.

Primary and Backup Protection

The main protection system for a given zone of protection is called the primary protection system. It operates in the fastest time possible and removes the least amount of equipment from service. On extra-

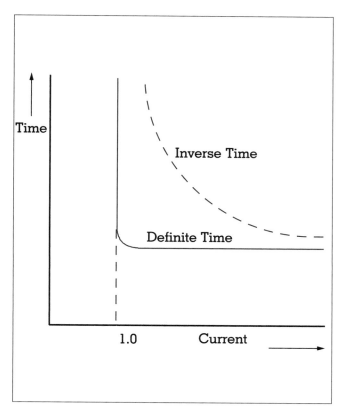

Figure 2.
Level detector relay.

high-voltage systems (230 kV and above) it is common to use duplicate primary protection systems in case any element in one primary protection chain fails to operate. This duplication is therefore intended to cover the failure of the relays themselves. One may use relays from a different manufacturer, or relays based on a different principle of operation, to avoid common-mode failures. The operating time and the tripping logic of both the primary and its duplicate system are the same.

It is not always practical to duplicate every element of the protection chain. Particularly on lower voltage systems, backup relaying is used. Backup relays are slower than the primary relays and, generally, remove more system elements than may be necessary to clear a fault. They may be installed locally, that is, in the same substation as the primary relays, or remotely.

RELAY OPERATING PRINCIPLES

In general, as faults (short circuits) occur, currents are increased and voltages decrease. Besides these magnitude changes, other changes may occur. Relay operating principles are based upon detecting these changes.

Level Detection

This is the simplest of all relay operating principles. Any current above, or voltage below, a set level may be taken to mean that a fault or some other abnormal condition exists inside the zone of protection. Figure 2 shows a definite time and an inverse time overcurrent relay.

Magnitude Comparison

This operating principle is based upon the comparison of one or more operating quantities. The relay will operate when the phasor division between the two or more circuits differs beyond the normal operating parameters. In Figure 3, I_A and I_B may be equal or at a fixed ratio to each other.

Differential Comparison

This is one of the most sensitive and effective methods of providing protection against faults and is shown in Figure 4. The algebraic sum of all currents entering and leaving the protected zone will be close to zero if no fault exists within the zone and will be the sum of I_1 and I_2 if a fault exists within the zone. A level detector can be used to detect the magnitude of this comparison or a special relay such as a percentage differential or harmonic restrained relay is applicable. This is the most common protective device used for generators, motors, buses, reactors, capacitors, etc. Its only drawback is that it requires currents from the extremities of a zone of protection which may require excessive cable lengths or a communication system.

Phase Angle Comparison

This type of relay compares the relative phase angle between two alternating-current quantities. It is commonly used to determine the direction of a current with respect to a reference quantity. Normal power flow in a given direction will result in the phase angle between the voltage and the current varying around the power factor angle (e.g., 30°) while power in the reverse direction will differ by 180°. Under fault conditions, since the impedance is primarily the inductance of the line, the phase angle of the current with respect to the voltage will be close to 90°.

Distance Measurement

This type of relay compares the local current with the local voltage. This is, in effect, a measurement of

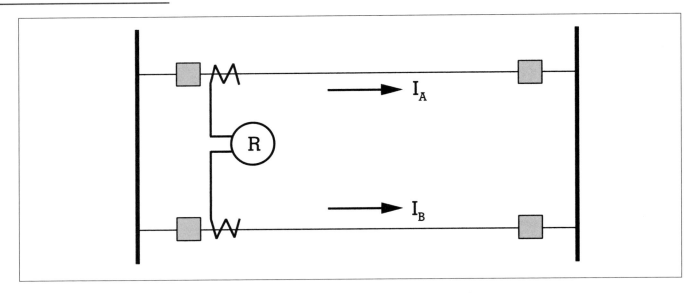

Figure 3.
Magnitude comparison relaying for two parallel transmission lines.

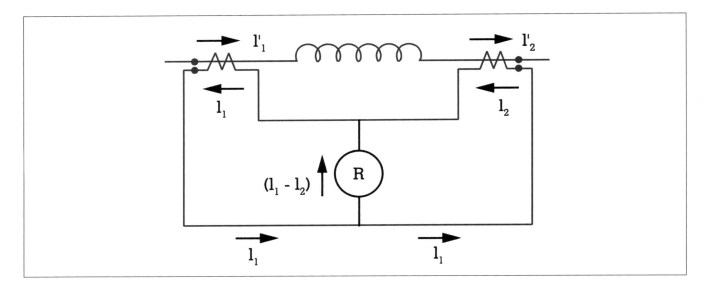

Figure 4.
Differential comparison principle applied to a generator winding.

the impedance as seen by the relay. An impedance relay depends on the fact that the length of the line (i.e., its distance) for a given conductor diameter and spacing determines its impedance. This is the most commonly used relay for the protection of high voltage transmission lines. As shown in Figure 5, zones can be identified as "zone one" which provides instantaneous protection to less than 100 percent of the associated line segment, and zones two and three which cover more than the line involved but must be delayed to provide coordination.

Harmonic Content

Currents and voltages in a power system usually have a sinusoidal waveform of the fundamental power system frequency plus other normal harmon-

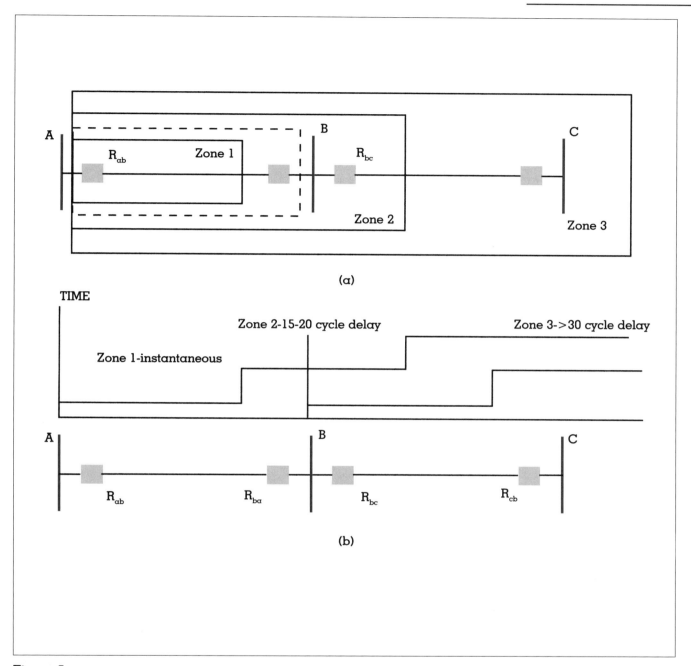

Figure 5.
Three-zone step distance relaying to protect 100 percent of a line and back up the neighboring line.

ics (e.g., the third harmonic produced by generators). Abnormal or fault conditions can be detected by sensing any abnormal harmonics that accompany such conditions.

Frequency Sensing

Normal power system operation is at 50 or 60 Hz depending upon the country. Any deviation from

these values indicates that a problem exists or is imminent.

RELAY DESIGN

The following discussion covers a very small sample of the possible designs. Specific details must be obtained from the manufacturers.

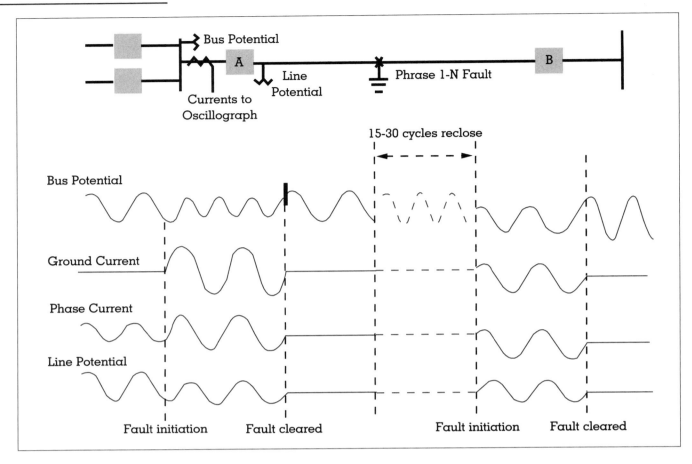

Figure 6.
Single phase-to-ground fault with unsuccessful high-speed reclose.

Fuse

The fuse is a level detector and is both the sensor and the interrupting device. It is installed in series with the equipment being protected, and it operates by melting a fusible element in response to the current flow.

Electromechanical Relays

The actuating forces are created by a combination of input signals, stored energy in springs, and dashpots. The plunger type relay consists of a moving plunger inside a stationary electromagnet. It is typically applied as an instantaneous level detector. The induction-type relay is similar to the operation of a single-phase ac motor in that it requires the interaction of two fluxes across a disc or cup. The fluxes can be produced by two separate inputs or by one input electrically separated into two components.

Depending on the treatment of the inputs (i.e. one current separated into two fluxes, two currents, or a current and a voltage), this design can be used for a time-delay overcurrent relay, a directional relay, or a distance relay.

Solid-State Relays

All of the functions and characteristics of electromechanical relays can be performed by solid-state devices, either as discrete components or as integrated circuits. They use low-power components, either analog circuits for fault-sensing or measuring circuits as a digital logic circuit for operation. There are performance and economic advantages associated with the flexibility and reduced size of solid-state devices. Their settings are more repeatable and hold closer tolerances. Their characteristics can be shaped by adjusting logic elements as opposed to the fixed characteristics of induction discs or cups.

Computer Relays

The observation has often been made that a relay is an analog computer. It accepts inputs, processes them, electromechanically or electronically, to develop a torque or a logic output resulting in a contact closure or output signal. With the advent of rugged, high performance microprocessors, it is obvious that a digital computer can perform the same function. Since the usual inputs consist of power system voltages and currents, it is necessary to obtain a digital representation of these parameters. This is done by sampling the analog signals and using an appropriate computer algorithm to create suitable digital representations of the signals.

PROTECTION SCHEMES

Individual types of electrical apparatus, of course, require protective schemes that are specifically applicable to the problem at hand. There are, however, common detection principles, relaying designs and devices that apply to all.

Transmission Line Protection

Transmission lines utilize the widest variety of schemes and equipment. In ascending order of cost and complexity they are fuses, instantaneous overcurrent relays, time delay overcurrent relays, directional overcurrent relays, distance relays, and pilot protection. Fuses are used primarily on distribution systems. Instantaneous overcurrent relays provide a first zone protection on low-voltage systems. Time delay overcurrent relays provide a backup protection on low-voltage systems. Directional overcurrent relays are required in loop systems where fault current can flow in either direction. Distance relays provide a blocking and tripping function for pilot relaying and first, second, and third zone backup protection on high-voltage and extra-high-voltage systems. Pilot protection provides primary protection for 100 percent of the line segment by transmitting information at each terminal to all other terminals. It requires a communication channel such as power line carrier, fiber optics, microwave, or wire pilot.

Rotating Apparatus

The dominant protection scheme for generators and motors is the differential relay. Access to all entry points of the protected zone is usually readily available, no coordination with the protection of other connected apparatus is required, and the faulted zone is quickly identified. Motor protection also includes instantaneous and time delay overcurrent relays for backup.

Substation Equipment

Differential relaying is the universal bus and transformer protection scheme. The inrush current associated with power transformers requires a special differential relay utilizing filters to provide harmonic restraint to differentiate between energizing current and fault current.

Instantaneous and time delay overcurrent relays are the most common protective devices used on shunt reactors, capacitors and station service equipment.

CONTROL

Transmission line faults are predominantly temporary, and automatic reclosing is a necessary complement to the protective relaying function. The reclose time must be greater than the time required to dissipate the arc products associated with the fault. This varies with the system voltage and ranges from 15–20 cycles at 138 kV to 30 cycles for the 800 kV systems. Automatic reclosing requires that proper safety and operating interlocks are provided.

Rotating equipment, transformers, and cables do not, in general, have temporary faults, and automatic reclosing is not provided.

MONITORING

The importance of monitoring the performance of power system and equipment has steadily increased over the years.

Oscillographs and other fault recorders such as sequence of events are, by nature, automatic devices. The time frame involved in recognizing and recording system parameters during a fault precludes any operator intervention. The most common initiating values are currents and voltages associated with the fault itself. Phase currents increase, phase voltages decrease, and there is normally very little ground current, so all of these are natural candidates for trigger mechanisms. There are transient components superimposed on the 60 Hz waveform that accompany faults and other switching events. They are revealed in the oscillographic records and are an essential element in analyzing performance. Figure 6 is a typical record of a single phase to ground fault and unsuccessful high-speed reclose.

With the advent of digital relays, the situation changed dramatically. Not only could the relays record the fault current and voltage and calculate the fault location, they could also report this information to a central location for analysis. Some digital devices are used exclusively as fault recorders.

Stanley H. Horowitz

See also: Electric Power, Generation of; Electric Power, System Reliability and; Electric Power Transmission and Distribution Systems.

BIBLIOGRAPHY

Blackburn, J. L. (1952) *Ground Relay Polarization.* AIEE Trans., Part III, PAS, Vol. 71, December, pp. 1088–1093.

Horowitz, S. H., and Phadke, A. G. (1996). *Power System Relaying.* New York: John Wiley & Sons, Inc.

IEEE Power Engineering Society. (1980). *Protective Relaying for Power Systems*, ed. Stanley H. Horowitz. New York: IEEE Press.

IEEE Power Engineering Society. (1992). *Protective Relaying for Power Systems II*, ed. Stanley H. Horowitz. New York: IEEE Press.

IEEE Power System Relaying Comm. (1979). *Protection Aspects of Multi-Terminal Lines.* IEEE Special Publication No. 79 TH0056-2-PWR. New York: IEEE Press.

Lewis, W. A., and Tippett, L. S. (1947). *Fundamental Basis for Distance Relaying on 3-Phase Systems.* AIEE Trans., Vol. 66, pp. 694–708.

Mason, C. R. (1956). *The Art and Science of Protective Relaying.* New York: John Wiley & Sons.

Westinghouse Electric Corp., Relay Instrument Division. (1979). *Applied Protective Relaying.* Coral Springs, FL: Author.

ELECTRIC POWER, SYSTEM RELIABILITY AND

An electric power system involves the production and transportation of electrical energy from generating facilities to energy-consuming customers. This is accomplished through a complex network of transmission lines, switching and transformer stations, and other associated facilities. The primary objective in the design of the delivery system is the creation and operation of a safe, reliable, and economic link between the generating supply and the customer load points. Reliability has always been a primary concern in the design and operation of electric power systems. In most of the developed countries throughout the world, a very high level of reliability has been the norm for many years. This is particularly true in North America, where customers have come to expect electricity supply to be available virtually on demand.

The Canadian Electricity Association (CEA) has collected Canadian service continuity statistics for many years and publishes these data on an overall basis. Canadian electric utility reliability performance statistics for 1997 are shown in Table 1.

The SAIFI, SAIDI, and IOR indices are defined in Table 1 as follows:

- SAIFI is the average number of interruptions per customer served per year: SAIFI = (Total Customer – Interruptions) ÷ Total Customers Served.
- SAIDI is the system average interruption duration for customers served per year: SAIDI = (Total Customer – Hours of Interruption) ÷ Total Customers Served.
- IOR is the per unit of annual customer-hours that service is available: IOR = (8760 hours/year – SAIDI) ÷ 8760 hours/year

These data show that in 1997 Canadian electric power customers experienced an average of 2.35 power failures and on average were without electricity for 3.70 hours. Their demands were satisfied 99.9578 percent of the time. The SAIFI, SAIDI, and IOR values shown are averages and vary widely across the country and for different segments of any given area. In general, urban customers experienced much higher levels of reliability than did rural customers. The CEA statistics also show that approximately 80 percent of the interruptions seen by the average customer occurred due to failures within the low-voltage distribution system, where the supply configuration is basically radial in nature. The high-voltage network, which links the generating facilities with the major load points, is known as the bulk supply network and accounts for approximately 20 percent of the power failures experienced by an average customer. Distribution system failures are usually local in nature and tend to affect a relatively small

System Average Interruption Frequency (SAIFI)	= 2.35 int/system customer
System Average Interruption Duration (SAIDI)	= 3.70 hours/system customer
Index of Reliability (IOR)	= 99.9578%

Table 1.
1997 Canadian Electric Utility Reliability Performance
SOURCE: Canadian Electricity Association

number of customers, while failures in the bulk system can affect many customers. This is not always the case, as extreme weather conditions such as the 1998 ice storms in Ontario, Quebec, and the eastern United States affected both the bulk power and the distribution systems in these regions. The ability to generate sufficient electrical energy has traditionally not been a major concern in North America for some time, as vertically integrated utilities have planned and constructed their generating facilities to stringent reliability criteria. Regulatory authorities have viewed the establishment of adequate generation facilities as a mandatory requirement of reliable supply. This may change in the near future as the vertically integrated utility structure is discarded and replaced by independent, market-driven generation, transmission, and distribution companies. A major requirement in the new unbundled electric power system structure is open access to the bulk delivery system. This will permit competition among generating companies as well as customer choice in regard to supply. The overall integrity of the bulk delivery system is the vital link in this structure in order to retain the high level of reliability that customers have become accustomed to.

THE NORTH AMERICAN ELECTRIC RELIABILITY COUNCIL (NERC)

The bulk system has long been seen as a vital element in the maintenance of an economic and reliable supply of electrical energy in North America, and a high degree of redundancy has been incorporated to achieve this objective. The North American Electric Reliability Council (NERC) was formed in 1968, following the November 9–10, 1965, blackout that affected the northeastern United States and Ontario,

Canada, in response to extreme concerns regarding the reliability of the rapidly developing interconnected power networks in North America. NERC is a not-for-profit corporation owned by ten regional councils, whose members come from virtually all segments of the electric power industry. Their parent companies account for virtually all the electrical energy supplied throughout the United States, Canada, and northern portions of Baja California, Mexico. The various regional councils are as follows:

- NPCC: Northeast Power Coordinating Council
- MAAC: Mid-Atlantic Area Council
- ECAR: East-Central Area Reliability Co-ordination Agreement
- MAIN: Mid-America Interconnected Network, Inc.
- MAPP: Mid-Continent Area Power Pool
- WSCC: Western Systems Co-ordinating Council
- SPP: Southwest Power Pool
- SERC: Southeastern Electric Reliability Council
- ERCOT: Electric Reliability Council of Texas
- FRCC: Florida Reliability Coordinating Council

The primary focus of NERC is on the bulk transmission systems spread throughout North America. The principal objective is to ensure that system disturbances is one area do not adversely affect other areas or regions. This is achieved through stringent and specific criteria; for example, a bulk power system should be able to withstand the loss of a large transmission component, a large generator, or a large area load. The reliability implications of customer supply in a specific or local area are not covered by NERC criteria unless that contingency event adversely affects a neighboring utility system. The 1965 blackout forcibly established recognition of the fact that reliable electric service is critical to the economic and social welfare of the millions of businesses and residents in the northeastern United States and eastern Canada. An immediate response to the blackout was the creation of the Northeast Power Coordinating Council (NPCC). The NPCC together with the nine other councils, form the present NERC:

- NPCC covers the northeastern United States and central and eastern Canada. NPCC's responsibility is to develop appropriate reliability criteria and guides, monitor the individual utility and participants' performance with these protocols, and thereby ensure that the individual bulk systems and therefore the overall bulk system has been planned and is operating reliably.

- MACC was established in 1967 and encompasses nearly 50,000 square miles from Virginia to New York. MAAC contains the PJM centrally dispatched electric control area, which is the largest such area in North America and the third largest in the world. PJM Interconnection became the first operational independent system operator in the United States on January 1, 1998.

- ECAR was established in 1967, and its current membership includes 29 major electricity suppliers in nine east-central states serving more than 36 million people. The current full members of ECAR are those utilities whose generation and transmission have an impact on the reliability of the overall region.

- MAIN regular and associate members include investor-owned utilities, cooperative systems, municipal power agencies, independent power producers, marketers, and two municipal systems that together serve 19 million people living in a 120,000-square-mile area.

- MAPP was formed in the mid-1960s to perform regional planning of transmission and generation and has a wide variety of members. The MAPP organization performs three functions. It is a reliability council under NERC, a regional transmission group, and a power and energy market.

- WSCC covers the largest geographic area of the ten regional councils. Its 1.8-million-square-mile service area covers more than half the conterminous area of the United States. WSCC was formed in 1967 by 40 electric power systems and in 1999 had more than 100 member organizations, which provide electric service to over 65 million people. The region is divided into four major areas that reflect varying, and sometimes extreme, geographic, and climatic conditions.

- SPP consists of 55 member entities and serves customers in 8 southwestern states with a population of more than 25 million. Initially formed in 1941, SPP has 17 control areas with over 40,000 MW of generation.

- SERC was created in 1970 by 22 electric systems. The initial NERC agreement in 1968 was signed by 12 regional and area organizations, which included from the Southeast, the CARVA Pool, the Tennessee Valley Authority, the Southern Company, and the Florida Electric Power Coordinating Group. Subsequent discussions indicated that the overall reliability of the southeastern bulk power system would be better served by the creation of a regional reliability council with broader membership. This lead to the formation of SERC.

- As noted earlier, FRCC became the 10th Reliability Region of NERC in 1996.

- ERCOT provides the overall electric system coordination in the state of Texas and it consists of six different market groups with more than 80 voting and nonvoting members.

- The Florida peninsula, previously part of SERC, became a separate NERC Region in 1996 and was designated as the FRCC. It currently has 34 members consisting of investor-owned utilities, cooperatives, municipals, power marketers, and independent power producers.

Overall coordination of the ten councils is the responsibility of NERC. The advent of deregulation, competition, and the unbundling of traditional vertically integrated utilities has necessitated considerable reevaluation of the role that NERC will play in the future. The ten regional councils and NERC are continually evolving in regard to their scope and mandates. NERC assembled the Electric Reliability Panel in 1997 to recommend the best ways to set, oversee, and implement policies and standards to ensure the continued reliability of North America's interconnected bulk electric systems in a competitive and restructured industry. The resulting report of the panel stated that the introduction of competition within the electric power industry and open access to transmission systems require creation of a new organization with the technical competence, unquestioned impartiality, authority, and respect of participants necessary to enforce reliability standards on bulk electric power systems. Since that time, NERC has been aggressively working to create a new structure, designated as the North American Electric Reliability Organization (NAERO), to achieve these objectives. The importance of the bulk delivery system will not diminish in the new utility environment and will if anything became even more important in the maintenance of reliable electric energy supply. The events that lead to the Northeast blackout of 1965 and the creation of NERC and its constituent councils are important reminders of the need to continually appraise the security and adequacy of the highly interconnected bulk power network in North America.

THE NORTHEAST BLACKOUT OF 1965

The massive power failure of November 9, 1965, affected approximately thirty million customers and initiated the most intensive examination of power system planning and operating practices in the history of the electric power industry. The overall blackout was the result of a series of cascading events initiated by a major power surge. The network connections to the south and the east were too weak to withstand this massive power surge, and the overall network became unstable and ceased to function as an integrated whole. Network islands, in which the load exceeded the generation, were created and then subsequently collapsed. Many of these individual utility systems did not have adequate black-start procedures and equipment; therefore, service restoration was delayed for significant periods of time.

The Federal Power Commission (FPC) report on the blackouts, published in December 1965, stated that the initial cause of the blackout was the operation of a backup relay on one of the five main transmission lines taking power to Toronto from hydro facilities on the Niagara River. Due to an improper setting, the relay disconnected one line. The other lines were then overloaded and consequently tripped out successively. Approximately 1,500 MW of power generated at these hydro facilities, which power had been flowing to Toronto, then attempted to take an alternate route through the one remaining U.S.-Canadian interconnection, at Massena, New York. When this subsequently tripped, the now approximately 1,700 MW previously going to Canada surged into the U.S. network and initiated the breakup of systems in the northeastern United States.

The FPC report details the three basic stages of the overall blackout event. The first was the initial system shock due to the massive power surge. There was a widespread separation of systems throughout New York and New England in only seconds. If this had been the only system reaction, the resulting blackout would have impacted about one-third of the actual customers affected and only those in the northern and northwestern areas of the affected region. The second stage was the subsequent collapse of those utilities in eastern New York and New England that were separated from their normal interconnected grids. These islands were basically left with insufficient generating capacity to meet the connected load, resulting in declining frequency and voltage levels and individual system collapse. The third stage of the overall blackout was the inability of many utilities to restore power due to inadequate or nonexistent black-start capability at the fossil-fired steam-generating stations. Difficulties were also encountered in reenergizing high-voltage underground oil-filled transmission cables.

The initiating and subsequent events associated with the Northeast blackout revealed many deficiencies in utility planning, operating, and maintenance practices. The criteria and procedures now embedded in the NERC Planning and Operating Standards stem from the many studies, investigations, and reports initiated because of the Northeast blackout. The electric power industry has been in continuous evolution since its inception some one hundred years ago. It is quite possible, however, that the changes occurring now are the most dramatic in its history. Electric energy supply in the United States, Canada, and many other parts of the world is moving from the traditional regulated industry of the past to a more open competitive, market-based environment. These changes are being driven by both market and legislative forces and could create significant obstacles to the maintenance of a high level of bulk system reliability. The traditional vertically integrated utility "obligation to supply" will not exist in the new electricity market, which will contain many new players who, at best, will have only an indirect responsibility for reliability. The transmission system of the future will be used in many different ways than was envisaged when it was planned and constructed. Firm transfers within and across systems coupled with both short-term and long-term economic transactions will affect emergency support from interconnected systems. These factors will reduce system flexibility and therefore tend to reduce system reliability. Open access will increase the uncertainty associated with the timing and availability of new generation resources. These concerns have already become evident with service disruptions in California, New York, and Alberta Traditional opposition to the construction of high-voltage transmission due to land use, aesthetics, and electromagnetic fields, and the need for judicial reviews will not diminish in the new electric utility environment and will place severe constraints on the use of existing transmission systems.

The reliability of a modern electric power system depends on continuous real-time control of power and energy production, transmission line flows, system frequency, and voltage. This complex task will get more involved in the new environment with increased market participation on both the supply and the demand sides.

CUSTOMER CONSIDERATIONS

Customers will have a choice of their energy provider in the new utility environment. It is unlikely, however, that most customers have any real appreciation for those aspects that will affect reliability of supply or how a highly integrated electric utility system operates. An electric transmission system is not a conventional transportation system in which a product is dispatched from a source to a receiving point. An electric transmission system is a total energy system in which energy generation and consumption must remain continuously and instantaneously in balance. A customer cannot buy a particular unit of energy from a given operating unit, as the electric power system is a total energy system. Power flows in the system are dictated by the laws of physics, and flows instantaneously and automatically respond to changes in network conditions by following the paths of least impedance. Third-party transactions superimposed on an existing transmission network will change voltage profiles and transmission losses and will affect individual load point reliability levels.

Virtually all residential, commercial, and small industrial customers are served through low-voltage transmission networks. Only a relatively small number of large industrial consumers are served directly from the high-voltage transmission network. As previously noted, Canadian data indicate that approximately 80 percent of all the electric power system outages that an average Canadian consumer experiences are due to failures in the low-voltage distribution system. These events will not be affected by unbundling traditionally vertical utilities into separate functional zones involving different companies. It is going to be extremely difficult to improve customer reliability by unbundling the three functional zones of generation, transmission, and distribution, but there is considerable potential for seriously lowering customer reliability.

CLIMATE CHANGE

Electric power transmission and distribution systems are extremely vulnerable to adverse weather conditions. It appears to many people that extreme weather is becoming more frequent and also more violent. It has been suggested that this is due to increased heat and humidity in Earth's environment, leading to increases in both rain and snowfall. The effect of

System Average Interruption Frequency (SAIFI)	= 3.68 int/system customer
System Average Interruption Duration (SAIDI)	= 31.35 hours/system customer
Index of Reliability (IOR)	= 99.6422%

Table 2.
1998 Canadian Electric Utility Reliability Performance
SOURCE: Ontario Hydro, May 1, 1998.

extremely adverse weather on both transmission and distribution facilities was clearly illustrated by "Ice Storm '98," which impacted eastern Ontario, southern Quebec, and parts of the northeastern United States. The weather condition was created by a combination of events that occurred over the North American continent. Moist, warm air from the Gulf of Mexico was pumped into southern Ontario and Quebec by a low-pressure system over the Texas panhandle. Coincident with this phenomenon, a large, stationary Arctic high-pressure area over Hudson Bay created a northeasterly circulation over central Quebec that moved very cold air into the Lawrence and Ottawa River valleys. The southerly warmer air current was unable to move the heavy, cold air and overrode the cold area at the contact surface, which resulted in considerable freezing rain. In the collision between the two air masses, the warm air was pushed upward and the cold air down. The snow that fell melted at the middle level but didn't have time to freeze again before hitting the ground. The ice that accumulated on the transmission and distribution facilities slowly increased as there were no periods of sunshine or thawing between the various periods of freezing rain.

The impact of "Ice Storm '98" was unprecedented in Canadian electric power system history. The Ontario Hydro Report provides a chronology of the twenty-three major storms that have impacted the system since 1942. The report also provides a detailed inventory of the effects, consequences, and mitigation measures taken due to the ice storm. The estimated direct total cost to Ontario due to the ice storm is $472 million (Canadian), which includes costs incurred by Ontario Hydro, local Ontario and federal governments, the Insurance Bureau of Canada and associated insurance companies, the Department of National Defence, and affected customers, including businesses, farms, and residents. The monetary impacts in Quebec are expected to be considerably higher. Frequent questions asked immediately after the ice storm were: Why not put the transmission and distribution facilities underground and avoid the consequences of freezing rain? Why not increase the redundancy built into the network? Why not build bigger and stronger transmission facilities? All of these are technically possible and would have different impacts on the reliability of electric power system supply. All would have considerable economic consequences. It was estimated that the cost to replace the transmission and distribution systems in eastern Ontario would be about $11 billion, resulting in electricity rate increases of more than 11 percent. The question of balancing reliability and economics is an ongoing requirement in electric power system planning, operations and decision-making.

The impact of "Ice Storm '98" on system reliability can be seen from the Canadian electric utility reliability performance statistics for 1998 in Table 2.

The impact of "Ice Storm '98" can clearly be seen by realizing that the number of customer interruptions and the customer hours of interruption for all of Canada in 1997 were 24,280,244 and 38,130,783, respectively. The ice storm alone resulted in 12,332,950 customer interruptions and 282,576,829 customer hours of interruption in the utilities affected. Removing the ice storm incidents from the 1998 Canada-wide data results in a SAIFI of 2.46 and a SAIDI of 3.40. The IOR is 99.9612 percent. "Ice Storm '98" had only a relatively moderate effect on SAIFI but a dominant effect on the customer hours of interruption and the SAIDI statistic due to the extremely long storm duration and the required restoration period. It is important to realize that the bulk transmission system retained its integrity according to NERC criteria, and the impact of the ice storm on Ontario Hydro facilities did not propagate into neighboring interconnected utilities.

CONCLUSION

Reliability of electric energy supply is an important requirement in modern society, and consumers in developed countries have grown to expect electricity to be available on demand. The electric utility industry in North America and throughout the rest of the developed world is undergoing considerable change as open transmission access and consumer choice are replacing traditional utility structures in which the obligation to meet customer requirements was a key component. One casualty in this move to the competitive marketplace may be the planning of adequate generating capacity to meet future load requirements. The bulk transmission system will become the focal point of power system operation and control in the new environment and the determining factor in retaining high levels of customer reliability and satisfaction. It is likely that judicial regulation will increase rather than decrease in the new environment, as regulatory authorities exercise increased vigilance to ensure that acceptable reliability levels are provided and competition is allowed to flourish. The utility industry has developed tremendously since the Northeast blackout of 1965 and has learned many lessons over the subsequent years. The requirement to exercise both flexibility and control in permitting open access and competition while maintaining an acceptable level of reliability will be the major challenge in the next decade.

Roy Billinton

See also: Climatic Effects; Consumption; Domestic Energy Use; Electric Motor Systems; Electric Power, Generation of; Electric Power, System Protection, Control, and Monitoring of; Electric Power Substations; Electric Power Transmission and Distribution Systems; Government and the Energy Marketplace; Regulation and Rates for Electricity.

BIBLIOGRAPHY

Canadian Electricity Association. (1997). *Service Continuity Report on Distribution System Performance in Canadian Electrical Utilities.* Montréal, Canada: Author.

"NERC Planning Standards." (1997). Princeton, NJ: North American Electric Reliability Council.

Ontario Hydro. (1998). "Ice Storm 98: Electricity Supply Impacts of the January 1998 Ice Storm in Eastern Ontario." Toronto: Author.

"Program Improvements in the CEA Service Continuity Report and Impact of the Ice Storm on CEA Data." (1999). CEA Electricity '99 Conference, Vancouver, B.C., March.

U.S. Federal Power Commission. (1965). *Northeast Power Failure, November 9 and 10, 1965.* Washington, DC: U.S. Government Printing Office.

U.S. Federal Power Commission. (1967). *Prevention of Power Failures: Volume I, Report of the Commission; Vol. II, Reliability of Electric Bulk Power Supply; Vol. III, Studies of the Task Groups on the Northeast Power Interruption.* Washington, DC: U.S. Government Printing Office.

ELECTRIC POWER MEASUREMENT

See: Units of Energy

ELECTRIC POWER SUBSTATIONS

An electric power substation is a facility that provides a junction between parts of the power grid. The substation's functions, critical for the proper operation of the power system, include the interconnection of power lines from different parts of the system; the monitoring and control of system operating conditions; and the protection of the power system equipment.

CLASSIFICATION AND GENERAL DESCRIPTION

Substations may be classified into one of several categories depending on their location and function within the system. Generator substations are located at the site of power generating stations and provide the connection to the transmission system. Bulk power substations link the transmission system to the subtransmission system, stepping the voltage down through a transformer (transformer substation), or linking high-voltage transmission lines from different parts of the system without changing the voltage (switching substation). A distribution substation provides the link between the subtransmission system and the much lower voltages of the distribution system. A converter station is a unique type of bulk power substation that provides a link between high-voltage alternating-current transmission lines and high-voltage direct-current transmission lines.

The siting of substations, electrical, geographic, economic, political, and aesthetic factors must be considered. The high voltages of the transmission system are utilized because the reduced currents result in more efficient power transmission. Therefore, substations are placed as close to the system loads as possible to minimize losses. This is con-

strained by the value and availability of real estate, as well as by the requirement that terrain be relatively level within the substation. Care is taken in substation placement, particularly in areas of dense population, that the location not obstruct scenic views or aesthetically depreciate commercial or residential developments. The physical size of substations can cover large areas because the high-voltage components are insulated from each other by air and thus must be separated by significant distances. Historically, these issues have limited the installation of large substations to areas of relatively sparse population. However, since the 1980s, substations have been insulated with pressurized sulfurhexafluoride gas (SF_6). Because of the highly insulating quality of SF_6, the size of these gas-insulated substations may be well under 25 percent of the size of an air-insulated substation with the same power-handling capability. In some applications, particularly those in proximity to population centers, the entire substation may be enclosed within buildings, reducing aesthetic concerns and deterioration by the environment. Nevertheless, air-insulated substations are still generally preferred because of the higher cost and environmental concerns regarding the release of SF_6 (which is being investigated as a greenhouse gas).

SYSTEM INTERCONNECTION

The primary function of substations is to provide an interconnection between transmission lines extending to other geographical areas and between parts of the system that may be operating at different voltages. A principal aspect of the substation design is the arrangement of connections through circuit breakers to common nodes called busses. Circuit breakers are large electrical switches that provide the ability to disconnect the transmission lines or transformers from the bus. Transformers provide a change in voltage.

Busses

Busses are typically made of aluminum or copper and are rigid bars in the substation, insulated from ground and other equipment through ample insulating material, typically air or sulfurhexafluoride. The arrangement of the busses in the substation may fall into a number of different categories; the most common are illustrated and explained in Table 1. The appropriate selection of configuration is made by carefully balancing cost, reliability, control, and space

Single Bus		• All connections are tied to a single bus, with one circuit breaker for each bus. This arrangement is favored for its simplicity and low cost, although it is least desirable with regards to reliability. Maintenance to substation equipment requires that connections be removed from service.	• This type of bus is usually the configuration of choice in substations at or below 130 kV.
Main and Transfer Bus		• As with the single bus arrangement, each connection is linked to the main bus through a circuit breaker, but the breaker may be bypassed using disconnect switches through a transfer bus and another breaker to the main bus. This permits isolation of the circuit breaker for maintenance without loss of service to the connection.	• Used in more critical applications at or below 130 kV, and occasionally at higher voltages.
Ring Bus		• This scheme has all circuit breakers linked in a closed loop, with connections entering at the junction between breakers. This way, any connection may be isolated or any single circuit breaker removed without interrupting the other connections. This provides a higher level of redundancy than the systems mentioned above. Control and protective relaying issues are somewhat more complicated for this arrangement.	• Usually found in substations above 130 kV, in smaller substations. Often installed with the expectation of future expansion to a breaker-and-a-half scheme.
Breaker-and-a-half Scheme		• This scheme has two equal busses, with three breakers connected between them. Each connection may be linked to one of the busses through one breaker, and in the event that one breaker is out of service or in need of maintenance, the connection may still be served through the two breakers to the other bus. The name of this arrangement comes from the fact that two connections are served by three breakers, so that there is an average of one and a half breakers per connection. This scheme is less complicated than the ring bus, with higher reliability, but is more costly.	• Most common on systems above 130 kV.
Double Bus		• A double bus, double breaker arrangement provides a link to each bus through an independent breaker for each connection. This provides full redundancy in case of malfunction, or the need to perform maintenance on a circuit breaker or bus, but is the most expensive configuration.	• Usually found in most critical transmission substations and in generator substations.
Bus	Line, Transformer, or Load	Disconnect Switch	Circuit Breaker

Table 1.
Most Common Arrangements of the Busses in a Substation

constraints. If the substation is providing service to critical loads, the need for high reliability may warrant the higher cost of a more complex bus arrangement, while for less critical loads, space constraints may dictate a minimal bus arrangement.

Disconnect Switches

For every piece of equipment in a substation, manual switches—called disconnect switches—are provided to enforce complete electrical isolation from equipment before any service is performed. Disconnect switches are placed in clearly visible locations so maintenance personnel can continuously confirm that the equipment is isolated. The disconnect switch cannot interrupt current, so it is opened only when the current has already been interrupted by an automatic switch such as a circuit breaker.

Circuit Breakers

Circuit breakers are switches that are operated by a signal, from a relay or from an operator. The circuit breaker is designed to interrupt the very large currents that may occur when the system experiences a fault, such as a lightning strike or arc to ground (e.g., a tree falling on a line, or a line falling to the ground). Because these extremely large currents can cause severe damage to equipment such as transformers or generators, and because these faults can disrupt the proper operation of the entire power system, the circuit breakers are designed to operate rapidly enough to prevent damage to equipment, often in 100 milliseconds or less.

The circuit breaker contacts consist of two pieces of metal that are able to move with respect to each other. When the circuit breaker is closed, the contacts are touching and current flows freely between them. When the circuit breaker opens, the two contacts are separated, typically by a high-strength spring or a pneumatic operator. As the contacts separate, current continues to flow through them, and the material between them is ionized, forming a conducting plasma. To provide isolation, the plasma must be eliminated and the contacts be separated a sufficient distance to prevent the reinitiation of an arc. Several different technologies are implemented to give four common types of circuit breakers.

Air blast circuit breakers are insulated by air, and the plasma is extinguished as a blast of compressed air is blown between the contacts. These are less common than the other types and generally are no longer applied in new installations because of size, and problems with the maintenance of the compressors. Oil-filled circuit breakers have the contacts enclosed within a sealed tank of highly refined oil, with oil ducts designed to force oil between the contacts to quench the arc when the contacts open. These are common, but decreasing in popularity due to the environmental concerns associated with the risk of an oil spill. Although breaker failures occur only rarely, hundreds of gallons of oil may be spilled in a single failure, requiring very costly remedial procedures. The more popular breakers for high-voltage systems are gas-filled breakers that have the contacts enclosed within a sealed tank of pressurized SF_6. These have proved highly reliable, although there have been some environmental concerns about the release of the SF_6 when maintaining the device or when the tank ruptures. For lower-voltage applications (less than 34 kV), vacuum breakers are often used. These eliminate arcing by enclosing the contacts within an evacuated chamber. Because there is no fluid to be ionized, there can be no plasma formed. Their major benefit is a very fast response time and elimination of environmental concerns.

In addition to circuit breakers, there are other classes of automatic switches that can be controlled or operated remotely, but with current-interrupting capability. These include circuit switchers, reclosers, and sectionalizers.

Transformers

Power transformers perform the very important function of linking parts of the power system that are at different voltages. They are found exclusively in substations, except in the distribution system, where they may be mounted on poles or pads close to the loads they are serving.

SYSTEM MONITORING AND PROTECTION

The substation provides a monitoring point for system operating parameters. The power system is a highly complex and sensitive conglomeration of parts that must all be coordinated to function properly. For this reason, the operating conditions must be very closely observed and controlled. This is done by using specialized sensors to acquire the information and then communication systems to convey the information to a central point. For immediate response to system faults (such as damaged conduc-

Pend Oreille Utility District workers at a control box in a power substation. (Corbis-Bettmann)

tors, arcs to ground, or other undesirable operating conditions), a system of protective relaying (consisting of sensors and automated switches) is used to operate circuit breakers.

Instrument Transformers

The high voltages and currents seen in a substation exceed the voltage and current ratings of monitoring equipment, so instrument transformers are used to convert them to lower values for monitoring purposes. Instrument transformers may be categorized as current transformers (CTs) or voltage transformers (VTs), which are also sometimes designated as potential transformers. CTs typically consist of a toroidal core of magnetic material wrapped with a relatively high number of turns of fine wire, with the current to be measured passing through the middle of the toroid. These devices are often located in the bushings of circuit breakers and transformers so as to be able to measure the current in those devices.

Bushings are the special insulated connections that allow the current to pass from the outside air into a sealed metal enclosure. VTs serve the function of stepping the voltage down to a measurable level. There is usually one connected to each of the substation busses. Most of the time VTs are constructed in essentially the same fashion as other transformers, although sometimes a capacitive coupling may enhance or replace the electromagnetics. Recent advances in technology have developed a new class of CTs and VTs that are optical devices that use specialized materials and advanced signal processing techniques to determine current based on the polarization of light as influenced by magnetic field strength, and voltage based on the polarization of light as influenced by electric field strength. While these devices are significantly more expensive than the traditional technologies, they provide higher accuracy and reliability and better electrical isolation.

Once the operating conditions have been measured,

the information is conveyed to a central location using a system known as SCADA (Supervisory Control and Data Acquisition). The SCADA system data are displayed in the regional dispatch center to assist operators to know what actions must be taken for the best operation of the system.

Protective Relaying

Instrument transformers provide inputs to the automatic protection system. To provide a quick response to faults, a group of devices called relays accept the voltage and current signals, determine when abnormal conditions exist, and open the circuit breakers in response to fault conditions. The protection system design opens only the circuit breakers closest to the problem so that all of the rest of the system may resume normal operation after the fault is isolated from the system. Historically, determining which breakers to open has been done using various electromechanical devices that had the necessary comparisons and delays built into their design. These include overcurrent relays, directional relays, distance relays, differential relays, undervoltage relays, and others. These electromechanical devices have proven rugged and reliable since the early 1900s. In the late 1950s a new class of relays, solid-state relays, using analog circuits and logic gates, provided basically the same performance, but without any moving parts and hence reduced maintenance requirements. With the advent of low-cost-high level microprocessors, a new generation of relays has been born in which a single microprocessor-based relay performs all of the functions of several different electromechanical or solid-state relays. The microprocessor provides the benefits of higher accuracy, improved sensitivity to faults, better selectivity, flexibility, ease of use and testing, and self-diagnostic capabilities. They can be integrated into the SCADA system to communicate the cause of breaker opening, and can be operated, reset, and updated through remote access. These advantages are why microprocessor-based relays are found in most new installations and are also being retrofitted into many existing substations.

In addition to protection against excessive currents, equipment must be protected against excessive voltages that commonly result from lightning strikes or switching transients. Because of the high speed of these surges, relays and circuit breakers are unable to respond in time. Instead, this type of protection is provided by surge arrestors, which are passive devices that prevent overvoltages without moving parts. An air gap was the earliest type of surge arrestor, in which a special set of contacts are set a distance apart specified by the maximum tolerable voltage. When the voltage exceeds that threshold an arc forms, essentially shorting out the overvoltage. The newer surge arrester technology is the metal-oxide varistor (MOV). This is a device that behaves like a very large resistor at voltages below the specified threshold, but at voltages above the threshold, the resistance of the device drops precipitously, effectively drawing enough current to limit the voltage, but without shorting it to ground.

SYSTEM VOLTAGE CONTROL

Another of the principal functions of a substation is to provide the means to control and regulate voltages and power flow. These functions are provided either by feedback from an automated system or by remote instruction from the dispatch center using an array of devices and systems within the substation.

A load tap changer, an integral part of a power transformer, is a special switch that adjusts the voltage ratio of the transformer up or down to keep the load side voltage at the desired level despite changing voltages on the source side. Capacitor banks are used to raise the voltage in a substation when it has dropped too low, particularly in areas of large industrial loads. Shunt reactors are used to lower voltages that have risen too high due to the capacitance in the transmission or distribution line.

Another class of devices used to control the voltage is operated using powered electronic switches to continuously adjust the capacitance and/or inductance in a substation to keep the voltage at precisely the voltage desired. These devices are relatively new in deployment, having been developed with the advent of inexpensive and robust power semiconductor components. These devices are part of a group broadly known as FACTS (Flexible AC Transmission System) devices and include static var compensators, static synchronous compensators, and dynamic voltage restorers.

John A. Palmer

See also: Capacitors and Ultracapacitors, Electric Motor Systems; Electric Powers, Generation of; Electric Powers, System Protection, Control, and Monitoring of; Electric Power, System Reliability and; Electric Power Transmission and Distribution Systems; Insulation; Transformers.

BIBLIOGRAPHY

Asea Brown Boveri. (1988). *Selection and Application of Gas Insulated Switchgear.* North Brunswick, NJ: Author.

Bosela, T. R. (1997). *Introduction to Electrical Power System Technology.* Englewood Cliffs, NJ: Prentice-Hall.

Faulkenberry, L. M., and Coffer, W. (1996). *Electrical Power Distribution and Transmission.* Englewood Cliffs, NJ: Prentice-Hall.

Glover, J. D., and Sarma, M. (1994). *Power System Analysis & Design*, 2nd ed. Boston: PWS.

ELECTRIC POWER TRANSMISSION AND DISTRIBUTION SYSTEMS

The North American electric power transmission system has been described as the largest, most complex machine ever built by humanity. It is a massive network of generating stations, transmission lines, substations, distribution lines, motors, and other electrical loads all interdependently linked for the conversion, transportation, and control of electrical energy. Approximately 60 percent of all energy utilized in the United States passes through the interconnected electric power system. The major goal of the system is to most efficiently and reliably deliver electric power from generating stations to residential, commercial, and industrial consumers.

A small portion of the power system is depicted in Figure 1. The flow of energy is as follows: at generating stations, mechanical, chemical, or some other form of energy is converted into electricity, most often using a synchronous generator. The electrical output from the generator is converted, through a transformer, to a very high voltage, to be conveyed through transmission lines to transmission substations. Within the transmission substations, the voltage is stepped down to the subtransmission system, through which the power is conveyed to the distribution substations and into the distribution system. The distribution system delivers the power to residential, commercial, and industrial users, where the

power is converted to light, heat, motion, or other desired forms of energy.

HISTORICAL CONTEXT AND TECHNOLOGICAL DEVELOPMENTS

The earliest commercial power system, believed to be Thomas Edison's Pearl Street station opened in 1882, consisted of a simple generator and a number of users. Within twenty years, more than 3,000 small electric generating stations were built in cities across the United States, each serving relatively local loads and with no interconnection. With the advent of the transformer, in about 1885, and the recognition that voltage drop and losses will be significantly reduced by stepping the voltage up and the current down, power transmission systems were created. The first demonstration of an ac power transmission line, in 1886, operated at 3,000 V over a distance of 4,000 ft (1,220 m). The first commercial transmission line in the United States was a 13-mile (21-km) transmission line operating at 3,300 V. As insulation systems improved and the technology of transformers was advanced, transmission voltage levels increased to 40 kV by 1907. This was a practical limit for the pin-type insulators (Figure 2) that were used to support the line on the towers, due to the structural stresses in the support. The voltage level was only able to increase further with the invention of the suspension insulator (Figure 2), which is in common use at the beginning of the twenty-first century. This increased the practical limit to about 150 kV, which was the limiting case because of corona. (Corona is a phenomenon in which the air in the vicinity of the energized surface is ionized because of the intensity of the electric field. It results in significant energy losses to the system.) The intensity of the electric field is reduced by having a larger-diameter conductor. The voltage was again increased in the 1960s after the realization that forming bundles of two, three, or four conductors could also mitigate the problem of corona. Power systems in the late 1990s operated at voltages as high as 765 kV in the United States, and as high as 1,100 kV in some parts of Europe.

SYSTEM INTERCONNECTIONS

Through the first several decades of commercial power systems, one generator or a small cluster of

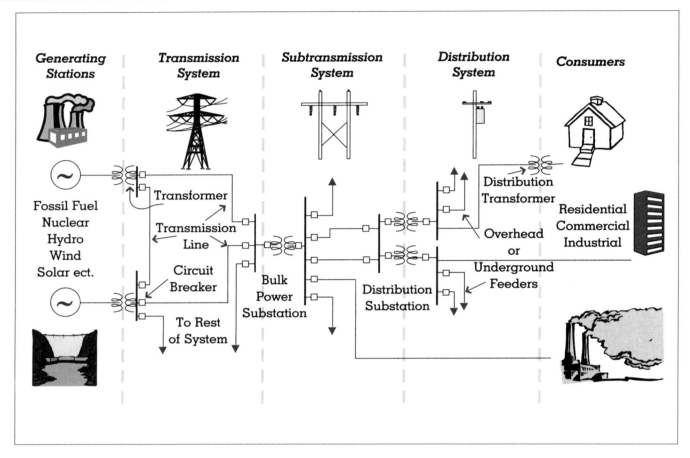

Figure 1.
A portion of the electric power transmission system.

generators would provide power to a group of users in a given region. In the 1930s, systems began to be interconnected for reliability and economic reasons. In the operation of any engineering system, equipment occasionally malfunctions due to degradation of the equipment itself or because of outside influences. When this happens in small systems, the load is no longer served and the region is subjected to blackout. Relatively minor changes to the system, such as the addition or removal of a single large load, also cause a significant impact on system operating frequency or other parameters. The interconnection of a large number of generators over a wide area avoids both types of problems. The loss of one generator is made up quickly by the controls on the other generators on the system. A change in load has a much smaller impact when the total load is many orders of magnitude larger. Additionally, the inter-connection of several local systems permits economic transactions, so that a utility requiring more power may purchase from another generating company rather than utilizing its own generation facilities, which may not be available or may be more expensive to operate.

The North American power system (covering continental Canada, the conterminous United States, and parts of northern Mexico) is made up of four independent power systems, with special connections among them. The independent systems are the Eastern interconnection, the Western interconnection, Quebec, and Texas. Within any of the four systems, there is a high level of connectivity, and all generators have coordinated control systems to enable them to work together fully. Among the systems are one or more high-voltage direct current (HVDC) links, which permit the flow of power for commerce and reliability. Several other smaller power systems also operate in North America but do not have any interconnection to the larger systems. There are a large number of other interconnections internationally in Europe and elsewhere.

Despite the very strong advantages to interconnection, there is an inherent weakness in the interconnection of large systems covering vast distances. Under some operating conditions, the system becomes unstable, and a small change may have a large impact. This occurs most frequently when the system is heavily loaded, such as during the summer, under heavy air-conditioning loads. Interarea oscillations may occur as the internal control systems for the various generators respond at slightly different times, so that power flow through transmission lines may widely fluctuate and even change direction. Careful monitoring and control must be maintained to avoid unstable operating regimes, and extensive efforts have gone into the development of simulation techniques to predict potentially unstable operating conditions so they may be avoided. Computational analyses are typically done that assess the impact of single and double contingencies—that is, the widespread effect of the failure of a single piece of equipment or generating unit or the loss of a single transmission line, or some combination of events. Naturally, inasmuch as the system has so many elements all working together, it is impossible to predict every possible contingency. Because of this, occasionally unstable operating regimes are entered, which may result in local or even widespread outages.

The most infamous outage in U.S. history was the Northeast blackout of November 9–10, 1965, in which thirty million people lost power for as long as thirteen hours. In this case, large quantities of power were being transmitted over long lines to New York City. The initiating event was the tripping of a single transmission line on the Ontario–New York border. This resulted in several other transmission lines having to pick up the load that had previously been carried by that line, and those lines overheated and tripped, removing 1,800 MW of generation (at Niagara Falls) from the system. The entire Northeast power system became unstable and separated into a number of different isolated systems, none of which had a balance between generation and loads. This resulted in the remaining generation tripping off line, and a widespread outage covering much of New

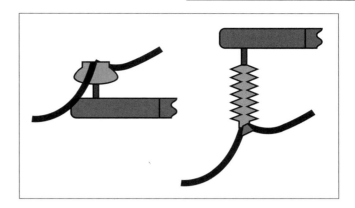

Figure 2.
Typical pin type insulator (left) and suspension insulator (right).

York, Ontario, New England, New Jersey, and Pennsylvania. The blackout was widespread, and the damages that resulted from looting and panic were extraordinary. In response to the outage, the National Electric Reliability Council (now the North American Electric Reliability Council—NERC) was created, with the responsibility to ensure system security and reliable operation. NERC is owned by ten regional coordinating councils consisting of various utilities, power producers, power marketers, and customers. While many other blackouts have occurred since the one in 1965, some affecting millions of customers, none has affected as many customers, and none has had an impact on the industry that was so widespread. This is largely due to the efforts of NERC to study and promote reliability and establish policies, guidelines, and standards conducive to reliable operation.

TRANSMISSION LINE CHARACTERISTICS AND THEIR IMPLICATIONS

The performance of a transmission line, and its limitations, are directly related to physical parameters that come from its design, construction, and even its location. Those parameters are common to many electrical circuits: resistance, inductance, and capacitance.

The series resistance of a transmission line is closely related to the losses that will be dissipated when current passes through the line (proportional to the square of the current magnitude). The resistance is proportional to the length of the line but inversely proportional to the cross-sectional area of the conductor.

The losses, and hence the effective resistance, are also increased by passing the line in close proximity to a noninsulating surface—for example, passing the line over seawater. The metal from which the conductor is made is also very important—for example, copper has a lower resistance, for the same geometry, than aluminum does. Also related to the losses of the transmission line is the shunt resistance. Under most circumstances, these losses are negligible because the conductors are so well insulated; however, the losses become much more significant as the insulators supporting the transmission line become contaminated, or as atmospheric and other conditions result in corona on the line.

The series inductance of the transmission line is a measure of the energy stored in the magnetic field of the conductor. High inductance is usually the limiting factor on the ability to transmit power over long distances, because the stability limit for power transfer is inversely proportional to the line inductance. Inductance increases for conductors that are farther apart, and decreases for conductors having a larger diameter.

The shunt capacitance of the transmission line is related to the energy stored in an electric field between conductors and/or earth. Capacitance negatively influences the operation of the transmission line by requiring higher currents from the generators (charging current), and by inducing a voltage rise (sometimes well in excess of safe operating limits) in a lightly loaded transmission line (the Ferranti effect). The capacitance increases when conductors are brought closer together, and it decreases for conductors having a smaller diameter.

The actual flow of power through the transmission system from one point to another is dictated by the operation of all of the generators and by the resistance, inductance, and capacitance of the system. Because those parameters are mainly characteristics of design, in the past very little flexibility in control was provided to change the path of power flow. In one well-known example, there is a lot of power generation that occurs in Niagara Falls, New York, and New York City consumes a lot of power. It has frequently happened that the flow of the majority of the power from Niagara Falls to New York City, rather than being along the most direct route, has been from Niagara Falls into Ontario, through Michigan, Ohio, Pennsylvania, and New Jersey before getting to New York City. Despite economic transactions and contractual agreements that would call for a direct route with minimized losses, the path of power flow is dictated by the laws of physics and the system parameters.

There are several techniques, however, that modify the system parameters to manipulate the power flow in localized regions. The one that has been in use the longest is modifying the parameters of generator operation, the quantity of power being produced, and the voltage at the terminals of the machine. This sometimes requires operation of less efficient and more costly generating stations, instead of optimizing efficiency and cost, and other limitations frequently come into play as well. The second approach is to install a very costly type of specialized transformer, a phase angle regulating transformer, at crucial points in the system. This device provides limited control, but can of modify power flows by imposing a change in phase angle in the voltage and current going through it. Switched series capacitor banks are sometimes used to reduce the effective inductance of a transmission line, permitting more power to flow through it, but these are not highly controllable devices either.

A fairly new technology that provides a high level of control for power flows and system stability is a class of devices known as flexible alternating current transmission systems (FACTS). A number of FACTS devices have been used at all different voltage levels. On the high-voltage system are devices such as static var compensators (SVC), thyristor controlled series capacitors (TCSC), static synchronous compensators (STATCOM), and universal power flow controllers (UPFC). These devices work by canceling out the inductance and/or capacitance of the transmission lines and the system loads. They are operated with high-power electronic devices and are fully controllable, either manually or automatically. Because of their dynamic nature and quick response time, they can respond to system disturbances and can provide an extraordinary increase to system stability.

Still another approach, which has been used since the 1960s to control power flow and transmit power over very long distances at high efficiencies, is high-voltage direct current transmission (HVDC). While they fill the same role as high-voltage ac transmission lines (bulk power transmission), HVDC lines are impervious to the effects of system inductance and capacitance, so power flow and system stability are not influenced by those parameters. Each end of the HVDC transmission line is located in a special converter station where high-power electronic devices convert alternating current to direct current at one end and reverse the process at the other end. HVDC power transmission is more efficient and more controllable, but the converter stations are very costly, so the planning and design of such a line includes a careful cost/benefit analysis in which typically the HVDC system is only chosen over a high-voltage ac line for very long distances or as an interconnection between independent systems.

TYPES OF TRANSMISSION LINES

For the ac or dc interconnections that span great distances, high-voltage transmission lines are almost exclusively overhead. These lines are typically constructed with aluminum wrapped around a reinforcing steel core. The conductors are connected to the support structures with suspension insulators that may be made of porcelain or polymers. The type of structure that supports the transmission line, the distance between structures, the arrangement of conductors, and so forth are functions of operating voltage, terrain, climate, right-of-way, aesthetic concerns, cost, etc. Common types of structures include treated wood, steel lattice, tubular steel, and concrete. Overhead power transmission lines are less expensive than underground transmission lines and are easier to install, inspect, and maintain. However, they are also more exposed to the hazards of environmental conditions such as severe winds, ice formation, and corrosion and contamination due to airborne pollutants.

Underground transmission lines are preferred in places where rights-of-way are severely limited because they can be placed much closer together than overhead lines. They are also favored for aesthetic reasons. They may be directly buried in the soil, buried in protective steel or plastic pipes, or placed in subterranean tunnels. The conductors are usually contained within plastic insulation encased in a thin metallic sheath. The conductors enclosed in steel pipes may be immersed in oil, which may be circulated for cooling purposes. For all types of underground lines, the capacitance is higher than for overhead lines, and the power transfer capability is usually limited by the resistive losses instead of the inductance. While not exposed to environmental

hazards, underground cables are at risk of damage by rodents, construction, or geological instabilities.

In the 1990s, significant efforts have gone into the development of transmission lines made from superconductors, which are materials that have essentially no resistance when operating at extremely low temperatures. While a significant amount of energy is required to operate the cryogenic systems, the reduction in power loss and the increase in power transfer capability have made the technology appealing. Several prototype installations in various stages of design and preparation have provided encouraging preliminary results, but as of the end of the twentieth century, no full-scale applications have been deployed.

ENVIRONMENTAL IMPACT OF TRANSMISSION SYSTEMS

While many people are concerned about the aesthetic impact of transmission lines, several other aspects of the transmission line environmental impact must also be considered. There is a danger to wildlife imposed by the presence of high-voltage surfaces such as overhead transmission lines. For small birds that land on transmission lines there is very little hazard because there is no path for current to flow through them. However, birds with large wingspans or climbing rodents will often reach between surfaces energized at different voltages, killing them and damaging system equipment.

Overhead transmission lines require that the area beneath them be cleared of trees or tall shrubs, which may result in erosion. When the transmission line right-of-way is not kept clear, the transmission line may come into contact with vegetation, causing a fault on the system and possibly starting a fire. Chemical contamination of soil may result from some types of transmission structures, such as treated wood. Burial of underground cables also can impact the environment due to erosion.

Another environmental concern that has been raised is the fear that electromagnetic fields (EMF) may cause negative physiological effects. Various epidemiological studies have purported to find an association between the presence of transmission lines and different types of cancer, especially in children. These studies have found at most a very weak association, and were based on estimated and not measured electromagnetic fields. While the issue has been given significant media attention, inciting grave public concern and much new research, the medical and scientific communities have been unable to definitively confirm that EMF has any measurable physiological impact.

TRANSMISSION SYSTEM ORGANIZATIONAL STRUCTURE

Until recently, all functions related to the generation, transmission, and distribution of electric power in the United States were executed under the umbrella of monopolies regulated by state utility boards (under a variety of titles). Through the 1990s, legislation has resulted in the deregulation of that industry and many changes in the nature of commerce in the energy industry. The details of the new system are still in development, but the deregulated system as presently designed consists of a plan to separate the utilities into independent entities consisting of regulated business-es and unregulated businesses. The regulated busi-nesses will include the transmission companies and the distribution companies. It is necessary to regulate these companies to utilize the assets that exist to pro-vide power to individual users and to transfer bulk power without the proliferation of an excessive quan-tity of new infrastructure. The deregulated businesses will consist of generating companies responsible for the production of power, and retail companies responsible for sale, billing, and other energy-related services. Sale and marketing transactions will be con-ducted by power marketers, with transactions being conducted on the power exchange. The entity respon-sible for observing and operating the overall system, ensuring that all functions are executed within the safe and reliable operating limits, will be the inde-pendent system operator (ISO).

John A. Palmer

See also: Capacitors and Ultracapacitors; Electric Motor Systems; Electric Power, Generation of; Electric Power, System Protection, Control, and Monitoring of; Electric Power, System Reliability and; Electric Power Substations; Environmental Problems and Energy Use; Insulation; Transformers.

BIBLIOGRAPHY

Faulkenberry, L. M., and Coffer, W. (1996). *Electrical Power Distribution and Transmission*, Englewood Cliffs, NJ: Prentice-Hall.

Glover, J. D., and Sarma, M. (1994). *Power System Analysis & Design,* 2nd ed. Boston: PWS.

Gyugyi, L., et al. (1995). "The Unified Power Flow Controller: A New Approach to Power Transmission Control." *IEEE Transactions on Power Delivery* 10:1085–1093.

National Research Council. (1995). *EMF Research Activities Completed Under the Energy Policy Act of 1992: Interim Report, 1995.* Washington, DC: National Academy Press.

ELECTRIC VEHICLES

Electric vehicle (EV) is an abbreviated term for bat-tery electric vehicle, which is a vehicle that uses energy from a battery to operate an electric motor that rotates the wheels. Throughout the twentieth century, a wide array of electric cars, trucks, buses, bikes, and scooters have been developed. However, except for the golf cart and special delivery vehicles, none has experienced any sustained commercial success.

THE BASICS

EVs use a variety of electric motors, dc, ac, and three-phase ac induction being the most common. No one electric motor is optimal for all vehicles. Motor selec-tion criteria include power needs, variable speed needs, operating voltage, and physical space limita-tions. Researchers at the turn of the twenty-first cen-tury are looking into directly attaching electric motors to the wheels to eliminate the drive shaft and differential. This will improve efficiency by reducing weight and by minimizing the mechanical losses between the motor and wheel so that propulsion to the wheels is generated at the wheels.

EVs deliver superior efficiency because of the elec-tric motor. Approximately 75 percent of the battery's chemical energy is used by the electric motor to gen-erate the mechanical energy rotating the wheels. In addition, little electricity is wasted during recharging. The extremely efficient rechargeable batteries of elec-tric vehicles result in somewhere between 70 to 90 percent of the electrical energy used to charge the bat-teries being available during discharging. In compari-son, the internal combustion engine (ICE) vehicle converts only about 20 percent. This figure drops even further when taking into account the energy losses through the transmission and idling. Because the electric motor does not consume any energy while the vehicle is stationary, the EV operates even more efficiently than the ICE vehicle in urban settings, especially for EVs with regenerative braking technology, which recovers braking energy that normally would be dissipated as heat to charge the batteries.

The key component of EVs remains the energy source making propulsion possible: batteries. Advances in batteries are necessary to match the performance of ICE vehicles, particularly the high power-to-weight ratio for acceleration and the high energy-to-weight ratio for range. Since the heavier the battery, the lower the power output per pound of weight, designers face a seemingly no-win tradeoff: additional batteries added to achieve greater acceleration result in more weight that takes away from vehicle range and performance. However, some of the weight problem can be offset by teaming high power density/low energy density devices, such as ultracapacitors or flywheels, to handle peak loads with the low power density/high energy density battery.

Unfortunately, often the EV's efficiency and cost advantages are wiped out by the battery's abbreviated life expectancy. When battery life expectancy is around 2,000 miles (reported by testers of General Motors's EV, the Impact), then the recharging and replacement battery costs of an EV (over $1,500 for the GM Impact) alone can exceed the total operating costs for an ICE vehicle. Energy costs per mile for the GM Impact ($0.02 electricity plus $0.42 for batteries) were greater than the total operating costs for an ICE vehicle ($0.42).

The energy costs of building vehicles must also be considered. For ICE vehicles, more energy is usually used in construction of the vehicle than will be consumed in fuel for driving 100,000 miles. For the EV, the dynamics are even worse since the material and energy costs of batteries are considerable. Batteries are expensive since they entail a substantial amount of material (added weight) and often involve multiple complex construction. For example, the thirty-two advanced lead-acid batteries for the 1995 GM Impact weighed over 850 pounds.

EARLY DEVELOPMENT

EVs are not new. They actually originated in the 1880s. In fact, it may be difficult to imagine, but at one time EVs were more popular than ICE vehicles. In the early 1900s, of the approximately 8,000 motor vehicles in the United States, 40 percent were powered by steam, 38 percent by battery, and 22 percent by gasoline. Since steam-powered vehicles needed a constant supply of water and high steam pressure, they proved expensive and difficult to maintain, which lead to their demise by around 1910. The problems with internal combustion engine (ICE) vehicles at that time were difficulties in starting, noisiness, and unreliability. In contrast, early EV owners enjoyed immediate starts, quiet operation, minimum maintenance, and impressive performance for the times: speeds of 15 to 20 miles per hour and 30 to 40 miles between charges.

The popularity of EVs did not last. By the 1920s, the performance of ICE vehicles improved dramatically, and the earlier major drawbacks had been solved. Ironically, the replacement of the dangerous hand crank with a battery-powered electric starter was a major innovation accelerating ICE vehicle sales at the expense of EV sales. At the same time, there were no concurrent solutions to the limitations of battery technology for EVs that addressed the demand to drive faster and farther.

REEMERGENCE DURING THE OIL SHORTAGES

From 1920 until the oil shortage crisis of the 1970s, the only major uses of the EV were for golf carts and neighborhood delivery vehicles. In the United Kingdom, the lower operating costs and fewer repairs of the EV more than offset the disadvantages of its modest range, speed, and acceleration for these short-range delivery vehicles.

Several EVs came on the market in the 1970s, both in the United States and abroad, as a transportation alternative in answer to soaring gasoline prices and supply disruptions causing the long lines at gasoline stations. But as the oil shortages ended in the early 1980s, interest in EVs again faded. EV technology advanced, but not nearly enough to seriously compete with ICE vehicles.

THE AIR QUALITY CRISIS

Regulations imposed on auto makers to address a perceived air pollution crisis once again renewed EV interest in the early 1990s. Perhaps the most aggressive regulations were imposed by the California Air Resources Board (CARB) low emission vehicle (LEV) program mandating that zero emission vehicles

(ZEV) comprise 2 percent of manufacturers sales by 1998 and 10 percent by 2003. The belief among regulators was that mandates would spur innovation, create the economies of scale to lower costs (as a result of greater vehicle output), and incite the sales momentum that would begin a large-scale transition to EVs. Considering the state of EV technology at the time, the auto industry was very reluctant to spend the millions necessary to develop an electric car that their marketing research told them people did not want.

General Motors made the most noteworthy foray into the market in 1995 with the Impact (later renamed the EV1). Unlike the offerings of other auto makers, the Impact was an all-electric design from the ground up. The advanced lead-acid battery series delivered 320 volts, and with an inverter that converted direct current to alternating current for transfer to two induction motors (one for each front wheel), the car had impressive performance: quick and smooth acceleration from 0 to 60 mph in eight seconds, and a top speed of 110 mph. Other innovative features included the lowest-drag exterior design, high-pressure tires that had half the rolling resistance of ordinary tires, and a regenerative braking system. However, there were problems. Compared to ICE vehicles, the Impact was considerably more expensive; its lightweight body design was not as safe; and its range was limited to 55 to 95 miles between charges (depending on terrain, driving habits, and temperature). A full charge took 6 hours, and it turned out the batteries needed to be replaced after only 2,000 to 3,000 miles, not 20,000 miles as hoped. Attempting to address the battery shortcomings, in 1999 GM began offering more expensive nickel-metal hydride batteries as an option. These batteries extended the range to 75 to 130 miles, but also took slightly longer to recharge.

The much higher initial cost remains the most significant objection to EVs. Economies of scale could solve the cost problem, yet range, change time, and battery life expectancy will continue to be major disadvantages. Even if the range of EVs was extended to 300 miles between charges, battery life extended to 40,000 miles, and full recharge time reduced to 30 minutes, these are still limitations that ICE vehicle owners do not have to endure. Moreover, EVs offer limited functionality at a much higher cost (even with federal and state subsidies), while the public keeps looking for more functionality. For example, many sports utility vehicle purchases are driven by desire for

their unmatchable multi-functionality. People use the vehicle for many things besides commuting: car pooling duties, hauling lumber, backroad touring, towing a boat, and plowing through the occasional snow storm, to name a few. The additional utility of these vehicles may cost the buyer a sizable premium, yet the buyer is happy to pay it for the security of having the added functionality, even though the advanced functionality may be terribly underutilized.

ICE innovations that resulted in significant emission reductions also hastened the demise of EVs. In fact, Toyota and Honda in 1999 announced the development of ICE vehicles that could meet California's Super Ultra Low Emission Standard proposed for 2004. It can be argued that these super ultra low emission vehicles (SULEVs) will generate levels of emissions comparable to EVs if the power plant emissions that generate the electricity to power the EV batteries are taken into account. However, figuring out EV emissions from the electric generation mix is a very imprecise science. Almost all of the electric energy will come from emission-producing coal and natural gas power plants, not emission-less hydroelectric and nuclear energy since these sources are "first through the meter." Thus EVs are not really zero emission vehicles. They are what Amory Lovins has called "elsewhere emission vehicles"—not occurring at the tailpipe but at a distant smokestack. Elsewhere emissions are still considered a worthwhile goal by many since the non-attainment of air quality standards is almost exclusively a problem for urban areas, and natural gas and coal power plants are usually all located outside of urban areas. Further, it is also easier to monitor and control emissions at a small number of stationary sources (power plants) than the tailpipe of every vehicle.

The combination of ICE innovation, the lack of EV advances, and the dismal acceptance of EVs by consumers forced California's ZEV program to suspend the vehicle sale mandates in 1995. The mandates were replaced by agreements with automakers to make a concerted effort and demonstrate technology. Nevertheless, declaring an inability to develop demand, Honda withdrew the EV Plus from the market in 1999, followed by General Motors withdrawing the EV1 (formerly Impact) in 2000. All automakers have shifted their efforts to developing the other more promising emerging electric vehicle options such as hybrid and fuel cell technologies.

Despite the repeated failures of EVs in the marketplace, there is widespread agreement that electric vehicles make sense for selected applications. For instance, city delivery vans have represented a good niche for EVs because they have sufficient space to be outfitted with large battery cases and do not need to have rapid acceleration or go long distances between charges.

As battery innovation continues, it is likely that the realm of applications for which EVs make sense will expand. A promising future market for electric vehicles may be electric scooters and electric motor-assisted bicycles. These types of vehicles will not only cut down on the urban air pollution problem, but also could help reduce congestion since the typical ICE vehicles being replaced are much larger. In particular, it is the hope in many highly populated urban areas of Asia that electric scooters and electric bicycles will become the personal transportation option of choice. In 2000, a nickel-metal hydride electric scooter developed by Ovonic Battery Company achieved a range of 73 miles with an efficiency equivalent to more than 300 miles per gallon of gasoline.

THE LARGE-SCALE RECHARGING CHALLENGE

By 1995 global passenger-car registrations surpassed 500 million, with the United States accounting for almost 160 million of the total. If significant performance improvements in battery technology occur that make EVs more attractive than ICE vehicles, major investments in the recharging infrastructure would be required to accommodate the millions of new EVs. This would present critical challenges for the electric utility industry, both at the bulk power system level and at the point of distribution or end use.

Those electric utility companies with low load growth and high generation capacity margins would benefit most from the increased power sales associated with having more electric vehicles. However, in the current transition to electric industry deregulation, generation capacity margins in many regions are declining to barely acceptable levels, since generating capacity is being added in response to market opportunities and not as part of a long range, comprehensive planning effort focusing on load growth. More importantly, a greatly increased number of EVs using charging systems located in residential areas could

vastly change current load profiles. Electric power engineers are concerned that adding these large "single-phase loads" may adversely affect power quality, safety, and system reliability. (Single-phase loads are lower voltage loads placed on distribution systems that may unbalance three-phase systems and may require large current flows.)

Power system planners need to consider how the costs associated with electric vehicles should be passed along to consumers. Fortunately from a load balancing perspective, it is likely that most EV charging will occur in off-peak periods. Charging in off-peak periods would reduce utility costs and therefore should allow utilities to reduce customer rates, but this would require "time of day metering" that is not available in most service areas.

There are also public safety concerns associated with electric vehicle charging devices. Single-phase charging systems for home use could require current flows that far exceed 100 to 200 amp service installations that are typical in homes, and could result in overloads, breaker operations, and blown fuses on distribution systems. Older homes were not designed for these high current flows and it is likely that their electrical systems would have to be updated. In addition, charging systems would pose problems for people who live in apartments, coops, and condominiums who are accustomed to street parking. Three-phase charging systems located at service stations would be better for charging, since they deliver constant (non-pulsating) power at higher voltages and avoid the high current and unbalancing effects associated with single-phase systems. These systems could reduce some of the public safety concerns, produce more power, and charge batteries more quickly. Also, the service station charging approach would make it easier to achieve the goals set by the United States Advanced Battery Consortium: 3 to 6 hours for normal recharge time, and a fast recharge time of 50 percent of capacity in less than 30 minutes.

Both single-phase charging systems and three-phase charging systems would contain power electronic devices that as a side effect introduce waveform distortion and create "power quality" problems. Filtering devices used in conjunction with residential charging systems could be used to reduce harmonics and other power quality problems, but the cost of such filtering devices is currently quite high.

Some engineers have suggested that electric vehicle

charging systems would have the least negative impact if integrated with emerging distributed resources technologies involving distributed generation and distributed storage. Distributed systems could help in addressing the new challenges of system operations in the open access, competitive, restructured electric power industry and relieve network congestion problems. In addition, the energy line losses could be significantly reduced by using distributed resources to meet electric vehicle charging requirements.

At present there is much speculation about the long-term future of EVs. Electric vehicles could dramatically lessen U.S. dependence on imported oil and help solve air quality problems, yet also will create new problems and challenges for electric utility system operators.

John Zumerchik
Fred I. Denny

See also: Batteries; Capacitors and Ultracapacitors; Electric Motor Systems; Emission Control, Vehicle; Environmental Problems and Energy Use; Flywheels; Fuel Cells; Fuel Cell Vehicles; Hybrid Vehicles; Materials; Transportation, Evolution of Energy Use and.

BIBLIOGRAPHY

Balzhiser, R. E., and Bryson, J. E. (1994). "The Strategic Role of Electric Vehicles." *Forum for Applied Research and Development* (Spring):31–34.

"Building the Electric Vehicle Future: EPRI's Vehicle Development Activities." (1987). Electric Power Research Institute Report EU.3017.11.87. Palo Alto, CA: Electric Power Research Institute.

Chachich, A. (1991). "Energy and Transportation: A Technological Update." In *The Energy Sourcebook*, eds. R. Howes and A. Fainberg. New York: American Institute of Physics.

"Charging Up Electric Vehicles." (1992). *EPRI Journal*, June.

D'Agostino. (1993). "The Electric Car." *IEEE Potentials* (February):28–32.

Fischetti, M. (1992). "Here Comes the Electric Car—It's Sporty, Aggressive and Clean." *Smithsonian* 23(1):34–43.

Flink, J. J. (1970). *America Adopts the Automobile, 1895–1910*. Cambridge, MA: MIT Press.

Greene, D. (1996). *Transportation and Energy*. Lansdowne, VA: Eno Transportation Foundation Inc.

Miller, K. (1996). "The Future Is Electric." *Life* 19(2):110–116.

Riezenmann, M. J. (1992). "Electric Vehicles." *IEEE Spectrum* 29(11):18–21.

Sperling, D. (1995). *Future Drive: Electric Vehicles and Sustainable Transportation*. Washington, DC: Island Press.

U.S. Department of Energy. (1995). "DOE'S Energy Partnerships for a Strong Economy: Electric Utilities Meet the Challenge." February/March, p. 3.